Optical Probes in Biology

Series in Cellular and Clinical Imaging

Series Editor
Ammasi Periasamy

PUBLISHED

Coherent Raman Scattering Microscopy
edited by Ji-Xin Cheng and Xiaoliang Sunney Xie

Imaging in Cellular and Tissue Engineering
edited by Hanry Yu and Nur Aida Abdul Rahim

Second Harmonic Generation Imaging
edited by Francesco S. Pavone and Paul J. Campagnola

The Fluorescent Protein Revolution
edited by Richard N. Day and Michael W. Davidson

Natural Biomarkers for Cellular Metabolism:
Biology, Techniques, and Applications
edited by Vladimir V. Gukasyan and Ahmed A. Heikal

Optical Probes in Biology
edited by Jin Zhang, Sohum Mehta, and Carsten Schultz

SERIES IN CELLULAR AND CLINICAL IMAGING
AMMASI PERIASAMY, SERIES EDITOR

Optical Probes in Biology

Edited by
Jin Zhang
Sohum Mehta
Carsten Schultz

CRC Press
Taylor & Francis Group
Boca Raton London New York

CRC Press is an imprint of the
Taylor & Francis Group, an **informa** business

CRC Press
Taylor & Francis Group
6000 Broken Sound Parkway NW, Suite 300
Boca Raton, FL 33487-2742

First issued in paperback 2017

ISBN-13: 978-1-4665-1011-1 (hbk)
ISBN-13: 978-1-138-19993-4 (pbk)

Library of Congress Cataloging-in-Publication Data

Optical probes in biology / editors, Jin Zhang, Sohum Mehta, Carsten Schultz.
p. ; cm. -- (Series in cellular and clinical imaging)
Includes bibliographical references and index.
ISBN 978-1-4665-1011-1 (hardcover : alk. paper)
I. Zhang, Jin (Molecular scientist), editor. II. Mehta, Sohum, editor. III. Schultz, Carsten, editor. IV. Series: Series in cellular and clinical imaging.
[DNLM: 1. Molecular Imaging--methods. 2. Optical Imaging--methods. 3. Biosensing Techniques--methods. WN 190]

RC78.7.D53
616.07'54--dc23 2014028661

Visit the Taylor & Francis Web site at
http://www.taylorandfrancis.com

and the CRC Press Web site at
http://www.crcpress.com

Contents

Series Preface

A picture is worth a thousand words.

This proverb says everything. Imaging began in 1021 with use of a pinhole lens in a camera in Iraq; later in 1550, the pinhole was replaced by a biconvex lens developed in Italy. This mechanical imaging technology migrated to chemical-based photography in 1826 with the first successful sunlight-picture made in France. Today, digital technology counts the number of light photons falling directly on a chip to produce an image at the focal plane; this image may then be manipulated in countless ways using additional algorithms and software. The process of taking pictures ("imaging") now includes a multitude of options—it may be either invasive or noninvasive, and the target and details may include monitoring signals in two, three, or four dimensions.

Microscopes are an essential tool in imaging used to observe and describe protozoa, bacteria, spermatozoa, and any kind of cell, tissue, or whole organism. Pioneered by Antoni van Leeuwenhoek in the 1670s and later commercialized by Carl Zeiss in 1846 in Jena, Germany, microscopes have enabled scientists to better grasp the often misunderstood relationship between microscopic and macroscopic behavior, by allowing for study of the development, organization, and function of unicellular and higher organisms, as well as structures and mechanisms at the microscopic level. Further, the imaging function preserves temporal and spatial relationships that are frequently lost in traditional biochemical techniques and gives two- or three-dimensional resolution that other laboratory methods cannot. For example, the inherent specificity and sensitivity of fluorescence; the high temporal, spatial, and three-dimensional resolution that is possible; and the enhancement of contrast resulting from detection of an absolute rather than relative signal (i.e., unlabeled features do not emit) are several advantages of fluorescence techniques. Additionally, the plethora of well described spectroscopic techniques providing different types of information, and the commercial availability of fluorescent probes such as visible fluorescent proteins (many of which exhibit an environment- or analytic-sensitive response), increase the range of possible applications, such as development of biosensors for basic and clinical research. Recent advancements in optics; light sources; digital imaging systems; data acquisition methods; and image enhancement, analysis, and display methods have further broadened the applications in which fluorescence microscopy can be applied successfully.

Another development has been the establishment of multiphoton microscopy as a three-dimensional imaging method of choice for studying biomedical specimens from single cells to whole animals with sub-micron resolution. Multiphoton microscopy methods utilize naturally available endogenous fluorophores—including NADH, TRP, FAD, and so on—whose autofluorescent properties provide a label-free approach. Researchers may then image various functions and organelles at molecular levels using two-photon and fluorescence lifetime imaging (FLIM) microscopy to distinguish normal from cancerous conditions. Other widely used non-labeled imaging methods are coherent anti-Stokes Raman scattering spectroscopy (CARS) and stimulated Raman scattering (SRS) microscopy, which allow imaging of molecular function using the molecular vibrations in cells, tissues, and whole organisms. These techniques have been widely used in gene therapy, single molecule imaging, tissue engineering, and stem cell research. Another non-labeled method is harmonic generation (SHG and THG), which is also widely used in clinical imaging, tissue engineering, and stem cell research. There are many more advanced technologies developed for cellular and clinical imaging including multiphoton tomography, thermal imaging in animals, ion imaging (calcium, pH) in cells, etc.

The goal of this series is to highlight these seminal advances and the wide range of approaches currently used in cellular and clinical imaging. Its purpose is to promote education and new research across a broad spectrum of disciplines. The series emphasizes practical aspects, with each volume focusing on a particular theme that may cross various imaging modalities. Each title covers basic to advanced imaging methods, as well as detailed discussions dealing with interpretations of these studies. The series also provides cohesive, complete state-of-the-art, cross-modality overviews of the most important and timely areas within cellular and clinical imaging.

Since my graduate student days, I have been involved and interested in multimodal imaging techniques applied to cellular and clinical imaging. I have pioneered and developed many imaging modalities throughout my research career. The series manager, Ms. Luna Han, recognized my genuine enthusiasm and interest to develop a new book series on Cellular and Clinical Imaging. This project would not have been possible without the support of Luna. I am sure that all the volume editors, chapter authors, and myself have benefited greatly from her continuous input and guidance to make this series a success.

Equally important, I personally would like to thank the volume editors and the chapter authors. This has been an incredible experience working with colleagues who demonstrate such a high level of interest in educational projects, even though they are all fully occupied with their own academic activities. Their work and intellectual contributions based on their deep knowledge of the subject matter will be appreciated by everyone who reads this book series.

Ammasi Periasamy, PhD, Series Editor
Professor and Center Director
W.M. Keck Center for Cellular Imaging
University of Virginia
Charlottesville, Virginia, USA

Preface

Optical probes—in particular the fluorescent varieties—are among the most sophisticated and valuable tools in cell biology and biochemistry, allowing researchers to observe cellular events in real time and often with great spatial resolution. With *Optical Probes in Biology*, we hope to provide a comprehensive resource that not only encapsulates the diverse capabilities of these probes but also pulls back the curtain on the various approaches used to design, develop, and implement these powerful and versatile tools. As such, this book is aimed at a wide audience of scientists in a variety of disciplines, from students to established investigators, who seek to incorporate optical probes and fluorescence imaging into their research.

A critical component of the success and popularity of optical probes has been the sheer breadth of cellular and molecular events that can be tracked using these tools; this versatility stems from the numerous ways that have been devised to construct optical probes. Younger readers ought to be reminded that the first optical probes suitable for use in living cells were in fact developed in the mid-1970s, mainly to monitor changes in membrane potential. This was soon followed by the development of the first small-molecule calcium indicators that were suitable for applications in intact cells. Indeed, such tools for scrutinizing intracellular calcium signaling may represent the single greatest success story to come out of the early history of optical probe development. The wide availability and application of these probes greatly strengthened entire fields in cell biology, and it was at this time that it also became clear how powerful ratiometric probes are for enabling reliable data acquisition. The early 1990s then saw the emergence of fluorescent proteins, thereby ushering in the era of genetically encoded fluorescence. Suddenly, it became possible to follow fluorescently tagged proteins in both space and time within single living cells, as well as in cell batches, owing to the ease with which proteins could be tagged using standard molecular biology techniques. Soon, genetically encoded sensors featuring two fluorescent proteins, useful for Förster/fluorescence resonance energy transfer (FRET) measurements, became available, and new FRET sensors are continually being developed to answer pressing biological questions through noninvasive live-cell experiments. In the meantime, numerous molecular beacons, including FRET-based reporters, have been introduced and have become of great use to study DNA and RNA both *in vitro* and in living cells. Furthermore, despite the prevalence of fluorescent protein–based probes, new chemical techniques continue to be developed, such as fluorogenic probes that become fluorescent when covalently reacting with the target enzyme.

All told, optical probes can be used to detect a truly remarkable variety of molecular processes in living cells, and we have included chapters detailing their use to track GTPase and kinase activities, membrane lipids, voltage, metal ions (e.g., calcium and zinc), metabolic signals, RNA, and histone modifications. Each chapter describes the different probe designs that have been and are being used, while also offering a frank discussion of the critical considerations associated with each. Our goal is to help readers develop a sense of which probe designs will be best suited to the desired application. Because fluorescent proteins play such a crucial role in the generation of many biosensors, we also delve into the many strategies that are being used to develop new fluorescent protein varieties with enhanced capabilities. Of course, these optical probes are only as good as the imaging techniques and equipment used to visualize and track them. A number of powerful and sophisticated imaging techniques have been developed over the years to allow a great deal of information to be coaxed out of the fluorescent signals from optical probes. Therefore, we have included a detailed discussion of fluorescence microscopy, covering several commonly used fluorescence imaging techniques, as a reference for selecting the appropriate imaging experiment and experimental set-up for answering a given biological question.

We should also point out that the use of optical probes is by no means limited to tracking events in single cells. Indeed, the range of biological insights that can be gleaned through the use of optical probes continues to grow steadily beyond single-cell tracking. A major focus of these efforts has been on the application of optical probes for following cellular events in intact living animals. Such *in vivo* imaging strategies carry their own particular requirements for delivering probes into the host organism, imaging the targeted tissue(s), obtaining robust signals, and so forth, and we have included chapters that go into some depth on the current state of *in vivo* imaging approaches. The growing use of activity-based probes for performing pharmacological drug screening directly in living cells and animals is also discussed. Numerous emerging techniques also promise to accelerate the expansion of optical probe-based approaches into new biological frontiers. For instance, super-resolution microscopy combines highly advanced imaging techniques with the sophisticated manipulation of specialized optical probes to observe biological phenomena in unprecedented spatial detail, while optogenetic techniques extend the role of optical probes to include both the detection and manipulation of intracellular events.

Although no book is able to cover all the exciting developments that are currently ongoing, we are confident that *Optical Probes in Biology* provides a sophisticated overview with illustrative examples that will help researchers enter or expand this field. We hope readers will enjoy every chapter in this volume, each of which has been contributed by outstanding authors.

Editors

Jin Zhang attended Tsinghua University for her undergraduate studies and pursued her graduate studies in chemistry at the University of Chicago. After completing her postdoctoral work at the University of California, San Diego, she joined the faculty of Johns Hopkins University School of Medicine, where she is currently a professor of pharmacology, neuroscience and oncology, with a joint appointment in the Department of Chemical and Biomolecular Engineering in the Whiting School of Engineering. She has received the American Heart Association National Scientist Development Award, the Biophysical Society Margaret Oakley Dayhoff Award, the National Institutes of Health (NIH) Director's Pioneer Award, the John J. Abel Award in Pharmacology from the American Society for Pharmacology and Experimental Therapeutics (ASPET), and the Pfizer Award in Enzyme Chemistry from the American Chemical Society.

Sohum Mehta received a BS degree in biology, with a minor in fine arts, from the George Washington University in 2003 before going on to pursue graduate studies at the Johns Hopkins University, where he received a PhD degree in biology in 2009. For his doctoral thesis, he studied calcineurin signaling in the yeast *Saccharomyces cerevisiae*, performing a structure–function study on the dual stimulatory and inhibitory behaviors of the evolutionarily conserved regulator of calcineurin (RCaN) family of proteins. After completing his graduate work, he began as a postdoctoral fellow in the lab of Jin Zhang at the Johns Hopkins University School of Medicine, using genetically encoded biosensors to continue investigating the spatiotemporal mechanisms of calcineurin in living cells while also developing novel genetically encoded tools to study intracellular signaling.

Carsten Schultz studied chemistry at the University of Bremen, Germany. After earning his PhD from the same university in 1989, he joined the lab of Roger Tsien at the University of California, San Diego. Carsten started his own group back in Bremen in 1993, received his habilitation in 1997, and in 2000 was appointed as a group leader at the Max-Planck-Institute for Molecular Physiology in Dortmund, Germany. In 2001, he moved to the European Molecular Biology Laboratory in Heidelberg, Germany, where he was promoted to senior scientist in 2008. Carsten is interested in intracellular signaling networks. With his expertise in preparative organic chemistry and chemical biology, his lab develops tools for imaging network components and probes to manipulate such networks noninvasively. Currently, the lab efforts focus mainly on lipid and membrane signaling. In more translational efforts, his lab contributes to better understand diseases such as cystic fibrosis (CF) and chronic obstructive pulmonary disease (COPD). Carsten is a member of the German Center for Lung Research. He is an author on more than 120 papers and has supervised more than 20 PhD students and close to 20 postdoctoral researchers.

Contributors

Walther Akemann
RIKEN Brain Science Institute
Saitama, Japan

Michal Arbel-Ornath
MassGeneral Institute of Neurodegenerative Disease
Massachusetts General Hospital
Charlestown, Massachusetts

Brian J. Bacskai
MassGeneral Institute of Neurodegenerative Disease
Massachusetts General Hospital
Charlestown, Massachusetts

Tamas Balla
Eunice Kennedy Shriver National Institute of Child Health and Human Development
National Institutes of Health
Bethesda, Maryland

Yury Belyaev
Advanced Light Microscopy Facility
European Molecular Biology Laboratory
Heidelberg, Germany

Daphne S. Bindels
Swammerdam Institute for Life Sciences
University of Amsterdam
Amsterdam, The Netherlands

Matthew Bogyo
Department of Pathology
and
Department of Microbiology and Immunology
Stanford University School of Medicine
Stanford, California

Peter Dedecker
Department of Chemistry
Katholieke Universiteit Leuven
Heverlee, Belgium

Theodorus W. J. Gadella
Swammerdam Institute for Life Sciences
University of Amsterdam
Amsterdam, The Netherlands

Elena Galea
Institució Catalana de Recerca i Estudis Avançats
Institut de Neurociències/Unitat de Bioquímica
Facultat de Medicina
Universitat Autònoma de Barcelona
Barcelona, Spain

Ambhighainath Ganesan
Department of Biomedical Engineering
The Johns Hopkins University
Baltimore, Maryland

Joachim Goedhart
Swammerdam Institute for Life Sciences
University of Amsterdam
Amsterdam, The Netherlands

Oliver Griesbeck
Max-Planck-Institut für Neurobiologie
Martinsried, Germany

Lindsay Haarbosch
Swammerdam Institute for Life Sciences
University of Amsterdam
Amsterdam, The Netherlands

Klaus M. Hahn
Department of Pharmacology
and
Lineberger Comprehensive Cancer Center
The University of North Carolina at Chapel Hill
Chapel Hill, North Carolina

Li He
Department of Genetics
Harvard Medical School
Boston, Massachusetts

Mark A. Hink
Swammerdam Institute for Life Sciences
University of Amsterdam
Amsterdam, The Netherlands

Felix Hövelmann
Department of Chemistry
Humboldt University Berlin
Berlin, Germany

Eloise Hudry
MassGeneral Institute of Neurodegenerative Disease
Massachusetts General Hospital
Charlestown, Massachusetts

Yin Pun Hung
Department of Neurobiology
Harvard Medical School
Boston, Massachusetts

Samie R. Jaffrey
Department of Pharmacology
Weill Medical College
Cornell University
New York, New York

Ksenia V. Kastanenka
MassGeneral Institute of Neurodegenerative Disease
Massachusetts General Hospital
Charlestown, Massachusetts

Kazuya Kikuchi
Immunology Frontier Research Center
and
Graduate School of Engineering
Osaka University
Osaka, Japan

Hiroshi Kimura
Graduate School of Bioscience and Biotechnology
Tokyo Institute of Technology
Yokohama, Japan

Thomas Knöpfel
RIKEN Brain Science Institute
Saitama, Japan

Toshiyuki Kowada
Immunology Frontier Research Center
Osaka University
Osaka, Japan

Wen-hong Li
Department of Cell Biology
and
Department of Biochemistry
University of Texas Southwestern Medical Center
Dallas, Texas

Hiroki Maeda
Graduate School of Engineering
Osaka University
Osaka, Japan

Gary C. H. Mo
Department of Pharmacology and Molecular Sciences
The Johns Hopkins University School of Medicine
Baltimore, Maryland

Denise Montell
Molecular, Cellular, and Developmental Biology Department
University of California, Santa Barbara
Santa Barbara, California

Hiroki Mutoh
RIKEN Brain Science Institute
Saitama, Japan

Robert K. Neely
Department of Chemistry
University of Birmingham
Birmingham, United Kingdom

Ellen C. O'Shaughnessy
Department of Pharmacology
and
Lineberger Comprehensive Cancer Center
The University of North Carolina at Chapel Hill
Chapel Hill, North Carolina

Yuko Sato
Graduate School of Bioscience and Biotechnology
Tokyo Institute of Technology
Yokohama, Japan

Ehud Segal
Department of Pathology
Stanford University School of Medicine
Stanford, California

Oliver Seitz
Department of Chemistry
Humboldt University Berlin
Berlin, Germany

Stefan Terjung
Advanced Light Microscopy Facility
European Molecular Biology Laboratory
Heidelberg, Germany

Thomas Thestrup
Max-Planck-Institut für Neurobiologie
Martinsried, Germany

Carol A. Vandenberg
Molecular, Cellular and Developmental Biology Department
University of California, Santa Barbara
Santa Barbara, California

Wim Vandenberg
Department of Chemistry
Katholieke Universiteit Leuven
Heverlee, Belgium

Laura van Weeren
Swammerdam Institute for Life Sciences
University of Amsterdam
Amsterdam, The Netherlands

Peter Varnai
Department of Physiology
Semmelweis University Faculty of Medicine
Budapest, Hungary

Xiaobo Wang
Laboratoire de Biologie Cellulaire et Moléculaire du Contrôle de la Prolifération
Université Paul Sabatier
Toulouse, France

Hong Xie
MassGeneral Institute of Neurodegenerative Disease
Massachusetts General Hospital
Charlestown, Massachusetts

Gary Yellen
Department of Neurobiology
Harvard Medical School
Boston, Massachusetts

Jason J. Yi
Department of Cell Biology and Physiology
The University of North Carolina at Chapel Hill
Chapel Hill, North Carolina

Jin Zhang
Department of Pharmacology and Molecular Sciences and the Solomon H. Snyder Department of Neuroscience
The Johns Hopkins University School of Medicine
Baltimore, Maryland

Section I

Introduction and Basics

Chapter 1

Engineering of Optimized Fluorescent Proteins

An Overview from a Cyan and FRET Perspective

Lindsay Haarbosch, Joachim Goedhart, Mark A. Hink, Laura van Weeren, Daphne S. Bindels, and Theodorus W. J. Gadella

CONTENTS

1.1 INTRODUCTION

1.1.1 The GFP Family

The green fluorescent protein (GFP) isolated from the jellyfish *Aequorea victoria* (avGFP) has revolutionized microscopy and cell biology. After the cloning of the gene (Prasher et al. 1992) and the demonstration that no specific jellyfish cofactor is needed for its production in other species (Chalfie et al. 1994), GFP soon became the most popular fluorescent tag in life sciences research. Starting in 1994, many research groups have worked on mutagenizing the GFP gene to isolate enhanced or spectroscopically altered variants of GFP. Besides random mutagenesis approaches, the crystal structure published in 1996 (Ormö et al. 1996) also allowed rationalized mutagenesis approaches. The crystal structure demonstrated that GFP consists of an outer β-barrel that is built up by 11 β-sheets. A short α-helix, to which the chromophore is attached, is positioned inside the β-barrel. The avGFP chromophore is formed via an autocatalytic reaction involving three consecutive amino acids: Ser65, Tyr66, and Gly67. Cyclization (imidazolinone formation) of the chromophore occurs by a nucleophilic attack of the amide of Gly67 (which is the best nucleophile because of the minimal steric hindrance) on the carbonyl of residue 65, followed by a dehydration step. The imidazolinone-conjugated system is then connected to the aromatic group by the oxidation of the α–β bond of residue 66 (Tsien 1998). Because the chromophore is buried inside the β-barrel, it is well protected from the extracellular environment, and GFP therefore shows remarkably similar spectroscopic properties in cells, in buffer solutions after purification, or even in protein crystals after crystallization (Chudakov et al. 2010).

Before the year 2000, the fluorescent protein (FP) family was limited to the avGFP-derived FPs (reviewed in Tsien 1998). Later, the FP family was extended after the cloning of the DsRed gene, encoding a red fluorescent protein (RFP) in *Discosoma* corals (Matz et al. 1999). From DsRed and related *Anthozoa*-derived

RFPs or chromoproteins that were subsequently identified and mutagenized, the color palette of FPs today spans the entire visible spectrum, from near ultraviolet (UV) to far-red (Shaner et al. 2004). In addition, several added features such as large Stokes shifts, monomerized versions, photochromic FPs, and photoswitchable FPs have become available (Shcherbakova et al. 2012). To date, all FPs known are 220–240 amino acids in length and adopt a similar β-barrel fold with a (monomeric) protein mass of about 27 kDa. For the coral-derived FPs, we refer to several extensive recent reviews for further information (Verkhusha and Lukyanov 2004; Chudakov et al. 2005; Alieva et al. 2008). Here we focus specifically on the avGFP-derived FPs, with a main emphasis on cyan fluorescent variants.

1.1.2 The avGFP-Derived Spectral FP Classes

From wild-type avGFP, eight classes of spectral FP variants (based on their excitation and emission properties) have been obtained through mutagenesis (Tsien 1998; Sawano and Miyawaki 2000).

The first class of avGFP is the wild-type chromophore (neutral phenol and anionic phenolate). This chromophore gives the most complicated spectrum, as it has a major excitation peak at 395 nm and a threefold smaller 475 nm peak, which yield emission peaking at 508 and 503 nm on excitation, respectively.

The second GFP class (phenolate anion) contains GFP variants with the S65T substitution. This mutation simplified the spectra (a single peak instead of multiple peaks) and resulted in red shifts of the excitation and emission peaks, which are 489 and 509 nm, respectively.

Substitution of the threonine at position 203 by an isoleucine resulted in the suppression of the 475 nm excitation peak and a small shift of the emission peak (511 nm) compared to wild-type GFP (λ_{em} 504 nm), which is observed in the third type of GFP chromophores (neutral phenol).

Inserting an amino acid with an aromatic ring (e.g., His, Trp, Phe, and Tyr) at position 203, in combination with the S65G substitution, showed a severe red shift in the spectra. As a consequence, this fourth spectral class was dubbed yellow fluorescent proteins (YFPs), although the fluorescence still appears more green than yellow. YFPs (phenolate anion with stacked π-electron) show only a small Stokes shift between excitation (peaking at 508–516 nm) and emission (peaking at 518–527 nm).

By replacing the phenolate with an indole moiety in the chromophore, another chromophore is formed in cyan fluorescent proteins (CFPs), which all have the Y66W substitution. Double excitation peaks (434 and 452 nm) and double emission peaks (476 and 505 nm) are found in the spectra of this type of chromophore.

The blue fluorescent proteins (BFPs), containing the Y66H mutation (imidazole in chromophore), are blue shifted compared to wild-type avGFP, with the excitation peak found at 384 nm and the highest emission within this class observed at a wavelength of 448 nm.

By replacing the phenolate moiety of the chromophore by a phenyl moiety through the Y66F mutation, a seventh class is obtained, which is still further blue shifted, with excitation and emission peaks at 360 and 442 nm, respectively. However, this class has no obvious practical use up until today.

The eighth and final class of GFP variants derived from avGFP is formed by cyan-green FP (CGFP). In CGFP, the T203Y mutation (known from the YFP class) is introduced (Sawano and Miyawaki 2000) in addition to the Y66W mutation as found in CFPs. This spectral class displays excitation and emission maxima at 455 and 506 nm, respectively.

GFP is used frequently in molecular cell biology for both structural (e.g., labeling of cells, organelles, or proteins) and functional (e.g., protein interactions, promoter activity, drug screening, and sensors) studies. For localization studies, GFP is attached to a protein of interest. This is accomplished by fusing together the encoding DNA, such that the GFP and the target protein are expressed and produced as a single protein in cells. Because each target protein has a GFP tag, both the temporal and spatial location of the protein can be followed by monitoring fluorescence. For gene expression studies, GFP is placed behind a promoter to monitor promoter activity. For both protein–protein interactions studies and for protease activity studies, Förster resonance energy transfer (FRET) pairs can be used. FPs can also be used as sensors, for example, to indicate the calcium concentration or the pH of the environment (Chudakov et al. 2010).

In this chapter, we do not focus on all the diverse applications of FPs or on describing in greater detail all the variants of each spectral class. Instead, we focus on the approaches used to optimize an FP, as well as on which features are important for optimization. We highlight these approaches and features by focusing on the CFP class. This class, with an indole moiety in the chromophore, does not occur in nature. Hence, no β-barrels have evolved to optimally accommodate this bulky noncharged chromophore. As a result, the route for engineered evolution to generate an optimal CFP variant has been quite long and many CFP variants with increasing brightness have been described. Another reason for highlighting CFPs is that they are the most frequently used donors in genetically encoded FRET sensors. We will demonstrate the benefits of using optimized CFPs in these sensors, as illustrated by specific examples.

1.2 OPTIMIZATION OF FPs

To improve the properties of FPs, their encoding DNA sequences need to be altered to introduce amino acid substitutions. This can occur either randomly (error-prone polymerase chain reaction [epPCR]) or by specifically mutating (a) codon(s) of interest (site-directed mutagenesis).

1.2.1 Random Mutagenesis

EpPCR requires the use of a low-fidelity DNA polymerase (e.g., *Taq* DNA polymerase) that allows the occurrence of mismatches in the newly synthesized DNA strand during the PCR. These mismatches occur at random positions in the DNA sequence; hence this method is also called "random mutagenesis." There are several aspects that have an influence on the mutation rate of the reaction. For example, introducing increasing amounts of Mn^{2+} in the PCR reaction mixture shows a linear correlation with the number of mutations found in the genetic sequence of the FP (i.e., doubling the [$MnCl_2$] results in approximately twice as many mutations in the DNA sequence) (Beckman et al. 1985; Koyanagi et al. 2008).

The mutation rate in a DNA sequence can also be controlled by varying the concentration of Mg^{2+}, using dNTPs in unequal concentrations (i.e., four times more dTTP and dCTP than dGTP and dATP), and by altering the total number of PCR cycles (Eckert and Kunkel 1990).

Controlling the mutation rate is important for FP improvement, as too many mutations might interrupt the folding process of the β-barrel and the chromophore, thereby causing the resulting protein to be non fluorescent. Too few alterations in the DNA sequence, on the other hand, might not show a clear improvement at all. In this respect, it should be noted that changing the third position of the codon often does not result in an amino acid change, in view of the degenerate genetic code (see Figure 1.1b).

1.2.2 Targeted Mutagenesis

Next to randomly changing the DNA sequences, it is also possible to make amino acid alterations at selected positions. For this site-directed mutagenesis, primers need to be designed that feature a mutation codon in their sequence. The mutation codons NNN or NNK will introduce all possible amino acid substitutions at a given position. However, in cases in which more background information is known (e.g., from previous studies and/or crystal structures), it is more efficient to insert only codons that encode a few amino acids. Next to the normal base symbols A, T, G, and C, there are 11 other symbols that encode two or more bases (see Figure 1.1) that can be combined to obtain a specific mutation codon. For instance, the codon TTW encodes either Phe or Leu. A single primer can contain one or more mutation codons, but the amount is limited, as the mismatches of the mutation codons interfere with the annealing of the primers to the template DNA sequence.

Bases	Description	Symbol
A or T	Weak	W
C or G	Strong	S
A or C	Amino	M
G or T	Keto	K
A or G	Purine	R
C or T	Pyrimidine	Y
C or G or T	Not A	B
A or G or T	Not C	D
A or C or T	Not G	H
A or C or G	Not T	V
A or C or G or T	Any base	N

(a)

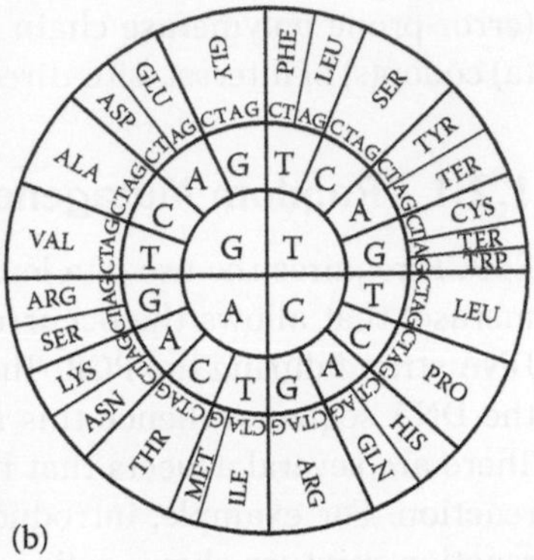

(b)

Figure 1.1 Creating site-directed mutation libraries using degenerate codons. For site-directed mutagenesis, codons can be made that encode only a specific set of amino acids. In the DNA primer, instead of one base, a mixture of bases is used at a specific elongation step during the synthesis of the DNA polymer. There are 11 combinations of two or more base mixtures, each represented by a unique symbol, as listed in (a). By comparing this with the genetic codes for amino acids (b), it follows, for example, that NNK and NNN encode for all possible amino acids but that NNK yields fewer stop codons (TER in b). Another example is the codon DYK, which encodes Ala, Ile, Leu, Met, Phe, Val, Ser, and Thr.

1.2.3 OmniChange

With regular site-directed mutagenesis, typically only one amino acid can be altered at a time by means of whole-vector PCR. A few approaches have emerged to introduce multiple (up to three) mutations at selected sites (Sawano and Miyawaki 2000) at once by whole-vector PCR, yet with varying saturation and efficiency. To introduce multiple amino acid substitutions in a single gene, usually several multicodon mutagenesis methods can be considered that differ in the number of altered amino acids produced and in the number of PCR rounds that are required. However, performing multiple PCRs on a single gene is very time consuming and decreases the variety of the library. The OmniChange method was introduced to overcome this inconvenience (Dennig et al. 2011). With this system, up to five NNK codons can be introduced simultaneously and independently of each other, without the use of restriction enzymes or ligases, by performing multiple PCRs of different lengths at the same time. OmniChange requires the use of primers that are partly phosphothioated (i.e., instead of oxygen, a sulfur molecule is inserted into the phosphate–sugar backbone of the DNA). These primers are designed in such a way that the forward and the reverse primers show an overlapping region of about 12 base pairs with phosphothioated bonds. Only the forward primers contain a mutation codon, positioned directly after the phosphothioated region. After all PCRs are performed, each overlapping region is found on two PCR products (see Figure 1.2). To reconstruct the vector, the PCR products are treated with an I_2/EtOH solution that reacts with the sulfur in the phosphothioated DNA backbones of the primers. This results in PCR products with (long) matching overhangs. Adding the PCR product to each other one by one makes it possible for hybridization to occur spontaneously at room temperature. These multiannealed vectors can then be directly transformed into bacteria by heat shock. A major advantage of this approach is that truly stochastic combinations of multiple targeted mutations can be made.

1.3 SCREENING METHODS

After mutagenesis is performed, a library of DNA molecules encoding new FP variants is obtained and needs to be screened to select the most improved variants. After the DNA library is transformed into bacteria, many bacterial colonies can be produced, each of which expresses a unique FP variant. Several possible screening methods for fluorescent proteins are highlighted in Sections 1.3.1–1.3.4 by illustrating their use for CFP optimization.

1.3.1 Intensity

The first and most straightforward screening approach is to check the fluorescence intensity of the bacterial colonies. For CFP, excitation at a wavelength of about 430 nm is required to induce maximal fluorescence emission. By comparing the intensity of the mutant colonies with those producing the nonmutated FP (i.e., the one that was used as the template for PCR), we aim to pick out colonies with increased intensity. However, this screening method has several serious drawbacks: The fluorescent intensity depends directly on gene expression level, excitation light intensity, optical thickness of the colony, protein folding efficiency, protein breakdown, and protein maturation. Hence, if a colony produces 10 times more of a suboptimal FP than another colony expressing an optimal FP, this screen will not identify the optimal FP, but perhaps the optimally transformed colony.

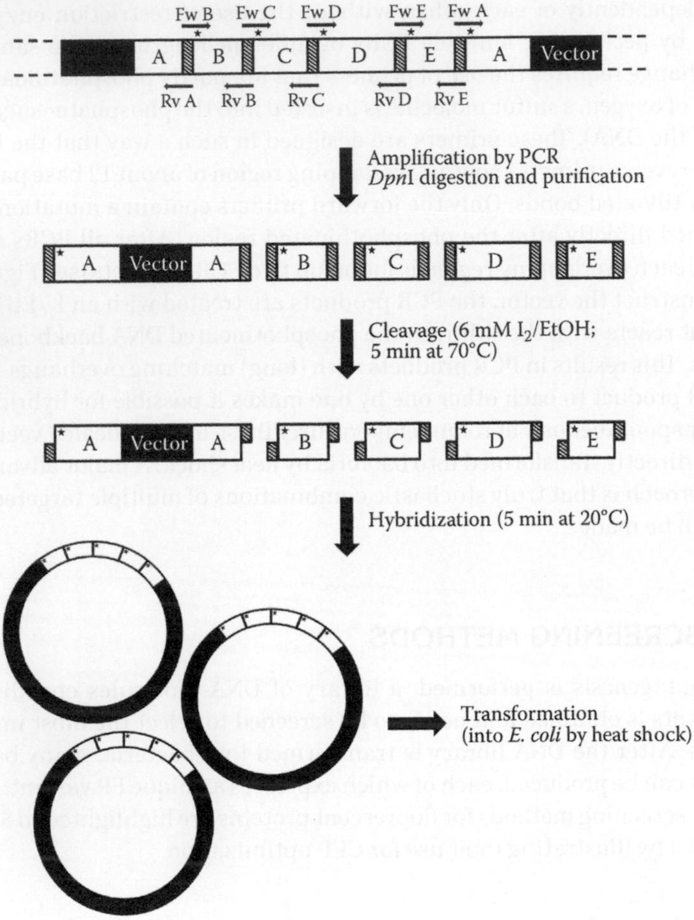

Figure 1.2 OmniChange: the simultaneous saturation of five independent codons. Primers are created with partly phosphothioated backbones. Each forward primer contains a certain mutation codon. Five independent polymerase chain reactions (PCRs) are performed on a chosen template FP. The PCR products are treated with *Dpn*I to destroy the template DNA. After purification, the PCR products are cleaved with an iodine solution for 5 min at 70°C to create overhangs (i.e., the iodine reacts with the sulfur in the phosphothioated backbones of the primers). Next, the pieces of DNA are hybridized one by one at room temperature, starting with the vector PCR product. Directly after the hybridization is completed, the resulting plasmid can be transformed into bacteria by heat shock. (Edited after Dennig, A. et al., *PLoS One* 6:e26222, 2011.)

1.3.2 Fluorescence Lifetime

A second way of highlighting improved FPs is by measuring the fluorescence lifetime (i.e., the average time an excited molecule spends in the excited state before decaying to the ground state). Importantly, the fluorescence lifetime is proportional to the fluorescence quantum yield but independent of excitation light intensity, detector efficiency, bacterial colony thickness, expression level, magnification used, and many other factors. Thus, fluorescence lifetime imaging microscopy (FLIM) can provide direct information on the intrinsic brightness of an FP, independent of its concentration, without the need for FP isolation and purification. Fluorescence lifetime measurements can be performed on bacterial colonies or after expressing the protein in mammalian cells.

There are two FLIM approaches: the time-domain approach and the frequency-domain approach. The first approach uses pulsed (fs–ps duration) excitation light and time-correlated detection of the induced decaying fluorescence emission after the pulse. The most common implementation is time-correlated single-photon counting on confocal microscopes (Becker 2012). In the frequency-domain method, the sample is excited with high-frequency (generally 40–80 MHz), sinusoidal intensity-modulated light. As a result, the fluorescence emission will be intensity modulated with the same frequency as the excitation light, but with a reduction in modulation depth and a shift in phase as compared to the excitation light. The modulation lifetime (τ_M) and the phase lifetime (τ_φ) are then calculated from the changes in the modulation depth and phase, respectively. In single-exponential FPs, the τ_φ and the τ_M are equal, but in FPs with a multiexponential (complex) decay, the phase and modulation lifetimes differ ($\tau_\varphi < \tau_M$).

1.3.3 Photobleaching

The photostability of FPs is substantially better than that of other fluorophores, because the chromophore is protected by the β-barrel. However, for longer experiments that require prolonged excitation, an even higher photostability can be desirable. To test the photostability of an FP, bleaching experiments are performed by exciting the FPs with high-intensity light sources. The time required to bleach the fluorescence intensity $1/e$-fold is called the photostability lifetime. Bleaching can be either monophasic or biphasic, meaning that a fluorescent protein bleaches at one constant rate (monophasic) or that it first shows fast bleaching followed by a decreased rate of bleaching after several seconds to minutes (biphasic) (Merzlyak et al. 2007). This method can be performed both on Petri dishes containing bacterial colonies (library bleaching) and on FPs (that were selected by other screening methods) expressed in mammalian cells. Besides irreversible photobleaching or photodestruction, short- or long-lived dark states can also be triggered inside FPs depending on the illumination power and light dose. This latter phenomenon can be especially pronounced with wide-field light sources. Reversible photobleaching (also called photochromism) can result from *cis–trans* isomerization or proton transfer reactions around the chromophore. Photochromism can be utilized for the development of reversibly switchable FPs (Shcherbakova et al. 2012), but for regular FPs, photochromic behavior is undesirable (Shaner et al. 2008).

1.3.4 Fluorescence-Activated Cell Sorting

The aforementioned screening methods all take some time if a large library of mutants FP needs to be screened. By using a fluorescence-activated cell sorting (FACS) machine, many thousands of bacteria expressing mutant FPs can be screened for fluorescence intensity within a few minutes (note that the expression vector is important for this method, as the amount of protein influences the intensity). After excitation with the 440-nm laser (for CFPs), the FACS machine measures the fluorescence intensity of a single droplet, and when the intensity is above the set threshold, the droplet is selected and sorted. To achieve reliable results, it is important that every drop contains only one bacterium. To obtain this, the OD_{600} of the bacterial culture needs to be measured and diluted to 0.01–0.03 in filtered phosphate-buffered saline (PBS) and mixed carefully before entering the FACS machine (Telford et al. 2012). Bacteria expressing high-intensity FPs can then be collected on Petri dishes or in multiwell plates (e.g., 96- or 384-well plates) containing the desired medium.

1.4 SPECTROSCOPIC CHARACTERIZATION/ PERFORMANCE

To determine the properties of FPs, these proteins are generally purified from *Escherichia coli*. Affinity tags are used to allow for easy purification with high yields. Often, the tags are not removed, as they are assumed not to interfere with fluorescence. Several of the parameters that are determined from purified proteins translate very well to *in vivo* behavior, such as the extinction coefficient (EC), quantum yield (QY), and fluorescence lifetime. Others, such as photostability, folding rate, and maturation, depend on the cellular environment. The latter parameters should therefore be verified in the experimental setting in which the FPs are used.

1.4.1 Extinction Coefficient

The extinction coefficient describes the amount of light that is absorbed by the chromophore. The higher the extinction coefficient is, the higher the chance of exciting a molecule will be. Therefore, high extinction coefficients are a desirable feature. The extinction coefficient is determined by measuring the amount of light that is absorbed by a known amount of purified protein. The major challenge is to obtain a pure FP sample in which all the FPs have a fully mature chromophore; contamination by immature FPs or other proteins will decrease the extinction coefficient. Another source of error is the protein concentration determination itself, as methods to determine protein concentrations are notoriously inaccurate. As for CFP variants, we have determined that the extinction coefficients between variants in this spectral class do not vary substantially. Typical CFP extinction values are 30,000 M^{-1} cm^{-1}, which is relatively low compared to green, yellow, and red FPs. Most likely, the EC is an intrinsic property of this chromophore structure, given the dimension of its absorption cross section, and it is unlikely that the EC can be significantly increased by mutations for FPs within the same spectral class.

1.4.2 Quantum Yield

The quantum yield is the ratio of the number of emitted photons (fluorescence) divided by the number of absorbed photons and is expressed as a number between 0 and 1. It determines how efficiently the excitation light is converted to fluorescence. The higher the QY, the brighter the FP. Quantum yields can be accurately determined using purified protein samples by comparing the fluorescence emission to a standard with a known QY. The QY of CFPs can vary between very low (0.2) and very high (0.9) and depends on mutations in the vicinity of the chromophore. In fact, the highest QY measured for a monomeric FP is determined for the CFP variant mTurquoise2. The origin of the extremely high QY is the stabilization of the chromophore by the surrounding amino acids. Consequently, the excitation energy is released almost exclusively via fluorescence rather than through nonradiative pathways owing to motions in the chromophore.

1.4.3 Theoretical Brightness

Choosing the best FP is often based on the theoretical brightness, which can be approximated by the product of the extinction coefficient and the quantum yield. The theoretical brightness is often used to judge the quality of an FP. Unfortunately, this parameter is heavily dependent on the extinction coefficient, which may have a large error margin. Several alternative methods exist to compare the relative brightness of FPs in the same spectral class. We have developed an *in vivo* assay (see Section 1.4.6) and an assay that measures the molecular brightness of purified proteins. To this end, we have used fluorescence correlation spectroscopy (FCS), which measures the brightness of individual molecules. Because FCS depends on fluorescence, only the molecules with a mature chromophore are measured. For CFPs, we have found that the brightness determined by FCS correlates with the QYs that we have determined, confirming the notion that the extinction coefficients for different CFPs are similar.

1.4.4 Fluorescence Lifetime

The fluorescence lifetime is the average time that an excited molecule spends in the excited state, before decaying back to the ground state. Fluorescence lifetimes are positively correlated with quantum yield, so longer lifetimes correlate with high QYs. Although the development of mCerulean and SCFP3A led to brighter CFPs, these variants were characterized by multiexponential decays (Kremers et al. 2006; Walther et al. 2011). Introducing the T65S mutation in CFPs yielded variants with monoexponential decay (Goedhart et al. 2010; Walther et al. 2011). The mTurquoise2 variant has a lifetime of 4 ns, which is among the highest lifetimes of all known FPs. Like the QY, the high fluorescence lifetime is caused by the stabilization of the chromophore.

1.4.5 Maturation

After the β-barrel folds, the three inner amino acids undergo a chemical reaction that requires other, catalytic residues. For avGFPs, chromophore formation occurs in two steps. First, cyclization/dehydration takes place, after which oxidation can occur. The latter step requires molecular oxygen and is necessary to yield a mature chromophore capable of absorbing and emitting light. Ideally, maturation is fast and complete. Several assays exist that measure maturation. One type of assay employs protein production and isolation under oxygen-free conditions, after which the proteins are exposed to oxygen to measure chromophore maturation. Another assay monitors the development of fluorescence in growing *E. coli* in real time. If cell growth and protein production are similar for different bacterial cultures producing FP variants, this assay can be used to measure differences in protein folding and chromophore maturation. Intriguingly, this assay highlighted a delayed maturation in T65S CFP mutants. This delay was repaired with the 146F mutation in mTurquoise2. The exact mechanism underlying the differences in maturation speed is currently unknown. In an attempt to increase the maturation speed of mTurquoise2 even further, we introduced four folding-enhancing mutations (S30R, Y39N, N105T, and I171V) that were described previously for the super-folder GFP (Pédelacq et al. 2006). Remarkably, introducing these super-folding mutations into mTurquoise2 yielded no apparent beneficial folding characteristics in either bacteria or mammalian cells, while the spectral properties were unchanged (Goedhart, unpublished observation).

1.4.6 Brightness in Cells

To scrutinize the brightness of FPs in eukaryotic cells, we have developed a novel, straightforward assay. The assay is based on quantitatively coproduced FPs. Because the production of an FP varies tremendously between cells on transient transfection with plasmids, we developed a coexpression system that uses an intrinsic control for the transfection, expression, and production efficiency. Two FP variants are used that can be spectrally separated. The FPs are expressed from a single plasmid, wherein one of the two proteins serves as an expression control, in our case an YFP, and is identical for all experiments. The other FP, for example, CFP, is varied. Plasmids are introduced into cells, and the CFP fluorescence is related to the YFP fluorescence (see Figure 1.4c). The brighter variants give higher signals. This assay determines *in vivo* brightness, which is determined by EC, QY, and expression/maturation efficiency (Goedhart et al. 2010).

1.4.7 Fusion Constructs

Assessing the performance in fusion constructs is an important first step to characterize the *in vivo* performance of an FP as a tag to visualize proteins of interest. Generally, a number of fusion constructs are made and expressed in eukaryotic cells. Most important are fusions that have a certain requirement for correct localization. For instance, it has been shown that tubulin or connexins do not localize correctly if the FP forms a dimer (Shaner et al. 2008). Conversely, proper localization has been shown with FPs that later turned out to have a tendency to dimerize. Therefore, the localization assay is rather qualitative, as it does not give a full representation of the *in vivo* performance. The biofunctionality of a particular fusion protein must always be verified under well-defined conditions. In our experience, CFP variants with the A206K mutation, which prevents dimerization, behave as monomeric FPs and can replace EGFP in all applications (Goedhart et al. 2010, 2012). Also, we have shown that the variant mTurquoise can be introduced into a loop of a protein, yielding a functional fusion protein (Adjobo-Hermans et al. 2011).

1.5 EVOLUTION OF ENGINEERED CFPs

1.5.1 An Overview of 18 Years of Work on the CFP Spectral Class

The first description of a CFP dates back to 1994, when Heim et al. (1994) demonstrated a blue-shifted mutant of the avGFP with the substitution Y66W (see Table 1.1 for amino acid changes). This mutant FP was not further characterized for spectroscopic properties. Heim and Tsien (1996) then published improved

TABLE 1.1 AMINO ACID MUTATIONS IN CFP VARIANTS

	64	65	66	72	123	145	146	147	148	153	163	166	167	168	169	175	206	224
WTGFP	F	S	Y	S	I	Y	N	S	H	M	V	K	I	R	H	S	A	V
GFP(Y66W)	F	S	*W*	S	I	Y	N	S	H	M	V	K	I	R	H	S	A	V
W2	F	S	*W*	S	*V*	*H*	N	S	*R*	**T**	**A**	K	I	R	H	S	A	V
W7	F	S	*W*	S	I	Y	*I*	S	H	**T**	**A**	K	I	R	H	S	A	V
ECFP	*L*	*T*	*W*	S	I	Y	*I*	S	H	**T**	**A**	K	I	R	H	S	A	V
mCerulean	*L*	*T*	*W*	*A*	I	*A*	*I*	S	*D*	**T**	**A**	K	I	R	H	S	**K**	V
Cerulean2	*L*	*T*	*W*	*A*	I	*A*	*I*	**H**	*G*	**T**	**A**	**G**	*L*	**N**	*C*	S	**K**	V
Cerulean3	*L*	S	*W*	*A*	I	*A*	*I*	**H**	*G*	**T**	**A**	**G**	*L*	**N**	*C*	S	**K**	V
Cerulean-T65S	*L*	S	*W*	*A*	I	*A*	*I*	S	*D*	**T**	**A**	K	I	R	H	S	**K**	V
SCFP2	*L*	*T*	*W*	*A*	I	Y	*I*	S	H	**T**	**A**	K	I	R	H	**G**	**K**	V
SCFP3A	*L*	*T*	*W*	*A*	I	Y	*I*	S	*D*	**T**	**A**	K	I	R	H	**G**	**K**	V
SCFP3B	*L*	*T*	*W*	*A*	I	*A*	*I*	S	*D*	**T**	**A**	K	I	R	H	**G**	**K**	V
mTurquoise	*L*	S	*W*	*A*	I	Y	*I*	S	*D*	**T**	**A**	K	I	R	H	**G**	**K**	V
mTurquoise GV	*L*	S	*W*	*A*	I	Y	*I*	S	*G*	**T**	**A**	K	I	R	H	**G**	**K**	V
mTurquoise GL	*L*	S	*W*	*A*	I	Y	*I*	S	*G*	**T**	**A**	K	I	R	H	**G**	**K**	*L*
mTurquoise2	*L*	S	*W*	*A*	I	Y	*F*	S	*D*	**T**	**A**	K	I	R	H	**G**	**K**	V
mTurquoise2-G	*L*	S	*W*	*A*	I	Y	*F*	S	*G*	**T**	**A**	K	I	R	H	**G**	**K**	V

Note: Bold are folding/monomerization mutations; italics are interior and/or QY changing.

CFPs. In a first round of mutagenesis, they introduced two folding mutations, M153T and V163A. After a second round, two variants were described: one called W2, with the additional mutations I123V, Y145H, and H148R; and another called W7, with one mutation N146I. In this study, published in 1996, the extinction coefficients and quantum yields of W2 and W7 were listed as 10,000 and 18,000 M^{-1} cm^{-1} and 0.72 and 0.67, respectively. In a later study, these numbers were revised to 23,900 M^{-1} cm^{-1} and QY 0.42 for W7 (Cubitt et al. 1999, reviewed in Tsien 1998). In the review of Tsien (1998), two additional cyan mutants were described called ECFP (also dubbed W1B) and W1C. ECFP differs from W7 by two amino acids (F64L, S65T) and shows improved maturation (extinction coefficient 32,500 M^{-1} cm^{-1}) but also a reduced quantum yield of 0.4. The other mutant (W1C) differs by just one mutation from W7, which is S65A, and shows quite similar properties as ECFP (EC 21,200 M^{-1} cm^{-1}, QY 0.39). In 2004, Rizzo et al. described several enhanced CFPs, of which the most enhanced was Cerulean. Three mutants were described that differ from ECFP: ECFP/H148D (EC 32,000 M^{-1} cm^{-1} and QY 0.68), a D10 variant ECFP/S72A/Y145G/H148D (EC 44,000 M^{-1} cm^{-1} and QY 0.46), and Cerulean (ECFP/S72A/Y145A/H148D) (EC 43,000 M^{-1} cm^{-1} and QY 0.62). Cerulean was notably enhanced as compared to ECFP and became widely used in cell biology and in FRET applications. However, the extent of QY and EC improvement, as well as the monoexponential decay, that was claimed in this paper was not reproduced in other laboratories. The corrected values for Cerulean are EC 33,000 M^{-1} cm^{-1}, QY 0.49 (see Table 1.2). A remarkable feature of the published mutants is the quantum yield enhancement of the H148D mutant of ECFP, which is less pronounced in both D10 and Cerulean. In 2006, Kremers et al. published a different branch of optimized CFPs derived from ECFP, with improved folding mutations, dubbed SCFP2 (= ECFP/S72A/S175G/A206K). The latter mutation also greatly reduces the dimerization tendency of the fluorescent protein. SCFP2 has a more or less identical EC and QY as that of ECFP. However, its folding and maturation are markedly enhanced in bacteria as compared to ECFP (Kremers et al. 2006). By introducing the largely beneficial mutation H148D (Rizzo et al. 2004), SCFP3A was made (= SCFP2/H148D). SCFP3A has an unaltered EC but displays an increased QY of 0.56. Introducing the additional Y145A (which is also found in Cerulean) reduced the QY to 0.5 for SCFP3B (= SCFP2/H148D/Y145A). Hence both studies (Rizzo et al. 2004; Kremers et al. 2006) indicate that H148D is seriously beneficial for the QY of CFP, but Y145A reverses the increased QY to some extent. In 2010, Goedhart et al. showed that the reversal of the S65T mutation in SCFP3A yielded a protein (dubbed mTurquoise = SCFP2/T65S/H148D) with a QY of 80% (without affecting the EC). Other mutants, mTurquoise-GV (= SCFP2/T65S/H148G) and mTurquoise-GL (= SCFP2/T65S/H148G/V224L), were also described with similarly high QYs but with slightly reduced maturation in bacteria. This demonstrates

TABLE 1.2 VERIFIED AND PUBLISHED PROPERTIES OF CYAN FLUORESCENT PROTEINS

CFP Variant	Reference	Extinction $M^{-1} cm^{-1}$	QY	EC*QY (% of mTq2)	Brightness *E. coli*	Brightness HeLa	Tau (phase)	Tau (mod)	Bleaching % in HeLa
ECFP	Kremers et al. 2006	28,000	0.36	36	20	37	2.3	3.0	13
	Cubitt et al. 1999	*32,500*	0.4	*47*				3.0[a]	
	Rizzo et al. 2004	29,000	0.37	38				3.2[a]	
	Lelimousin et al. 2009	30,000	0.3	32				*2.5*[a]	
	Fredj et al. 2012	29,000	0.4	42					
Cerulean	Kremers et al. 2006	33,000	0.49	56	46	57	2.3	3.1	6
	Rizzo et al. 2004	*43,000*	*0.62*	*96*				3.3[a]	
	Markwardt et al. 2011	*43,000*	0.48	*74*				3.17[a]	
	Lelimousin et al. 2009	31,000	0.44	49				*3.8*[a]	
	Fredj et al. 2012	29,200	*0.67*	*70*				3.05[a]	
SCFP3A	Kremers et al. 2006	30,000	0.56	60	52	66	2.7	3.2	7
SCFP3B	Kremers et al. 2006	30,000	0.50	54	—	—	2.3	3.1	—

mTurquoise	Goedhart et al. 2010 Markwardt et al. 2011 Fredj et al. 2012	30,000 *34,000* *36,200*	0.84 0.84 0.85	90 *102* *110*	72	85	3.7	3.8 4.04[a] 4.06[a]	3
mTurquoise-GV	Goedhart et al. 2010	23,000	0.82	68	—	79	3.9	4.0	9
mTurquoise-GL	Goedhart et al. 2010	27,000	0.82	80	—	83	3.9	4.1	7
Cerulean2	This study Markwardt et al. 2011	29,000 *47,000*	0.6 0.6	62 *101*	19	41	2.5	3.1 3.04[a]	33
Cerulean3	Goedhart et al. 2012 Markwardt et al. 2011	30,000 *40,000*	0.80 *0.87*	86 *124*	77	83	3.7	3.8 4.10[a]	39
Cerulean-T65S	Fredj et al. 2012	*34,800*	0.84	*104*	—	—	—	3.96[a]	—
mTurquoise2	Goedhart et al. 2012	30,000	0.93	100	100	100	3.8	4.0	3
mTurquoise2-G	Goedhart, pers. comm.	—	0.92	—	—	—	3.8	4.0	4

[a] Fluorescence lifetime determined by time-correlated single-photon counting FLIM.

that not only H148D but also H148G can be beneficial. After crystallization of mTurquoise and SCFP3A, the structural basis for the improvement became apparent (Goedhart et al. 2010). Very recently, the T65S mutation was shown to be beneficial not only for SCFP3A but also for ECFP (Erard et al. 2013) and for mCerulean (Fredj et al. 2012). Remarkably, adding H148D or H148G to the T65S mutation in ECFP yielded CFPs with fluorescence lifetimes similar to that of mTurquoise (Erard et al. 2013). Goedhart et al. (2012) also found one suboptimal residue around the chromophore (I146); mutating this residue to I146F yielded the brightest CFP variant published to date, dubbed mTurquoise2 (= SCFP2/T65S/I146F/H148D). This CFP displays an unchanged EC but a markedly improved fluorescent QY of 93%. In parallel, Markwardt et al. (2011) attempted to optimize Cerulean. They published Cerulean2 (= Cerulean/S147H/D148G/K166G/I167L/R168N/H169C), with an increased QY of 0.6, but with reduced photostability. After the introduction of the mTurquoise mutation T65S into mCerulean2, they obtained the markedly improved CFP variant Cerulean3 (= Cerulean/T65S/S147H/D148G/K166G/I167L/R168N/H169C). In this study, substantial improvements in the EC and photostability were claimed for mCerulean3 as compared to mTurquoise, but our detailed side-by-side comparison indicates that mTurquoise is slightly brighter than mCerulean3. The fluorescence QY and fluorescence lifetime of mCerulean3 are also slightly lower than for mTurquoise. In addition, experiments showed that the ECs of all Cerulean variants are the same (about 30,000 M^{-1} cm^{-1}) and that the photostability of Cerulean3 in mammalian cells is far worse as compared to mTurquoise (see Figure 1.3). To test whether a glycine instead of the aspartate at position 148 could be beneficial for mTurquoise2, we introduced this mutation and crystallized this protein along with the mTurquoise–D148G variant (von Stetten et al., PDB entry 4B5Y). It was found that in mTurquoise the D148G mutation caused increased van der Waals contact with the isoleucine at position 146. Introducing the D148G mutation into mTurquoise2 with a phenylalanine at position 146 demonstrated that no further stabilization of the contacts between the chromophore and Phe146 could be achieved. Instead, this study revealed that the more flexible β-strand 7, with the glycine at position 146, now introduced a slightly higher susceptibility toward photobleaching in mTurquoise2-G as compared to mTurquoise2 (von Stetten et al., PDB entry 4B5Y). Hence, the current most optimal CFP remains mTurquoise2 (see Table 1.2).

1.5.2 Zooming in on Photostability in mCerulean2, mCerulean3, and mTurquoise2

The strikingly decreased photostability of mCerulean3, as compared to mTurquoise or mCerulean, prompted us to further analyze the properties of mCerulean2 that differ from mCerulean3 only by the T65S mutation. This mutation caused a serious enhancement of mTurquoise as compared to SCFP3A. Given the remarkable photostability of mTurquoise, our assumption was that the decreased photostability of mCerulean3 could have been introduced during the mutagenesis of mCerulean that yielded mCerulean2. Indeed, we observed a serious reduction in the photostability of mCerulean2 as compared to mCerulean. The photobleaching kinetics of mCerulean2 were more complex than that of mCerulean3, as an additional fast bleaching component was found (see Figure 1.3). Next, we tested the brightness of mCerulean2 in bacteria and in mammalian cells and found that this protein was less bright than mCerulean

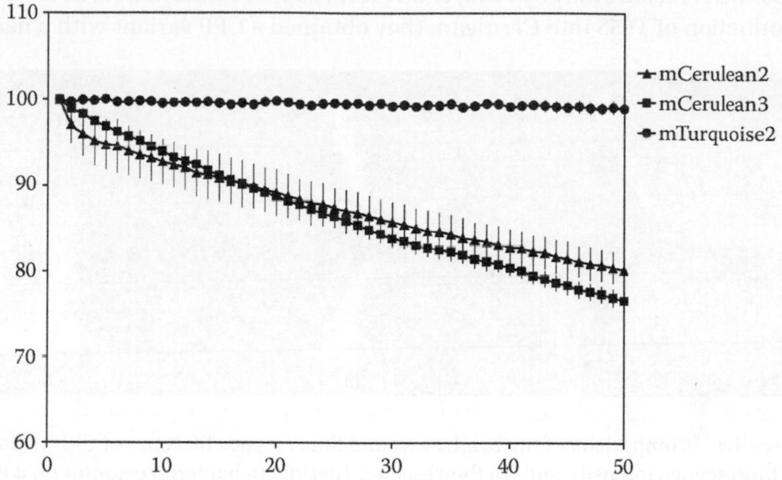

Figure 1.3 Photostability of mCerulean2, mCerulean3, and mTurquoise2 in HeLa cells using wide-field illumination. mCerulean2 was generated after introducing the S65T mutation into mCerulean3 (which was obtained from Dr. R.N. Day). mCerulean2, mCerulean3 and mTurquoise2 were cloned into a C1 vector and transfected into HeLa cells as described (Goedhart et al. 2012). At 24 h after transfection, the cells were mounted in microscopy medium (140 mM NaCl, 5 mM KCl, 1 mM $MgCl_2$, 1 mM $CaCl_2$, 10 mM glucose, 20 mM HEPES, pH 7.4) at room temperature. The cells were illuminated with light from a continuous 100-mW Hg lamp that was filtered with an ND 1.3 gray filter and a 436/20 excitation filter and reflected onto the sample with a 455 DCLP dichroic mirror (Chroma Technology). A 63× NA 1.4 oil immersion objective (Zeiss, Plan Apochromat) and a Zeiss Axiovert 200M inverted microscope were used, resulting in a light intensity of 1.4 W/cm^2 at the sample. On continuous illumination, the fluorescence emission from CFP produced in the cells was filtered through a 480/40 nm emission filter (Chroma), and images were captured with a Princeton Instruments CCD camera (with a Kodak 1300 KAF chip and ST133 controller). The average intensities of several cells were measured as a function of time and corrected for background intensity using an Image J processing routine. The intensity was normalized to the initial fluorescence intensity.

in both cell systems and yielded results comparable to ECFP (see Table 1.2 and Figure 1.4). We conclude that the mutagenesis of mCerulean into mCerulean2 was a step backward rather than forward. Many of the effects of this step backward could be rescued by the T65S mutation identified previously in mTurquoise, but the decreased photostability remains in mCerulean3. It is possible that removing the bulk from the side of the β-barrel by introducing the 148G and 166G mutations into mCerulean2 and mCerulean3 increased the susceptibility of the chromophore toward solvent oxidizing agents. Also, the positioning of a redox-sensitive cysteine 169 near the chromophore may cause problems. A recent study by Fredj et al. (2012) supports this notion, as after the introduction of T65S into Cerulean, they obtained a CFP variant with a nearly

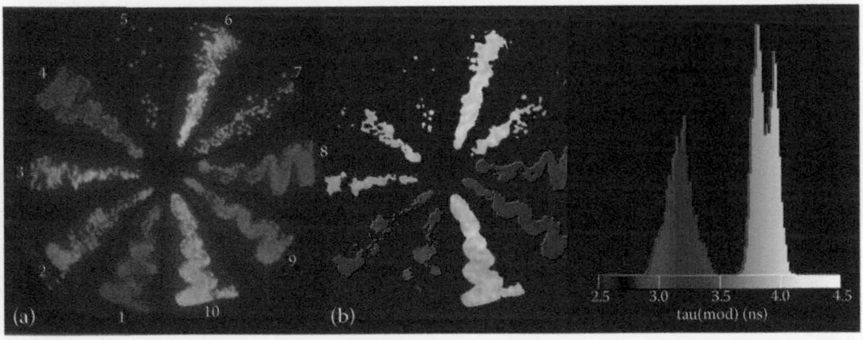

Figure 1.4 Comparison of the brightness and fluorescence lifetimes of CFP variants. (a) Fluorescence intensity and (b) fluorescence lifetime in bacterial colonies on a Petri dish. 1, ECFP; 2, SCFP3A; 3, mTurquoise; 4, mTurquoise-GV; 5, mTurquoise-GL (with just a few colonies); 6, mTurquoise2; 7, mTurquoise2-G; 8, mCerulean; 9, mCerulean2; 10, mCerulean3. The modulation fluorescence lifetime image in (b) is pseudocolored according to the histogram of lifetime values indicated at the right. Note the slightly higher lifetime of mTurquoise2 as compared to mTurquoise and mCerulean3 in bacteria. For (a) and (b), constructs encoding the diverse CFP variants were cloned into an RSET expression vector, and this was transformed into BL21 (DE3) *E. coli* bacteria as described (Goedhart et al. 2010, 2012). mTurquoise2-G was generated after introducing the E148G mutation into mTurquoise2. All other constructs were described previously (Goedhart et al. 2010, 2012). All constructs were verified by sequencing. Subsequently, bacteria were spread out on a Petri dish with LB-Agar medium. At 48 h after transformation, the fluorescence intensity of the bacteria was measured using the Hg lamp, microscope, filter set, and camera described in Figure 1.3, but without the ND 1.3 filter, and the 63× lens was replaced by a low-magnification (80-cm focal distance) glass plano-convex lens (Melles-Griot, LPX-25.0-415.0-C-SLMF-400-700). The Petri dish was mounted using a 70-cm poster tube, placing the Petri dish at the focal distance of this lens (Goedhart et al. 2010). Lifetime imaging of the Petri dish at 440 nm excitation was performed as described (Goedhart et al. 2010).

(*Continued*)

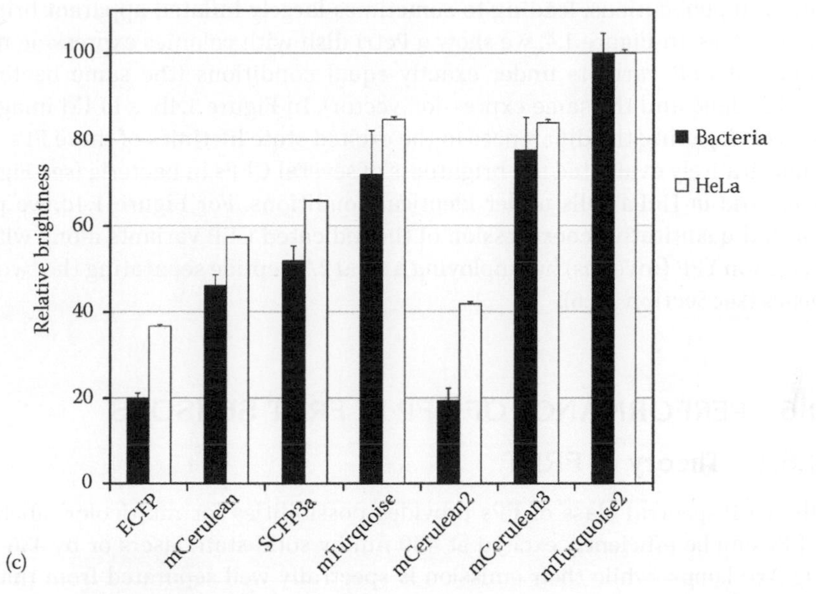

Figure 1.4 (Continued) Comparison of the brightness and fluorescence lifetimes of CFP variants. (c) Brightness of the indicated constructs in bacterial liquid cultures or in HeLa cells 24 h after transfection. For (c), the diverse CFP variants were expressed from an RSET expression vector in BL21 (DE3) *E. coli* bacteria in liquid LB medium, and the maximum fluorescence intensity was measured after 18 h at identical cell densities as described (Goedhart et al. 2010, 2012). The average brightness of three independent experiments (+stdev) was normalized to that observed for mTurquoise2 (black bars). Brightness in HeLa cells was measured as described by Goedhart et al. (2010, 2012) using a CFPX-mVenus construct separated by a viral 2A sequence, resulting in 1:1 stoichiometric coproduction of CFPX and mVenus. The relative brightness of CFPX (indicated in the figure by white bars) over YFP fluorescence, normalized to that for mTurquoise2, is shown.

identical QY and fluorescence lifetime to that of mTurquoise (see Table 1.2), as well as increased photostability and lower reversible photobleaching as compared to Cerulean.

1.5.3 Parallel Comparison of CFPs

We present an overview of the diverse mutations found in CFP variants in Table 1.1, as well as an overview of their published spectral properties in Table 1.2, which also includes our own independent reevaluation of the published values after detailed cross-examination. In Table 1.2, a few numbers are printed in italics; these represent the overestimation of the spectral properties in some original publications, leading to sometimes largely inflated apparent brightness values. In Figure 1.4, we show a Petri dish with colonies expressing nine different CFP variants under exactly equal conditions (the same bacterial background and the same expression vector). In Figure 1.4b, a FLIM image is shown depicting the differences in the excited state lifetimes of these FPs. We quantitatively evaluated the brightness of several CFPs in bacteria (see Figure 1.4c) and in HeLa cells under identical conditions. For Figure 1.4c, we performed quantitative coexpression of the indicated CFP variants along with a common YFP (mVenus) by employing a viral 2A peptide separating the two FP genes (see Section 1.4.6).

1.6 PERFORMANCE OF CFP IN FRET SENSORS

1.6.1 Theory of FRET

The CFP spectral class of FPs provides possibilities for multicolor labeling. CFPs can be efficiently excited at 440 nm by solid-state lasers or by 436-nm Hg Arc lamps, while their emission is spectrally well separated from that of the YFP and RFP spectral classes. Still, their most important application is in genetically encoded FRET sensors and in FRET-based protein–protein interaction studies. This is because the CFP–YFP FRET pair outcompetes the GFP–RFP, YFP–RFP, and OFP–RFP combinations for ratiometric FRET, even if these other pairs show a higher FRET efficiency (*E*; as determined via donor quenching). This is due to the lower quantum yield of the RFP spectral class as compared to YFP. To understand this better, we first introduce some basic equations that quantitatively describe FRET.

FRET is the nonradiative transfer of a quantum of energy from an excited molecule (called the donor) to a neighboring acceptor chromophore (called the acceptor). The efficiency of FRET is defined as the number of quanta transferred between a donor and an acceptor divided by the quanta absorbed by the donor (see Equation 1.1). As a result, fewer quanta of energy are available for donor fluorescence, and the probability that the donor will decay to the ground state is also increased. This decreases the duration of the excited state and hence decreases the donor fluorescence lifetime. In the event the acceptor is also a fluorophore (as in a CFP–YFP FRET pair), FRET will enhance the acceptor fluorescence, which is referred to as sensitized emission. The degree of sensitized emission, as compared to directly excited acceptor fluorescence (i.e., in the absence of FRET), is also proportional to the FRET efficiency, *E*, according to

$$E = \frac{\text{number of transferred quanta}}{\text{number of absorbed quanta}} = 1 - \frac{I_{DA}}{I_D} = 1 - \frac{\tau_{DA}}{\tau_D} = \frac{\varepsilon_A}{\varepsilon_D}\left(\frac{I_{AD}}{I_A} - 1\right) \quad (1.1)$$

in which I_{DA} and I_D represent donor fluorescence intensity in the presence and absence of the acceptor; τ_{DA} and τ_D are the donor fluorescence lifetime in the presence and absence of the acceptor; I_{AD} and I_A are the acceptor fluorescence intensity in the presence and absence of the donor; and ε_A and ε_D are the acceptor and donor ECs at the chosen (donor) wavelength used to excite the FRET pair.

The energy transfer efficiency, E, is related to the separation between the donor fluorophore and the acceptor chromophore and their intrinsic Förster radius:

$$E = \left(1 + \frac{R^6}{R_0^6}\right)^{-1} \tag{1.2}$$

The very steep (inverse sixth power) distance dependence depicted in Equation 1.2 is key to the FRET phenomenon and underlies the ability to probe intermolecular distances and conformational changes within sensors. The Förster radius (R_0) is related to the overlap between donor fluorescence emission and acceptor absorbance according to

$$R_0^6 = Cn^{-4} Q_D \kappa^2 J \tag{1.3}$$

$$J = \int f_D(\lambda) \cdot \varepsilon_A(\lambda) \cdot \lambda^4 \, d\lambda \tag{1.4}$$

in which R_0^6 has units nm^6, $C = 8.79 \times 10^{-11}$ (units M cm nm^2), n is the refractive index, Q_D is the donor fluorescence quantum yield; κ^2 is the orientation factor (dependent on the relative orientation of the acceptor dipole moment to the orientation of the dipole field of the donor; it is usually assumed that $\kappa^2 = \frac{2}{3}$, representing random orientations for both); and J is the spectral overlap integral. This integral is the wavelength-integrated multiplication of the donor fluorescence spectrum $f_D(\lambda)$, the acceptor absorbance spectrum $\varepsilon_A(\lambda)$, and the wavelength to the fourth power, where $f_D(\lambda)$ is normalized so that the wavelength-integrated donor fluorescence spectrum equals unity (Equation 1.4) (Clegg 2009). From inspecting Equations 1.3 and 1.4, it can be seen that the Förster radius for a FRET pair (and hence also the FRET efficiency E) increases with increasing donor quantum yield, Q_D, and increasing acceptor absorption, ε_A (overlap with the donor fluorescence).

From the equations, it is obvious that FRET measurements utilizing only donor fluorescence intensity or lifetime are dependent only on E (see Equation 1.1) and that higher donor QYs provide higher E values (see Equations 1.2 and 1.3). In fact, E and donor-based FRET measurements are completely independent of the acceptor QY. Even a nonfluorescent acceptor can be used for donor-based FRET measurements (Ganesan et al. 2006). However, not only E but also the acceptor brightness (or QY) is important when considering ratiometric FRET, as the latter is proportional to the extent of sensitized acceptor emission. Hence, high quantum yield acceptors, preferably ones with good spectral separation from the donor fluorescence, are important for ratiometric FRET. Further, optimal acceptors will exhibit low absorbance at the wavelength used for donor excitation (providing low direct acceptor excitation) and high absorbance overlapping with the donor fluorescence.

1.6.2 Simulated FRET Spectra of a CFP–YFP Fusion

The different requirements for donor-based FRET methods (e.g., FLIM) and ratiometric FRET methods have often led to confusion about which FP pairs are the best to use for FRET. Clearly the best FRET (highest value of E) is obtained for pairs that have the highest R_0, so YFP–RFP and OFP–RFP FRET pairs that display higher R_0 values (6.1–6.4 nm) than mTurquoise2-mVenus, with an R_0 of 5.9 nm, should also be considered (Goedhart et al. 2007, 2012; Shcherbo et al. 2009; Lam et al. 2012). However, the best ratiometric FRET responses can be observed from FRET pairs that show even suboptimal FRET (lower E values) but a high acceptor/donor QY and donor/acceptor fluorescence separation. In Figure 1.5, this is worked out for a theoretical CFP–YFP ratiometric FRET sensor. The sensor consists of a 1:1 fusion of CFP–YFP, and here the CFP QY is varied systematically, as is the distance between the donor and the acceptor. $F(\lambda)$ is the total FRET spectrum:

$$F(\lambda) = \varepsilon_D(1 - E)Q_D f_D(\lambda) + (\varepsilon_A + E\varepsilon_D)Q_A f_A(\lambda) \tag{1.5}$$

where $f_D(\lambda)$ and $f_A(\lambda)$ are the normalized donor and acceptor fluorescence spectra, respectively (of which the wavelength-integrated intensity equals unity), and ε_D and ε_A are the ECs of the donor and the acceptor at the (donor) excitation wavelength used. We used an excitation wavelength of 435 nm in the simulation, yielding a CFP extinction coefficient (ε_D) of 29,300 M^{-1} cm^{-1}, a YFP extinction coefficient (ε_A) of 2900 M^{-1} cm^{-1}, and a constant acceptor quantum yield (Q_A) of 0.64 (the latter two values corresponding to the spectral properties of mVenus). The $f_D(\lambda)$ and $f_A(\lambda)$ spectra used for the simulation are derived from the measured corrected spectra of mTurquoise and mVenus (Nagai et al. 2002; Kremers et al. 2006; Goedhart et al. 2010). In Figure 1.5a, the FRET spectra are shown a case in which the distance between CFP and YFP is held constant at 5 nm but with an increasing donor QY. Noticeably, a strong sensitized YFP emission is produced. From a careful inspection of Figure 1.5a, it is clear that the separation of the lines at the acceptor peak is larger than at the donor peak, which means (considering the lower donor fluorescence at the acceptor peak) that the sensitized emission increases with increasing donor quantum yield. Furthermore, it is obvious from looking at the donor and acceptor peaks that

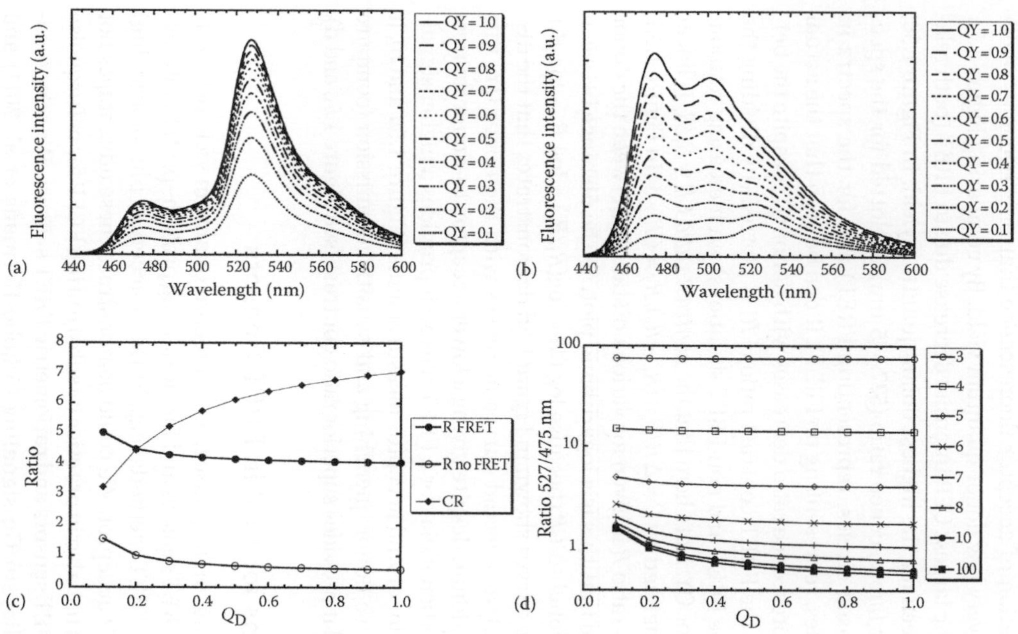

Figure 1.5 FRET simulation for a 1:1 CFP-mVenus fusion construct. (a) Fluorescence emission spectra for a FRET situation at 5 nm CFP-mVenus separation as a function of CFP quantum yield. (b) Fluorescence emission spectra for a non-FRET situation (at 100 nm CFP-mVenus separation) as a function of CFP quantum yield. (c) Ratio of mVenus/CFP fluorescence as a function of CFP quantum yield, at 5 nm separation (R FRET, closed circles) and at 100 nm separation (R no FRET, open circles), and the contrast ratio (RFRET/R no FRET). (d) Ratio of mVenus/CFP fluorescence emission at the indicated distances between CFP and mVenus as a function of CFP quantum yield. For the simulation, we used a CFP extinction coefficient (ε_D) of 29,300 $M^{-1} cm^{-1}$, a YFP extinction coefficient (ε_A) of 2900 $M^{-1} cm^{-1}$, and a constant acceptor quantum yield (Q_A) of 0.64. For more details see the text.

the signal-to-noise ratio increases owing to increased fluorescence at every wavelength upon increasing the donor QY. Figure 1.5b shows the spectra at (infinite) large distance between CFP and YFP (hence $E = 0$) as a function of the donor quantum yield. Here, it is shown that increasing the donor quantum yield proportionally increases the donor fluorescence, and it demonstrates a constant, direct excitation of acceptor fluorescence that only becomes really significant in cases of very low donor quantum yield. By comparing Figure 1.5a and b, it is clear that the largest CFP intensity decrease due to FRET (both relative and absolute) is seen for the highest donor quantum yield. In Figure 1.5c, the acceptor-to-donor fluorescence ratio (527:475 nm) is plotted for the spectra in Figure 1.5a (closed circles, representing FRET) and for the spectra in Figure 1.5b (open circles, representing no FRET). It can be seen that the actual acceptor-to-donor fluorescence ratio decreases with donor QY, despite the better FRET. If, however, the FRET contrast ratio (*CR*) is plotted (by dividing the YFP/CFP ratio *R* for the FRET and non-FRET situations), an increasing relationship is found with donor QY. Note here that the *contrast ratio* $CR = R_1/R_2$ is linear with *ratio contrast* as defined by $C = \Delta R/R = (R_1 - R_2)/R_2 = CR - 1$. In Figure 1.5d, the YFP/CFP intensity ratio *R* is given for various distances between the donor and the acceptor. Note that this is a logarithmic plot, so the distance between two curves is proportional to *CR* (because $\log CR = \log(R_1/R_2) = \log R_1 - \log R_2$). Close inspection of the curves shown in Figure 1.5d demonstrates that the distance between any of two selected curves increases with the donor QY. This proves that a high-QY donor, despite giving a lower acceptor-to-donor fluorescence ratio *R*, is always beneficial for a FRET sensor: it provides a higher signal-to-noise ratio for the donor and acceptor measurements (Figure 1.5a and b), it provides more extensive donor quenching and sensitized emission (compare Figure 1.5a and b), and it provides superior ratio contrast (see Figure 1.5c and d).

1.6.3 Conclusions for CFP in FRET Sensors

In view of the preceding, mTurquoise2, with a record quantum yield of 93%, is the preferred CFP for FRET sensors and for measuring protein–protein interactions via donor-based FRET methods (e.g., FLIM or acceptor photobleaching) when coupled to a YFP acceptor. We could demonstrate these advantages not only in simulated data but also in practice by substituting ECFP for mTurquoise in both multimeric FRET sensors and ratiometric FRET sensors. We refer readers to published studies on Gq signaling (Adjobo-Hermans et al. 2011) and cAMP signaling (Klarenbeek et al. 2011) to see the respective improvements.

For ratiometric FRET sensors, it can be beneficial to engineer physical contacts between the donor/acceptor β-barrels in one of the sensor states. These so-called sticky FPs then give rise to extremely high FRET in this state. Especially for the yellow cameleon (YCAM) FRET sensor, this phenomenon is responsible for a high dynamic range (Nagai et al. 2004; Laptenok et al. 2012). This is why replacing the sticky ECFP with mTurquoise or mTurquoise2 gives rise to a lower ratio contrast in this sensor, owing to the monomerization and altered β-barrel exteriors of both mTurquoise variants as compared to ECFP (not shown).

ACKNOWLEDGMENTS

We acknowledge Richard N. Day (Department of Cellular and Integrative Physiology, Indiana University School of Medicine, Indianapolis) for kindly donating the plasmid with cDNA encoding mCerulean3. We thank Koen Oost for performing the D148G point mutagenesis on mTurquoise2 to produce mTurquoise2-G. We are grateful to Dr. Gert-Jan Kremers (Erasmus Medical Centre, Rotterdam, The Netherlands) for his important contribution to the development of SCFP3A in our laboratory. We thank Linda Joosen for her important contribution in developing mTurquoise2. This research was funded by an STW-perspectief program grant P10-15 and by an NWO-CW ECHO grant (711.01.01812).

REFERENCES

Adjobo-Hermans, M. J., Goedhart, J., van Weeren, L. et al. (2011). Real-time visualization of heterotrimeric G protein Gq activation in living cells. *BMC Biol.* 9:32.

Alieva, N. O., Konzen, K. A., Field, S. F. et al. (2008). Diversity and evolution of coral fluorescent proteins. *PLoS One* 3:e2680.

Becker, W. (2012). Fluorescence lifetime imaging—Techniques and applications. *J. Microsc.* 247:119–136.

Beckman, R. A., Mildvan, A. S., and Loeb, L. A. (1985). On the fidelity of DNA replication: Manganese mutagenesis in vitro. *Biochemistry* 24:5810–5817.

Chalfie, M., Tu, Y., Euskirchen, G., Ward, W. W., and Prasher, D. C. (1994). Green fluorescent protein as a marker for gene expression. *Science* 263:802–805.

Chudakov, D. M., Lukyanov, S., and Lukyanov, K. A. (2005). Fluorescent proteins as a toolkit for in vivo imaging. *Trends Biotechnol.* 23:605–613.

Chudakov, D. M., Matz, M. V., Lukyanov, S., and Lukyanov, K. A. (2010). Fluorescent proteins and their applications in imaging living cells and tissues. *Phys. Rev.* 90:1103–1163.

Clegg, R. M. (2009). Förster resonance energy transfer—FRET what is it, why do it, and how it's done. *Lab. Tech. Biochem. Mol. Biol.* 33:1–57.

Cubitt, A. B., Woollenweber, L. A., and Heim, R. (1999). Understanding structure-function relationships in the *Aequorea victoria* green fluorescent protein. *Methods Cell Biol.* 158:19–30.

Dennig, A., Shivange, A. V., Marienhagen, J., and Schwaneberg, U. (2011). OmniChange: The sequence independent method for simultaneous site-saturation of five codons. *PLoS One* 6:e26222.

Eckert, K. A., and Kunkel, T. A. (1990). High fidelity DNA synthesis by the Thermus aquaticus DNA polymerase. *Nucleic Acids Res.* 18:3739–3744.

Erard, M., Fredj, A., Pasquier, H. et al. (2013). Minimum set of mutations needed to optimize cyan fluorescent proteins for live cell imaging. *Mol. Biosys.* 9:258–67.

Fredj, A., Pasquier, H., Demachy, I. et al. (2012). The single T65S mutation generates brighter cyan fluorescent proteins with increased photostability and pH insensitivity. *PLoS One* 7:e49149.

Ganesan, S., Ameer-Beg, S. M., Ng, T. T., Vojnovic, B., and Wouters, F. S. (2006). A dark yellow fluorescent protein (YFP)-based resonance energy-accepting chromoprotein (REACh) for Förster resonance energy transfer with GFP. *Proc. Natl. Acad. Sci. U. S. A.* 103:4089–4094.

Goedhart, J., Vermeer, J. E., Adjobo-Hermans, M. J., van Weeren, L., and Gadella, T. W. Jr. (2007). Sensitive detection of p65 homodimers using red-shifted and fluorescent protein-based FRET couples. *PLoS One* 2:e1011.

Goedhart, J., van Weeren, L., Hink, M. A., Vischer, N. O., Jalink, K., and Gadella, T. W. Jr. (2010). Bright cyan fluorescent protein variants identified by fluorescence lifetime screening. *Nat. Methods* 7:137–139.

Goedhart, J., von Stetten, D., Noirclerc-Savoye, M. et al. (2012). Structure-guided evolution of cyan fluorescent proteins towards a quantum yield of 93%. *Nat. Commun.* 3:751.

Heim, R., Prasher, D. C., and Tsien, R. Y. (1994). Wavelength mutations and posttranslational autoxidation of green fluorescent protein. *Proc. Natl. Acad. Sci. U. S. A.* 91:12501–12504.

Heim, R., and Tsien, R. Y. (1996). Engineering green fluorescent protein for improved brightness, longer wavelengths and fluorescence resonance energy transfer. *Curr. Biol.* 6:178–182.

Klarenbeek, J. B., Goedhart, J., Hink, M. A., Gadella, T. W., and Jalink, K. (2011). A mTurquoise-based cAMP sensor for both FLIM and ratiometric read-out has improved dynamic range. *PLoS One* 6:e19170.

Koyanagi, T., Yoshida, E., Minami, H., Katayama, T., and Kumagai, H. (2008). A rapid, simple, and effective method of constructing a randomly mutagenized plasmid library free from ligation. *Biosci. Biotechnol. Biochem.* 72:1134–1137.

Kremers, G. J., Goedhart, J., van Munster, E. B., and Gadella, T. W. Jr. (2006). Cyan and yellow super fluorescent proteins with improved brightness, protein folding, and FRET Förster radius. *Biochemistry* 45:6570–6580.

Lam, A. J., St-Pierre, F., Gong, Y. et al. (2012). Improving FRET dynamic range with bright green and red fluorescent proteins. *Nat. Methods* 9:1005–12. [Epub ahead of print].

Laptenok, S. P., van Stokkum, I. H., Borst, J. W., van Oort, B., Visser, A. J., and van Amerongen, H. (2012). Disentangling picosecond events that complicate the quantitative use of the calcium sensor YC3.60. *J. Phys. Chem. B* 116:3013–3020.

Lelimousin, M., Noirclerc-Savoye, M., Lazareno-Saez, C. et al. (2009). Intrinsic dynamics in ECFP and Cerulean control fluorescence quantum yield. *Biochemistry* 48:10038–10046.

Markwardt, M. L., Kremers, G. J., Kraft, C. A. et al. (2011). An improved cerulean fluorescent protein with enhanced brightness and reduced reversible photoswitching. *PLoS One* 6:e17896.

Matz, M. V., Fradkov, A. F., Labas, Y. A. et al. (1999). Fluorescent proteins from nonbioluminescent *Anthozoa* species. *Nat. Biotechnol.* 17:969–973.

Merzlyak, E. M., Goedhart, J., Shcherbo, D. et al. (2007). Bright monomeric red fluorescent protein with an extended fluorescence lifetime. *Nat. Methods* 4:555–557.

Nagai, T., Ibata, K., Park, E. S., Kubota, M., Mikoshiba, K., and Miyawaki, A. (2002). A variant of yellow fluorescent protein with fast and efficient maturation for cell-biological applications. *Nat. Biotechnol.* 20:87–90.

Nagai, T., Yamada, S., Tominaga, T., Ichikawa, M., and Miyawaki, A. (2004). Expanded dynamic range of fluorescent indicators for Ca($^{2+}$) by circularly permuted yellow fluorescent proteins. *Proc. Natl. Acad. Sci. U. S. A.* 101:10554–10559.

Ormö, M., Cubitt, A. B., Kallio, K., Gross, L. A., Tsien, R. Y., and Remington, S. J. (1996). Crystal structure of the *Aequorea victoria* green fluorescent protein. *Science* 273:1392–1395.

Pédelacq, J. D., Cabantous, S., Tran, T., Terwilliger, T. C., and Waldo, G. S. (2006). Engineering and characterization of a superfolder green fluorescent protein. *Nat. Biotechnol.* 24:79–88.

Prasher, D. C., Eckenrode, V. K., Ward, W. W., Prendergast, F. G., and Cormier, M. J. (1992). Primary structure of the *Aequorea victoria* green-fluorescent protein. *Gene* 111:229–233.

Rizzo, M. A., Springer, G. H., Granada, B., and Piston, D. W. (2004). An improved cyan fluorescent protein variant useful for FRET. *Nat. Biotechnol.* 22:445–449.

Sawano, A., and Miyawaki, A. (2000). Directed evolution of green fluorescent protein by a new versatile PCR strategy for site-directed and semi-random mutagenesis. *Nucleic Acids Res.* 28:E78.

Shaner, N. C., Campbell, R. E., Steinbach, P. A., Giepmans, B. N., Palmer, A. E., and Tsien, R. Y. (2004). Improved monomeric red, orange and yellow fluorescent proteins derived from *Discosoma* sp. red fluorescent protein. *Nat. Biotechnol.* 22:1567–1572.

Shaner, N. C., Lin, M. Z., McKeown, M. R. et al. (2008). Improving the photostability of bright monomeric orange and red fluorescent proteins. *Nat. Methods* 5:545–551.

Shcherbakova, D. M., Subach, O. M., and Verkhusha, V. V. (2012). Red fluorescent proteins: Advanced imaging applications and future design. *Angew. Chem. Int. Ed. Engl.* 51:2–17.

Shcherbo, D., Souslova, E. A., Goedhart, J. et al. (2009). Practical and reliable FRET/FLIM pair of fluorescent proteins. *BMC Biotechnol.* 9:24.

Telford, W. G., Hawley, T., Subach, F., Verkhusha, V., and Hawley, R. G. (2012). Flow cytometry of fluorescent proteins. *Methods* 57:318–330.

Tsien, R. Y. (1998). The green fluorescent protein. *Annu. Rev. Biochem.* 67:509–544.

Verkhusha, V. V., and Lukyanov, K. A. (2004). The molecular properties and applications of Anthozoa fluorescent proteins and chromoproteins. *Nat. Biotechnol.* 22:289–296.

Walther, K. A., Papke, B., Sinn, M. B., Michel, K., and Kinkhabwala, A. (2011). Precise measurement of protein interacting fractions with fluorescence lifetime imaging microscopy. *Mol. BioSyst.* 7:322–336.

Chapter 2

Fluorescent Imaging Techniques

FRET and Complementary Methods

Stefan Terjung and Yury Belyaev

CONTENTS

2.1 INTRODUCTION

Most optical probes in biology are based on a fluorescent readout and are designed to report biological or even physiological parameters in living samples. Effectively, this enables researchers to gain information on a molecular scale, well below the resolution limit of light microscopy. Förster resonance energy transfer (FRET) is one method for gaining information on a molecular level, and it is frequently utilized in fluorescent biosensor design. FRET involves the nonradiative transfer of energy between two fluorophores, and it can occur only if a fluorescent molecule—the donor—is in very close proximity to another aromatic system such as a dye or another chromophore. The maximal distance is usually below 10 nm, which is conveniently close to the size of an average protein. The efficiency with which energy transfer occurs depends very strongly on the orientation and distance between the donor and the acceptor. The efficiency of energy transfer is inversely proportional to the sixth power of the donor and acceptor distance, rendering it very useful for readouts of, for example, protein conformational changes or protein cleavage (Schultz 2007). In addition, FRET can be exploited to determine interactions between molecules (Jares-Erijman and Jovin 2003) and as a molecular ruler on a single-molecule level to measure distances in molecules or complexes (Stryer 1978). This chapter provides an overview of FRET and complementary techniques that are useful in combination with optical probes or FRET experiments. Chapter 1 introduced fluorescent proteins (FPs) and their properties, which are widely exploited in the design of fluorescent biosensors. The use of genetically encoded constructs based on FPs is often advantageous, because they are easily expressed in living cells and investigated *in vivo*. The design and application of probes for biomolecules are described in subsequent chapters in Part II. A key prerequisite to gaining physiological readouts with these probes is the use of the appropriate imaging technique while keeping the environment of the specimen at physiological conditions. Because elevated light intensity can influence the physiological state of cells, it is important to use an imaging setup with high sensitivity (Gräf et al. 2005; Brown 2007) to minimize bleaching and phototoxicity. These imaging requirements are discussed in the first part of this chapter. The second part focuses on advanced microscopy methods, which are interesting in the context of biosensors. The main topic in this respect is FRET, as a large number of fluorescent biosensors depend on FRET, which is used to report the biosensor status due to conformational changes on binding of a ligand (e.g., calcium in case of cameleon sensors). In addition to biosensor readout, the mobility of molecules is another important parameter to measure. Fluorescence recovery after photobleaching (FRAP) and fluorescence cross-correlation spectroscopy (FCCS) are techniques used to determine protein dynamics, including binding and diffusion, and are complementary methods to FRET.

2.2 MICROSCOPE TECHNIQUES AND SETUPS

In general, only light efficient and sensitive microscopy techniques are useful for imaging living specimens to avoid photobleaching and photodamage. This is important for the readout of physiological parameters by optical probes to avoid any influence that the imaging conditions may have on the detected values. Imaging conditions that induce very little bleaching are also very important for probes such as GCaMP (Zhao et al. 2011), which indicate concentration fluctuations through corresponding fluorescence changes in a single channel, as the readout will be masked by the bleaching rate of the dye. To measure physiological parameters in thicker samples such as embryos, optical slicing techniques such as laser scanning microscopy, spinning disk microscopy, or light-sheet microscopy, including selective plane illumination microscopy (SPIM) and digital scanned light-sheet microscopy (DSLM), are required (Krzic et al. 2012; Keller 2013; Pitrone et al. 2013; Schmid et al. 2013; Swoger et al. 2014).

2.2.1 How to Reduce Phototoxicity

Being a quantitative method, FRET requires that the fluorescence signal be acquired with the highest possible accuracy. Unfortunately, the signal is limited not only by low fluorophore concentration but also by the fact that the live sample should be subjected to as low an illumination intensity as possible to minimize phototoxicity. When high acquisition speed is necessary, the requirements on detector sensitivity are even higher. Thus, the general suggestion for FRET measurements is to use the most sensitive detectors possible, especially if employed for live-cell imaging. In this section, we present a detailed discussion of microscope components affecting sensitivity.

2.2.1.1 Detector Types

The detector is a major component of the imaging setup that strongly influences the sensitivity of the microscope. Detectors can be divided roughly into two main classes depending on their usage: area sensors and point detectors. Area sensors are typically cameras comprising an array of sensitive elements allowing the simultaneous detection of an entire field of view. Currently, two types of cameras are most often used for FRET measurements: electron multiplication charge-coupled device (EMCCD) cameras and scientific complementary metal-oxide-semiconductor (sCMOS) cameras. Cameras are used for widefield microscopy, SPIM, and spinning disk confocal microscopy. Point detectors, also referred to as confocal detectors, such as photomultipliers (PMTs) and avalanche photodiodes (APDs), allow light intensity measurements at a given point in the sample and are used for confocal and two-photon laser scanning microscopes.

The detector quantum efficiency (QE, a ratio between the number of produced electrons and the number of photons arriving at the detector) is of paramount importance. The QEs of detectors used in microscopy differ significantly. PMTs, which are still regularly used as confocal detectors, offer the lowest QE. PMTs with multi-alkali photocathodes typically reach QE values in the range of 10%–20%. PMTs integrating more sensitive GaAsP photocathodes offer up to 50% quantum efficiency. These so-called GaAsP detectors were recently introduced and are used in several commercial confocal microscopes (e.g., Olympus FV1200, Zeiss LSM 780). Among confocal detectors, APDs have the highest QE, reaching up to 90%. A combination of working principles of PMT and APD are used in so-called hybrid detectors (HyD). These were introduced by Leica in its SP5 and SP8 confocal laser scanning microscopes (LSMs) and feature a QE equal to that of GaAsP detectors. The reason why PMTs are still used in LSMs is the higher dynamic range and robustness: GaAsP, HyD, and APD detectors can be destroyed by too much light, which is largely prevented by precautions such as shutter or automatic switch off at too high count rates. APDs have a lower dynamic range than GaAsP and HyD detectors and are used mostly for photon counting applications (e.g., fluorescence correlation spectroscopy).

The QE of widefield detectors is typically higher than the QE of confocal detectors. EMCCD cameras have the highest QE among cameras. Back-illuminated EMCCD cameras can reach QE values of up to 90%, whereas front-illuminated EMCCD cameras have a slightly lower QE value in the range of 70%. The sCMOS cameras are roughly on par with front-illuminated EMCCD in terms of their QE.

The QE is wavelength dependent and specific for each detector. This spectral dependence should be taken into account when a FRET experiment is planned. Commonly used detectors are optimized for the visible range of light (400–750 nm). If ultraviolet (UV) or near-infrared (NIR) detection is necessary, special detector types should be chosen that have enough sensitivity in the required spectral range.

Another factor affecting detector sensitivity is read noise, an additional stray signal added to the image during detector readout. Read noise is normally negligible for point detectors, whereas it can be important for the widefield detectors.

This noise type is relevant only for a low signal level, which is often the case in FRET experiments. For sCMOS cameras, a typical value for read noise is in the range of one electron per pixel. EMCCD cameras, due to the additional multiplication register amplifying the signal before the readout electronics, have a very low effective read noise, less than a fraction of an electron per pixel.

2.2.1.2 Light Sources and Filters

The light source is the first element needed for a fluorescence readout. It has to offer sufficient intensity in the spectral range in which the used fluorophores absorb. In widefield fluorescence setups, lamps delivering a broad spectrum are usually used in combination with filters to select the appropriate light for excitation. Common light sources are mercury (HBO) or xenon (XBO) short-arc lamps as well as metal halide bulbs. Mercury and metal halide lamps have output spectra with several distinctive peaks of higher intensity, with lower intensity levels elsewhere. Xenon lamps emit a relatively uniform intensity distribution over the entire visible spectrum and are thus very well suited to ratio imaging. All lamps have in common that a relatively large fraction of the emitted spectrum is in the UV range, which is useful if dyes such as 4′,6-diamidino-2-phenylindole (DAPI) are used. On the other hand, this can be a strong disadvantage for live-cell imaging if the filter sets used do not efficiently suppress this UV contribution. A modern variant for fluorescence illumination are light-emitting diodes (LEDs), which have several advantages. First, the lifetime of LEDs (in the range of 10,000 h) is longer than those of HBO (typically 200–500 h), XBO (about 1500 h), and even metal halide (up to 2000 h) lamps. During the entire LED lifetime, no service is required. Second, the LED can be switched in the sub-millisecond range, which is faster than any conventional mechanical shutter. This allows for precise synchronization of sample illumination and image acquisition and can improve the viability of the live samples during long time-lapse experiments. LEDs have no strong UV or IR components in the spectrum, unless specifically designed for this purpose, which ensures favorable conditions for live-cell imaging without the need for additional IR and UV cutoff filters in the microscopy system. In addition, LEDs offer good short- and long-term intensity stability. Some of the LED systems available for microscopy (e.g., XLED1 from Lumen Dynamics) allow for intensity modulation with a minimum pulse duration of 10 μs. There are some indications that pulsed illumination is more favorable for live specimens because of lower sample photobleaching and phototoxicity (Nishigaki et al. 2006).

Excitation and emission filters should be matched as closely as possible to the spectra of the fluorophores used. Optimal filters should not only have high transmission—above 90% within the required spectral range—but should also strongly suppress the signal outside the transmission range, ideally on the order of OD 6.

2.2.1.3 Objective Lenses

The objective lens determines the magnification, potential resolution, and brightness of the system by the numerical aperture (NA). For biosensors (e.g., calcium sensors), optimizing for sufficient acquisition speed and readout intensity of the optical probe is favorable over the highest spatial resolution. A good flat field and color correction of the objective lens are important for FRET experiments. Objectives with ideal color correction (so-called apochromats) are in general very well suited for live-cell imaging. For FRET sensor ratio imaging, it is essential that the detected signal of the acquired channel originates from exactly the same *z*-plane of the sample, and the use of apochromatic lenses is therefore preferred. Especially for live-cell experiments, it is advantageous to maximize the amount of fluorescence light collected by the lens. The image brightness depends on the transmission of the objective lens, as well as its NA and magnification. The high transmission of a conventional objective lens (80%–90%) is maximized only in the visible range. If an extended transmission range is required by the experimental conditions, a special objective lens is required, for example, an IR-corrected multiphoton objective lens.

The brightness of the objective lens depends on how much of the sample can be illuminated (defined by objective magnification M) and which amount of excitation light is projected and how much fluorescence light is collected (defined by objective numerical aperture NA). For fluorescence or epi-fluorescence imaging, the brightness B is normally defined as

$$B = 10^4 \cdot (NA^2/M)^2 \tag{2.1}$$

Objective lenses with higher brightness values are preferable because they are more light efficient and allow imaging under lower light conditions. The brightness values of some objective lenses, calculated using Equation 2.1, are listed in Table 2.1.

2.2.2 Acquisition Speed, Sample Thickness, and Optical Sectioning

FRET experiments can be performed on different microscopy systems ranging from a simple widefield to a state-of-the-art multiphoton or confocal microscope. The system of choice depends on the desired acquisition speed, sensitivity, and whether optical sectioning is necessary.

2.2.2.1 Widefield Microscopy

Widefield epifluorescence microscopes are typically light efficient and equipped with a very fast camera. Here, the choice of a detector is defined by the fluorescence signal. For low-fluorescence signal applications, an EMCCD camera is required. For less demanding applications with respect to sensitivity, an

TABLE 2.1 AS AN EXAMPLE, THE BRIGHTNESS (*B*) VALUES FOR SEVERAL COMMON OBJECTIVE LENSES OF OLYMPUS ARE CALCULATED ACCORDING TO EQUATION 2.1

Objective Type	Magnification	NA	Immersion	Brightness
UPlanSApo	10	0.4	Air	2.56
UPlanApo	20	0.75	Air	7.91
LUCPlanFLN	20	0.45	Air	1.03
UPLSApo	30	1.05	Sil	13.51
UPlanApo	40	0.95	Air	5.091
UPLSAPO	60	1.3	Sil	7.93
UPlanApo	60	1.2	W	5.76
UPlanApo	60	1.35	Oil	9.23
PlanApo	60	1.4	Oil	10.67

sCMOS camera is usually a good choice. In addition, sCMOS cameras can reach frame rates of up to 100 frames per second and are recommended for maximizing acquisition speed. Widefield acquisition works well only for relatively thin samples (e.g., cell culture).

2.2.2.2 Spinning Disk Microscopy

If thicker samples have to be imaged (e.g., tissue sections or embryos), an appropriate optical sectioning technique should be used. Spinning disk microscopes are often the method of choice for acquiring optical sections of relatively thin live samples. With an array of pinholes swiping the field of view continuously and the signal detected by a sensitive camera, high acquisition rates with relatively low phototoxicity are possible (Gräf et al. 2005). Similar to widefield microscopy, EMCCD or sCMOS cameras are best suited to spinning disk microscopy. Owing to the relatively short distances between the pinholes on the spinning disk, out-of-focus light may increase the background level by detection through neighboring pinholes.

2.2.2.3 Confocal and Two-Photon Laser Scanning Microscopy

LSMs are well suited for thicker samples. Two-photon LSMs are especially well suited if the specimen produces strong scattering or is thicker than approximately 50–100 μm. Because of the scanning concept, these techniques are inherently slower than camera-based techniques. LSMs with resonant galvo-scanners (e.g., optionally available for Leica SP8) can reach speeds close to those of spinning disk microscopes, but owing to very short pixel dwell times, this comes at the cost of increased noise levels. To increase image brightness, it is advisable to use signal accumulation in combination with a moderate increase in laser power when resonant scanners are used. For confocal microscopy, there is a relatively wide detector choice including PMT, HyD, GaAsP, and APD. The PMT is still used frequently and is sufficient for applications with high fluorescence levels where the sensitivity is not critical. The HyD and GaAsP detectors are more sensitive than traditional PMTs and are thus preferable for more light-limited applications. The APDs are traditionally used for photon counting, where ultimate sensitivity is necessary, for example, in FCS as explained in Section 2.5. APDs are used for imaging only in exceptional cases, as they are damaged by too strong illumination and offer a lower dynamic range.

In two-photon microscopy, the use of non-descanned detectors can in addition increase the sensitivity of the system, as the fluorescence signal in this case is guided directly to the detector, without passing through additional optical elements in the microscope scan head, including the pinhole, which would block the detection of photons scattered on the path between the focus and objective lens. Confocal and two-photon LSMs are used for imaging both live and fixed thicker samples, such as embryos.

2.2.2.4 Selective Plane Illumination Microscopy

SPIM is another optical sectioning technique combining selective illumination of the acquired plane with efficient widefield detection (Huisken et al. 2004). In most cases, speed, for example, to acquire large stacks quickly, and a large field of view are important, and therefore, sCMOS cameras are usually best suited. SPIM is an outstanding method for long-term imaging of thick live samples (Krzic et al. 2012). Excitation light will be absorbed and scattered along the specimen, resulting in unevenly illuminated planes for thick, scattering, and strongly absorbing samples. This can be compensated by rotating the sample and acquiring stacks from different angles. These stacks are subsequently fused to gain a more complete view, leading to isodiametric resolution (Preibisch et al. 2010). SPIM has already been used in a few FRET studies (Greger et al. 2011; Costa et al. 2013). However, stacks created by the fusion of multiangle data are not suitable for FRET measurement because the fusion itself is not quantitative. Further, SPIM is not well suited for strongly scattering samples. Sample preparation is also relatively complicated and is not compatible with all specimens.

2.2.2.5 Total Internal Reflection Fluorescence Microscopy

Total internal reflection fluorescence microscopy (TIRF) is a rather specialized technique that illuminates only a very thin optical slice directly above the cover slip (Axelrod 2001). The thickness of the optical slice can be tuned by adjusting the incidence angle of the laser and is typically in the range of 70–250 nm. Because this optical slice can be obtained only in the direct vicinity of the cover slip, it is suitable only if the features of interest are close enough to the cover slip. For example, TIRF can be very useful for studies of processes localized at the plasma membrane, such as endocytosis or exocytosis or focal adhesions, as well as the visualization of *in vitro* experiments. TIRF is often used for single-molecule detection studies, which require the highest detector sensitivity possible. The preferred detector in this case is often a back-thinned EMCCD camera. This offers the highest possible QE and extremely low read noise. Recently, it has been shown that sCMOS cameras may also be used (Saurabh et al. 2012).

An overview of the different microscopy techniques and their applicability for FRET measurements is shown in Table 2.2. The use of emerging super-resolution techniques is also covered in Chapter 16.

2.3 FÖRSTER ENERGY RESONANCE TRANSFER

The readout of the majority of optical probes in biology is based on fluorescence, or more correctly, FRET. Here, we give a short introduction of the underlying theory of FRET and cover the most common methods to measure it.

2.3.1 FRET Fundamentals

The effect of fluorescence resonance energy transfer was first described by Theodor Förster (Förster 1946), and FRET is referred to as Förster resonance energy transfer in his honor. In a nutshell, the energy of an excited fluorophore (donor) can be transferred to an accepting molecule (acceptor) through a nonradiative process (i.e., without the emission of a photon). The acceptor needs to be in close proximity to the donor, usually closer than 10 nm. As a result, the acceptor is elevated to an excited state and subsequently emits a photon in its typical emission range, while the donor returns to the ground state without fluorescence emission (Figure 2.1). It is not mandatory that the acceptor is a fluorophore, especially for fluorescent lifetime imaging microscopy (FLIM) measurements, where the acceptor might as well be nonfluorescent.

The prerequisites for FRET to occur are as follows (Vogel et al. 2006):

1. Proximity
   The donor and acceptor need to be in very close proximity. The FRET efficiency is inversely proportional to the sixth power of the distance between the donor and the acceptor (Equation 2.2). The distance at which FRET occurs with 50% efficiency is defined as the Förster radius R_0. Values of R_0 for common FRET pairs are in the range between 2

TABLE 2.2 OVERVIEW OF DIFFERENT MICROSCOPY TECHNIQUES AND THE APPLICABILITY FOR FRET MEASUREMENTS

Technology	Advantages	Disadvantages	Optical Sectioning	Speed	Sensitivity	Optimal for These FRET Techniques	Area of Application
Widefield epifluor.	High speed and efficiency	Out of focus fluorescence	—	+++	+++ Can be increased with EM CCD or sCMOS	Ratiometric detection Sensitized emission Additional equipment needed for FLIM	Adherent cells and thin samples, if no optical sectioning is required
Confocal laser scanning	Good optical sectioning, same timing of channels in line-by-line sequential	Slow, usually high excitation intensity in focus, low sensitivity, bleaching not only in focus plane	++	+	+ Can be increased with HyD or GaAsP detectors	Ratiometric detection Sensitized emission Acceptor and donor photobleaching Additional equipment needed for FLIM	Thick samples that need good optical sectioning, but speed is not limiting. Combination with other techniques like FRAP or FCS
Spinning disk microscopy	Optical sectioning, fast, low photobleaching	Less sectioning and less flexible compared to CLSM	+	++	++ Can be increased with EM CCD or sCMOS detectors	Ratiometric detection Sensitized emission Additional equipment needed for FLIM	Specimen of medium thickness, optical sectioning with low bleaching and fast acquisition needed

Multiphoton laser scanning	Good optical sectioning, imaging of thick and scattering samples, bleaching only in focus volume same timing of channels in line-by-line sequential	Very expensive, lower resolution than confocal laser scanning	+++	+	+ Can be increased with HyD or GaAsP detectors	Ratiometric detection Sensitized emission Additional equipment needed for FLIM	Thick and scattering samples, if speed is not limiting
Total internal reflection microscopy	Very thin section, very high signal to background	Imaging only one slice on top of the coverslip	++	+++	+++ Can be increased with EM CCD or sCMOS	Ratiometric detection Sensitized emission	Processes that can be imaged very close to the coverslip like plasma membrane, focal adhesions
Selective plane illumination microscopy (SPIM)/DSLM	Illuminating only the imaging plane, very light efficient, isodiametric resolution possible, if multiple angles are fused	Stripe artifacts possible, sample preparation usually more demanding	++	+++	+++ Can be increased with EM CCD or sCMOS	Ratiometric detection Sensitized emission Additional equipment needed for FLIM	Thicker and light sensitive samples that can be mounted for SPIM acquisition

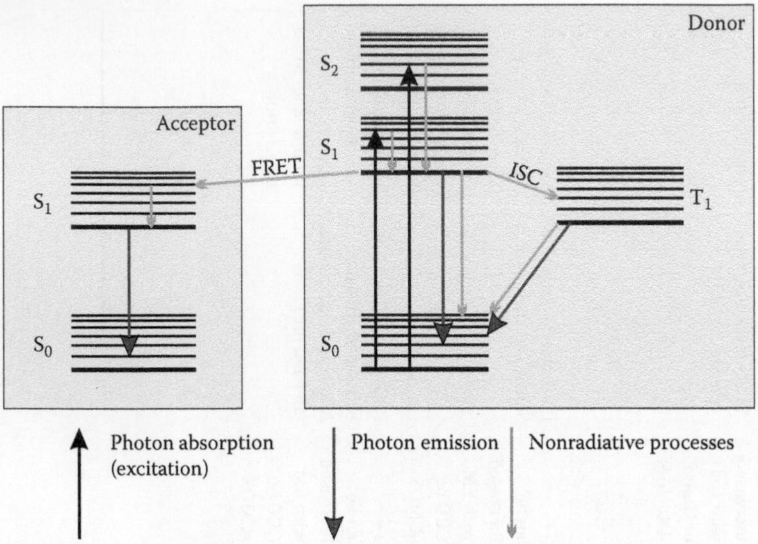

Figure 2.1 A simple Jablonski diagram including FRET.

and 7 nm, which effectively limits FRET to distances below 10 nm (Figure 2.2).

$$E = \frac{R_0^6}{R_0^6 + r^6} \tag{2.2}$$

where

r = distance between the donor and the acceptor

R_0 = Förster radius, distance with 50% FRET efficiency

2. Spectral overlap
The emission spectrum of the donor has to overlap significantly with the absorption spectrum of the acceptor to facilitate FRET (Figure 2.3). Higher overlap increases the FRET efficiency of the pair.
3. Favorable spatial orientation
Energy transfer strongly depends on the spatial orientation of the FRET fluorophores. Donor energy is best transferred to the acceptor if the excitation dipole of the acceptor and the emission dipole of the donor are oriented in parallel to each other. The transfer efficiency decreases for other orientations and is zero if the dipoles are oriented perpendicularly. In practice, it is often assumed that the dipoles are randomly oriented, but in the case of tandem sensors, in which both

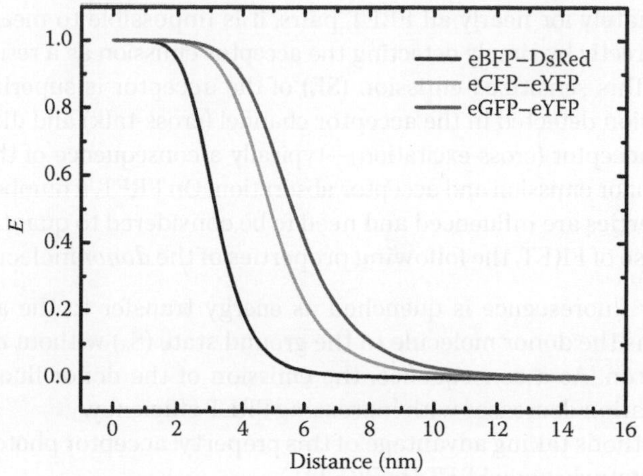

Figure 2.2 Example of the dependence of FRET efficiency on distance, calculated for the Förster radius of three FRET pairs. (Förster radius values from Patterson, G. H. et al., *Anal. Biochem.* 284(2):438–440, 2000.)

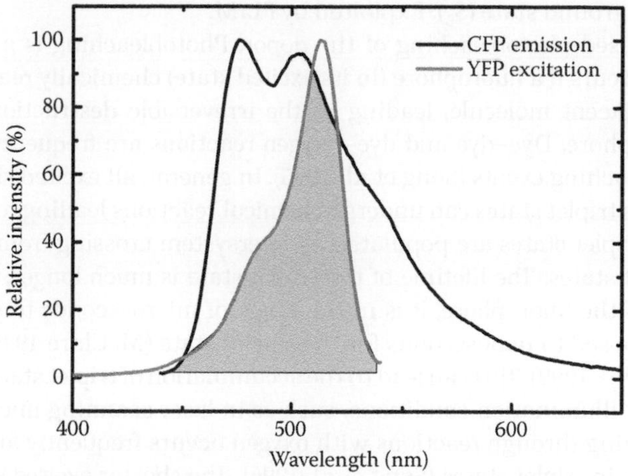

Figure 2.3 Overlap of eCFP emission with eYFP absorption.

the donor and the acceptor are in the same molecule, it is possible that the orientation is partially constrained by the protein conformation, for example, if only short and stiff linker sequences are used. A (designed) fixed orientation between the donor and the acceptor might be an advantage, for example, for a cleavage sensor to gain a high FRET signal in the intact construct before cleavage.

Unfortunately for nearly all FRET pairs, it is impossible to measure FRET efficiency directly by simply detecting the acceptor emission as a result of donor excitation. This sensitized emission (SE) of the acceptor is superimposed by donor emission detected in the acceptor channel (cross-talk) and direct excitation of the acceptor (cross-excitation)—typically a consequence of the essential overlap of donor emission and acceptor absorption. On FRET, a number of fluorophore properties are influenced and need to be considered to quantify FRET.

In the case of FRET, the following properties of the *donor* molecule change:

1. Donor fluorescence is quenched as energy transfer to the acceptor returns the donor molecule to the ground state (S_0) without emitting a photon. As a consequence, the emission of the donor fluorophore population decreases with increasing FRET efficiency.

   Methods taking advantage of this property: acceptor photobleaching, photochromic FRET (pcFRET).
2. Fluorescence lifetime of the donor is shortened. The average time spent by donor molecules in the excited state (S_1), termed the fluorescence lifetime, is shortened because FRET is an additional pathway for the donor to dispose of excited-state energy (Figure 2.1) and return to the ground state (S_0). Exploited by FLIM.
3. Decreased photobleaching of the donor. Photobleaching is a process that occurs if a fluorophore (in its excited state) chemically reacts with an adjacent molecule, leading to the irreversible destruction of the fluorophore. Dye–dye and dye–oxygen reactions are frequent reasons for bleaching events (Song et al. 1995). In general, all excited singlet as well as triplet states can undergo chemical reactions leading to bleaching. Triplet states are populated by intersystem crossing from excited singlet states. The lifetime of the triplet state is much longer; depending on the fluorophore, it is in the range of microseconds to seconds, as opposed to nanoseconds for the singlet state (McClure 1949; Visser and Hink 1999). This can lead to the accumulation of triplet states under strong illumination conditions, such as in laser scanning microscopy. Bleaching through reactions with oxygen occurs frequently for fluorophores in triplet states (Song et al. 1996). The shorter excited state (S_1) lifetime of donor molecules due to FRET also leads to a lower probability of populating the excited triplet state (T_1). This decreases the chances that the excited donor molecule will react and bleach; thus, the rate of donor photobleaching is decreased. Method: donor photobleaching.

In the case of FRET, the following property of the *acceptor* molecule changes:

1. *Sensitized emission.* Describes the emission of photons from an acceptor molecule that received energy from the donor by FRET.

Unfortunately, sensitized emission is superimposed by cross-excitation (acceptor directly excited by donor excitation) and cross-talk of donor fluorescence into the acceptor channel (bleed-through) for most FRET pairs.

Methods based on this property: sensitized emission, ratiometric imaging.

FRET is used mainly in three different types of experiments:

1. *Biological sensors.* Optical probes containing both a donor and an acceptor on the same molecule deliver physiological readouts via FRET changes due to conformational changes, for example, on ligand binding. Although the proper design of probes and the verification of their function under physiological conditions are very tedious, the usage of a validated probe is relatively straightforward, for example, with sensors measuring ion concentrations such as calcium (Miyawaki et al. 1997) or the activity of kinases like aurora B (Fuller et al. 2008); see Figure 2.5.
2. *Molecular interaction.* FRET is employed to test the interaction of two molecules. One molecule of interest has to be tagged with the donor fluorophore, the second with the accepting molecule. Usually, fluorescent proteins such as cyan fluorescent proteins (CFPs) and yellow fluorescent proteins (YFPs) are genetically linked to the two proteins of interest. Because the position of the attached fluorophore might interfere with the binding properties of the tagged protein, the design of the FRET pair needs to be determined and tested carefully. Further, the orientation and distance between a donor and an acceptor, as well as the maturation time of the fluorescent proteins, have to be considered. The stoichiometry of the donor and acceptor molecules also strongly influences the apparent FRET efficiency (Thaler et al. 2005; Piston and Kremers 2007). Meaningful FRET experiments can be performed with a ratio of the donor to the acceptor ranging between 10:1 and 1:10 (Berney and Danuser 2003). If the relative number of interacting molecules is low (e.g., only 1% of the donor molecules are interacting), it is difficult to measure reliable FRET values because the FRET signal is masked by the majority of fluorescent molecules not displaying FRET.

   Labeled proteins can influence the results in an unwanted manner in many ways:

   a. If the number of labeled molecules is much higher than the number of interacting molecules, the FRET signal contributes only a small fraction of the measured fluorescence and is often not clearly distinguishable from background due to noise.

b. Too high levels of fluorescent molecules, due to massive overexpression, are likely to influence the FRET equilibrium and can induce nonphysiological effects in cells.
c. In most cases, endogenous molecules interact more efficiently than their artificially introduced, fluorescently tagged counterparts, leading to a lower apparent FRET efficiency.

   It is possible to reduce the number of endogenous molecules by siRNA treatment to avoid this competition if the siRNA targets an endogenous sequence that is not present in the construct encoding the FP-tagged protein (Szymborska et al. 2013).

Owing to the challenges mentioned previously for obtaining a proper FRET signal from heterogeneous FRET pairs, negative results do not necessarily rule out an interaction between the probed molecules. In contrast, detecting a verified FRET signal clearly indicates interaction. In the case of negative results, it is possible to optimize the conditions. For example, the amount and relative concentration of the donor and acceptor molecules, or the FP constructs themselves, may be improved. One opportunity to improve the constructs is to permutate the position of the fluorescent protein in the construct. Another option is to exchange the FP with an optimized variant (see Chapter 1). Often, this process is time consuming and tedious, especially with a low fraction of interacting molecules, as those will be masked by the excess of noninteracting molecules. Alternative and complementary methods such as FCCS (Bacia et al. 2006), bimolecular fluorescence complementation (Kerppola 2009), and dimerization-dependent fluorescence (Alford et al. 2012a) are covered later in this chapter.

3. *Molecular ruler—Determination of intramolecular distance*. FRET is also used at the single-molecule level. By measuring the FRET efficiency per molecule, it is possible to calculate the distance between a donor and an acceptor based on the knowledge of the Förster radius R_0 (Gansen et al. 2009; Preus and Wilhelmsson 2012; Schuler and Hofmann 2013). In environments with multiple molecules per pixel, calculating the donor–acceptor distance does not lead to meaningful results, as the real FRET efficiency is masked by the contribution of noninteracting molecules and the measurement averages multiple events with different distances in the probed volume. Nevertheless, this effect can be exploited to design sensor molecules to measure tension (Borghi et al. 2012).

The signal generated by FRET is usually not very bright and is masked by direct donor and acceptor emission. In any case, the best-suited FRET readout method depends on the experiment as well as the available instrumentation.

2.3.2 Detection Methods for FRET

As described in the preceding text, the donor and acceptor molecules exhibit altered properties when they are engaged in FRET. These properties are employed to measure the (apparent) FRET efficiency. An overview of FRET techniques is shown in Table 2.3. Multiple methods have been established to measure FRET, but all of them have in common that control experiments are essential to make sure that the experimental conditions are set up correctly. As a positive control, a well-characterized FRET sample containing the same fluorophores as in the intended experiment should be measured. Often, constructs with the two fluorophores connected by a short flexible linker are used, such as eCFP–linker–eYFP (Zimmermann et al. 2002), if eCFP and eYFP are the FRET pair of choice. With this positive control, it is important to reach a value close to the apparent FRET efficiency reported previously (using the same FRET technique) to verify that the conditions being used produce a reliable readout. If the result is significantly lower, this is an indication of suboptimal imaging conditions. In the next step, negative controls are tested using the same acquisition conditions as used for the positive control: a donor only, an acceptor only, and a sample containing both the donor and the acceptor without interactions. Negative controls containing both fluorophores include, for example, a double transfection of plasmids encoding both the donor and acceptor molecules without any interacting domain or linker or a control construct separating the donor and the acceptor sufficiently to avoid FRET, such as mEGFP–MBP–mCherry (Huet et al. 2010). Ideally, no "FRET" signal is detected in these negative controls. However, depending on the readout method and the detection noise, there is usually a signal detectable in the "FRET" channel. Very low values in the negative control can be due to errors introduced by noise and can be subtracted from the acquired apparent FRET efficiency values of the sample of interest. Larger results in the negative control point toward systematic problems that need to be solved before meaningful FRET values can be measured.

2.3.2.1 Sensitized Emission—Quantifying the Acceptor Emission Induced by FRET

The term *sensitized emission* is used in different contexts in the literature. Here we refer to the amount of acceptor fluorescence that is caused by energy transfer from the donor to the acceptor molecule (Gordon et al. 1998; Rheenen et al. 2004; Zal and Gascoigne 2004). As described previously, the sensitized emission fluorescence is usually overlaid by cross-talk of the donor and cross-excitation of the acceptor. One method to measure the apparent FRET efficiency by sensitized emission is similar to linear unmixing (Zimmermann et al. 2002). The signal acquired on excitation of the donor and detection of acceptor emission

TABLE 2.3 COMPARISON OF DIFFERENT FRET MEASUREMENT TECHNIQUES

FRET Technique	Advantages	Disadvantages	Area of Use
Acceptor photobleaching	Available on standard widefield and confocal microscopes Relatively easy method	Not (well) suited for living samples, because of artifacts due to movement and possible photodamage Measures only apparent FRET efficiency	Fixed samples
Donor photobleaching	Easy method available on standard widefield and confocal microscopes	Not quantitative due to potential acceptor bleaching Measures only apparent FRET efficiency	Fixed samples in combination with acceptor photobleaching
Sensitized emission	Dynamic readout with living samples, possible on standard confocal laser scanning microscopes and widefield systems with appropriate filter wheels	Artifacts possible in frame-by-frame sequential mode Measures only apparent FRET efficiency	Fixed and live samples
Ratiometric imaging	Dynamic readout with living samples, possible on standard confocal laser scanning microscopes and widefield systems with appropriate filter wheels	Possible only if the donor and the acceptor are on the same molecule (fixed ratio)	Optical probes, live samples
FLIM	Can be used to determine FRET efficiency and amount of interacting molecules	Specialized equipment needed, relatively slow (especially time domain FLIM)	Fixed and live samples
Polarization anisotropy imaging	Determination of homo-FRET	Polarization filter and control needed	Live samples to determine multimerization, e.g., of receptor molecules
pcFRET	Donor emission with and without FRET can be measured multiple times	Need for photochromic acceptor molecules	Live samples

("FRET" channel) has to be corrected for cross-talk and cross-excitation. Three channels are needed to gain sufficient information (Jalink and van Rheenen 2009):

Donor channel (D): donor excitation, donor emission
Acceptor channel (A): acceptor excitation, acceptor emission
"FRET" or sensitized emission channel (S): donor excitation, acceptor emission

The acquisition of these channels is relatively straightforward. Widefield microscopes with the appropriate filter sets, as well as spinning disk or confocal LSMs, are all well suited to the task. In the case of widefield and spinning disk systems, the channels usually have to be acquired sequentially, which is not a big problem because these systems have fast acquisition rates. One possibility for acquiring these images is to use three filter cubes (three-cube FRET) with the appropriate filters; the donor and acceptor cubes are standard cubes. Fast filter wheels for excitation and emission filters, parallel detection using multiple cameras, or splitting the signal side by side onto separate areas of the camera chip are employed to optimize further the acquisition rate for fast processes. Confocal LSMs usually offer the option to acquire two channels excited with the same laser simultaneously (e.g., to acquire both donor and acceptor emission under donor excitation). The third channel can be added in line-by-line sequential setting to record the acceptor channel under acceptor excitation. In this case, the images are all acquired with the same frame timing (only the lines have slightly different timing), but still usually with a lower frame rate compared to a widefield system. The acquisition of all three channels at the same time can be an advantage, especially if fast processes are investigated, as motion between the sequential frames will create artifacts in the calculation of FRET signals.

Correction factors need to be derived from the acquisition of samples containing donor-only and acceptor-only probes. These images need to be acquired under exactly the same settings used for the acquisition of the FRET sample. It is also possible to mix cells expressing the donor only and the acceptor only into the sample, thereby creating an internal control, instead of acquiring these separately. This internal control is especially useful for the FRET analysis of time-lapse sequences, as problems during acquisition, such as too strong bleaching of one fluorescent protein, can be detected easily (Jalink and van Rheenen 2009).

A number of correction factors have to be determined to calculate the sensitized emission (Rheenen et al. 2004):

A donor-only data set is needed to determine donor cross-talk:

$\beta = S/D$ donor cross-talk in the acceptor channel

An acceptor-only data set is needed to calculate several correction factors:

$\alpha = D/A$ cross-excited acceptor detected in the donor channel
$\gamma = S/A$ cross-excitation of the acceptor
$\delta = D/S$ cross-talk of sensitized emission back into the donor channel

Using these correction factors, the amount of sensitized emission is determined as follows (Rheenen et al. 2004):

$$F_{SE} = \frac{S - \beta D - A(\gamma - \alpha\beta)}{1 - \beta\delta} \tag{2.3}$$

To convert the amount of sensitized emission, which is only the relative amount of energy transfer detected per pixel, into apparent FRET efficiency, this value is normalized to either the donor or acceptor intensity.

Apparent FRET efficiency normalized to acceptor fluorescence:

$$E_A = \frac{F_{SE}}{A} \tag{2.4}$$

Apparent FRET efficiency normalized to donor fluorescence:

$$E_D = \frac{F_{SE}}{D} \tag{2.5}$$

Sensitized emission is a technique that is well suited to acquiring time-lapse sequences for investigating the apparent FRET efficiency of two proteins of interest over time. Starting the experiment is a bit tedious as the donor-only and acceptor-only control samples have to be acquired every time, but the data analysis can be performed more or less automatically, e.g., with FiJi (http://fiji.sc) plugins or macros such as PixFRET (Feige et al. 2005) and FluoQ (Stein et al. 2013).

To investigate optical probes containing a FRET pair on a single molecule, the method of sensitized emission described previously is also suitable, but it can be simplified because the donor/acceptor ratio is fixed (Figure 2.4).

2.3.2.2 Ratiometric FRET Imaging

Most optical probes based on FRET contain both donor and acceptor fluorophores fused with a reporter element that senses the physiological parameter of

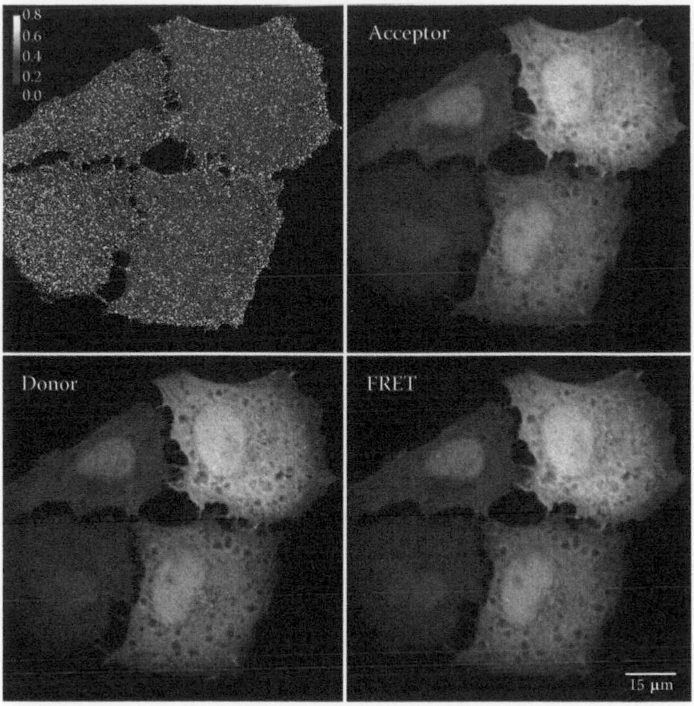

Figure 2.4 Sensitized emission of eCFP–linker–eYFP.

interest (see the following discussion on probe design for details). The fixed intramolecular ratio between the donor and the acceptor can be exploited to measure conformational changes, for example, on ligand binding, by FRET using only two channels, without the need for very specialized equipment (Wouters et al. 2001; Jalink and van Rheenen 2009). In most cases, donor excitation is used to detect both donor and acceptor emission. In the majority of experiments, qualitative changes during the time course are sufficient, for example, calcium oscillations triggered by the addition of ATP or the relative readout of aurora B activity (Figure 2.5) during mitosis (Fuller et al. 2008). For more quantitative analyses, the measured values can be calibrated, provided minimum and maximum levels can be reliably established. For calcium sensors, a common procedure is to calibrate at the end of the experiment by adding ionomycin to reach the calcium concentration of the medium and to subsequently add chelators like BAPTA (1,2-bis(*o*-aminophenoxy)ethane-*N*,*N*,*N*′*N*′-tetraacetic acid) and EGTA (ethylene glycol tetraacetic acid) to determine the minimum response (Miyawaki et al. 1997). Conventional widefield fluorescence

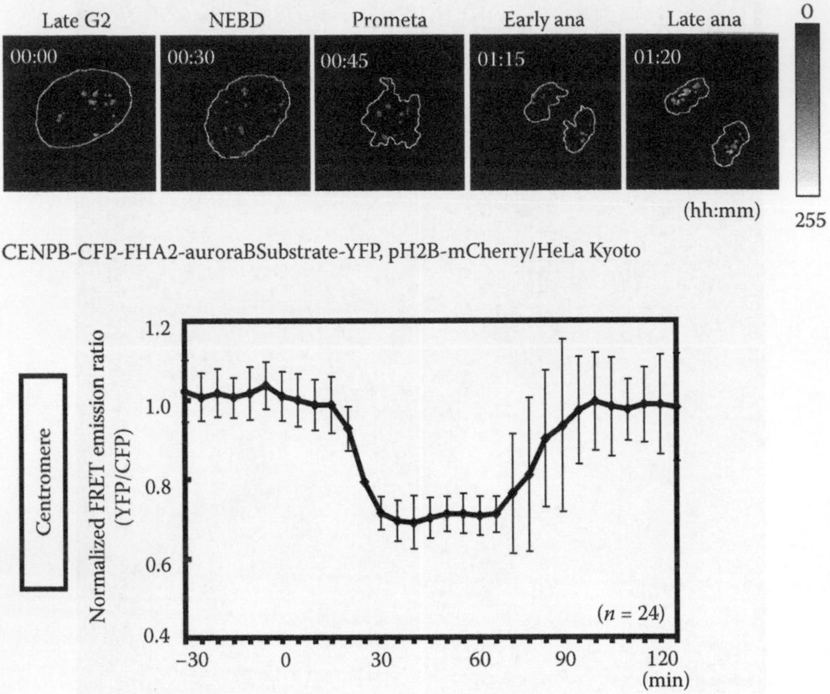

Figure 2.5 Ratiometric FRET measurement of an aurora B sensor (Fuller et al. 2008) targeted to kinetochores in a stable cell line. (Courtesy of Mayumi Isokane, EMBL Heidelberg.)

microscopes with excitation and emission filter wheels, as well as spinning disk or confocal laser scanning microscopes, are well suited for ratiometric FRET imaging. The amounts of donor cross-talk and direct acceptor excitation are constant and are cancelled out by calculating the ratio between the donor and acceptor emission, both acquired using donor excitation. The ratio calculation can be performed with several software packages. These are frequently integrated into the microscope software. FiJi (http://fiji.sc) is freeware software that is also well suited for this, especially because it can be extended via macros and plugins shared by an active community, such as the FluoQ macro (Stein et al. 2013).

In an ideal case, the donor and acceptor emission channels are acquired simultaneously after donor excitation. This option is available on most confocal LSMs that are equipped with a minimum of two detectors in the correct detection range for the fluorescent proteins used. For simultaneous acquisition on camera-based systems like spinning disk confocal or widefield microscopes, the signal is usually separated by a beam splitter either projected onto two cameras (e.g., AndorTucam) or projected side by side onto the same camera chip (e.g., Optosplit from Cairn). It is important to align the images before calculating ratio values, as the two images are usually slightly shifted (Spiering et al. 2013). If the investigated process is relatively slow, a less expensive but slower option is to acquire the two channels sequentially; however, the acquisition rate needs to be fast enough to avoid artifacts from the lag time during changes between filter cubes. This sequential acquisition may be accelerated by using fast filter wheels to minimize the time delay between the channels. It has to be noted that the sequential acquisition of the two channels also leads to more light exposure of the specimen, as the donor has to be excited twice. This may cause more photobleaching and photodamage. Note that widefield and spinning disk microscopes collect light more efficiently than confocal LSMs, and the effective light dose is highly dependent on the sensitivity and the settings of the equipment used.

2.3.2.3 Acceptor Photobleaching—Quantifying Donor Quenching

Under FRET conditions, interacting donor molecules partially transfer absorbed energy to acceptor molecules, leading to decreased fluorescence emission by the donor (quenching). This reduced fluorescence emission is not immediately quantifiable without further information because the donor fluorescence intensity strongly depends on the donor concentration, but only an unknown fraction of donor molecules is interacting and quenched due to FRET. The amount of quenching is determined by acquiring a series of images in the donor and acceptor channel before and after bleaching the acceptor. On confocal LSMs, which are able to bleach by scanning a selected area, preferably only a region of interest (ROI) is bleached in the acceptor channel to control for artifacts in the nonbleached area. The FRET efficiency is then calculated using the donor image before and after acceptor bleaching by the following formula (Wouters et al. 2001):

$$E_{\mathrm{D}}(i)=1-\frac{F^{\mathrm{D}}(i)}{F^{\mathrm{D}}_{\mathrm{pb}}(i)}=E\cdot\alpha_{\mathrm{D}}(i) \tag{2.6}$$

FiJi plugins such as AccPbFRET (Roszik et al. 2008) are available and facilitate the analysis of the acquired data.

Unfortunately, the measured apparent FRET efficiency acquired with this simple version of acceptor photobleaching very much depends on the fraction of bleached acceptor molecules. To get the maximal amount of unquenching, 100% of the acceptor molecules have to be bleached. For most FRET pairs, the light dose needed to bleach all the acceptor molecules will also bleach donor molecules to a certain extent, thereby decreasing the measured apparent FRET efficiency. One solution for this problem, which leads to more precise values, is a method based on repetitive acceptor photobleaching (Amiri et al. 2003; Van Munster et al. 2005). Similar to a fluorescence loss in photobleaching (FLIP) experiment (Rabut and Ellenberg 2005), a time course of donor and acceptor channels, interleaved with low-dose acceptor bleaching events, is acquired to measure the amount of donor unquenching with respect to the degree of acceptor bleaching (Figure 2.6a and b). The calculated partial FRET efficiency is then plotted against the normalized amount of acceptor. The apparent FRET efficiency at maximum unquenching is calculated by linear regression analysis of the partial FRET efficiency determined for different acceptor bleaching levels (Figure 2.6c). Without significant donor bleaching, the intersection with the y-axis (apparent FRET efficiency) indicates the desired value (Amiri et al. 2003). If donor bleaching occurs, data points at low amounts of unbleached acceptor molecules will systematically deviate from the linear regression, indicating

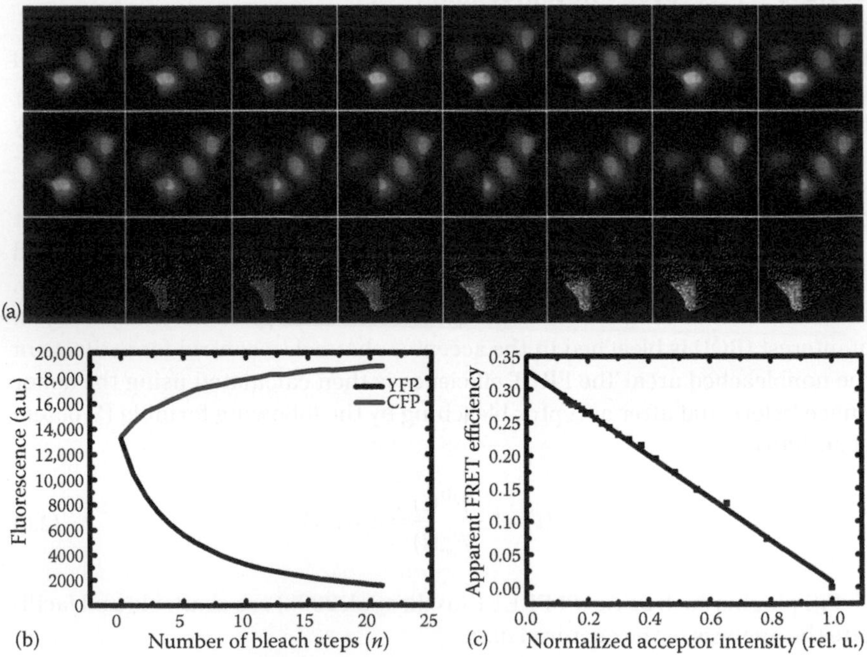

Figure 2.6 Example of repetitive acceptor photobleaching. (a) Montage of donor (upper row) and acceptor (center row) images and the calculated apparent FRET efficiency (lower row). (b) Mean fluorescence intensity of donor and acceptor channels in the ROI repetitively bleached. (c) Apparent FRET efficiency normalized to the acceptor intensity.

lower values than expected. In this case, only the linear part of the data should be used for the linear regression to exclude the influence of donor bleaching.

Acceptor photobleaching is also possible with lamp-based widefield systems (Kenworthy and Edidin 1999). Depending on the fluorophore and the available laser or lamp intensity, the bleaching event can take from a few seconds up to several minutes. During this time, focal drift might occur, for example, induced by temperature fluctuations. Even a slight drift of the sample leads to differences between the prebleach and postbleach donor images in regions without bleaching. The differences are usually most pronounced at the borders of the ROI or cells, as these areas experience the strongest relative difference between the shifted images. This problem can be minimized by aligning the prebleach and postbleach donor images by cross-correlation, if the shift was predominantly in the *xy*-direction. Temperature stabilization or hardware autofocus can be used to prevent drift in the *z*-direction. Unbleached areas outside the ROI serve as an internal control regarding drift, but for acceptor photobleaching, it is important to acquire additional controls with the same settings: a positive control and acceptor-only and donor-only as negative controls. The sensitivity to movement and the strong bleach pulse needed are the reasons why acceptor photobleaching is predominantly used with fixed samples, although it has been used, for example, in combination with FRAP (Dinant et al. 2008). With the positive control, it is important to reach a value close to the apparent FRET efficiency reported previously (using the same FRET technique), to verify that the conditions being used enable a reliable readout. If the result is significantly lower, this might be an indication that the acceptor is insufficiently bleached or bleached too much, leading to unwanted donor bleaching and thereby reducing the measured apparent FRET efficiency. In the next step, an acceptor-only sample has to be measured using the same conditions. Ideally no "FRET" signal is detected in this negative control. Any signal in the donor channel might be due to drift (as explained above), or the acceptor (most prominent for FPs) might be partially converted into a "donor-like species" and emit in the donor instead of the acceptor channel (Valentin et al. 2005; Kirber et al. 2007). In most cases, the contribution of photoconverted acceptors to the FRET signal is very low. It may be reduced by choosing more stable acceptor fluorophores (such as Citrine or Venus instead of eYFP), or it may be quantified and corrected if the acceptor is not very abundant (Seitz et al. 2012).

2.3.2.4 Photochromic FRET—Reversible Donor Quenching

Acceptor photobleaching is an established method to measure FRET by quantifying the donor quenching due to FRET. However, acceptor photobleaching has a big drawback: it is not well suited for live cell imaging because it requires high laser power to bleach the acceptor, which can give rise to artifacts if the donor molecules move during the time the acceptor is being bleached. An emerging FRET method is pcFRET. A photochromic acceptor that switches reversibly between on and off states on irradiation with the appropriate wavelengths can be used to measure donor quenching at relatively high frequencies (Giordano et al. 2002; Mao et al. 2008). Recently, a red fluorescent protein (rsTagRFP) has been engineered that is suitable for pcFRET in living cells (Subach et al. 2010). The reversible photochromic effect works like an inbuilt control. This permits the use of a nonfluorescent photochromic protein as the acceptor in pcFRET (Don Paul et al. 2013).

2.3.2.5 Donor Photobleaching

Energy transfer to acceptor molecules is an additional pathway through which donor molecules can dispose of excitation energy and return to the ground state. One consequence of this is that the chance to bleach is reduced because the donor molecules populate less excited singlet and triplet states and bleaching occurs mainly via chemical reactions with these excited states (Song et al. 1995, 1996; Widegren and Rigler 1996). This is also exploitable to detect FRET (Gadella and Jovin 1995), although donor photobleaching is less suited for quantitative measurements because of the bleaching of the acceptor molecules by cross-excitation or even energy transfer. Nevertheless, donor photobleaching may be combined with acceptor photobleaching by bleaching the acceptor in an ROI and subsequently bleaching the donor in the whole field of view (Bastiaens et al. 1996). If FRET is occurring, the bleaching rates of the donor will differ between areas with and without intact acceptor molecules.

2.3.2.6 FLIM—Reduced Fluorescence Lifetime of the Donor

FLIM is a method for the accurate measurement of the lifetime of fluorescent molecules (Borst and Visser 2010). In the context of FRET, FLIM is considered as one of the methods that quantitatively measures FRET efficiency, as the lifetime of the donor is a direct indication of the FRET efficiency. According to the Jablonski diagram (Figure 2.1), the fluorescence lifetime of the donor molecule will be shorter if it has an additional pathway to relax from the excited state (e.g., due to nonradiative energy transfer, as in the case of FRET). The fluorescence lifetime of the donor decreases proportional to the FRET efficiency. Thus, the difference between the lifetime of a pure donor τ_d and the lifetime of a donor interacting with an acceptor τ_{da} is related to the real FRET efficiency E. E can be determined via FLIM according to the following formula (Bastiaens and Pepperkok 2000):

$$E = 1 - \frac{\tau_{da}}{\tau_d} \tag{2.7}$$

FLIM may therefore be exploited for calibrating the apparent FRET efficiencies obtained in experiments performed by other methods, for example, sensitized emission.

FLIM is a relatively challenging method and requires dedicated experimental equipment. Two distinct approaches to measuring the fluorescence lifetime are currently available, operating in either the time domain or the frequency domain.

Time-domain FLIM provides direct and very accurate values for the fluorescence lifetime, but it is technically challenging, and compared to frequency-domain FLIM, it is relatively slow (approximately 10–30 s with time-domain FLIM vs. 1–3 s with frequency-domain FLIM at comparable quality). A typical setup of this type is based on a confocal microscope and requires a pulsed laser and additional electronic equipment for time-correlated measurements. A sample is excited with a pulsed laser, and the intensity time trace of the resulting fluorescence signal is collected with subnanosecond precision. The obtained information on how the fluorescence intensity changes relative to the time of excitation is used to derive the value of the lifetime via fitting an exponential decay model. The accuracy of the fitting indicates whether the lifetime of a given fluorophore has a monoexponential or multiexponential character. The use of pulsed interleaved excitation (Müller et al. 2005) makes it possible to distinguish between molecules with different lifetimes. In the case of FRET, this allows the determination of the percentage of interacting donor molecules.

Frequency-domain FLIM determines the lifetime values based on an indirect calculation, which renders it less accurate, but it requires less sophisticated equipment and reaches higher frame rates in time-lapse experiments (Pietraszewska-Bogiel and Gadella 2011). The method is typically implemented in widefield mode or on a spinning disk confocal microscope. Fluorescence is excited using a modulated light source such as an LED or a laser at a frequency of several tens of megahertz. The resulting fluorescence signal is sampled at different phase positions by a modulated intensified camera. The change of the modulation depth and the phase shift with respect to the excitation light is determined by a frequency-domain cross-correlation function. Based on these values, the modulation and phase lifetimes are calculated. If these two values are close, a monoexponential lifetime of the fluorophore is indicated. A large discrepancy between phase and modulation lifetime is a sign of a multiexponential decay. In this case, more thorough investigation using multifrequency modulation should be performed to obtain additional information about the lifetime components. In the case of FRET, the phasor polar plot can be used to determine the ratio of interacting and free donor (Redford and Clegg 2005; Digman et al. 2008).

2.3.2.7 Polarization Anisotropy Imaging

On stimulation of fluorescence by a polarized light source, typically a laser, the fluorescent molecules whose dipoles are oriented in the direction of the polarization axis are predominantly exited and exhibit polarized emission directly after excitation. The polarization anisotropy of the excited molecules decreases by rotation and can be used to measure rotational diffusion (Axelrod 1979). FRET is another possible reason why the polarization of the excited fluorophores can decrease, for example, via the transfer of energy to an acceptor that is in a random orientation (Lidke et al. 2003). This is especially useful for measuring energy transfer between identical fluorophores (homo-FRET), for example, to investigate the oligomerization state of interacting proteins (Blackman et al. 1998; Squire et al. 2004).

2.3.3 FRET Method Selection

The large variety of available FRET techniques is quite overwhelming, as all the methods have their advantages and disadvantages (Table 2.3). As a first step, it is advisable to start with easy and available techniques. For example, measuring intramolecular FRET sensors is relatively easy using ratiometric FRET detection on a large number of standard microscope systems and should be preferred for qualitative measurements over more complicated FLIM measurements. Proper controls for the selected technique, as described in Section 2.3.2, are as important as the FRET detection method. A general suggestion for novices is to get in contact with more experienced colleagues, for example, via a light microscopy core facility or by collaborating with an expert. A more detailed FRET microscopy method selector has been compiled by Pietraszewska-Bogiel and Gadella (2011).

2.4 FLUORESCENCE RECOVERY AFTER PHOTOBLEACHING

FRAP is a method used to measure the mobility of molecules and their binding characteristics. FRAP is in some ways a complementary method to FRET. Whereas FRET allows one to determine if two entities interact, FRAP can be used to measure diffusion rates and to determine the binding coefficients of complexes (Houtsmuller and Vermeulen 2001; Sprague et al. 2006; Wachsmuth 2014). The basic principle of FRAP is very easy. A fluorophore in the ROI of a certain shape within the sample is quickly bleached using a strong light dose, typically a laser. The recovery of fluorescence intensity in this ROI is then monitored until the intensity once again reaches a plateau level. Based on the time and degree of recovery, the apparent diffusion coefficient and the level of the bound fraction ("immobile" fraction—on the time scale of the experiment) in the sample are determined by fitting the recovery curve with an appropriate function describing the recovery. Three cases can be distinguished:

1. *Recovery limited only by diffusion.* If the investigated molecules move freely without binding events, the recovery reflects pure diffusion. The effective diffusion coefficient D_{eff} is determined by fitting the recovery curve with a formula describing diffusion into a region corresponding to the bleached area (Axelrod et al. 1976; Soumpasis 1983). Cytoplasm and nucleoplasm are not simple fluids and instead are packed with a large number of different biomolecules. Diffusion in membranes is limited along its folded two-dimensional structure. Especially for larger molecules, FRAP-derived diffusion coefficients will be influenced, for example, by molecular sieving effects (Sprague and McNally 2005) and, at least for larger bleach areas, by constriction due to organelles such as the endoplasmic reticulum, vesicles, or the Golgi apparatus (Weiss et al. 2003; Bancaud et al. 2010).
2. *Equilibration dominated by binding events.* The average time the investigated molecules are bound is high and the major fraction of molecules is bound. The time to equilibrate the bleached area with freely moving molecules is much faster than the dissociation events. The dynamics of recovery are determined only by the dissociation rate k_{off} (Sprague et al. 2004).
3. *Combination of diffusion and binding contribution.* In many cases, the investigated molecules will interact only with a low number of binding sites or on a fast time scale, decreasing the speed of recovery. The recovery curve can still be very similar to a diffusion-limited case. If a diffusion coefficient is calculated, this is often indicated by using the term effective diffusion coefficient D_{eff} (Sprague et al. 2004). To determine more details, such as binding rates, residence times, and the diffusion coefficient itself, it is necessary to analyze the FRAP data in combination with modeling approaches (Houtsmuller 2005; Forster et al. 2006; McNally 2008; Bancaud et al. 2010).

   A simple method to determine if the recovery speed depends on diffusion is to apply FRAP to ROIs of different sizes. In the case of binding-dominated recovery, the rate will not depend on the shape and size of the ROI (Bancaud et al. 2010). For diffusion-limited recovery, plotting the square of the ROI radius against the halftime of recovery should give a straight line with a slope proportional to the diffusion coefficient.

FRAP experiments can be performed on virtually every microscope system. The most straightforward way is to use a modern confocal LSM, as it already has a scanner and does not require any system modification to selectively bleach an ROI. On most of modern commercial confocal systems, a FRAP software module is implemented, which makes FRAP experiments technically very accessible. Olympus provides a very interesting technical solution by implementing a second scanner in its FV1200 microscope to make it very flexible for FRAP applications, allowing users to fine-tune the timing between bleaching and the first postbleach acquisition. One of the disadvantages of confocal FRAP is the scan speed limitation. For samples with very fast recovery rates (recovery by diffusion only), the time resolution can be increased by scanning a smaller area of the sample. The size and the shape of the bleaching ROI also influence the time needed for bleaching and switching between imaging and bleaching mode; often a horizontal line or strip ROI is the fastest option for bleaching. The speed problem is less severe when FRAP is performed on a spinning disk microscope. Currently, several commercial FRAP add-ons for spinning disk systems are available (e.g., PK unit from PerkinElmer or FRAPPA from Andor). In this case, a limiting factor is the switching time between the imaging and bleaching modes. The first FRAP experiments were performed on custom-modified widefield systems, usually with the laser spot fixed in the center of the image field (Koppel et al. 1976). Today, several commercial options to equip widefield microscopes with a scanning FRAP module are available (e.g., Rapp Optoelectronic UGA-40, Roper iLas^2). FRAP experiments in widefield mode offer high time resolution but can be limited by the thickness of the sample, as out-of-focus light deteriorates the dynamic range of FRAP recovery. Another aspect of FRAP that should be kept in mind is that the sample is bleached not only in the focal plane but also above and below it along the optical axis. The only solution to confine the bleached region to the focal volume is the use of a two-photon LSM (Waharte et al. 2005). Measuring of the mobility of molecules by FRAP can potentially be combined with acceptor photobleaching (Royen et al. 2009).

2.5 FLUORESCENCE CORRELATION SPECTROSCOPY

Fluorescence correlation spectroscopy (FCS) is another method that is complementary to both FRET and FRAP (Stasevich et al. 2010). Whereas FRAP depends on rather high dye concentrations and measures a larger area of interest, FCS works only at low concentrations of fluorescent molecules and measures only in the focal volume. It is a very sensitive method to study the diffusion dynamics of fluorescently labeled particles moving in a liquid environment, with living cells being a special case (Weidemann 2014). With proper calibration, FCS allows very accurate measurements of particle concentrations. A typical FCS system is built on the basis of a confocal microscope.

Laser light is focused into a diffraction-limited spot using a high-quality water immersion objective lens, and the fluorescence signal of the particles diffusing through the focal spot is detected with an APD, HyD, or GaAsP detector set to photon counting mode.

The entry of a fluorescent particle into the focal volume increases the fluorescence signal, and its exit correspondingly decreases the signal. Thus, this method is based on the fact that fluorescence intensity fluctuations are high for a smaller number of molecules (low concentration) in the focal spot. In a typical experiment, the fluorescence signal is recorded for a period of several tens of seconds. Subsequently, the acquired time trace is analyzed by calculating the autocorrelation function, indicating how the signal fluctuations are correlated over time. The amplitude of the autocorrelation curve is inversely proportional to the number of molecules in the focal volume and is thus an indication of the concentration. The characteristic time of decrease in the autocorrelation function represents the time required for the particles to diffuse through the focal volume. If the size of the focal volume is known (e.g., calibrating with different dilutions of a dye), the diffusion coefficient of the investigated molecules can be accurately calculated.

The requirement for a low number of fluorescent particles in the focal volume (between 0.1 and 1000) imposes a limitation on the possible particle concentration range in which FCS is useful. The concentrations of fluorescent molecules that can be faithfully measured by FCS are in the subnanomolar to micromolar range. Luckily enough, this is exactly the concentration range typical for most proteins in living cells under physiological conditions. This fact justifies the very wide use of FCS as an experimental method to study protein dynamics and interactions within the living cell.

If two types of particles are marked with different fluorophores, not only the correlation of the signal for each fluorophore but also the cross-correlation between the signals (i.e., FCCS) can be determined. If the particles move independently, there will be no cross-correlation, provided that there is no bleed-through between the fluorescence channels. If the particles interact, which implies that they move together, it will lead to a significant cross-correlation between the signals. Thus, the degree of cross correlation can be an indication for both interaction and the percentage of interacting species. In this respect, FCCS is a complementary method to FRET, not only to probe if binding takes place but also to measure additional information regarding the concentrations and stoichiometry of the donor, acceptor, and interacting particles. It has to be mentioned that FCCS measurements will be influenced if FRET is occurring, and this fact needs to be taken into account by suitable corrections or adapted FCCS measurement methods (Sahoo and Schwille 2011).

Recently, several successful attempts to extend FCCS from a single spot measurement to a two-dimensional method have been made by combining

FCCS with SPIM and camera-based detection (Wohland et al. 2010; Capoulade et al. 2011). This enables the imaging of protein concentration and interaction maps within the living cell and provides a deeper understanding of protein interactions within the cell.

Another approach for concentration and diffusion measurements in a complex sample is the so-called image correlation spectroscopy (ICS). In addition to high spatial and sufficient temporal resolution, this method allows diffusion to be distinguished from binding, which might be useful for studying protein interactions in living samples, as well as for several FRET applications. This method is based on the fact that a confocal microscope scans the image line by line, such that the pixels of the image are acquired sequentially. Thus, for a given scan speed, the position of a pixel in the image defines at which time point it was imaged. By taking several consecutive images and applying correlation procedures equivalent to FCCS but in the spatial domain on the scale of the whole image, information about molecule diffusion in the sample can be derived. Originally, this method was developed for imaging with confocal systems, but it can also be extended to the widefield detection modality. Several implementations of this technique have been reported, among them raster image correlation spectroscopy (RICS; Rossow et al. 2010) and spatiotemporal image correlation spectroscopy and cross-correlation spectroscopy (STICS and STICCS; Wiseman 2013).

2.6 BIMOLECULAR FLUORESCENCE COMPLEMENTATION

For a low percentage of interacting molecules, the available FRET methods are often not sensitive enough. The main problem here is the high level of fluorescence from the noninteracting molecules, which renders the tiny FRET signal indistinguishable from noise. To detect small fractions of interacting molecules, it is desirable to induce fluorescence only in the event of an interaction, leading to a signal from interacting molecules above the dark background. Bimolecular fluorescence complementation (BiFC) is one such method. The different molecules of interest are labeled with complementary, nonfluorescent fragments of a fluorescent protein derivative. Noninteracting molecules remain dark, whereas interacting molecules bring the fused complementary fragments into close proximity, inducing complementation to form a mature fluorophore (Kerppola 2009). This method is used to detect whether interactions occur, but because the complementation of the fluorescent protein is not reversible, it is not possible to investigate the dynamics and levels of such events. In addition, the strong and irreversible molecular interaction has the tendency to induce dimerization.

2.7 DIMERIZATION-DEPENDENT FLUORESCENCE

Dimerization-dependent fluorescence is a recent development (Alford et al. 2012a) that promises a more dynamic readout than BiFC. This method is based on fluorescent proteins that are very dim as monomers but whose brightness increases significantly on dimerization, for example, dimerization-dependent GFP and YFP (ddGFP/ddYFP (Alford et al. 2012b). As the dimerization process is reversible, this opens the door to investigating the dynamics of weak interactions.

2.8 CONCLUSION AND OUTLOOK

Thanks to continuous improvements in fluorophores (especially FP variants), microscopy equipment, and techniques, FRET and complementary techniques are increasingly applied to investigate biological processes. The growing repository of optical probes (see Sections II and III) offers readouts of a wide range of physiological parameters and contributes substantially to this success. Multiparameter fluorescence imaging to investigate the interplay between different pathways at once is now becoming feasible using existing tools (Piljic and Schultz 2008; Carlson and Campbell 2009; Welch et al. 2011; Woehler 2013). Automating these techniques in high-throughput microscopy (Pepperkok and Ellenberg 2006; Verissimo and Pepperkok 2013; Robinson et al. 2014) also opens the door to systems biology approaches. Finally, the use of automatic cell detection for the functional imaging of rare events by FRET and complementary techniques offers increased efficiency and more objective data generation (Conrad et al. 2011). However, regardless of whatever new developments may improve our view of the microscopic world, FRET will always be on the menu.

REFERENCES

Alford, S. C., Abdelfattah, A. S. et al. (2012a). A fluorogenic red fluorescent protein heterodimer. *Chem. Biol.* 19(3):353–360.

Alford, S. C., Ding, Y. et al. (2012b). Dimerization-dependent green and yellow fluorescent proteins. *ACS Synth. Biol.* 1:569–575.

Amiri, H., Schultz, G. et al. (2003). FRET-based analysis of TRPC subunit stoichiometry. *Cell Calcium* 33(5–6):463–470.

Axelrod, D. (1979). Carbocyanine dye orientation in red cell membrane studied by microscopic fluorescence polarization. *Biophys. J.* 26(3):557–573.

Axelrod, D. (2001). Total internal reflection fluorescence microscopy in cell biology. *Traffic* 2:764–774.

Axelrod, D., Koppel, D. E. et al. (1976). Mobility measurement by analysis of fluorescence photobleaching recovery kinetics. *Biophys. J.* 16:1055–1069.

Bacia, K., Kim, S. A. et al. (2006). Fluorescence cross-correlation spectroscopy in living cells. *Nat. Methods* 3(2):83–89.

Bancaud, A., Huet, S. et al. (2010). Fluorescence-perturbation techniques to study mobility and molecular dynamics of proteins in live cells: FRAP, photoactivation, photoconversion, and FLIP. In R. D. Goldman, J. R. Swedlow, and D. L. Spector (Eds.), *Live Cell Imaging: A Laboratory Manual*, 2nd ed. Cold Spring Harbor, NY: Cold Spring Harbor Laboratory Press.

Bastiaens, P. I., Majoul, I. V. et al. (1996). Imaging the intracellular trafficking and state of the AB5 quaternary structure of cholera toxin. *EMBO J.* 15(16):4246–4253.

Bastiaens, P. I. H., and Pepperkok, R. (2000). Observing proteins in their natural habitat: The living cell. *Trends Biochem. Sci.* 25(12):631–637.

Berney, C., and Danuser, G. (2003). FRET or no FRET: A quantitative comparison. *Biophys. J.* 84(6):3992–4010.

Blackman, S. M., Piston, D. W. et al. (1998). Oligomeric state of human erythrocyte band 3 measured by fluorescence resonance energy homotransfer. *Biophys. J.* 75(2):1117–1130.

Borghi, N., Sorokina, M. et al. (2012). E-cadherin is under constitutive actomyosin-generated tension that is increased at cell–cell contacts upon externally applied stretch. *Proc. Natl. Acad. Sci. U. S. A.* 109(31):12568–12573.

Borst, J. W., and Visser, A. J. W. G. (2010). Fluorescence lifetime imaging microscopy in life sciences. *Meas. Sci. Technol.* 21(10):102002.

Brown, C. M. (2007). Fluorescence microscopy—Avoiding the pitfalls. *J. Cell Sci.* 120(10):1703–1705.

Capoulade, J., Wachsmuth, M. et al. (2011). Quantitative fluorescence imaging of protein diffusion and interaction in living cells. *Nat. Biotechnol.* 29:835–839.

Carlson, H. J., and Campbell, R. E. (2009). Genetically encoded FRET-based biosensors for multiparameter fluorescence imaging. *Curr. Opin. Biotechnol.* 20(1):19–27.

Conrad, C., Wunsche, A. et al. (2011). Micropilot: Automation of fluorescence microscopy-based imaging for systems biology. *Nat. Methods* 8(3):246–249.

Costa, A., Candeo, A. et al. (2013). Calcium dynamics in root cells of *Arabidopsis thaliana* visualized with selective plane illumination microscopy. *PLoS One* 8(10):e75646.

Digman, M. A., Caiolfa, V. R. et al. (2008). The phasor approach to fluorescence lifetime imaging analysis. *Biophys. J.* 94(2):L14–L16.

Dinant, C., van Royen, M. E. et al. (2008). Fluorescence resonance energy transfer of GFP and YFP by spectral imaging and quantitative acceptor photobleaching. *J. Microsc.* 231(Pt 1):97–104.

Don Paul, C., Kiss, C. et al. (2013). Phanta: A non-fluorescent photochromic acceptor for pcFRET. *PLoS One* 8(9):e75835.

Feige, J. N., Sage, D. et al. (2005). PixFRET, an ImageJ plug-in for FRET calculation that can accommodate variations in spectral bleed-throughs. *Microsc. Res. Tech.* 68(1):51–58.

Forster, R., Weiss, M. et al. (2006). Secretory cargo regulates the turnover of COPII subunits at single ER exit sites. *Curr. Biol.* 16(2):173–179.

Förster, T. (1946). Energy transport and fluorescence [in German]. *Naturwissenschaften* 33:166–175.

Fuller, B. G., Lampson, M. A. et al. (2008). Midzone activation of aurora B in anaphase produces an intracellular phosphorylation gradient. *Nature* 453(7198):1132–1136.

Gadella, T. W., and Jovin, T. M. (1995). Oligomerization of epidermal growth factor receptors on A431 cells studied by time-resolved fluorescence imaging microscopy. A stereochemical model for tyrosine kinase receptor activation. *J. Cell Biol.* 129(6):1543–1558.

Gansen, A., Valeri, A. et al. (2009). Nucleosome disassembly intermediates characterized by single-molecule FRET. *Proc. Natl. Acad. Sci. U. S. A.* 106(36):15308–15313.

Giordano, L., Jovin, T. M. et al. (2002). Diheteroarylethenes as thermally stable photoswitchable acceptors in photochromic fluorescence resonance energy transfer (pcFRET). *J. Am. Chem. Soc.* 124(25):7481–7489.

Gordon, G. W., Berry, G. et al. (1998). Quantitative fluorescence resonance energy transfer measurements using fluorescence microscopy. *Biophys. J.* 74(5):2702–2713.

Gräf, R., Rietdorf, J. et al. (2005). Live cell spinning disk microscopy. *Adv. Biochem. Eng. Biotechnol.* 95:57–75.

Greger, K., Neetz, M. J. et al. (2011). Three-dimensional fluorescence lifetime imaging with a single plane illumination microscope provides an improved signal to noise ratio. *Opt. Express* 19(21):20743–20750.

Houtsmuller, A. B. (2005). Fluorescence recovery after photobleaching: Application to nuclear proteins. *Adv. Biochem. Eng. Biotechnol.* 95:177–199.

Houtsmuller, A. B., and Vermeulen, W. (2001). Macromolecular dynamics in living cell nuclei revealed by fluorescence redistribution after photobleaching. *Histochem. Cell Biol.* 115:13–21.

Huet, S., Avilov, S. V. et al. (2010). Nuclear import and assembly of influenza A virus RNA polymerase studied in live cells by fluorescence cross-correlation spectroscopy. *J. Virol.* 84(3):1254–1264.

Huisken, J., Swoger, J. et al. (2004). Optical sectioning deep inside live embryos by selective plane illumination microscopy. *Science* 305(5686):1007–1009.

Jalink, K., and van Rheenen, J. (2009). FilterFRET: Quantitative imaging of sensitized emission. In T. W. J. Gadella (Ed.), *Laboratory Techniques in Biochemistry and Molecular Biology*, Vol. 33, pp. 289–349. Amsterdam: Elsevier.

Jares-Erijman, E. A., and Jovin, T. M. (2003). FRET imaging. *Nat. Biotechnol.* 21(11):1387–1395.

Keller, P. J. (2013). Imaging morphogenesis: Technological advances and biological insights. *Science* 340(6137):1234168.

Kenworthy, A. K., and Edidin, M. (1999). Imaging fluorescence resonance energy transfer as probe of membrane organization and molecular associations of GPI-anchored proteins. *Methods Mol. Biol.* 116:37–49.

Kerppola, T. K. (2009). Visualization of molecular interactions using bimolecular fluorescence complementation analysis: Characteristics of protein fragment complementation. *Chem. Soc. Rev.* 38:2876–2886.

Kirber, M. T., Chen, K. et al. (2007). YFP photoconversion revisited: Confirmation of the CFP-like species. *Nat. Methods* 4(10):767–768.

Koppel, D. E., Axelrod, D. et al. (1976). Dynamics of fluorescence marker concentration as a probe of mobility. *Biophys. J.* 16(11):1315–1329.

Krzic, U., Gunther, S. et al. (2012). Multiview light-sheet microscope for rapid in toto imaging. *Nat. Methods* 9(7):730–733.

Lidke, D. S., Nagy, P. et al. (2003). Imaging molecular interactions in cells by dynamic and static fluorescence anisotropy (rFLIM and emFRET). *Biochem. Soc. Trans.* 31(Pt 5):1020–1027.

Mao, S., Benninger, R. K. P. et al. (2008). Optical lock-in detection of FRET using synthetic and genetically encoded optical switches. *Biophys. J.* 94(11):4515–4524.

McClure, D. S. (1949). Triplet-singlet transitions in organic molecules: Lifetime measurements of the triplet state. *J. Chem. Phys.* 17(10):905–913.

McNally, J. G. (2008). Quantitative FRAP in analysis of molecular binding dynamics in vivo. In F. S. Kevin (Ed.), *Methods in Cell Biology*, Vol. 85, pp. 329–351. San Diego, CA: Academic Press.

Miyawaki, A., Llopis, J. et al. (1997). Fluorescent indicators for Ca^{2+} based on green fluorescent proteins and calmodulin. *Nature* 388(6645):882–887.

Müller, B. K., Zaychikov, E. et al. (2005). Pulsed interleaved excitation. *Biophys. J.* 89(5): 3508–3522.

Nishigaki, T., Wood, C. D. et al. (2006). Stroboscopic illumination using light-emitting diodes reduces phototoxicity in fluorescence cell imaging. *BioTechniques* 41(2): 191–197.

Patterson, G. H., Piston, D. W. et al. (2000). Forster distances between green fluorescent protein pairs. *Anal. Biochem.* 284(2):438–440.

Pepperkok, R., and Ellenberg, J. (2006). High-throughput fluorescence microscopy for systems biology. *Nat. Rev. Mol. Cell. Biol.* 7(9):690–696.

Pietraszewska-Bogiel, A., and Gadella, T. W. J. (2011). FRET microscopy: From principle to routine technology in cell biology. *J. Microsc.* 241(2):111–118.

Piljic, A., and Schultz, C. (2008). Simultaneous recording of multiple cellular events by FRET. *ACS Chem. Biol.* 3(3):156–160.

Piston, D. W., and Kremers, G.-J. (2007). Fluorescent protein FRET: The good, the bad and the ugly. *Trends Biochem. Sci.* 32(9):407–414.

Pitrone, P. G., Schindelin, J. et al. (2013). OpenSPIM: An open-access light-sheet microscopy platform. *Nat. Methods* 10(7):598–599.

Preibisch, S., Saalfeld, S. et al. (2010). Software for bead-based registration of selective plane illumination microscopy data. *Nat. Methods* 7(6):418–419.

Preus, S., and Wilhelmsson, L. M. (2012). Advances in quantitative FRET-based methods for studying nucleic acids. *ChemBioChem* 13(14):1990–2001.

Rabut, G., and Ellenberg, J. (2005). Photobleaching techniques to study mobility and molecular dynamics of proteins in live cells: FRAP, iFRAP, and FLIP. In R. D. Goldman and D. L. Spector (Eds.), *Live Cell Imaging—A Laboratory Manual.* Cold Spring Harbor, NY: Cold Spring Harbor Laboratory Press.

Redford, G., and Clegg, R. (2005). Polar plot representation for frequency-domain analysis of fluorescence lifetimes. *J. Fluoresc.* 15(5):805–815.

Rheenen, J. v., Langeslag, M. et al. (2004). Correcting confocal acquisition to optimize imaging of fluorescence resonance energy transfer by sensitized emission. *Biophys. J.* 86:2517–2529.

Robinson, K. H., Yang, J. R. et al. (2014). FRET and BRET-based biosensors in live cell compound screens. *Methods Mol. Biol.* 1071:217–225.

Rossow, M. J., Sasaki, J. M. et al. (2010). Raster image correlation spectroscopy in live cells. *Nat. Protoc.* 5(11):1761–1774.

Roszik, J., Szöllősi, J. et al. (2008). AccPbFRET: An ImageJ plugin for semi-automatic, fully corrected analysis of acceptor photobleaching FRET images. *BMC Bioinform.* 9(1):1–6.

Royen, M. E., Dinant, C. et al. (2009). FRAP and FRET methods to study nuclear receptors in living cells. *Methods Mol. Biol.* 505:69–96.

Sahoo, H., and Schwille, P. (2011). FRET and FCS—Friends or foes? *ChemPhysChem* 12(3):532–541.

Saurabh, S., Maji, S. et al. (2012). Evaluation of sCMOS cameras for detection and localization of single Cy5 molecules. *Opt. Express* 20(7):7338–7349.

Schmid, B., Shah, G. et al. (2013). High-speed panoramic light-sheet microscopy reveals global endodermal cell dynamics. *Nat. Commun.* 4:2207.

Schuler, B., and Hofmann, H. (2013). Single-molecule spectroscopy of protein folding dynamics—Expanding scope and timescales. *Curr. Opin. Struct. Biol.* 23(1):36–47.

Schultz, C. (2007). Molecular tools for cell and systems biology. *HFSP J.* 1(4):230–248.

Seitz, A., Terjung, S. et al. (2012). Quantifying the influence of yellow fluorescent protein photoconversion on acceptor photobleaching-based fluorescence resonance energy transfer measurements. *J. Biomed. Opt.* 17(1):011010.

Song, L., Hennink, E. J. et al. (1995). Photobleaching kinetics of fluorescein in quantitative fluorescence microscopy. *Biophys. J.* 68:2588–2600.

Song, L., Varma, C. A. et al. (1996). Influence of the triplet excited state on the photobleaching kinetics of fluorescein in microscopy. *Biophys. J.* 70(6):2959–2968.

Soumpasis, D. M. (1983). Theoretical analysis of fluorescence photobleaching recovery experiments. *Biophys. J.* 41:95–97.

Spiering, D., Bravo-Cordero, J. J. et al. (2013). Quantitative ratiometric imaging of FRET-biosensors in living cells. *Methods Cell Biol.* 114:593–609.

Sprague, B. L., and McNally, J. G. (2005). FRAP analysis of binding: Proper and fitting. *Trends Cell Biol.* 15(2):84–91.

Sprague, B. L., Muller, F. et al. (2006). Analysis of binding at a single spatially localized cluster of binding sites by fluorescence recovery after photobleaching. *Biophys. J.* 91(4):1169–1191.

Sprague, B. L., Pego, R. L. et al. (2004). Analysis of binding reactions by fluorescence recovery after photobleaching. *Biophys. J.* 86(6):3473–3495.

Squire, A., Verveer, P. J. et al. (2004). Red-edge anisotropy microscopy enables dynamic imaging of homo-FRET between green fluorescent proteins in cells. *J. Struct. Biol.* 147(1):62–69.

Stasevich, T. J., Mueller, F. et al. (2010). Cross-validating FRAP and FCS to quantify the impact of photobleaching on in vivo binding estimates. *Biophys. J.* 99(9):3093–3101.

Stein, F., Kress, M. et al. (2013). FluoQ: A tool for rapid analysis of multiparameter fluorescence imaging data applied to oscillatory events. *ACS Chem. Biol.* 8(9):1862–1868.

Stryer, L. (1978). Fluorescence energy transfer as a spectroscopic ruler. *Annu. Rev. Biochem.* 47(1):819–846.

Subach, F. V., Zhang, L. et al. (2010). Red fluorescent protein with reversibly photoswitchable absorbance for photochromic FRET. *Chem. Biol.* 17(7):745–755.

Swoger, J., Pampaloni, F. et al. (2014). Light-sheet-based fluorescence microscopy for three-dimensional imaging of biological samples. *Cold Spring Harb. Protoc.* 2014(1):1–8.

Szymborska, A., de Marco, A. et al. (2013). Nuclear pore scaffold structure analyzed by super-resolution microscopy and particle averaging. *Science* 341(6146):655–658.

Thaler, C., Koushik, S. V. et al. (2005). Quantitative multiphoton spectral imaging and its use for measuring resonance energy transfer. *Biophys. J.* 89(4):2736–2749.

Valentin, G., Verheggen, C. et al. (2005). Photoconversion of YFP into a CFP-like species during acceptor photobleaching FRET experiments. *Nat. Methods* 2(11):801.

Van Munster, E. B., Kremers, G. J. et al. (2005). Fluorescence resonance energy transfer (FRET) measurement by gradual acceptor photobleaching. *J. Microsc.* 218(3): 253–262.

Verissimo, F., and Pepperkok, R. (2013). Imaging ER-to-Golgi transport: Towards a systems view. *J. Cell Sci.* 126(22):5091–5100.

Visser, A. W. G., and Hink, M. (1999). New perspectives of fluorescence correlation spectroscopy. *J. Fluoresc.* 9(1):81–87.

Vogel, S. S., Thaler, C. et al. (2006). Fanciful FRET. *Sci. STKE* 2006(331):re2.

Wachsmuth, M. (2014). Molecular diffusion and binding analyzed with FRAP. *Protoplasma* 251(2):373–382.

Waharte, F., Brown, C. M. et al. (2005). A two-photon FRAP analysis of the cytoskeleton dynamics in the microvilli of intestinal cells. *Biophys. J.* 88(2):1467–1478.

Weidemann, T. (2014). Application of fluorescence correlation spectroscopy (FCS) to measure the dynamics of fluorescent proteins in living cells. *Fluoresc. Spectrosc. Microsc.* 1076:539–555.

Weiss, M., Hashimoto, H. et al. (2003). Anomalous protein diffusion in living cells as seen by fluorescence correlation spectroscopy. *Biophys. J.* 84(6):4043–4052.

Welch, C. M., Elliott, H. et al. (2011). Imaging the coordination of multiple signalling activities in living cells. *Nat. Rev. Mol. Cell Biol.* 12(11):749–756.

Widegren, J., and Rigler, R. (1996). Mechanisms of photobleaching investigated by fluorescence correlation spectroscopy. *BioImaging* 4:149–157.

Wiseman, P. W. (2013). Image correlation spectroscopy: Mapping correlations in space, time, and reciprocal space. *Methods Enzymol.* 518:245–267.

Woehler, A. (2013). Simultaneous quantitative live cell imaging of multiple FRET-based biosensors. *PLoS One* 8(4):e61096.

Wohland, T., Shi, X. et al. (2010). Single plane illumination fluorescence correlation spectroscopy (SPIM-FCS) probes inhomogeneous three-dimensional environments. *Opt. Express* 18(10):10627–10641.

Wouters, F. S., Verveer, P. J. et al. (2001). Imaging biochemistry inside cells. *Trends Cell Biol.* 11:203–211.

Zal, T., and Gascoigne, N. R. J. (2004). Photobleaching-corrected FRET efficiency imaging of live cells. *Biophys. J.* 86:3923–3939.

Zhao, Y., Araki, S. et al. (2011). An expanded palette of genetically encoded Ca^{2+} indicators. *Science* 333(6051):1888–1891.

Zimmermann, T., Rietdorf, J. et al. (2002). Spectral imaging and linear un-mixing enables improved FRET efficiency with a novel GFP2–YFP FRET pair. *FEBS Lett.* 531:245–249.

Section II

Tracking

Sensors for Tracking Biomolecules

Chapter 3

Protein-Based Calcium Sensors

Thomas Thestrup and Oliver Griesbeck

CONTENTS

3.1 INTRODUCTION

Free calcium ions (Ca^{2+}) are important and omnipresent second messengers for intracellular signal transduction with key regulatory functions in numerous biological processes. Ca^{2+} signals are involved in diverse processes ranging, for example, from gene expression, apoptosis and cell death, bacterial chemotaxis, cell division, and fertilization, to the activation of cells of the immune system. As such, Ca^{2+} signals are extremely versatile owing to their complex spatiotemporal regulation. In neurons, Ca^{2+} is among the most important intracellular signals driving essential mechanisms in neurobiological communication. Action potentials (APs) can trigger and release bursts of Ca^{2+} through voltage-gated channels, causing rapid changes in intracellular Ca^{2+} concentrations. Similarly, Ca^{2+} influxes can be triggered by the release of neurotransmitters, leading to the activation of neurotransmitter-gated ion channels such as *N*-methyl-D-aspartate (NMDA) receptors (Müller and Connor 1991; Jaffe et al. 1992; Berridge et al. 2000). Notably, the processes regulated by Ca^{2+} signaling span a wide time frame from microsecond release of neurotransmitters in synapses to gene transcription lasting up to several hours (Berridge et al. 2003). The magnitude of activity-induced changes in neurons makes Ca^{2+} a unique ion to monitor neuronal activity (Hille 1992). Monitoring changes in free [Ca^{2+}] in spines, dendrites, axons, or somas can be used as a reliable representation of neuronal activity by providing an indirect measure of action potential frequency (Denk et al. 1996; Svoboda et al. 1996, 1997). Thus, Ca^{2+} imaging has become an important method to report the activity of various neuronal cell types and to understand the connectivity and signaling pathways in neuronal networks.

The last three decades have delivered significant leaps in the visualization and quantification of intracellular Ca^{2+}. From a historical perspective, the measurement of free Ca^{2+} in cells and intracellular environments started with the microinjection of bioluminescent proteins (Johnson and Shimomura 1972) and the use of Ca^{2+}-sensitive electrodes (Rink et al. 1980). With the introduction of fluorescent synthetic Ca^{2+} dyes like Fura-2 or Quin-2 (Tsien et al. 1982; Grynkiewicz et al. 1985), the seed for the era of Ca^{2+} imaging was planted. Using synthetic dyes for labeling is often a tedious job, as these dyes require loading either via pipettes or as membrane-permeable acetoxymethyl esters. There is also a strong cell type to cell type variability in the loading efficiency, and loading is practically impossible in some cells and tissues. Owing to the leakage of synthetic dyes, the chronic *in vivo* imaging of neurons is not possible—nor is the targeting of dyes to specific cell types and organelles. New technologies, in particular the use of two-photon fluorescence microscopy (Denk et al. 1990) and the significant evolution of genetically encoded calcium indicators (GECIs), have advanced the capabilities of *in vitro* and *in vivo* imaging (Kleinfeld and Griesbeck 2005).

3.2 GENETICALLY ENCODED CALCIUM INDICATORS

GECIs offer a noninvasive approach for studying the real-time spiking activity of individual neurons or neuronal populations. This is possible because GECIs do not require loading into specific tissues, like synthetic fluorescent dyes, and instead are constitutively expressed by the cells of interest. They are encoded by stretches of DNA and composed only of amino acids, with no need for external cofactors or synthetic compounds to function. Consequently, the encoding DNA can be manipulated in any way modern recombinant molecular biology has to offer. In combination with specific promoters, appropriate targeting sequences, and transgenic approaches, the GECI will be expressed *in situ* and targeted to specific tissues or cell organelles, offering a perfect noninvasive method for Ca^{2+} imaging.

Today, GECIs follow two major design dogmas to provide sufficient fluorescent signal output on Ca^{2+} binding: single fluorescent protein (FP)- and Förster resonance energy transfer (FRET)-based indicators (Figure 3.1; Zhang et al. 2002; Mank and Griesbeck 2008). Probes based on the Ca^{2+}-induced subcellular redistribution of a protein domain are less prominent. Single FP (also often referred to as single-wavelength or intensity-based) indicators are often designed using variants of circularly permutated FPs. The signal output arises

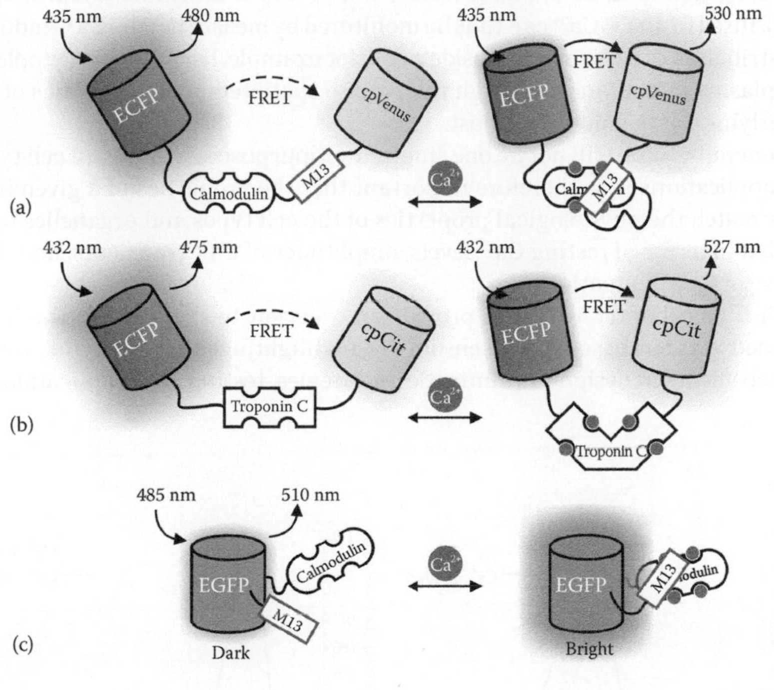

Figure 3.1 Schematics of protein-based calcium sensors. Protein-based Ca^{2+} sensors depend on either changes in Förster resonance energy transfer (FRET) (a, b) or changes in the fluorescence intensity of a single fluorescent protein (c). (a) Schematic representation of YC3.60, a member of the Yellow Cameleon class of sensors. Here, CaM and M13 are inserted between variants of CFP (donor) and YFP (acceptor). The binding of Ca^{2+} to CaM leads to a conformational change and the formation of a CaM–M13 complex, thereby bringing the donor and the acceptor into close proximity to promote FRET. (From Nagai, T. et al., *Proc. Natl. Acad. Sci. USA* 101(29):10554–10559, 2004.) (b) Different family of FRET sensors relies on conformational changes in Troponin C (TnC) on Ca^{2+} binding. In TN-XXL, two TnC C-terminal lobes are fused together, forming the Ca^{2+} sensing domain between the ECFP and cpCitrine FRET pair. (From Mank, M. et al., *Nat. Methods*, 5(9):805–811, 2008.) (c) Schematics of the G-CaMP–type Ca^{2+} sensors. Much like the Yellow Cameleons, G-CaMPs rely on the formation of a CaM–M13 complex on Ca^{2+} binding, which increases the fluorescence intensity of a circularly permutated GFP protein. (From Akerboom, J. et al., *J. Neurosci.* 32(40):13819–13840, 2012.)

through the modulation of chromophore protonation, which is induced by Ca^{2+} binding (Figure 3.2; Baird et al. 1999; Nagai et al. 2001). For FRET-based indicators, the binding of calcium changes the conformation of the sensor, thereby altering the donor and acceptor emission and ultimately the fluorescence ratio between the two (Figures 3.2 and 3.3; Miyawaki et al. 1997; Heim and Griesbeck 2004). There is a third variety, however: calcium-sensitive probes that translocate, for instance, C2 domains (Sakai et al. 1997) or annexins (Clemen et al. 2001), fused to an FP. Ca^{2+} can thus be monitored by measuring the Ca^{2+}-induced redistribution of fluorescence inside a cell, for example, between the cytoplasm and plasma membrane, although the precise spatiotemporal dynamics of the underlying Ca^{2+} signal may be lost.

Generally, there will not be one single, multipurpose sensor for all cell types and applications. It is therefore important that the properties of a given indicator match the physiological properties of the cell types and organelles to be studied in terms of resting Ca^{2+} levels, amplitudes of Ca^{2+} transients, and time course of Ca^{2+} fluctuations.

With this chapter, we try to provide an overview to the field of genetically encoded protein-based Ca^{2+} sensors by highlighting historical and recent advancements in designs, optimization strategies, testing, and applications.

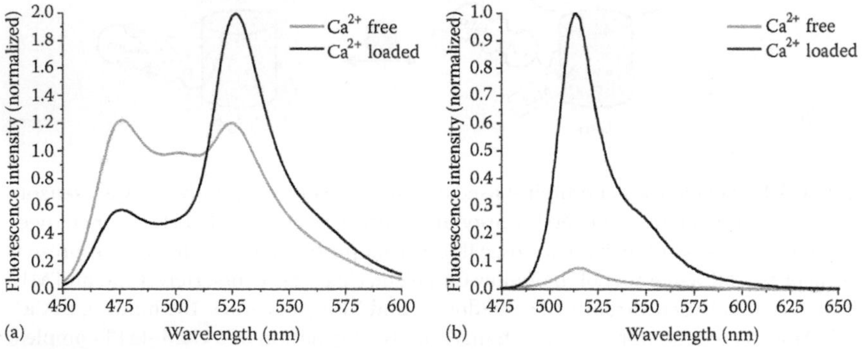

Figure 3.2 *In vitro* emission spectra of ratiometric and single-fluorophore sensors. (a) Emission spectrum of the FRET sensor TN-XXL at 0 Ca^{2+} (gray line) and Ca^{2+} saturation (black line; e.g., 432 nm). (b) Emission spectrum of the single-fluorophore sensor G-CaMP5G showing low baseline fluorescence emission in the Ca^{2+}-free state (gray line) and a significant increase in fluorescence upon Ca^{2+} binding (black line). Spectra were obtained using purified recombinant proteins. (G-CaMP5G data kindly provided by J. Akerboom.)

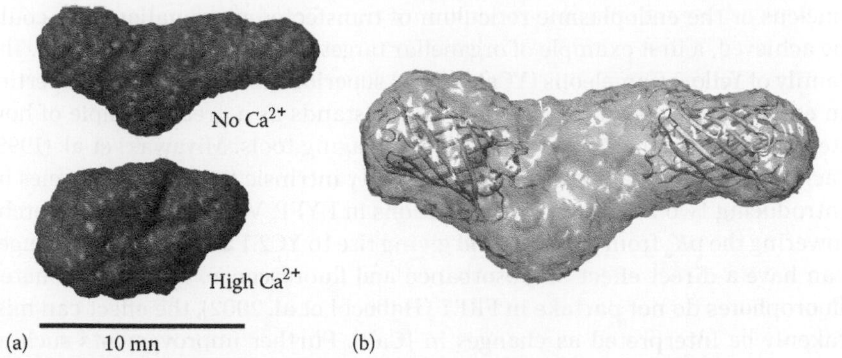

Figure 3.3 Shape reconstruction of the FRET sensor TN-XXL using SAXS. (a) Shape reconstruction of TN-XXL in the Ca^{2+}-free and saturated states. (b) To illustrate the possible positioning of the two β-barrel domains of the donor and acceptor FPs in TN-XXL, the crystal structures for ECFP and citrine (PDB 1CV7 and 1HUY) are superimposed over the shape envelope of the Ca^{2+}-free form. (From Geiger, A. et al., *Biophys. J.* 102(10):2401–2410, 2012.)

3.3 STRATEGIES FOR DESIGNING GENETICALLY ENCODED CALCIUM INDICATORS

3.3.1 Calmodulin-Based FRET Calcium Indicators

The very first protein-based GECIs materialized in the groups of A. Persechini (Romoser et al. 1997) and R. Tsien (Miyawaki et al. 1997). FIP-CB_{SM}, designed by Romoser and colleagues, was based on a ratiometric FRET signal output by employing the smooth muscle myosin light chain kinase (smMLCK) M13 subunit/peptide between a pair of modified green fluorescent proteins (GFPs)—the donor and acceptor BGFP and RGFP, respectively. Through a calmodulin (CaM)-dependent modulation, the orientation and the distance between the donor and acceptor pair are altered, relaying a Ca^{2+}-induced change in the emission spectrum of FIP–CB_{SM}. The FRET-based Cameleons, developed by Miyawaki and co-workers, utilized a similar principle, while including both the M13 CaM-binding peptide and CaM in combination with BFP/GFP (Cameleon-1), EBFP/EGFP (Cameleon-2), and ECFP/EYFP (Yellow Cameleon-2) for enhanced responses (Miyawaki et al. 1997). In these sensors, the binding of Ca^{2+} to CaM triggers an interaction between CaM and the M13 peptide, reducing the distance between the donor and acceptor pair. By fusing appropriate targeting sequences to the sensor, locally restricted expression in the nucleus or the endoplasmic reticulum of transfected mammalian cells could be achieved, a first example of organellar targeting (Miyawaki et al. 1997). The family of Yellow Cameleons (YCs) showed superior cellular imaging properties in comparison to Cameleons-1 and -2 and stands as a great example of how iterative improvements can evolve GECI imaging tools. Miyawaki et al. (1999) tackled the significant issue of pH sensitivity intrinsic to the EYFP species by introducing two adjacent point mutations in EYFP, V68L, and Q69K, thereby lowering the pK_a from 6.9 to 6.1 and giving rise to YC2.1 and YC3.1. pH changes can have a direct effect on absorbance and fluorescence, and as protonated fluorophores do not partake in FRET (Habuchi et al. 2002), the effect can mistakenly be interpreted as changes in [Ca^{2+}]. Further improvements such as increased photostability, reduced sensitivity to chloride, and improved folding efficiency at 37°C consequently came from the evolution of YFP into "Citrine" via the Q69M mutation (Griesbeck et al. 2001). In similar attempts to improve YFP, Nagai et al. (2002) identified the beneficial F46L mutation, which greatly improved maturation time by accelerating the oxidation process of the chromophore at 37°C. Because of its brightness, this variant of YFP was given the name "Venus" and was incorporated into the Cameleon variant YC2.12.

Henceforth, the evolution of the FRET-based Cameleons took two separate directions, with the design of the noteworthy YC2.6 and YC3.6 (Nagai et al. 2004), as well as the computationally redesigned "Design" series (Palmer et al. 2006). Despite substantial efforts to improve YC dynamic range and low signal-to-noise ratios, variants like YC2.12, 2.3, and 3.3 still did not show more than a 120% Ca^{2+}-induced ratio change *in vitro* and suboptimal performance when targeted to organelles. To overcome these limitations, Nagai and co-workers performed circular permutation using Venus, as initially described by Baird et al. (1999), by interchanging and reconnecting the original N- and C-termini using a short spacer peptide (Nagai et al. 2004). Using YC3.12 as the parental construct, the result was a new group of YC members, with YC3.6 outperforming the lot by achieving a 560% FRET ratio change (i.e., five- to sixfold larger dynamic range) and a Ca^{2+} affinity (K_d) of 250 nM (Nagai et al. 2004).

A slightly different approach was undertaken in the design of an indicator targeted to the endoplasmic reticulum (ER), the Cameleon D1ER. The goal was to redesign the binding interface between CaM and M13 to generate specific protein–peptide pairs with less sensitivity to endogenous wild-type calmodulin (Palmer et al. 2004). Palmer and co-workers utilized available nuclear magnetic resonance (NMR) solution structure data to target six possible salt–bridge interactions between CaM and the M13 peptide, resulting in the D1ER variant with a low Ca^{2+} affinity of K_d = 60 μM. To further address the problem of perturbed interaction with endogenous CaM, the binding interface of CaM and the CaM-binding peptide M13 was targeted with a new round of computational reengineering (Palmer et al. 2006). Steric bumps were introduced in the CaM-binding peptide, along with complementary holes in CaM, to create a series of indicators with a wide range of Ca^{2+} affinities, D2cpV, D3cpV, and D4cpV, which utilize circularly permutated Venus (cpV). Of these, D3cpV is the most widely used, showing good performance in reporting Ca^{2+} transients in the cytosol and mitochondria of HeLa cells, as well as in hippocampal neurons (Palmer et al. 2006).

The most recent attempt at optimizing the Cameleon family yielded a series of high-affinity indicators (K_d = 15–140 nM), named Yellow Cameleon-nano (YC-nano), by altering the peptide linker between CaM and M13 (Horikawa et al. 2010).

3.3.2 Troponin C–Based FRET Calcium Indicators

Despite the great effort to improve the Cameleons with respect to their optical performance and sensitivity (K_d), the concern of cross-reactivity with the endogenous CaM signaling pathway spawned a parallel approach to FRET-based GECI design.

The initial attempt to overcome these limitations came with the creation of a family of GECIs based on Troponin C (TnC) (Heim and Griesbeck 2004), the Ca^{2+}-sensing element of the muscle tropomyosin complex (Vassylyev et al. 1998; Gordon et al. 2000; Mercier et al. 2000). The first of these GECIs, TN-L15, used a truncated version of chicken skeletal muscle Troponin C (csTnC) inserted between CFP and YFP (Heim and Griesbeck 2004). One of the promising aspects of the TN-L15 TnC-based sensor was its minimal interference with cellular biochemistry compared with Cameleons.

The original TN-L15 was the first of a series of TnC sensors. Whereas it lacked the ability to detect small Ca^{2+} transients, particularly those caused by single action potentials, when expressed in neurons, its lack of endogenous ligands made TnC a viable Ca^{2+}-binding motif for generating GECIs suitable for *in vivo* neuronal imaging (Heim et al. 2007). The next generation of TnC-based indicators resulted in TN-XL (Mank et al. 2006). TN-XL showed FRET ratio changes at a maximal range of 400%, faster kinetics, and a high selectivity for Ca^{2+} over magnesium (Mg^{2+})—a problem often found in Ca^{2+}-binding protein motifs. However, the lower Ca^{2+} affinity (K_d = 2.2 μM) made it less suitable for detecting small cytosolic changes in free [Ca^{2+}]. TN-XXL was therefore created with the goal of shifting the Ca^{2+} affinity in the right direction while retaining properties such as low Mg^{2+} selectivity and fast kinetics. The fundamental engineering principle behind TN-XXL was a doubling of the highly Ca^{2+}/Mg^{2+}-sensitive C-terminal domain while also abolishing Mg^{2+}-induced conformational changes. The final construct consisted of the two TnC C-terminal domains sandwiched between ECFP and the circularly permuted citrine variant "citrine cp174" (Mank et al. 2008). *In vitro* experiments with TN-XXL showed a significantly lower K_d of 0.8 μM for Ca^{2+} in comparison with TN-XL. *In vivo* experiments in the motor neuron boutons in *Drosophila melanogaster* expressing TN-XXL showed promising applicability for GECIs like TN-XXL. By using two-photon laser scanning microscopy, changes in intracellular free [Ca^{2+}] could be measured with a maximal FRET ratio change of 150% in the *D. melanogaster* neuromuscular junction (Mank et al. 2008). TN-XXL allowed for the first chronic imaging of neuronal response properties in mouse cortex after sensory stimulation over repeated imaging sessions spread over periods of up to 3 weeks, although it reported a relatively low number of responsive cells *in vivo* (22%) in comparison to the synthetic dye indicator OGB-1 (Mank et al. 2008). Recently, the Ca^{2+}-dependent conformational changes in TN-XXL could be visualized using NMR spectroscopy and small-angle x-ray scattering (SAXS; Figure 3.3) (Geiger et al. 2012). The shape reconstruction showed a large conformational change. In the absence of Ca^{2+}, TN-XXL was found to have a flexible, elongated rod-like shape with the FP β-barrels most likely forming both ends of the rod, whereas on Ca^{2+} binding, the protein becomes essentially rigid and globular. The biocompatibility of this sensor, in particular for applications in the brain, appears to be a very favorable aspect of these types of indicators (Direnberger et al. 2012).

3.3.3 Single-Wavelength/Single Fluorescent Protein Indicators

The emergence of single FP sensors was initiated by the discovery that large protein fragments could be inserted into certain sites within the β-barrel of FPs without destroying their folding. In short, this led to three main groups of single FP Ca^{2+} sensors using CaM as the Ca^{2+}-binding domain: Camgaroos, Pericams, and G-CaMPs (Baird et al. 1999; Nakai et al. 2001; Nagai et al. 2002). The first successful attempt to generate a single-fluorophore calcium sensor was made by Baird et al. (1999) by introducing *Xenopus* CaM to replace residue tyrosine 145 of EYFP. The resulting Ca^{2+} indicator, the EYFP-based camgaroo-1, displayed a change in its absorbance spectrum from a peak at 400 nm at 0 Ca^{2+} to 490 nm through the deprotonation of the YFP fluorophore on Ca^{2+} binding. Overall, an eightfold increase in brightness was recorded (Baird et al. 1999; Griesbeck et al. 2001).

To date, the most popular family of single-wavelength Ca^{2+} sensors, the G-CaMPs, rely on a similar functional principle as the camgaroos; however, they are using a circularly permuted EGFP instead of EYFP. The very first G-CaMP variant was reported by Nakai et al. (2001). A strategy of linker manipulation and point mutations in the M13–cpGFP–CaM construct gave rise to a series of 26 variants, which were screened in HEK293 for expression efficiency as well as Ca^{2+} performance using ATP and carbachol. The best variant, "G85" (G-CaMP), showed a 1.5-fold increase in fluorescence intensity on ATP stimulation and up to a fourfold increase upon the addition of ionomycin (Nakai et al. 2001). The Ca^{2+}-induced change in the fluorescence intensity of G-CaMP arises from the interaction between Ca^{2+} and CaM at the C terminus of cpGFP and the N-terminal M13 peptide (Tallini et al. 2006). As with Camgaroos, G-CaMP1 suffered from poor fluorescence when expressed under physiological temperatures. Thus, GFP-stabilizing mutations were introduced to improve the maturation of the GFP-based G-CaMP1. The resulting new indicator, G-CaMP1.6, displayed increased brightness as well as less sensitivity to lower pH (Ohkura et al. 2005). Nevertheless, G-CaMP and G-CaMP1.6 provided only a fraction of the brightness of the original eGFP molecule and still showed maturation problems at temperatures above 30°C. To overcome the limited possibility of applying these indicators for *in vivo* experiments under physiological conditions, a series of targeted and random alterations were carried out to improve both brightness and stability (Tallini et al. 2006). Two new mutations (D180Y and V93I) that improved the brightness of cpGFP were identified during this process. Furthermore, it turned out that the addition of the N-terminal polyHis sequence from RSET linked to the M13 sequence was essential for thermal stability at 37°C. G-CaMP2 displays a four- to fivefold increase in signal between no Ca^{2+} and Ca^{2+} saturation, and most importantly, retains high brightness at 37°C (Tallini et al. 2006).

Recently, the structural and functional mechanisms of G-CaMP2 were elucidated in back-to-back articles published by Wang et al. (2008) and Akerboom et al. (2009). The x-ray crystal structure of G-CaMP2 in the Ca^{2+}-free and Ca^{2+}-bound states provided knowledge of the Ca^{2+} binding mechanism, and thus allowed for systematic targeted mutagenesis around the crucial GFP–CaM interface and chromophore. The crystal structure of G-CaMP2 (Wang et al. 2008; Akerboom et al. 2009) revealed a crucial N-terminal arginine residue, known to destabilize the fluorescent protein, adding to lower baseline fluorescence (Varshavsky 2008; Tian et al. 2009). Small libraries of G-CaMP2.1 lacking the N-terminal arginine were created through site-directed mutagenesis at specific sites near the EGFP chromophore and at superfolder GFP positions. Improvements were thus achieved in brightness, signal-to-noise ratio, Ca^{2+}-response kinetics, and fluorescence dynamic range, which finally led to G-CaMP3. With an ~12-fold change in fluorescence intensity upon Ca^{2+} binding, which is threefold larger than that of G-CaMP2, and a Ca^{2+} affinity of 660 nM, G-CaMP3 is in many ways a superior indicator compared to G-CaMP2 and other previously described single-wavelength Ca^{2+} indicators. The increase in dynamic range results from a twofold decrease in fluorescence in the Ca^{2+}-free state and a 1.5-fold fluorescence increase in the Ca^{2+}-saturated state (Tian et al. 2009). A series of other attempts were made to improve G-CaMP2, for example, by introducing GFP "superfolder" mutations (Pédelacq et al. 2005) into the G-CaMP2 scaffold to create G-CaMP-HS (Muto et al. 2011). Another variant is the more recent G-CaMP4 (Shindo et al. 2010). The engineering of G-CaMP5 stands as a great example of how complex the engineering of protein-based sensors can be. Here, crystal structures of both G-CaMP2 and G-CaMP3 were used to identify further positions for site-directed mutagenesis at the cpGFP/CaM proto-interface and targeted library screening at the region of the M13 peptide/cpGFP and cpGFP/CaM linkers. In addition, mutations introduced directly within the M13 peptide, as well as in the Ca^{2+}-binding loops of CaM, were also tested (Akerboom et al. 2012). The color palette of single-wavelength GECIs has also been expanded (Zhao et al. 2011) by mutating the G-CaMP3 scaffold and performing bacterial colony screening. Simple error-prone polymerase chain reaction (PCR) of G-CaMP3 resulted in G-GECO. The incorporation of chromophore mutations into G-CaMP3, followed by random mutagenesis, produced a blue indicator B-GECO1, and using the red fluorescent protein mApple in place of GFP as the fluorescent backbone led to the creation of R-GECO1 (Zhao et al. 2011) (Figure 3.1b). Intriguingly, during the development of B-GECO1 and improved versions of G-CaMP3 (G-GECO1.1 and G-GECO1.2), variants were discovered with ratiometric blue/green emission or excitation (GEM–GECO1 and GEX–GECO1, respectively) (Zhao et al. 2011). However, initial attempts at applying GECOs in neurophysiology were not encouraging (Yamada and Mikoshiba 2012).

3.4 FRET OR SINGLE-WAVELENGTH PROBES?

The choice of using either a ratiometric, FRET-based GEGI or a single-wavelength, intensity-based GECI often depends on the experimental setup available and on the type of experiment being performed.

Typically, FRET-based GECIs are brighter and thus generate more photons at basal Ca^{2+} levels, making it easier to identify fluorescently labeled target cells. In addition, FRET-based GECIs are less sensitive to motion artifacts and nonuniform levels of expression or illumination, as the signal readout is calculated from the ratio of donor and acceptor fluorescence. Furthermore, FRET-based indicators are better for quantifying $[Ca^{2+}]$ over long time periods, as over repeated imaging sessions, changes in the optical path length, fluctuations in the excitation intensity, or, during longer intervals, even changes in indicator expression levels may occur. Here, intensity-based GECIs are at a disadvantage. On the other hand, FRET-based GECIs require data to be collected from two fluorescence channels using multiple filters, which can slow down imaging processing and place constraints on temporal speed if filter wheels need to be used. The use of a beam splitter to split the emission into separate donor and acceptor channels can solve these issues, however. Other complications include the loss of photons through the use of bandpass filters and splitters. Conversely, single-wavelength GECIs require the collection of data from only one channel, allowing for simpler instrumentation. Another advantage of single-wavelength GECIs is their smaller size, which makes them more suitable as fusion proteins for specific labeling. Moreover, they provide for the possibility of using additional labeling probes with different spectral properties, as only a smaller area of the visual spectrum is occupied by each indicator.

3.5 BIOCOMPATIBILITY, EXPRESSION LEVELS, AND BUFFERING OF CELLULAR CALCIUM

GECIs are expressed inside cells using various gene transfer techniques. Consequently, the precise indicator concentration within the cell is difficult to titrate and depends, among other factors, on the promoters and methods used for gene transfer. In addition, the indicators reside within the cells of interest for a relatively long period of time, leaving ample time and opportunity for potential interactions with the biochemical machinery of the host cell. What are typical GECI concentrations inside cells? Initial attempts to titrate the concentration of Yellow Cameleons (YCs) inside cells resulted in relatively high values of 50–300 μM (Miyawaki et al. 1999) when expressed in HeLa cells using liposome-mediated gene transfer. Later studies, however, arrived at smaller values that ranged from approximately 3 to 10 μM indicator concentrations inside living cells (Tian et al. 2009; Direnberger et al. 2012). Still, it bears keeping in mind that each of the GECIs used in these studies (TN-XXL, G-CaMP3) harbors four Ca^{2+} binding sites per sensor molecule, and the Ca^{2+} buffering properties of GECIs do have detectable effects. There are obvious effects of sensor concentrations on the measured Ca^{2+} transient profile, as known from work with synthetic dyes (see, e.g., Helmchen et al. 1996). Similar effects were reported in early work with YC (Miyawaki et al. 1999) and the TnC-based TN-L15 (Heim and Griesbeck 2004) in HeLa and HEK293 cells, respectively. Following histamine-induced calcium oscillations in YC3.1-expressing HeLa cells, Miyawaki et al. (1999) observed significant differences between cells with low and high protein concentrations; oscillations were clearly attenuated in highly expressing cells. Similar findings were reported with HEK293 cells expressing different levels of TN-L15 (Heim and Griesbeck 2004). Are there any potential *in vivo* effects of Ca^{2+} buffering on the physiology of a GECI transgenic organism? In a recent study, the sensor TN-XXL was expressed in transgenic mice using the ubiquitous CAG promoter, and the effects of sensor expression on the organism were studied in detail using methods ranging from anatomy and physiology to gene expression profiling and behavioral testing (Direnberger et al. 2012; Figure 3.4).

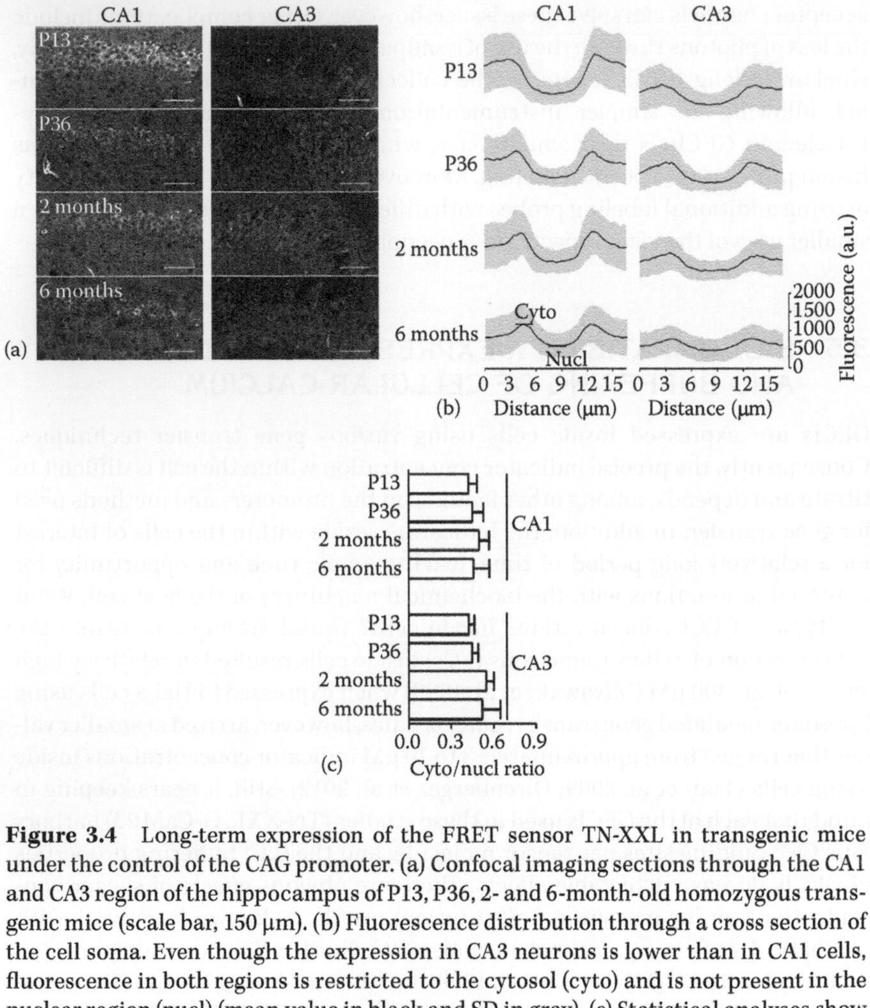

Figure 3.4 Long-term expression of the FRET sensor TN-XXL in transgenic mice under the control of the CAG promoter. (a) Confocal imaging sections through the CA1 and CA3 region of the hippocampus of P13, P36, 2- and 6-month-old homozygous transgenic mice (scale bar, 150 μm). (b) Fluorescence distribution through a cross section of the cell soma. Even though the expression in CA3 neurons is lower than in CA1 cells, fluorescence in both regions is restricted to the cytosol (cyto) and is not present in the nuclear region (nucl) (mean value in black and SD in gray). (c) Statistical analyses show that even expression over 6 months does not significantly change the cellular distribution of the indicator, which could be taken as a sign of cytomorbidity. Values are means ±SD. (From Direnberger, S. et al., *Nat. Commun.* 3: 1031, 2012.)

In these mice, mild signatures of indicator expression, presumably due to Ca^{2+} buffering, were indeed detected. These effects consisted primarily of cardiac pathologies, because maximal heart rates were lower in TN-XXL transgenic mice, and the chronotropic competence of the heart was somewhat compromised (Direnberger et al. 2012). However, the majority of parameters tested in these mice were not changed.

3.6 CONCLUSIONS

With more GECIs becoming available, researchers will have an expanded set of indicators to choose from, with increased performance, different colors, and varying response properties for specific questions. Although this is highly desirable, it also stresses the need for some common guidelines to evaluate GECIs. To facilitate this choice, standardized testing of GECI performance should be performed in selected preparations under standard conditions, using identical stimuli, cell types, and expression systems, as was done, for example, for GECIs expressed at the *Drosophila* neuromuscular junction (Hendel et al. 2008). Finally, new techniques to engineer GECI performance via evolutionary library screening approaches are currently being put to practice in a number of labs and will most likely lead to the development of sensors with unprecedented performance. Thus, the future of GECIs looks good.

REFERENCES

Akerboom, J., Chen, T.-W., Wardill, T. J., Tian, L., Marvin, J. S., Mutlu, S., Calderón, N. C. et al. (2012). Optimization of a GCaMP calcium indicator for neural activity imaging. *J. Neurosci.* 32(40):13819–13840.

Akerboom, J., Vélez Rivera, J. D., Rodríguez Guilbe, M. M., Alfaro Malavé, E. C., Hernandez, H. H., Tian, L., Hires, S. A., Marvin, J. S., Looger, L. L., and Schreiter, E. R. (2009). Crystal structures of the GCaMP calcium sensor reveal the mechanism of fluorescence signal change and aid rational design. *J. Biol. Chem.* 284(10):6455–6464.

Baird, G. S., Zacharias, D. A., and Tsien, R. Y. (1999). Circular permutation and receptor insertion within green fluorescent proteins. *Proc. Natl. Acad. Sci. U. S. A.* 96(20):11241–11246.

Berridge, M. J., Bootman, M. D., and Roderick, H. L. (2003). Calcium signalling: Dynamics, homeostasis and remodelling. *Nat. Rev. Mol. Cell Biol.* 4(7):517–529.

Berridge, M. J., Lipp, P., and Bootman, M. D. (2000). The versatility and universality of calcium signalling. *Nat. Rev. Mol. Cell Biol.* 1(1):11–21.

Clemen, C. S., Herr, C., Lie, A. A., Noegel, A. A., and Schröder, R. (2001). Annexin VII: An astroglial protein exhibiting a Ca^{2+}-dependent subcellular distribution. *NeuroReport* 12(6):1139–1144.

Denk, W., Strickler, J. H., and Webb, W. W. (1990). Two-photon laser scanning fluorescence microscopy. *Science (NY)* 248(4951):73–76.

Denk, W., Yuste, R., Svoboda, K., and Tank, D. W. (1996). Imaging calcium dynamics in dendritic spines. *Curr. Opin. Neurobiol.* 6(3):372–378.

Direnberger, S., Mues, M., Micale, V., Wotjak, C. T., Dietzel, S., Schubert, M., Scharr, A. et al. (2012). Biocompatibility of a genetically encoded calcium indicator in a transgenic mouse model. *Nat. Commun.* 3:1031.

Geiger, A., Russo, L., Gensch, T., Thestrup, T., Becker, S., Hopfner, K.-P., Griesinger, C., Witte, G., and Griesbeck, O. (2012). Correlating calcium binding, Förster resonance energy transfer, and conformational change in the biosensor TN-XXL. *Biophys. J.* 102(10):2401–2410.

Gordon, A. M., Homsher, E., and Regnier, M. (2000). Regulation of contraction in striated muscle. *Physiol. Rev.* 80(2):853–924.

Griesbeck, O., Baird, G. S., Campbell, R. E., Zacharias, D. A., and Tsien, R. Y. (2001). Reducing the environmental sensitivity of yellow fluorescent protein: Mechanism and applications. *J. Biol. Chem.* 276(31):29188–29194.

Grynkiewicz, G., Poenie, M., and Tsien, R. Y. (1985). A new generation of Ca^{2+} indicators with greatly improved fluorescence properties. *J. Biol. Chem.* 260(6):3440–3450.

Habuchi, S., Cotlet, M., Hofkens, J., Dirix, G., Michiels, J., Vanderleyden, J., Subramaniam, V., and De Schryver, F. C. (2002). Resonance energy transfer in a calcium concentration-dependent cameleon protein. *Biophys. J.* 83(6):3499–3506.

Heim, N., Garaschuk, O., Friedrich, M. W., Mank, M., Milos, R. I., Kovalchuk, Y., Konnerth, A., and Griesbeck, O. (2007). Improved calcium imaging in transgenic mice expressing a troponin C-based biosensor. *Nat. Methods* 4(2):127–129.

Heim, N., and Griesbeck, O. (2004). Genetically encoded indicators of cellular calcium dynamics based on troponin C and green fluorescent protein. *J. Biol. Chem.* 279(14): 14280–14286.

Helmchen, F., Imoto, K., and Sakmann, B. (1996). Ca^{2+} buffering and action potential-evoked Ca^{2+} signaling in dendrites of pyramidal neurons. *Biophys. J.* 70(2): 1069–1081.

Hendel, T., Mank, M., Schnell, B., Griesbeck, O., Borst, A., and Reiff, D. F. (2008). Fluorescence changes of genetic calcium indicators and OGB-1 correlated with neural activity and calcium in vivo and in vitro. *J. Neurosci.* 28(29):7399–7411.

Hille, B. (1992). *Ionic Channels of Excitable Membranes*. Sunderland, MA: Sinauer Associates.

Horikawa, K., Yamada, Y., Matsuda, T., Kobayashi, K., Hashimoto, M., Matsu-ura, T., Miyawaki, A., Michikawa, T., Mikoshiba, K., and Nagai, T. (2010). Spontaneous network activity visualized by ultrasensitive Ca^{2+} indicators, yellow cameleon-nano. *Nat. Methods* 7(9):729–732.

Jaffe, D. B., Johnston, D., Lasser-Ross, N., Lisman, J. E., Miyakawa, H., and Ross, W. N. (1992). The spread of Na^{+} spikes determines the pattern of dendritic Ca^{2+} entry into hippocampal neurons. *Nature* 357:244–246.

Johnson, F. H., and Shimomura, O. (1972). Preparation and use of aequorin for rapid microdetermination of Ca^{2+} in biological systems. *Nature* 237(78):287–288.

Kleinfeld, D., and Griesbeck, O. (2005). From art to engineering? The rise of in vivo mammalian electrophysiology via genetically targeted labeling and nonlinear imaging. *PLoS Biol.* 3(10):e355.

Mank, M., Santos, A. F., Direnberger, S., Mrsic-Flogel, T. D., Hofer, S. B., Stein, V., Hendel, T. et al. (2008). A genetically encoded calcium indicator for chronic in vivo two-photon imaging. *Nat. Methods* 5(9):805–811.

Mank, M., and Griesbeck, O. (2008). Genetically encoded calcium indicators. *Chem. Rev.* 108(5):1550–1564.

Mank, M., Reiff, D. F., Heim, N., Friedrich, M. W., Borst, A., and Griesbeck, O. (2006). A FRET-based calcium biosensor with fast signal kinetics and high fluorescence change. *Biophys. J.* 90:1790–1796.

Mercier, P., Li, M. X., and Sykes, B. D. (2000). Role of the structural domain of troponin C in muscle regulation: NMR studies of Ca^{2+} binding and subsequent interactions with regions 1-40 and 96-115 of troponin I. *Biochemistry* 39(11):2902–2911.

Miyawaki, A., Griesbeck, O., Heim, R., and Tsien, R. Y. (1999). Dynamic and quantitative Ca^{2+} measurements using improved cameleons. *Proc. Natl. Acad. Sci. U. S. A.* 96(5):2135–2140.

Miyawaki, A., Llopis, J., Heim, R., McCaffery, J. M., Adams, J. A., Ikura, M., and Tsien, R. Y. (1997). Fluorescent indicators for Ca^{2+} based on green fluorescent proteins and calmodulin. *Nature* 388(28):882–887 (Letter).

Müller, W., and Connor, J. A. (1991). Dendritic spines as individual neuronal compartments for synaptic Ca^{2+} responses. *Nature* 354(6348):73–76.

Muto, A., Ohkura, M., Kotani, T., Higashijima, S., Nakai, J., and Kawakami, K. (2011). Genetic visualization with an improved GCaMP calcium indicator reveals spatiotemporal activation of the spinal motor neurons in zebrafish. *Proc. Natl. Acad. Sci. U. S. A.* 108(13):5425–5430.

Nagai, T., Ibata, K., Park, E. S., Kubota, M., Mikoshiba, K., and Miyawaki, A. (2002). A variant of yellow fluorescent protein with fast and efficient maturation for cell-biological applications. *Nat. Biotechnol.* 20(1):87–90.

Nagai, T., Sawano, A., Park, E. S., and Miyawaki, A. (2001). Circularly permuted green fluorescent proteins engineered to sense Ca^{2+}. *Proc. Natl. Acad. Sci. U. S. A.* 98(6):3197–3202.

Nagai, T., Yamada, S., Tominaga, T., Ichikawa, M., and Miyawaki, A. (2004). Expanded dynamic range of fluorescent indicators for Ca^{2+} by circularly permuted yellow fluorescent proteins. *Proc. Natl. Acad. Sci. U. S. A.* 101(29):10554–10559.

Nakai, J., Ohkura, M., and Imoto, K. (2001). A high signal-to-noise Ca^{2+} probe composed of a single green fluorescent protein. *Nat. Biotechnol.* 19(2):137–141.

Ohkura, M., Matsuzaki, M., Kasai, H., Imoto, K., and Nakai, J. (2005). Genetically encoded bright Ca^{2+} probe applicable for dynamic Ca^{2+} imaging of dendritic spines. *Anal. Chem.* 77(18):5861–5869.

Palmer, A. E., Giacomello, M., Kortemme, T., Hires, S. A., Lev-Ram, V., Baker, D., and Tsien, R. Y. (2006). Ca^{2+} indicators based on computationally redesigned calmodulin-peptide pairs. *Chem. Biol.* 13(5):521–530.

Palmer, A. E., Jin, C., Reed, J. C., and Tsien, R. Y. (2004). Bcl-2-mediated alterations in endoplasmic reticulum Ca^{2+} analyzed with an improved genetically encoded fluorescent sensor. *Proc. Natl. Acad. Sci. U. S. A.* 101(50):17404–17409.

Pédelacq, J.-D., Cabantous, S., Tran, T., Terwilliger, T. C., and Waldo, G. S. (2005). Engineering and characterization of a superfolder green fluorescent protein. *Nat. Biotechnol.* 24(1):79–88.

Rink, T. J., Tsien, R. Y., and Warner, A. E. (1980). Free calcium in *Xenopus* embryos measured with ion-selective microelectrodes. *Nature* 283(5748):658–660.

Romoser, V. A., Hinkle, P. M., and Persechini, A. (1997). Detection in living cells of Ca^{2+}-dependent changes in the fluorescence emission of an indicator composed of two green fluorescent protein variants linked by a calmodulin-binding sequence: A new class of fluorescent indicators. *J. Biol. Chem.* 272(20):13270–13274.

Sakai, N., Sasaki, K., Ikegaki, N., Shirai, Y., Ono, Y., and Saito, N. (1997). Direct visualization of the translocation of the γ-subspecies of protein kinase C in living cells using fusion proteins with green fluorescent protein. *J. Cell Biol.* 139(6):1465–1476.

Shindo, A., Hara, Y., Yamamoto, T. S., Ohkura, M., Nakai, J., and Ueno, N. (2010). Tissue-tissue interaction-triggered calcium elevation is required for cell polarization during *Xenopus* gastrulation. *PLoS One* 5(2):e8897.

Svoboda, K., Denk, W., Kleinfeld, D., and Tank, D. W. (1997). In vivo dendritic calcium dynamics in neocortical pyramidal neurons. *Nature* 385(6612):161–165.

Svoboda, K., Tank, D. W., and Denk, W. (1996). Direct measurement of coupling between dendritic spines and shafts. *Science* 272(5262):716–719.

Tallini, Y. N., Ohkura, M., Choi, B.-R., Ji, G., Imoto, K., Doran, R., Lee, J. et al. (2006). Imaging cellular signals in the heart in vivo: Cardiac expression of the high-signal Ca^{2+} indicator GCaMP2. *Proc. Natl. Acad. Sci. U. S. A.* 103(12):4753–4758.

Tian, L., Hires, S. A., Mao, T., Huber, D., Chiappe, M. E., Chalasani, S. H., Petreanu, L. et al. (2009). Imaging neural activity in worms, flies and mice with improved GCaMP calcium indicators. *Nat. Methods* 6(12):875–881.

Tsien, R. Y., Pozzan, T., and Rink, T. J. (1982). Calcium homeostasis in intact lymphocytes: Cytoplasmic free calcium monitored with a new, intracellularly trapped fluorescent indicator. *J. Cell Biol.* 94(2):325–334.

Varshavsky, A. (2008). The N-end rule at atomic resolution. *Nat. Struct. Mol. Biol.* 15(12):1238–1240.

Vassylyev, D. G., Takeda, S., Wakatsuki, S., Maeda, K., and Maeda, Y. (1998). Crystal structure of troponin C in complex with troponin I fragment at 2.3-A resolution. *Proc. Natl. Acad. Sci. U. S. A.* 95(9):4847–4852.

Wang, Q., Shui, B., Kotlikoff, M. I., and Sondermann, H. (2008). Structural basis for calcium sensing by GCaMP2. *Structure* 16(12):1817–1827.

Yamada, Y., and Mikoshiba, K. (2012). Quantitative comparison of novel GCaMP-type genetically encoded Ca^{2+} indicators in mammalian neurons. *Front. Cell. Neurosci.* 6:41.

Zhang, J., Campbell, R. E., Ting, A. Y., and Tsien, R. Y. (2002). Creating new fluorescent probes for cell biology. *Nat. Rev. Mol. Cell Biol.* 3(12):906–918.

Zhao, Y., Araki, S., Wu, J., Teramoto, T., Chang, Y.-F., Nakano, M., Abdelfattah, A. S. et al. (2011). An expanded palette of genetically encoded Ca^{2+} indicators. *Science* 333(6051):1888–1891.

Chapter 4

Monitoring Membrane Lipids with Protein Domains Expressed in Living Cells

Peter Varnai and Tamas Balla

CONTENTS

4.1 INTRODUCTION

Almost every important molecular event in a cell takes place on the surface of biological membranes, and the membrane lipid composition has a great impact on how protein signaling complexes interact and function in the context of these membranes. In addition to the structural lipids, there is a class of phospholipids, the phosphoinositides, that are present in small amounts but show very rapid changes and high metabolic turnover. These lipids regulate a vast majority of cellular processes via interactions with peripheral or integral membrane proteins, and a plethora of synthesizing and metabolizing enzymes ensure tight temporal and spatial control over their changes. The highly localized and rapid changes in phosphoinositide levels demand experimental tools that are capable of detecting and following these changes, preferably in live cells. Moreover, to understand better the impact of phosphoinositide changes on any cellular process, it is also desirable to develop tools that allow for the artificial manipulation of these lipids with proper spatial and temporal control. This chapter briefly summarizes the developments that have fueled this field in recent years.

Inositol lipids are minor constituents of the biological membranes of all eukaryotic cells, which are built from a diacylglycerol backbone linked to an inositol ring via a phosphodiester linkage. The inositol ring can be phosphorylated at three different positions (positions 3, 4, or 5), giving rise to the seven known phosphoinositide isomers (Balla 2013). A whole host of inositol lipid kinases and phosphatases controls the phosphorylation state of these lipids and ensures their tight metabolic control (Fruman et al. 1998; Pirruccello and De Camilli 2012). These lipids were first identified as precursors of important second messengers, inositol 1,4,5-trisphosphate [Ins(1,4,5)P_3] and diacylglycerol (DG), that inform cells on the presence of stimuli that activate specific kinds of cell surface receptors (Berridge 1984). It has become apparent in the last 20 years that these lipids also serve as focal points in cellular membranes, marking compartments for the assembly of protein signaling complexes whose functions are also regulated by these lipids. Through these actions, phosphoinositides regulate almost every biological process, such as actin cytoskeleton dynamics, Ca^{2+} signaling, vesicular transport, lipid transfer, ion channels, and transporters. Phosphoinositides also regulate a whole set of enzymes that control cell cycle progression, proliferation, and apoptosis. It is not surprising, therefore, that phosphoinositides are linked to a variety of diseases, and several drugs that target phosphoinositide kinase enzymes are under development or have advanced to clinical trials (Wong et al. 2010). Naturally, none of these developments would have transpired without the efforts of basic science that focused on understanding the biology of these lipids. Advances in methods and approaches have always had a major impact on the progression of this field. In this chapter, we attempt to highlight the newest methods used in phosphoinositide research.

The metabolic labeling of cells with *myo*-[^{3}H]inositol or ^{32}P-phosphate, followed by lipid extraction and separation by thin-layer chromatography (TLC), has long been the method of choice to follow phosphoinositide changes in a variety of cells (e.g., Christy et al. 1998). These studies have taught us a great deal and established the inositol lipid research field. However, these techniques reached their limits in that they require millions of cells to obtain a sufficient signal, and they did not give any information on the subcellular location of inositides. The latter required cell fractionation procedures, with lingering questions as to whether the observed distribution really reflects what was present in the intact cell. To overcome the uncertainty of metabolic labeling efficiency and specific activity changes, the total cellular mass of inositides can also be measured based on the quantification of the inositide headgroup that is liberated from the extracted lipid species (Chilvers et al. 1991) or by mass measurements of lipids separated on high-performance liquid chromatography (HPLC) and quantified with metal dye detection (Pittet et al. 1989) or suppressed conductivity detection (Nasuhoglu et al. 2002). Recently, measurements of the total lipid mass by mass spectrometry have been achieved with great sensitivity (Di Paolo et al. 2004; Clark et al. 2011). However, these approaches still suffer from the aforementioned lack of spatial resolution.

Another line of research focused on using fluorescently labeled lipids (phosphoinositides or other lipid classes) (Lipsky and Pagano 1983; Epand et al. 2004; Golebiewska et al. 2011). The great advantage of these tools was that they could follow spatial and compartmentalized changes in living cells. Their major limitation is that the fluorophores are usually attached to shortened acyl side chains, which grossly alters the hydrophobicity and therefore the distribution and movements of these probes inside the cell (Chattopadhyay and London 1987). Moreover, once in the cell, the lipid headgroup can quickly change,

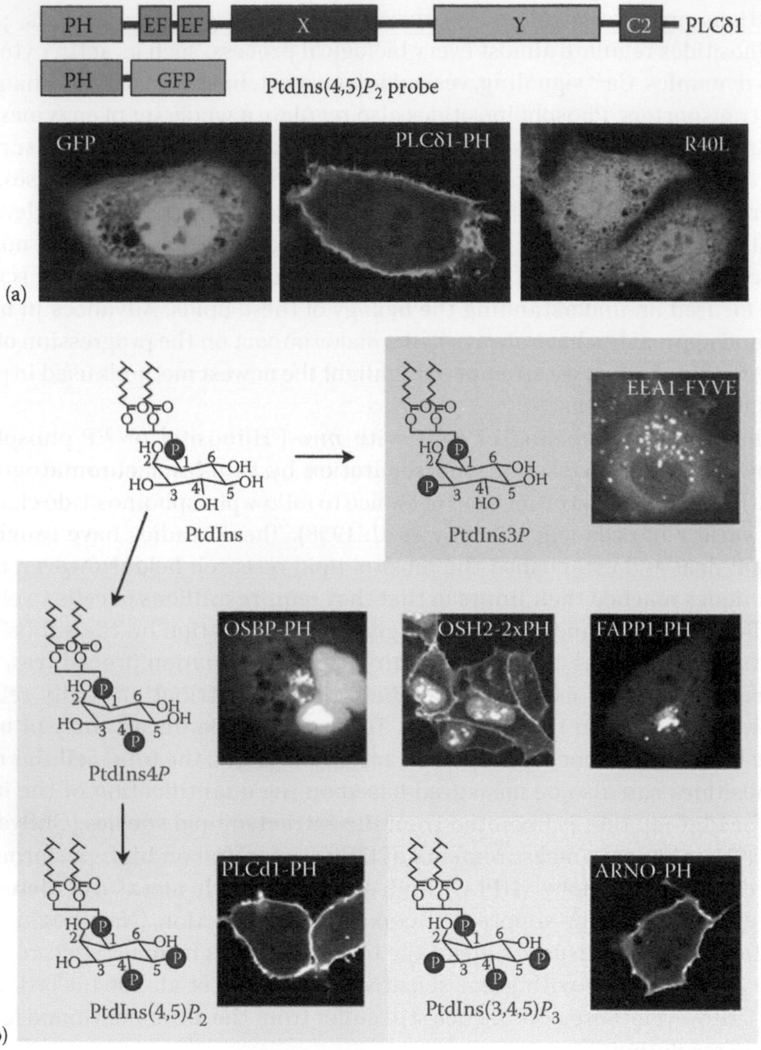

Figure 4.1 The principle and application of fluorescent inositol lipid probes. (a) To create probes that can be used to visualize and measure individual phosphoinositide isomers, lipid-binding domains from various proteins were fused to fluorescent proteins (e.g., GFP). On expression of these proteins in living cells, the lipids can be visualized by confocal microscopy. The R40L mutation renders the PH domain of the PLCδ1 protein incapable of binding lipids, resulting in the loss of membrane localization for this probe. (b) Intracellular localizations of selected lipid probes recognizing specific configurations of the phosphorylated inositol ring as found in phosphoinositides.

and one does not know what changes occur because the label remains on the backbone regardless of what happens to the headgroup.

One would think that the best approach used in fixed cells is to use antibodies raised against specific isomers of inositol lipids (Miyazawa et al. 1988). Obviously, the specificity of the antibody is a paramount issue, but there are additional problems associated with this approach. Fixation procedures, developed over the years for immunofluorescence (IF) studies on proteins, have not been thoroughly tested for their ability to preserve lipids unchanged in their locations. In fact, thorough studies addressing the impact of fixation on the preservation and access of inositol lipids suggest that the fixation problem should not be underestimated (Hammond et al. 2006). While these questions are still being sorted out, it is also important to remember that only fixed cells can be studied with the antibody approach.

The appearance of the green fluorescent protein (GFP) has revolutionized cell biology and made live-cell imaging of genetically encoded material possible (Lippincott-Schwartz et al. 1999). The challenge for lipid research was to find ways to detect lipids with GFP, as GFP can be fused only to proteins and not lipids. The idea of how this can be done came from Mother Nature herself, as proteins have evolved to interact with and respond to changes in specific inositol lipid isomers. Protein modules capable of specific inositol lipid recognition, such as pleckstrin homology (PH; Cohen et al. 1995), FYVE (Kutateladze et al. 1999), or Phox homology (PX; Kanai et al. 2001) domains, are present in a variety of proteins. These modules usually recognize the soluble inositol phosphate headgroups of phosphoinositides, although additional factors may also determine their membrane-binding properties, as will be detailed in Section 4.5.4. Nevertheless, these modules can be genetically isolated and fused to GFP to create probes capable of finding inositol lipids inside living cells and following the distribution and dynamic changes that occur under a variety of experimental conditions (Figure 4.1) (Halet 2005; Shen et al. 2009). This chapter is mainly devoted to questions related to the generation and use of these molecular tools.

4.2 WHAT ARE THE CRITERIA OF A USEFUL LIPID PROBE?

4.2.1 Specificity

The criteria for a good lipid probe are not too different from those required for a good antibody. The probe has to display a decent level of specificity toward the inositide lipid species. This is an interesting question where our views have been changing as we have learned more about the ways in which inositides regulate cellular functions. The first question relates to the *in vitro* specificity. A few methods have been used to determine lipid-binding specificity, but sometimes they give conflicting results. The most commonly used and simplest methods are fat blots, wherein various lipid isomers are spotted and dried on membranes and incubated with recombinant lipid-binding modules to see which lipids bind the protein (Kavran et al. 1998; Dowler et al. 2000). Although this method is great as a first approximation, many protein domains show only weak interactions with not just one but several lipid species (Narayan and Lemmon 2006). Because fat blots do not reproduce the lipids in their natural membrane environment, the results they show are not always very reliable (Dowler et al. 2000). It is better to measure the binding of the proteins to liposomes produced from lipid mixtures that mimic the natural membrane composition (e.g., Kavran et al. 1998) or to polymers coated with the particular inositide (Ferguson et al. 2005). Surface plasmon resonance (SPR) is a widely used method for determining lipid binding to proteins (e.g., Komander et al. 2004), but here again, it does matter how the lipid or the protein is attached to the Biacore surface. Therefore, it is important that more than one method is used to determine the *in vitro* inositide binding features of protein modules. A critical evaluation of these methods has recently been published (Narayan and Lemmon 2006).

Many inositol binding modules, such as PH domains, show relatively poor specificity *in vitro*, yet they seem to report lipid changes with surprising specificity when expressed in cells (Yu et al. 2004). This apparent discrepancy is due to the fact that the lipid-recognizing modules also interact with proteins that contribute to their membrane binding, and this dual (or even multiple) lipid and protein binding localizes these modules to specific compartments where both the lipid and protein binding partner are present (Lemmon 2008). This notion has a few consequences. First, the probe does not necessarily equally detect all the lipid pools present within the cell and will show a bias. Second, the release of the probe from the membrane is not always due to lipid changes but can also be caused by changes in the amount or conformation of the protein binding partner. For example, most PtdIns4*P* binding probes also require Arf1 for membrane recruitment and are therefore biased for detection of Golgi PtdIns4*P* and can be released from the membrane by treatment with brefeldin A (Balla et al. 2005). At the same time, this restricted lipid recognition could in fact be useful, as various reporters may be able to help understand the regulation and importance of specific inositide pools within the cell.

It should also be mentioned that we understand very little about whether the fatty acid side chain makes any difference in the inositide recognition behavior of these protein domains. Based on static views of membranes, as often depicted in textbook cartoons or in our slideshow presentations, the fatty acid side chains are part of the hydrophobic center of membranes and are not accessible to these peripheral membrane proteins. However, increasing evidence supports a more dynamic view in which proteins may lift lipids (even if partially) out of the membrane, exposing some of their fatty acid side chains. Moreover, several inositide-recognizing domains have structural features consistent with membrane-penetrating segments that can interact with the fatty acid chains (see Lenoir et al. 2010 for an example).

4.2.2 Affinity

The probe has to have a reasonably high affinity (K_d: ~0.3–1 μM), with rapid association and dissociation rates. Too high of an affinity is not desirable, as such a probe would bind tightly to the lipid and not allow access to enzymes that normally modify lipid levels. Also, a tightly bound probe usually has a slow dissociation rate and hence is unable to follow rapid changes. A high affinity probe can also be entircly bound to the membrane and not leave enough "free" unbound probe in the cytosol to report on increased lipid levels in the membrane. The true affinity of the probe to the membrane within the cell is not identical to the lipid-binding affinity measured through the various *in vitro* lipid-binding assays listed in Section 4.2.1. This again is partly due to the way the endogenous lipid is presented in the membrane environment and also to the presence of other binding components, such as proteins or nonspecific hydrophobic or electrostatic interactions with the membrane. Using *in vitro*-measured lipid-binding constants to model what happens inside the cell may lead to incorrect conclusions regarding the behavior of the probe during lipid changes inside the cell. Sometimes, the lipid affinity of an otherwise specific probe is too low to detect lipids inside the cell. In this case, a tandem construct, in which two such domains are fused together, may still produce a useful probe, as was shown for PtdIns3*P* (Gillooly et al. 2000) and PtdIns4*P* detection in the plasma membrane (Roy and Levine 2004).

4.2.3 The Choice of Fluorophore

There has been enormous progress in the development of fluorescent proteins (FPs) with various excitation and emission properties, quantum yields, and photostability (Shaner et al. 2005; Stepanenko et al. 2011). Some are based on the original *Aequoria victoria* GFP (Giepmans et al. 2006; Goedhart et al. 2012), while others have been isolated from corals (Alieva et al. 2008). Although the wealth of proteins is extremely useful, as it turns out, some FPs significantly change the behaviors of lipid probes. This is mostly due to differences in the reactive groups present on the surface of the FP. Therefore, one should not assume that simply changing the color by changing the fluorophore will not affect the properties of the lipid probe. Also, it should be recognized that a monomeric FP variant should be used, especially in cases where the natural FP form exhibits strong dimerization or tetramerization tendencies.

4.2.4 Conformational Change or Change in Localization

Single molecular probes that contain a donor–acceptor pair of fluorescent molecules (such as cyan fluorescent proteins [CFPs] and yellow fluorescent proteins [YFPs] or their improved versions) are very useful to detect conformational changes that affect the distance or dipole orientation of the two fluorophores, enough so that the change in the energy transfer (FRET; see Chapters 1 and 2) between the molecules can be detected (Wallrabe and Periasamy 2005). These probes have been successfully applied to measure changes in Ca^{2+} (see Chapter 3), cAMP, Ins(1,4,5)P_3, and the GDP–GTP transition of a variety of small GTP binding proteins (see Chapter 5), or the activity of protein kinases (see Chapter 9) (Balla 2009). Unfortunately, in our experience, the conformational change in the case of lipid binding by inositide binding domains is too small to generate an easily measurable change in the FRET efficiency of single-molecule sensors, although they have been introduced and used with some success (Nishioka et al. 2008). Therefore, we prefer to use single molecular sensors that can still be used in FRET (van Der Wal et al. 2001) or BRET (Toth et al. 2012) mode when coexpressing the same protein module tagged with the two fluorophores separately. This is a convenient way of monitoring membrane association, especially when the changes are relatively small to be easily detected by other methods.

4.3 STUDYING SPECIFIC FORMS OF PHOSPHOINOSITIDES AND OTHER LIPID SPECIES

Table 4.1 lists the various protein domains that have been successfully used for imaging specific inositol lipid pools. In Sections 4.3.1–4.3.10, we describe the methods used for different inositol lipids. Both Table 4.1 and the descriptions that follow are updated versions of our previously published summaries (Varnai and Balla 2007; Balla and Varnai 2009).

4.3.1 PtdIns(4,5)P_2

PtdIns(4,5)P_2 is the most abundant polyphosphoinositide, which is present primarily in the plasma membrane and regulates many biological processes (Halstead et al. 2005). To image this lipid, most studies use the PLCδ1PH–GFP

TABLE 4.1 PHOSPHOINOSITIDE BINDING MODULES IN USE FOR LIVE-CELL IMAGING PURPOSES

Lipid Protein Domain	References for *In Vitro*	Live-Cell Localization	References
	PtdIns(4,5)P_2		
PLCδ_1–PH	Lemmon et al. (1995)	PM	Stauffer et al. (1998); Várnai and Balla (1998)
Tubby domain	Santagata et al. (2001)	PM	Santagata et al. (2001)
		+ cleavage furrow	Field et al. (2005)
	PtdIns(3,4,5)P_3		
GRP1–PH	Klarlund et al. (1997); Rameh et al. (1997a)	PM	Venkateswarlu et al. (1998a); Klarlund et al. (2000)
ARNO–PH	Klarlund et al. (2000)	PM	Venkateswarlu et al. (1998b)
Cytohesin-1–PH	Klarlund et al. (2000)	PM	Nagel et al. (1998); Venkateswarlu et al. (1999)
Btk–PH	Rameh et al. (1997a); Salim et al. (1996)	PM	Varnai et al. (1999)
	PtdIns(3,4,5)P_3/PtdIns(3,4)P_2		
Akt–PH	Franke et al. (1997)	PM	Kontos et al. (1998); Watton and Downward (1999); Servant et al. (2000)
PDK1–PH	Komander et al. (2004)	PM	Komander et al. (2004)
CRAC	Huang et al. (2003)	*Dictyostelium*	
		PM	Dormann et al. (2002)
	PtdIns(3,4)P_2		
TAPP1–PH	Dowler et al. (2000)	PM	Kimber et al. (2002)

(*Continued*)

TABLE 4.1 (CONTINUED) PHOSPHOINOSITIDE BINDING MODULES IN USE FOR LIVE-CELL IMAGING PURPOSES

Lipid Protein Domain	References for *In Vitro*	Live-Cell Localization	References
	PtdIns(3,5)P_2		
Ent3p-ENTH[a]	Friant et al. (2003)	Yeast pre-vacuole	Friant et al. (2003)
Svp1p[a]	Dove et al. (2004)	Yeast vacuole	Dove et al. (2004)
Tup1[a]	Han and Emr (2011)	Yeast vacuolar-endosomal compartment, nucleus	Han and Emr (2011)
Cti6[a]	Han and Emr (2011)	Yeast nucleus	Han and Emr (2011)
TRPML-N-term	Li et al. (2013)	Vacuoles, yeast PM	Li et al. (2013)
	PtdIns3*P*		
FYVE (Hrs, EEA1)	Burd and Emr (1998); Simonsen et al. (1998)	Early endosome	Gillooly et al. (2000)
FYVE (Vps27)		Yeast vacuole	Burd and Emr (1998)
P40phox–PX	Ellson et al. (2001); Kanai et al. (2001)	Early endosome	Ellson et al. (2001)
	PtdIns4*P*		
OSH2–2xPH[b]	Yu et al. (2004)	PM	Roy and Levine (2004); Yu et al. (2004); Balla et al. (2007)
OSBP–PH	Dowler et al. (2000); Levine and Munro (1998)	Golgi	Levine and Munro (1998, 2002)
		+ PM	Balla et al. (2005)
FAPP1–PH	Dowler et al. (2000)	Golgi	Levine and Munro (2002); Godi et al. (2004)
		+ PM	Balla et al. (2005)
SidC	Ragaz et al. (2008); Weber et al. (2006)		

(Continued)

TABLE 4.1 (CONTINUED) PHOSPHOINOSITIDE BINDING MODULES IN USE FOR LIVE-CELL IMAGING PURPOSES

Lipid Protein Domain	References for *In Vitro*	Live-Cell Localization	References
		PtdIns4*P*	
SidM	Brombacher et al. (2009)		Hammond et al. (2014)
		PtdIns5*P*	
3xPHD (ING2)	Gozani et al. (2003)	Nucleus?	
		PM	Gozani et al. (2003); Pendaries et al. (2006)
		DAG	
C1 domains	Sharkey et al. (1984); Burns and Bell (1991); Dries et al. (2007)	PM	Meyer and Oancea (1998); Oancea et al. (1998)
		Nuclear membrane	Stahelin et al. (2005)
		Golgi	Lehel et al. (1995); Schultz et al. (2004)
		PtdOH	
Spo20–PABD	Nakanishi et al. (2004)	PM	Zeniou-Meyer et al. (2007); Kassas et al. (2012); Zhang et al. (2014)
DOCK2–PABD	Nishioka et al. (2010)	PM	Nishioka et al. (2010)

Note: PM, plasma membrane.

[a] The usefulness of these domains for imaging purposes is questionable (see Michell et al. 2005).

[b] The OSH2–PH domain shows little discrimination between PtdIns4*P* and PtdIns(4,5)P_2 based on *in vitro* binding (Yu et al. 2004), and it is likely that it actually reports on both of these molecules to some degree.

construct developed independently by both the Meyer lab and our group (see references in Table 4.1). PLCδ1PH–GFP expresses very well and decorates the plasma membrane, as well as some vesicular structures, but no other organelles. This has raised the question of whether PtdIns(4,5)P_2 is present at detectable levels only in the plasma membrane or whether the probe is biased toward the plasma membrane pool of the lipid. Few reliable works have compared PLCδ1PH–GFP distribution with PtdIns(4,5)P_2 antibody staining; some of these found the lipid in the Golgi using antibody staining but not PLCδ1PH–GFP (Matsuda et al. 2001), while others saw no discrepancy as well as no Golgi staining with the PtdIns(4,5)P_2 antibody (Hammond et al. 2006). Electron microscopic (EM) studies showed some PtdIns(4,5)P_2 in the Golgi using GST-fused PLCδ1PH post-fixation (Watt et al. 2002). Our unpublished observation is that PLCδ1PH–GFP can detect PtdIns(4,5)P_2 in the Golgi when we acutely recruit a PIP 5-kinase enzyme to this organelle, making it quite unlikely that the probe has a constraint that allows it to only see the plasma membrane pool of PtdIns(4,5)P_2 (Z. Szentpetery and T. Balla, unpublished).

The plasma membrane pool of PtdIns(4,5)P_2 can be monitored by PLCδ1PH–GFP, as the probe nicely reports changes in this lipid either after phospholipase C (PLC) activation or after degradation by a phosphoinositide-directed 5-phosphatase (Stauffer et al. 1998; Várnai and Balla 1998; Varnai et al. 2006). There is, however, another complicating issue when following the PLC-mediated hydrolysis of these lipids. PLCδ1PH–GFP also binds the corresponding soluble inositol phosphate, Ins(1,4,5)P_3, which can compete with membrane PtdIns(4,5)P_2 for the binding of the PH-domain–GFP fusion protein. Because of this competition, changes in the PLCδ1PH–GFP membrane localization could be exaggerated by InsP_3 elevations (Nash et al. 2001). This topic was covered in great detail in one of our previous reviews (Varnai and Balla 2006), but the bottom line is that inositol phosphates can compete for lipid binding by PH-domain constructs, and the effects on localization changes could be quite significant under certain conditions and therefore cannot be ignored. At the same time, we believe that it is misleading to treat the PLCδ1PH–GFP translocation response as an index of InsP_3 changes. This has been demonstrated recently by simultaneous measurements of InsP_3 and PLCδ1PH–GFP translocation (Matsu-ura et al. 2006).

Another PtdIns(4,5)P_2 binding module was described in the Tubby protein and called the Tubby domain (Santagata et al. 2001; Field et al. 2005; Yaradanakul and Hilgemann 2007). The Tubby domain also binds PtdIns(4,5)P_2, but it shows very little affinity to Ins(1,4,5)P_3 (Szentpetery et al. 2009), and hence it does not suffer from the "InsP_3 effect" described above. Several studies using the full-length Tubby protein (Nelson et al. 2008) or a mutant form of the Tubby domain showed that the Tubby domain is less sensitive to InsP_3 changes than PLCδ1PH–GFP inside cells (Nelson et al. 2008; Quinn et al. 2008;

Szentpetery et al. 2009). Our experience, in agreement with reports by the Tinker group (Quinn et al. 2008), is that the Tubby domain binds more tightly to PtdIns(4,5)P_2 and is more resistant to PLC-mediated PtdIns(4,5)P_2 hydrolysis. Curiously enough, we did not observe any such masking effect against a 5-phosphatase enzyme (Szentpetery et al. 2009). Other PtdIns(4,5)P_2-binding PH domains described in the yeast genome and characterized previously (Yu et al. 2004) have also been tested for their utility as PtdIns(4,5)P_2 probes, but none has proven to perform as well as PLCδ1PH-GFP or the Tubby domain (Szentpetery et al. 2009).

One important point to remember when using any type of sensor in living cells is the inhibitory effect of the expressed domain on the cellular responses regulated by inositol lipids. The binding of PH domain–GFP reporters to PtdIns(4,5)P_2 is expected to inhibit lipid-mediated cellular processes because the reporter competes with lipid binding by endogenous effectors. High PLCδ1PH–GFP expression causes morphological changes that include rounding of the cells, due to the loss of attachment to the matrix, and the development of intracellular vesicles. These side effects can be explained by the known importance of the lipid in linking the cytoskeleton to the plasma membrane (Raucher et al. 2000; Zhang et al. 2012) and the interference of the PH domain with this process. However, PLCδ1PH–GFP expression also enlarges the PtdIns(4,5)P_2 pool, and because of the rapid dissociation and reassociation of the PH domain with the lipid (van Der Wal et al. 2001; Hammond et al. 2009b), there is a larger pool of the lipid that can interact with various effectors. This may explain why the overexpression of PLCδ1PH–GFP generates intracellular vesicles that are very reminiscent of those seen after PIP 5-kinase overexpression. All of these problems can be minimized by using cells that express low levels of the protein. This, however, requires a sensitive microscope that can still detect the weak signals from such cells.

4.3.2 PtdIns(3,4,5)P_3

PtdIns(3,4,5)P_3 is far less abundant than PtdIns(4,5)P_2 and is hardly detectable in quiescent cells. It is produced by PI3Ks after the stimulation of either receptor tyrosine kinases (RTKs) (in the case of class IA PI3Ks) or G-protein–coupled receptors (GPCRs) (in the case of the class IB PI3K). Because of the high interest in PI 3-kinase signaling and its importance in polarized cell movements such as chemotaxis, as well as in carcinogenesis, a large number of studies have been making use of imaging tools to follow PtdIns(3,4,5)P_3 dynamics. One of the earliest reports came from studies in *Dictyostelium discoideum*, a widely used model for cell polarization and chemotaxis, where the PH domain of the CRAC protein (cytosolic regulator of adenyl cyclase) was used as a reporter of PtdIns(3,4,5)P_3 distribution (Dormann et al. 2002; Huang et al. 2003). In mammalian cells, three PH domains have been popular as PtdIns(3,4,5)P_3 detectors: the Akt–PH domain, the PH domain of the tyrosine kinase Btk, and the PH domain of the ARNO/Grp1 family of proteins (see Table 4.1 for references).

Each of these probes has advantages and disadvantages. For example, AktPH–GFP may not discriminate between PtdIns(3,4,5)P_3 and PtdIns(3,4)P_2. This is not as big of a problem as often stated, especially in certain cases where it is not so critical whether the product is PtdIns(3,4,5)P_3 or PtdIns(3,4)P_2. Yet AktPH–GFP is the most sensitive sensor of the three and gives a very robust signal on stimulation. Not surprisingly, it has been the most widely used. The ARNO/Grp1 PH domain has two splice variants that only differ in the number of Gly residues at the beginning of the loop between the β1 and β2 strands (Ogasawara et al. 2000; Cronin et al. 2004). The 3G variant is a very poor PtdIns(3,4,5)P_3 binder, and it also binds PtdIns(4,5)P_2 (Klarlund et al. 2000), although it does not bind either lipid strongly enough to be pulled to the plasma membrane. The 2G variant is a better PtdIns(3,4,5)P_3 detector and does not bind PtdIns(3,4)P_2 (Gray et al. 1999), but it is highly enriched in the nucleus (unrelated to PtdIns(3,4,5)P_3 binding), and its plasma membrane localization depends on the amount of Arf6–GTP in the plasma membrane (Cohen et al. 2007). These shortcomings make these probes less useful in cellular studies. On the other hand, the Btk PH domain is a good compromise: It shows specificity for PtdIns(3,4,5)P_3, it responds to stimulation with a reasonable amount of translocation (although less so than the Akt–PH), and it is only moderately nuclear. A recent study compared several PtdIns(3,4,5)P_3-binding PH domains for their *in vitro* binding specificity and cellular localization response (Manna et al. 2007). This study found strikingly different membrane recruitment kinetics between the various domains in PDGF-stimulated NIH3T3 cells, suggesting that in addition to inositide lipid binding, membrane penetration and possibly protein–protein interactions have a role in the membrane association of these domains. Based on all of the measurements presented in that study, the Btk–PH appeared to be the best probe for the selective detection of PtdIns(3,4,5)P_3 in intact cells.

These data already suggest that in addition to PtdIns(3,4,5)P_3, protein–protein interactions probably also play a role in the effective recruitment of these PH domains to the plasma membrane. This conclusion was also supported by our study in which we compared the dominant-negative inhibitory effects of these PH domains on various cellular responses, all of which were dependent on PtdIns(3,4,5)P_3. Notably, we found that the PH domains exerted the strongest inhibitory effects on the cellular responses in which their parent molecules played important roles. However, PtdIns(3,4,5)P_3 binding was always important for the inhibitory effects of these constructs. These results suggested that the lipid-dependent recruitment of the domains to the membrane led to the sequestration of specific proteins that were unique to the PH domain used (Varnai et al. 2005).

4.3.3 PtdIns4*P*

Finding a suitable reporter for PtdIns4*P* has been a high priority in our group, given our long-standing interest in PI 4-kinases. PtdIns4*P* is a precursor of PtdIns(4,5)P_2 in the plasma membrane but is also a regulatory lipid in the Golgi and endosomes, where the PI4Ks are also located (De Matteis et al. 2005). PtdIns4*P* should therefore be present in all of these membranes, although their relative amounts there are unknown. PH domains showing PtdIns4*P* binding, such as those from the oxysterol binding protein (OSBP), the four-phosphate-adaptor protein (FAPP1), and ceramide transfer protein (CERT), have been described in yeast and mammalian cells (Levine and Munro 1998, 2001, 2002). The two most popular domains used in cellular imaging studies are the PH domains of the OSBP and FAPP1, which were first identified as specific PtdIns4*P* binders by fat blots (Dowler et al. 2000). Based on lipid vesicle binding assays, however, these domains were shown to bind not only PtdIns4*P* but also PtdIns(4,5)P_2 (Levine and Munro 1998; Roy and Levine 2004; Yu et al. 2004; He et al. 2011). These PH domains, as well as their close relative found in the CERT protein, localize to the Golgi in both yeast and mammalian cells (Levine and Munro 2002), but only the OSBP PH domain binds slightly to the plasma membrane, where PtdIns4*P* is present (Hammond et al. 2009a, 2012) and PtdIns(4,5)P_2 is highly abundant. These data suggest that these PH domains do not recognize PtdIns(4,5)P_2 efficiently within cells. By now, it has become clear that both the OSBP- and FAPP1-PH domains require active (i.e., GTP-bound) Arf1 for their Golgi localization, and their Golgi targeting requires binding to both PtdIns4*P* and Arf1–GTP (Levine and Munro 2002; He et al. 2011). Neither interaction alone is sufficient for efficient membrane recruitment. This explains why brefeldin A treatment, which prevents the formation of Arf1–GTP in the Golgi, releases these PH domains from their Golgi location. The limited lipid-binding specificity and the need for an additional protein interaction for membrane targeting make these probes less than optimal for PtdIsn4*P* imaging in live cells if the goal is to see all the PtdIns4*P* inside cells. However, they have been used successfully to obtain information about Golgi PtdIns4*P* and are in use in many laboratories including ours (see Table 4.1 for references).

Based on these results, PtdIns4*P*-binding domains that do not depend on Arf1 would be highly desirable. Two additional PH domains, found in the yeast Osh1 and Osh2 proteins (both are *Saccharomyces cerevisiae* OSBP homologues), showed cellular localization consistent with PtdIns4*P* binding (Roy and Levine 2004; Yu et al. 2004). Remarkably, the Osh2–PH (used in tandem to increase its apparent affinity) revealed both the plasma membrane and the Golgi pool of PtdIns4*P* in yeast, whereas the Osh1 PH domain detected only the Golgi pool (Roy and Levine 2004; Yu et al. 2004). This spatial confinement was surprising given the poor ability of these PH domains to discriminate between PtdIns4*P* and several other inositides, including PtdIns(4,5)P_2, in various *in vitro* lipid binding assays (Yu et al. 2004). Based on sequence comparisons and residue swapping between the Osh1 and Osh2 proteins, it was shown that the latter did not interact with Arf1 (or at least interacted to a lesser extent) (Roy and Levine 2004). In our hands, using these two PH domains in mammalian cells yielded somewhat different results: while the Osh1–PH is a very good marker for PtdIns4*P* in the Golgi (as it was in yeast) and shows minimal plasma membrane localization, the Osh2–2xPH (or the single Osh2–PH domain) construct localizes only to the plasma membrane and does not show Golgi localization. The OSH2–2xPH has been useful in studies where the source of the plasma membrane pool of PtdIns4*P* was investigated (Balla et al. 2007), but it has to be noted that Osh2–PH also binds PtdIns(4,5)P_2, which should be kept in mind when interpreting data obtained with this probe. This dual specificity may explain why this probe does not show an increased signal when PtdIns(4,5)P_2 is acutely converted to PtdIns4*P* using a recruitable 5-phosphatase system (Balla et al. 2007; Korzeniowski et al. 2009). The extent of discrimination between these two lipids, as well as possible interactions with other proteins that could modify the relative affinity of Osh2–PH to PtdIns(4,5)P_2 and PtdIns4*P*, needs to be investigated further. Recent reports have shown the structure of the FAPP1 PH domain (Lenoir et al. 2010; He et al. 2011), revealing that the PtdIns4*P*- and Arf1-binding regions are located on different parts of the molecule, allowing the generation of mutants that show much lower affinity to Arf1 (He et al. 2011).

Recent studies in the bacterium *Legionella pneumophila* identified several virulence factor proteins that show PtdIns4*P* recognition (Ragaz et al. 2008; Brombacher et al. 2009; Schoebel et al. 2010; Zhu et al. 2010). Both the SidC and SidM proteins exhibit high-affinity PtdIns4*P* binding (Hilbi et al. 2011). However, these proteins also interact with specific Rab proteins, and it has so far been difficult to separate the lipid and Rab interactions to generate an unbiased PtdIns4*P* reporter. Our recent efforts suggest that a probe made from the SidM protein may just be the long-sought-after "unbiased" probe that detects different PtdIns4*P* pools simultaneously (Hammond et al. 2014).

Taking all these data together, it is clear that only specific pools of PtdIns4*P* can be monitored using some of these domains but the single SidM PtdIns4*P* binding domain identified in Legionella can detect all the different PtdIns4*P* pools within cells. This latter probe finally recognizes the PtdIns4*P* produced by type II PI 4-kinases on endosomes. Within the Golgi, PtdIns4*P* is produced by different PI 4-kinases (De Matteis et al. 2005), and it is possible that the different PH domains do not detect these pools equally. There is also an effect of the overexpressed domains on the Golgi itself. For example, the FAPP1–PH localizes primarily to the *trans*-side of the Golgi (Godi et al. 2004), but its localization between the *cis* and *trans* sides depends on the expression level (Weixel et al. 2005). In COS-7 cells, increased FAPP1 and OSBP expression levels cause morphological changes that are quite distinct (Balla and Varnai 2009), suggesting that they interact with distinct proteins (in addition to PtdIns4*P* and Arf1) and indicating that even though they may appear in the same Golgi compartment at low expression levels, they still detect functionally distinct pools of lipids. These details are all important signs indicating that not all PtdIns4*P* pools are created equal, and it may be challenging to image them using a single probe.

4.3.4 PtdIns3*P*

PtdIns3*P* is produced primarily in endosomes. In yeast, this lipid is critical for regulating the vesicular pathways by which newly synthesized or endocytosed molecules reach the vacuole (Schu et al. 1993). In mammalian cells, PtdIns3*P* is important in the early endosomal pathway (Simonsen et al. 1998). PtdIns3*P* was the first inositol lipid shown to bind a recognition domain other than a PH domain. The FYVE domain (the acronym originates from the first letter of the first four proteins in which the domain was described: Fab1p, YOTB, Vac1p, and EEA1) is responsible for PtdIns3*P* recognition and has been shown to be sufficient to recognize PtdIns3*P* when fused to GFP (Burd and Emr 1998). The strictly defined FYVE domain recognizes PtdIns3*P* quite specifically *in vitro*, but it detects PtdIns3*P* poorly in cells because of its limited affinity. To make this domain an effective PtdIns3*P* probe, the Stenmark lab created a tandem FYVE domain from the Hrs protein, which has become the most widely used tool for this purpose (Gillooly et al. 2000). We also made a construct from the EEA1 (Early Endosome-Associated Antigen) FYVE domain that is slightly longer, containing a few adjacent residues that assist in its natural dimerization (Hunyady et al. 2002). Both of these constructs decorate an early endosomal compartment and will fall off of vesicles on inhibition of the class III PI 3-kinase (the mammalian Vps34p homologue), which constitutively generates PtdIns3*P* on early endosomes.

Another domain that recognizes PtdIns3*P* is the phox (PX) domain, which is found in several proteins, the best characterized of which is the one from the p40phox protein (Kanai et al. 2001). When expressed as a GFP fusion protein, the p40phox PX domain also decorates the early endosomes, but this construct leads to the accumulation of very bright aggregated vesicles in cells that express somewhat higher amounts of the probe. These bright vesicles are more resistant to the effects of PI 3-kinase inhibitors, suggesting that the PX domain in these structures becomes part of a more stable protein complex that cannot disassemble and is probably responsible for the aggregation of vesicles. This is yet another example of how protein–protein interactions that must differ between the FYVE and PX domains may alter the behavior of these probes and why low expression levels should be used in these studies.

4.3.5 PtdIns(3,4)P_2

There is much less information available on this lipid than on the previous classes. PtdIns(3,4)P_2 is an intermediate of PtdIns(3,4,5)P_3 degradation by the SHIP 5-phosphatase enzymes, but it also can activate the protein kinase Akt (Franke et al. 1997). The relative increases between PtdIns(3,4)P_2 and PtdIns(3,4,5)P_3 vary from cell to cell depending on the activities of SHIP and the 4-phosphatase enzymes that degrade PtdIns(3,4)P_2 (Fedele et al. 2010). Larger increases in PtdIns(3,4)P_2 can be achieved via H_2O_2 treatment, which inactivates the phosphatase and tensin homolog (PTEN) enzyme (Leslie et al. 2003), directing PtdIns(3,4,5)P_3 degradation toward PtdIns(3,4)P_2, and H_2O_2 perhaps also inhibits the 4-phosphatases. The PH domains of the TAPP1 and TAPP2 proteins showed *in vitro* binding specificity to PtdIns(3,4)P_2 (Dowler et al. 2000). The crystal structure of the TAPP1 PH domain revealed structural features that explain its binding specificity (Thomas et al. 2001). The GFP-tagged TAPP1 PH domain detects the lipid in the plasma membrane in living cells under conditions in which PtdIns(3,4)P_2, but not PtdIns(3,4,5)P_3, is elevated (Kimber et al. 2002). A GST-fused TAPP1 PH domain also labeled the membranes of internal vesicles and the multivesicular body in fixed cells in EM analysis (Watt et al. 2004). Moreover, the TAPP1–PH domain does not show membrane association when PtdIns(3,4,5)P_3 is selectively elevated, indicating that it can discriminate between these two otherwise closely related lipid products. Other studies also found this domain to be useful in detecting the formation of PtdIns(3,4)P_2 in phagocytic cups in macrophages (Horan et al. 2007). The PX domain of p47phox has also been claimed as a PtdIns(3,4)P_2-recognizing module (Zhan et al. 2002) and was used to detect the lipid as a GFP fusion protein (Stahelin et al. 2003). However, in our hands the p47phox PX domain–GFP chimera does not show any indication of binding to membranes in a lipid-dependent manner (Balla and Varnai 2009), and its binding to other phospholipids, such as phosphatidic acid (Karathanassis et al. 2002), and proteins (Zhan et al. 2002) raises questions about the usefulness of this construct as an imaging tool.

4.3.6 PtdIns(3,5)P_2

This lipid has gained enormous interest because of its role in controlling trafficking to the vacuole in yeast and to the multivesicular body in higher eukaryotes (Gary et al. 1998; Ikonomov et al. 2003; Jefferies et al. 2008). Therefore, there is high demand for a tool to image PtdIns(3,5)P_2. Unfortunately, no reliable reporter is known for the visualization of PtdIns(3,5)P_3. Naturally, one would look for proteins that are downstream effectors of PtdIns(3,5)P_3, and there have indeed been new developments in this area. Previously, two proteins were described as targets of PtdIns(3,5)P_2: the yeast proteins Ent3p (Friant et al. 2003) and Svplp/Atg18 (Dove et al. 2004). The inositide binding site is located within the ENTH domain of Ent3p, and it was attributed to a cluster of basic residues on a beta propeller within Svplp. To our knowledge, there has been no systematic analysis of the intracellular distribution of any isolated domains extracted from these molecules with the aim of localizing PtdIns(3,5)P_2 in live cells. More recent studies identified another two yeast proteins, Tupl and Cti6, which act as transcription factors, as highly specific PtdIns(3,5)P_2 interactors (Han and Emr 2011). These proteins localize mostly to the nucleus, but the full-length Tupl protein also showed vesicular localization consistent with PtdIns(3,5)P_2 binding in yeast cells (Han and Emr 2011). At its C-terminus, the Tupl protein contains a WD40-like, seven-blade propeller motif that shows similarities to Atg18, and this 400-residue fragment is responsible for binding PtdIns(3,5)P_2 (Han and Emr 2011). To our knowledge, no studies have reported the distribution of this isolated fragment in mammalian cells. The Cti6 protein contains a PHD domain that also exhibits inositol-lipid-binding properties, but this fragment alone did not show the same specificity for PtdIns(3,5)P_2 that was observed with the full-length molecule (Han and Emr 2011). Another PtdIns(3,5)P_2 effector identified in mammalian cells was the TRPML1/mucolipin cation channel located in lysosomes, which requires PtdIns(3,5)P_2 for channel activity (Dong et al. 2010). The PtdIns(3,5)P_2-binding region was mapped to a basic stretch at the N-terminus of the channel, and a very recent study reported that a tandem version of this domain shows promise as a PtdIns(3,5)P_2 lipid probe (Li et al. 2013). More studies should clarify the utility of this promising new research tool.

4.3.7 PtdIns5P

PtdIns5*P* was first identified as an obligate substrate of the type II PIP kinases (PI5P 4-kinase) (Rameh et al. 1997b). This lipid is present in small amounts, and its metabolism and functions are poorly understood. The main route(s) of its production in cells is still highly debated, but it is most likely the result of the dephosphorylation of polyphosphoinositides (Coronas et al. 2007; Zou et al. 2007) rather than the direct 5-phosphorylation of PtdIns. To date, the only domain that recognizes PtdIns5*P* was found in the nuclear ING2 protein and was mapped to the PHD domain (PHD as for plant homeodomain), which binds PtdIns5*P*, and to a lesser degree PtdIns3*P*, *in vitro* (Gozani et al. 2003). A GFP fusion protein made with the 3xPHD domain of ING2 detected PtdIns5*P* in the plasma membrane in response to the overexpression of a bacterial 4-phosphatase, IpgD (Gozani et al. 2003; Pendaries et al. 2006), and did not show endosomal localization, suggesting that it does not bind PtdIsn3*P* in cells. More experience with this domain will be needed to determine where PtdIns5*P* is found in cells and whether its role in the nucleus is associated with detectable changes in its nuclear level.

4.3.8 Diacylglycerol

Diacylglycerol (DAG) is one of the most important lipid second messengers generated directly from phosphoinositides on PLC activation or indirectly from phosphatidylcholine by phospholipase D (PLD) and phosphatidic acid phosphohydrolase. The best known protein domain that recognizes DAG is the C1 domain, which shows some structural similarities to FYVE domains (Zhang et al. 1995; Misra and Hurley 1999). DAG imaging is based on C1 domains isolated from several proteins, such as PKC, PKD, and many others (Newton 2009), and several such sensors have been introduced (Meyer and Oancea 1998; Oancea et al. 1998; Dries et al. 2007; Kim et al. 2011). Most of these report on DAG production through their recruitment from the cytosol to the site of DAG formation, mostly the plasma membrane, but FRET-based sensors that also contain C1 domains and are targeted to different membranes have also been described (Sato et al. 2006). It is worth pointing out that C1 domains can have unique properties that manifest in protein–protein interaction and can restrict where they will detect DAG (Schultz et al. 2004). Accordingly, C1 domains isolated from various PKCs were shown to localize to different membranes other than the plasma membrane (see Table 4.1 for citations).

4.3.9 Phosphatidic Acid

Several protein domains have been postulated to confer phosphatidic acid (PtdOH) regulation to proteins (Wang et al. 2006). The PX domain of p47phox has a second shallow binding grove that binds PtdOH (Karathanassis et al. 2002), and there is a PtdOH-recognizing domain in the Raf-1 protein (Rizzo et al. 2000; Ghosh et al. 2003) as well as in protein phosphatase-1 (Jones et al. 2005). However, there are no reports of any globular domains that recognize PtdOH via a deeper binding pocket. Instead, there are shorter amphipathic helices containing basic residues that seem to be able to detect PtdOH in living cells. Given this structural simplicity, it is somewhat mysterious how these sequences achieve their lipid binding specificity. Live-cell imaging of PtdOH distribution has been performed using the GFP-fused PtdOH-binding domains of the yeast Spo20p protein (Zeniou-Meyer et al. 2007; Kassas et al. 2012) and Raf1 (Kassas et al. 2012). In comparative studies, the Spo20-based probe appeared to be a much better sensor than the one based on the Raf1–PtdOH-binding domain (Kassas et al. 2012). Because of the high nuclear accumulation of the Spo20-based probe, a recent study generated a modified version with an added nuclear export signal (NES) that has greatly increased its utility (Zhang et al. 2014). Another study reported a FRET sensor for PtdOH that relies on the recognition of this lipid by the C-terminal basic amino acid–rich binding domain of DOCK2 (Nishioka et al. 2010). This study showed a great variety of PtdOH patterns and changes relative to DAG in various cell types and in response to different forms of stimulation, which, based on appropriate controls, seemed to reflect true PtdOH changes. Although we need many more studies to evaluate the accuracy of these PtdOH probes, this is surely an important direction to pursue, as many investigators have expressed a desire to find a reliable probe to monitor PtdOH changes inside cells.

4.3.10 Phosphatidylserine

The appearance of phosphatidylserine (PtdSer) on the outer surface of cellular membranes observed during platelet activation (Zwaal et al. 1998) and as an early sign of apoptosis has generated great interest in probes that could detect this lipid on the outer surface of cells (Williamson and Schlegel 2002). Annexin V, which binds PtdSer at millimolar Ca^{2+} concentrations, has been the most widely used probe for this purpose (Koopman et al. 1994). These methods, however, could not be used to study the intracellular distribution and dynamics of PtdSer. Recent studies pioneered in the Grinstein laboratory filled this void when they introduced a new tool to follow PtdSer within the cell (Yeung et al. 2008; Fairn et al. 2011; Kay et al. 2012). Because the C2 domain of lactadherin binds PtdSer in a Ca^{2+}-independent manner (Andersen et al. 2000), the Grinstein group fused this domain to GFP (Yeung et al. 2008). The resulting probe was indeed able to detect PtdSer on the cytoplasmic leaflet of the plasma membrane, and when targeted to the lumen of the ER, it also detected PtdSer on the luminal side of the ER membrane (Kay et al. 2012). These authors also characterized a new fluorescent PtdSer analogue (TopFluor–PS) that mimics native PtdSer better than the previously used 7-nitrobenz-2-oxa-1,3-diazol-4-yl-conjugates and was useful in determining the diffusion properties of this molecule both in model membranes and inside living cells (Kay et al. 2012). These tools will be extremely useful in future studies on the cell biology of PtdSer.

4.4 DATA ACQUISITION AND ANALYSIS

These fluorescent reporters allow a wide array of methods to be employed for the analysis of lipid dynamics and distribution. Even simple widefield fluorescence microscopes can give very satisfactory pictures of the distribution of these reporters, especially in flat cells such as COS-1 or COS-7 cells. Confocal microscopes are undoubtedly the best for obtaining high levels of detail, but they usually lack the speed required to follow fast-moving objects. For this purpose, spinning-disc confocal microscopes are the best choice, although flat cells can be imaged at high speed almost as efficiently in a widefield scope equipped with a sensitive high-speed camera. As discussed in Chapter 2, TIRF microscopy is especially useful for analyzing events that occur in the plane of the plasma membrane. Recent developments in super-resolution imaging, such as photoactivated localization microscopy (PALM), stimulated emission depletion (STED), and structured illumination microscopy (SIM), also offer new possibilities to learn more about the molecular organization of lipids, but they also pose important technical challenges regarding fixation and data analysis. More details about these latter methods can be found elsewhere (Hess et al. 2006; Kner et al. 2009; Lippincott-Schwartz and Manley 2009) and in Chapter 16 of this book. Unfortunately, there is no single high-resolution microscope setup that can fulfill all the needs of live-cell experiments.

4.4.1 Changes That Can Be Seen by Eye

One can extract a great deal of information from using fluorescent lipid reporters. Some of these are visible and very obvious for even novice users, whereas others require observation by trained eyes. Moreover, a great deal of information is not appreciable by visual inspection and requires more sophisticated data analysis. Nonetheless, even the obvious and visible changes need quantification, and because of individual cell-to-cell variations, these need to be subjected to averaging and statistical analysis.

A good example is the question of how to analyze phosphoinositide dynamics monitored by PH domain GFP probes for quantification and statistics. The simplest way to monitor a change in the amount of cytoplasmic vs. membrane-bound fluorescent probe is to measure the average cytoplasmic intensity. This will increase when the probe falls off the membrane, and conversely, it will decrease when the probe gets recruited to the membrane (Figure 4.2a). However, cell shrinkage or the redistribution of the fluorescent probe (such as slow diffusion in and out of the nucleus) unrelated to lipid changes may complicate this simple approach. Still, this method is a good first approximation. Another method is to measure plasma membrane/cytoplasmic fluorescence ratios calculated from line intensity histograms. The fact that these values may change depending on where the line is drawn across the cell may entail multiple readings from each cell, which is quite a labor-intensive process. It is important to note that intensity increases in the membrane can be caused by intense membrane ruffling as new membranes are added to the plasma membrane, and these intensity increases—which mostly require PtdIns(3,4,5)P_3—do not necessarily represent actual membrane recruitment of the fluorescent

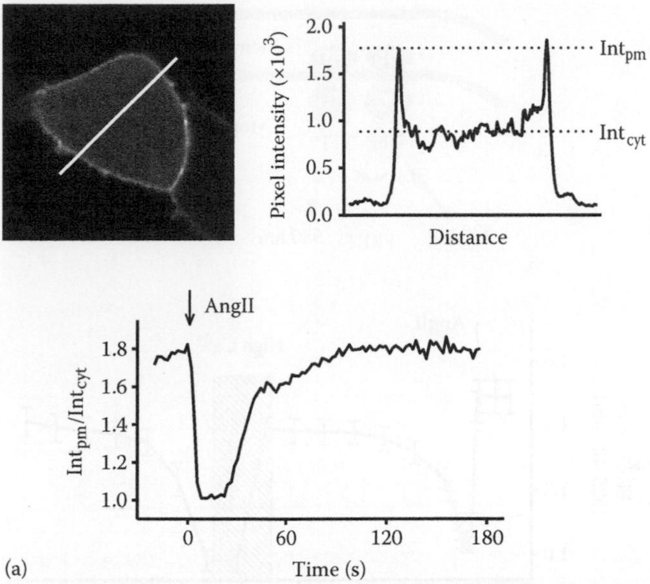

Figure 4.2 Quantifying the distribution of fluorescent probes between the membranes and the cytosol. (a) From a confocal picture taken of a cell expressing the domain in question, ratio values are calculated using line intensity histograms. The intensity of the fluorescence in the membrane (Int_{pm}) and the cytoplasm (Int_{cyt}) is used to assign a numerical value to the distribution of the lipid probe between the membrane and the cytoplasm. In live-cell experiments, plotting this ratio value as a function of time will show changes in probe distribution that follow the changes in membrane lipid concentration as a consequence of stimulation. In this example, the representative curve shows the transient angiotensin II (AngII)-evoked decrease of PtdIns(4,5)P_2 in the plasma membrane, as monitored with the PLCδ1PH–GFP probe in HEK293–AT1R cells. *(Continued)*

probes. To overcome this problem, it is highly recommended that fluorescent membrane markers be used as reference signals when assessing recruitment during intense membrane activity.

A better way to measure membrane association is to use Förster resonance energy transfer (FRET) measurements between proper fluorophore pairs (usually CFP and YFP or YFP and mRFP or their improved versions) (van Der Wal et al. 2001; Balla et al. 2005). Coexpressed CFP- and YFP-fused PH domains produce high FRET signal when they are both bound to the membrane. This signal decreases as soon as the probes dissociate from the membrane and diffuse into the cytosol (Figure 4.2b). FRET can be calculated using sensitized emission, in which case the microscope has to be tested for bleed-through between the donor and acceptor channels. This procedure has been described

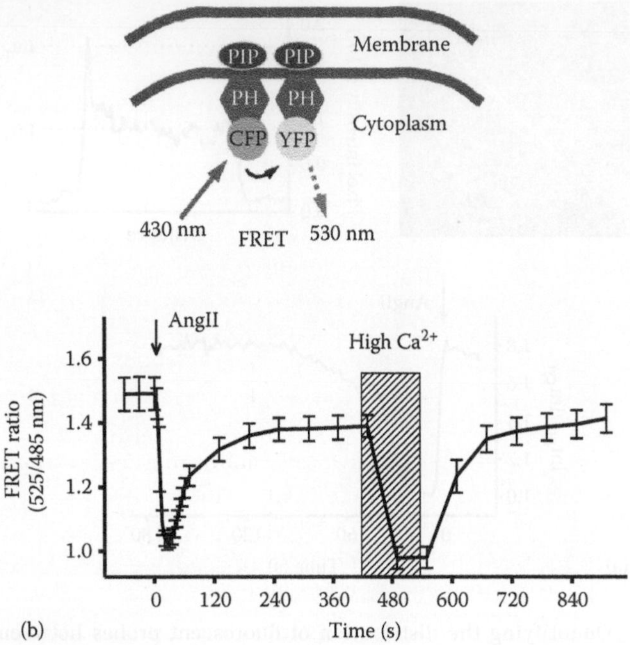

Figure 4.2 (Continued) Quantifying the distribution of fluorescent probes between the membranes and the cytosol. (b) Probes fused to appropriate fluorescent proteins for FRET measurements can also be used to follow membrane lipid levels. In this example, the PLCδ1PH domain was fused either to CFP or YFP and expressed in equimolar amounts in HEK293-AT1R cells. Cells were analyzed on a widefield fluorescence microscope and illuminated at 430 nm, and their emission was recorded at 530 nm and 485 nm. Simple FRET ratios were calculated from the emission intensities (530/485 nm). Because energy transfer is higher between the probes when they are found in the membrane, the changing FRET ratio reflects the change in the distribution of the probes between the membrane and the cytoplasm. This signal reports on changes in the PtdIns(4,5)P_2 content of the plasma membrane. It is important to note that this simple FRET ratio cannot be used to calculate the absolute FRET efficiency between the molecules, only the change that occurs during stimulation.

in detail elsewhere (van Rheenen et al. 2004) (see also Chapter 2). If the absolute FRET value between the fluorophores is not so important, a "quick and dirty" way to detect if any change occurs is to simply monitor the emission ratios for the two fluorophores on donor excitation (van Rheenen et al. 2004; Varnai and Balla 2006). This simplified method, however, is not really quantitative with regard to the absolute FRET value between the two molecules. A disadvantage of this approach is that at low expression levels (which would be desirable to minimize the ill effects of the presence of the probes), the FRET efficiency may not be very high, and even at high probe concentrations, there is low FRET signal if the density of the lipids is below a certain level.

A more accurate way to measure FRET is by analyzing changes in the lifetime of the donor fluorophore. Fluorescence lifetime imaging microscopy (FLIM) can monitor the duration of the excited state of the donor fluorophore, which will decrease upon FRET, and this change can be reliably monitored during stimulation (van Munster and Gadella 2005). FLIM is a very elegant and accurate method, but it requires special instrumentation. Detailed technical and theoretical background relating to FRET measurements using either sensitized emission or FLIM, including corrections for bleed-through and uneven illumination, can be found elsewhere in very comprehensive publications (Sekar and Periasamy 2003; van Rheenen et al. 2004; Thaler et al. 2005) and are also discussed in Chapter 2.

When spatial resolution is not essential, bioluminescent resonance energy transfer (BRET) is a good choice that does not require a microscope. BRET is based on similar principles as FRET, except that the donor molecule is not fluorescent but instead generates photons by bioluminescence. These photons will excite the fluorescent acceptor molecule when the two are in close proximity. The advantage of this method include its high sensitivity and low background, as well as the fact that it reports changes from a cell population that can be read in plate readers with luminescence reading capabilities (Toth et al. 2012).

A fourth method to monitor membrane recruitment or dissociation is total internal reflection fluorescence (TIRF) microscopy (Tengholm et al. 2003), which detects fluorescence only in the membrane region of the cell attached to the coverslip (Haugh et al. 2000; Schneider and Haugh 2004). This method is fairly simple and straightforward (although it needs a special instrument), but it is also sensitive to artifacts originating from changes in the cell shape and footprint or to membrane additions during the experiment. The use of a fluorescent membrane marker can also help to correct for changes unrelated to the altered membrane localization of the lipid probe.

4.4.2 Changes Not Seen by Visual Inspection

The next level of analysis is to obtain information on the behavior of lipid molecules that is not appreciable by visual inspection but requires more sophisticated data analysis. For example, determining the diffusion coefficient of lipid molecules using the lipid probes described above (in Section 4.3) is complicated by the fact that the reporters constantly associate with and dissociate from the lipids, and their diffusion therefore has two components: one related to when they bound to the lipids and another related to when they are unbound. These components of diffusion with the lipids and the dissociation/association can be calculated from fluorescence recovery after photobleaching (FRAP) data, but this requires additional mathematical analysis (see Hammond et al. 2009b for details). Single particle tracking in a TIRF microscope can also provide useful information about the spatial constraints of the movement of molecules in the membrane plane (see Kay et al. 2012 for an example). Another important question to be answered is whether the phosphoinositides are found in clusters within the plasma membrane, either by sequestration by proteins (Wang et al. 2002) or by partitioning into special lipid domains (often called rafts). This question has been addressed by analyzing the molecular proximity of the PLCδ1PH–GFP reporter using FRET. This study exploited the principle that clustered molecules will undergo FRET even when diluted. In contrast, when the molecules are randomly distributed, FRET will gradually decrease with decreasing molecular density (van Rheenen et al. 2005). The behavior of the lipid molecules in the membrane can also be analyzed by fluorescence correlation spectroscopy (FCS) (He and Marguet 2011; Heinemann et al. 2012; Wang et al. 2012). However, these methods require complex instrumentation and mathematical analyses that are often challenging even for experts and are intimidating for cell biologists. Doing FCS in membranes only adds to the complexity of an already difficult approach.

4.5 CRITICAL EVALUATION OF RESULTS

4.5.1 Inhibitory Effects of Expressed Lipid-Binding Protein Modules

The expression of lipid-binding protein modules has numerous effects on the biology of the cell (Figure 4.3a). These modules will generate a dynamic pool of the lipid by sequestering and protecting it from both effectors and metabolizing enzymes. In addition, protein binding by the lipid-binding modules is a major point of consideration. The localization of a lipid-binding protein that bind inositides with fewer phosphate groups requires other interactions, mostly with small GTP-binding proteins. For example, the FAPP1– and OSBP–PH–GFP fusion proteins are kept at the Golgi membrane by the combined presence of PtdIns4*P* and the GTP-bound form of Arf1 (Levine and Munro 2002). Therefore, changing the Arf1–GTP concentration will lead to the release of the construct from the membrane without any changes in the inositol lipid levels. Conversely, one cannot recruit more FAPP1–PH to the Golgi membrane by producing extra PtdIns4*P* without additional Arf1–GTP molecules. These

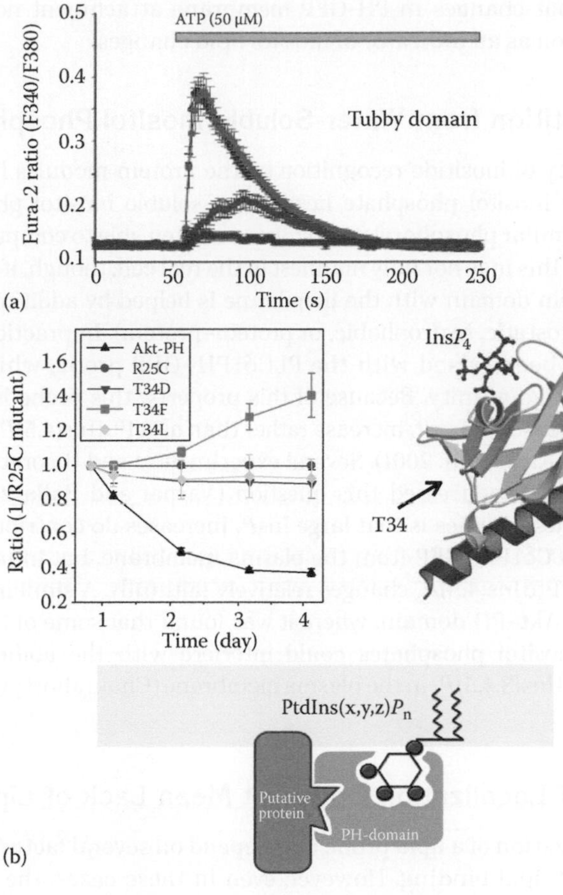

Figure 4.3 Intracellular expression of lipid probes can modify signaling events. (a) Expression of the Tubby domain in COS-7 cells inhibits the ATP-evoked Ca^{2+} signal measured with Fura-2. The binding of the domain to PtdIns(4,5)P_2 progressively inhibits the Ca^{2+} signals with increasing concentration of the domain (from pink through red to brown indicating increasing expression levels). (b) The PtdIns(3,4,5)P_3-binding PH domain of Akt inhibits cell growth in COS-7 cells. The growth of the transfected cells was measured by following the fluorescent intensities of various mutants of the GFP-tagged Akt–PH probe. The intensity values were normalized to those of the non-lipid binding R25C mutant for each day. Therefore, a decrease in the ratio corresponds to an inhibitory effect, as seen for the wide-type Akt–PH domain. Other mutations introduced at the T34 position, which do not affect PtdIns(3,4,5)P_3 binding, do not inhibit (T34D, T34L) or even facilitate (T34F) cell growth. This suggests that the inhibitory effect of the PH domain is not solely due to PtdIns(3,4,5)P_3 sequestration but must also involve another, probably protein–protein interaction at the surface where T34 is found.

examples show that changes in PH-GFP membrane attachment need to be treated with caution as an indicator of inositol lipid changes.

4.5.2 Competition from Water-Soluble Inositol Phosphates

Since the specificity of inositide recognition by the protein modules lies in the recognition of the inositol phosphate headgroup, soluble inositol phosphates with the same or similar phosphorylation status are often able to compete for the same binding site. This may not fully manifest in the real cell, though, if the interaction of the protein domain with the membrane is helped by additional interactions (e.g., electrostatic, hydrophobic, or protein–protein). In practical terms, this problem has been raised with the PLCδ1PH–GFP probe, which binds Ins(1,4,5)P_3 with high affinity. Because of this property, this probe is used by some studies as a sensor of InsP_3 increase rather than as a PtdIns(4,5)P_2 reporter (Hirose et al. 1999; Nash et al. 2001). Several experimental and theoretical modeling approaches have addressed this question (Varnai and Balla 2006). The conclusion from these studies is that large InsP_3 increases do contribute to the translocation of PLCδ1PH–GFP from the plasma membrane, but in most cells, this probe follows PtdIns(4,5)P_2 changes relatively faithfully. A similar problem was raised for the Akt–PH domain, where it was found that some of the highly phosphorylated inositol phosphates could interfere with the ability of this domain to bind PtdIns(3,4,5)P_3 in the plasma membrane (Chakraborty et al. 2010; Prasad et al. 2011).

4.5.3 Lack of Localization Does Not Mean Lack of Lipids

The cellular localization of a lipid probe may depend on several factors in addition to its inositol lipid binding. However, even in these cases, the presence of the lipids is essential for probe localization, and lipid changes can be followed as changes in localization. However, it has to be remembered that the lipid changes may be monitored only in the context of a molecular complex in a specific membrane compartment as opposed to an overall change in every membrane where the lipid is found. The best examples for this membrane-specific lipid reporting are the PH domains that recognize PtdIns4*P*. Some of these probes report on PtdIns4*P* only in Golgi/TGN membranes, where their localization also depends on Arf1–GTP (see the description of these probes above under Section 4.3). These probes will detect PtdIns4*P* in the plasma membrane only under special conditions, such as after recovery from a massive PLC activation (Balla et al. 2005), even though there is a significant amount of PtdIns4*P* in the plasma membrane in quiescent cells (Hammond et al. 2009a, 2012). This plasma membrane pool is seen by the OSH2 PH domain (Roy and Levine 2004), which then does not see the Golgi pool of the same lipid in mammalian cells.

These examples clearly illustrate that a lack of probe localization does not necessarily mean that the lipid is not present in a particular membrane.

Another point to consider is that more and more studies are raising the possibility that there can be a very high turnover rate of inositol lipid phosphorylation–dephosphorylation cycles regulating effector molecules without an easily detectable change in the steady-state level of the phosphorylated inositol lipid intermediate. For example, there are several lines of evidence suggesting that PtdIns(4,5)P_2 is present in the Golgi, yet none of the PH domains that recognize this lipid decorate the Golgi (De Matteis et al. 2005). More studies will be needed to address these questions with alternative methods such as EM analyses of fixed tissues or cells with PH domains, which should be less dependent on protein–protein interactions of the probes (Watt et al. 2002).

4.5.4 Binding and Sequestering Inositides Distorts Biology

By definition, overexpressing protein domains that bind inositides is expected to alter the availability of these lipids to enzymes and effectors. The only question is the extent to which these aberrations occur and limit our conclusions based on the use of these probes. FRAP experiments have shown that the steady-state membrane localization of the PLCδ1PH–GFP is the result of very rapid cycling between its membrane (PtdIns(4,5)P_2)-bound and cytosolic states (van Der Wal et al. 2001). PLCδ1PH–GFP overexpression increases the amount of PtdIns(4,5)P_2, most of which is bound to the PH domain at any given time, and this pool represents "extra" PtdIns(4,5)P_2 that is in dynamic equilibrium with the rest of the lipid pool. Owing to the rapid on- and off-rates of the PH domain, this extra PtdIns(4,5)P_2 pool is available for PLC-mediated hydrolysis and for 5-phosphatases, as demonstrated by numerous studies (Gill et al. 2007). This also means that the sequestration of a PtdIns(4,5)P_2 pool by the PH domain does not have very prominent functional effects. Nevertheless, the overexpression of this probe is clearly toxic to cells, as evidenced by rounding and detachment, as well as the accumulation of large vesicles inside cells expressing high levels of the protein. This effect is partly due to the interruption of membrane and cytoskeletal contacts (Raucher et al. 2000).

When we analyzed the dominant-negative effects of several PtdIns(3,4,5)P_3-recognizing PH domains on cellular responses known to depend on PtdIns(3,4,5)P_3, we found an unexpected specificity in their inhibitory profiles. Some of the PH domains preferentially inhibited PtdIns(3,4,5)P_3-dependent pathways that were related to the function of the proteins from which the PH domain originated (Figure 4.3b). The conclusion drawn from these studies was that the dominant-negative effects were in many cases more likely related to the sequestration of protein-binding partners rather than to the sequestration of PtdIns(3,4,5)P_3 itself, as the lipids can be produced easily in extra amounts

(Varnai et al. 2005). In a subsequent study, the protein-binding partner for the Grp1/ARNO PH domain was identified as Arf6–GTP (Cohen et al. 2007). It is important to emphasize that at moderate expression levels, the presence of these probes is well compensated by the cells, as was shown most convincingly by the successful creation of transgenic mice expressing PH-domain GFP reporters without any obvious functional defects (Nishio et al. 2007). Identifying protein-binding partners for the individual inositide-binding domains will largely help with our understanding of their biology as well as with the critical evaluation of the lipid data obtained through their use.

4.6 METHODS TO MANIPULATE INOSITOL LIPIDS IN LIVING CELLS

The ability to follow localized lipid changes in living cells is only one aspect of the study of spatially restricted phosphoinositide signaling. Of equal importance is the ability to induce changes in the lipid levels within cells. Artificially changing lipid levels can be used to determine the accuracy of our lipid probes, but more importantly, this allows us to study the downstream effectors that respond to these lipid changes. One approach is to add lipids exogenously to the cells in such a way that the lipid enters the cells and reaches internal membranes. Shuttling lipids into cells has been a popular way to test their effects (Ozaki et al. 2000). However, these techniques have several disadvantages: First, they flood the cell with the lipid without controlling where the lipid ends up. This takes away the spatially restricted aspect of the lipid changes and therefore may not be informative enough. Second, it is hard to know how the lipid is degraded while traveling inside the cell, and we therefore do not know whether the added lipid or its metabolites are exerting any effect.

The most widely used approach has been to express an inositide kinase or phosphatase enzyme to generate or eliminate lipids in the locations where these enzymes are naturally targeted. The reverse of this method is to use RNAi-mediated gene silencing to eliminate the kinase or phosphatase and study the consequences on the signaling process in question. Although these techniques are still the most widely used and most informative regarding the specific role of the enzymes in a particular cellular process, they have a big caveat, namely, that the time the cells are exposed to the lipid changes before they can be studied. Because phosphoinositides are key in the control of membrane flow and traffic between organelles, cells will undergo many changes that are secondary to the altered inositides. This makes it very difficult to determine the exact processes controlled by the particular lipid.

The best solution would be to use specific inhibitors that selectively block the effects of the individual enzymes acting on phosphoinositides. This is clear when reviewing the literature on the successful use of PI 3-kinase inhibitors (Wortmannin and LY 294002) or the inhibition of inositol phosphatase by Li^+ ions. Unfortunately, these inhibitors still inhibit too many inositide kinase enzymes, and no specific inhibitor of the phosphatases has surfaced as of yet. Recent progress in identifying subtype-specific PI 3-kinase inhibitors will tremendously help this field (Knight et al. 2006), and PI 4-kinases have also become important pharmaceutical targets because of their critical involvement in viral RNA replication (Altan-Bonnet and Balla 2012). We can certainly anticipate having a larger variety of lipid kinase inhibitors to selectively block certain pathways. However, until this happens, we needed an alternative way to locally manipulate inositide levels in an acutely regulated fashion.

4.6.1 Chemically Induced Recruitment of Enzymes to Subcellular Compartments

The rapamycin-induced heterodimerization of the FRB domain of mTOR and FKBP12 has been used to recruit proteins and induce molecular complex formation (Muthuswamy et al. 1999; Terrillon and Bouvier 2004). This heterodimerization system has been offered by Ariad (http://www.ariad.com, now sold by Clontech), and recently was adapted to alter phosphoinositide changes in the plasma membrane or in endosomes in a rapidly inducible manner (Heo et al. 2006; Suh et al. 2006; Varnai et al. 2006). In this approach, the FRB fragment of mTOR, which is a <9 kDa-sized module, is targeted to the plasma membrane by an N-terminal palmitoylation/myristoylation sequence derived from Lyn kinase or by the palmitoylation sequence of GAP43. An FP is also added to track its expression and localization. Simultaneous expression of an inositol lipid 5-phosphatase, truncated to eliminate its own localization signals and fused to FKBP-12 and a red fluorescent protein, yields a protein that remains in the cytosol and therefore has only limited (although not negligible) impact on membrane phosphoinositides. The addition of rapamycin rapidly recruits the phosphatase to the membrane and causes the dephosphorylation of PtdIns(4,5)P_2, as shown by the rapid decrease in the plasma membrane localization of the PLCδ1PH–GFP reporter (Figure 4.4a). This method allows for very quick reduction of PtdIns(4,5)P_2 (and also PtdIns(3,4,5)P_3) levels, the speed and extent of which depend on the concentration of the heterodimerizer (Varnai et al. 2006).

This method has for the first time allowed the unequivocal demonstration in intact cells that PtdIns(4,5)P_2 regulates KCNQ potassium channels (Suh et al. 2006) as well as Trpm8 channels (Varnai et al. 2006), and that the depletion of plasma membrane PtdIns(4,5)P_2 abolishes the endocytosis of transferrin receptors (Varnai et al. 2006; Abe et al. 2008) and leads to the elimination of endocytic clathrin-coated pits (Zoncu et al. 2007). A similar approach was used to target the phosphoinositide 3-phosphatase, myotubularin, to Rab5-positive

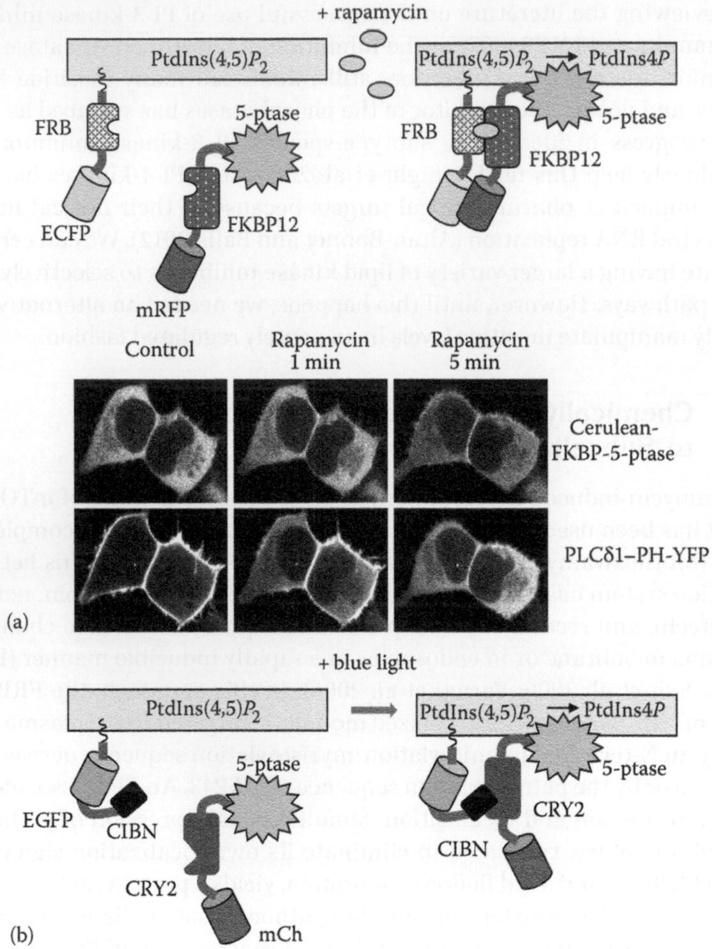

Figure 4.4 Molecular tools exploiting rapidly inducible protein heterodimerization to recruit inositol lipid-modifying enzymes to membranes in order to acutely modify inositol lipid levels. (a) Plasma membrane PtdIns(4,5)P_2 depletion was achieved by applying the rapamycin-induced heterodimerization of FKBP12 and the FRB domain of mTOR using an FKBP-fused INPP5E enzyme. Lipid depletion is demonstrated in COS-7 cells expressing the components of the depletion system (Cerulean FKBP-5-ptase, plasma membrane-targeted FRB [not shown], and the PtdIns(4,5)P_2 probe [PLCδ1–PH-YFP]). Translocation of the probe upon rapamycin treatment corresponds to the acute decrease in plasma membrane PtdIns(4,5)P_2 levels. (b) A similar result is achieved using a pair of plant protein modules, CIBN and CRY2, as a heterodimerizing pair. Here, the interaction can be induced by blue light, and the effects are reversible and allow very fine special control.

early endosomes to demonstrate the role of PtdIns3*P* in the morphogenesis of endosomes (Fili et al. 2006). This technique was extended to the targeting of other enzymes (Hammond et al. 2012) and other compartments, such as the Golgi (Komatsu et al. 2010; Szentpetery et al. 2010) and mitochondria (Csordas et al. 2010; Komatsu et al. 2010), allowing the analysis of the inositide-dependent regulation of several other processes in intact cells.

Because rapamycin is a known inhibitor of mTORC1, efforts have been made to create a system that does not affect mTOR. Ariad has developed an analogue (AP21967) that does not act on endogenous mTOR but acts on a mutant FRB that can be easily made in the FRB construct. However, the AP21967 analogue is too slow to mimic the physiological recruitment and activation process, and at concentrations at which it evokes a robust recruitment, it also acts on endogenous mTOR (Inoue et al. 2005). Therefore, another analogue (iRap) has been developed that induces much faster recruitment (Inoue et al. 2005), but this analogue also suffers from some of the same problems listed for AP21967.

To further confine the site of enzymatic activity, a photoactivatable rapamycin analogue was developed that could be used to locally liberate rapamycin via optical illumination and hence generate a very local activation of lipids or small GTP-binding proteins (DeRose et al. 2012). Another heterodimerization system utilizing fragments of two plant proteins, GAI and GID1, and the cell-permeable acetoxymethyl-derivative of the plant hormone gibberellin to induce their dimerization was also reported (Miyamoto et al. 2012), making it possible to switch on two processes independently. Future studies will determine how widely these tools can be applied to studies on the inositol lipid regulation of cellular processes.

4.6.2 Nonchemical Approaches to Alter Phosphoinositides

The rapamycin-inducible system has been very useful, especially in studies of plasma membrane-associated processes such as channels and transporters (Suh et al. 2006; Varnai et al. 2006; Lukacs et al. 2007; Klein et al. 2008; Yudin et al. 2011). However, this method has several limitations. Most importantly, it is not easily reversible owing to the high affinity and slow dissociation of the FRB–FKBP–rapamycin trimeric complex. Also, it is not known yet what effects the extra FRB and FKBP12 have on the biology of the cell. The discovery of a voltage-sensing inositol lipid phosphatase from *Ciona intestinalis* (CiVSP) by the Okamura group (Murata et al. 2005) has offered an alternative method to rapidly and reversibly alter plasma membrane PtdIns(4,5)P_2 levels (Murata and Okamura 2007; Iwasaki et al. 2008; Halaszovich et al. 2009). Homologues of the Ciona enzyme have since been identified in *Xenopus laevis*, and versions with altered voltage sensitivity have been created (Ratzan et al. 2011). Several studies utilized these new tools to study the role of plasma membrane PtdIns(4,5)P_2

in ion channel regulation (Lindner et al. 2011; Xie et al. 2011) as well as the kinetics of phosphoinositide fluxes (Falkenburger et al. 2010). The success of this approach has prompted efforts to generate voltage-sensing phosphoinositide phosphatase enzymes with different specificities. The similarity of the VSP enzymes to the PtdIns(3,4,5)P_3 3-phosphatase PTEN offered such a possibility, yielding a voltage-sensing PTEN (Lacroix et al. 2011). This technique has several advantages over the rapamycin-inducible system (see Section 4.6.1), but it can be applied only to the plasma membrane, and it will not be easily expandable to altering other phosphoinositide isoforms.

Lastly, an alternative method of heterodimerization driven by light exposure has recently been introduced (Figure 4.4b). This system uses two plant proteins, the PHR domain of cytochrome 2 (CRY2) and the CRY2-binding domain of the transcription factor CIBN, that form a heterodimer on exposure to blue light. By targeting one of these domains to the plasma membrane and using the other in a fusion construct containing a 5-phosphatase, the authors were able to control membrane recruitment by illuminating only a fraction of the cell membrane, thereby generating lipid gradients (Idevall-Hagren et al. 2012). This elegant new technique not only allows more refined spatial control but is also reversible, although not at the speed provided by the voltage-sensing phosphatases. Moreover, it requires neither the expression of proteins that can associate with endogenous protein partners nor the use of chemicals that can affect mTOR signaling. A small caveat is that only red and far-red fluorescence can be used to observe the cells, but this technique certainly has great potential to become a widely used experimental tool.

4.7 CONCLUDING REMARKS

Phosphoinositide research has undergone several transformations. It began with a few devoted scientists studying the curious phenomenon of the increased turnover of ^{32}P-labeled PtdIns4P and PtdOH, lipids that are highly abundant in the brain and that can be further phosphorylated. The discovery of the PtdIns(4,5)P_2–InsP_3–Ca^{2+} signaling cascade as a major signaling route from cell surface receptors (Michell 1975; Berridge and Irvine 1984) raised interest tremendously, which was then exponentially increased on the discovery of PI 3-kinases and their link to cancer (Yuan and Cantley 2008). Today, we know that phosphoinositides regulate vesicular trafficking, lipid transport, and channel activities and are implicated in a number of human diseases. These advances could not have occurred without the development of new experimental methods. Whether we think of the lipid extraction method introduced by Jordi Folch (Folch 1949), isotope labeling methods (Michell and Hawthorne 1965), or the separation of inositol phosphates by Dowex columns (Downes and

Michell 1981) or later by HPLC (Irvine et al. 1985), all of these advances opened up new worlds to be explored. Continuing this series of advances, the imaging of inositol lipids has earned its place as a major driving force in understanding the spatial aspects of the signaling roles of these lipid regulators. All of the aforementioned methods have their faults and limitations. However, the combination of various techniques and the desire to improve their accuracy has already produced a new level of understanding and has raised new questions for which further refinement of the tools are required. This chapter should serve as an inspiration for future scientists to continue this quest to find new ways to explore this exciting research field.

REFERENCES

Abe, N., T. Inoue, T. Galvez, L. Klein, and T. Meyer. 2008. Dissecting the role of PtdIns(4,5)P_2 in endocytosis and recycling of the transferrin receptor. *J. Cell Sci.* 121:1488–94.

Alieva, N.O., K.A. Konzen, S.F. Field, E.A. Meleshkevitch, M.E. Hunt, V. Beltran-Ramirez, D.J. Miller, J. Wiedenmann, A. Salih, and M.V. Matz. 2008. Diversity and evolution of coral fluorescent proteins. *PLoS One* 3:e2680.

Altan-Bonnet, N., and T. Balla. 2012. Phosphatidylinositol 4-kinases: Hostages harnessed to build panviral replication platforms. *Trends Biochem. Sci.* 37:293–302.

Andersen, M.H., H. Graversen, S.N. Fedosov, T.E. Petersen, and J.T. Rasmussen. 2000. Functional analyses of two cellular binding domains of bovine lactadherin. *Biochemistry* 39:6200–6.

Balla, A., Y.J. Kim, P. Varnai, Z. Szentpetery, Z. Knight, K.M. Shokat, and T. Balla. 2007. Maintenance of hormone-sensitive phosphoinositide pools in the plasma membrane requires phosphatidylinositol 4-kinase IIIα. *Mol. Biol. Cell* 19:711–21.

Balla, A., G. Tuymetova, A. Tsiomenko, P. Varnai, and T. Balla. 2005. A plasma membrane pool of phosphatidylinositol 4-phosphate is generated by phosphatidylinositol 4-kinase type-III alpha: Studies with the PH domains of the oxysterol binding protein and FAPP1. *Mol. Biol. Cell* 16:1282–95.

Balla, T. 2009. Green light to illuminate signal transduction events. *Trends Cell Biol.* 19:575–86.

Balla, T. 2013. Phosphoinositides: Tiny lipids with giant impact on cell regulation. *Physiol. Rev.* 93:1019–1137.

Balla, T., and P. Varnai. 2009. Visualization of cellular phosphoinositide pools with GFP-fused protein-domains. *Curr. Protoc. Cell Biol.* Chapter 24:Unit 24 4.

Berridge, M.J. 1984. Inositol trisphosphate and diacylglycerol as intracellular messengers. *Biochem. J.* 220:345–60.

Berridge, M.J., and R.F. Irvine. 1984. Inositol trisphosphate, a novel second messenger in cellular signal transduction. *Nature* 312:315–21.

Brombacher, E., S. Urwyler, C. Ragaz, S.S. Weber, K. Kami, M. Overduin, and H. Hilbi. 2009. Rab1 guanine nucleotide exchange factor SidM is a major phosphatidylinositol 4-phosphate-binding effector protein of *Legionella pneumophila*. *J. Biol. Chem.* 284:4846–56.

Burd, C.G., and S.D. Emr. 1998. Phosphatidylinositol(3)-phosphate signaling mediated by specific binding to RING FYVE domains. *Mol. Cell* 2:157–62.

Burns, D.J., and R.M. Bell. 1991. Protein kinase C contains two phorbol ester binding domains. *J. Biol. Chem.* 266:18330–8.

Chakraborty, A., M.A. Koldobskiy, N.T. Bello, M. Maxwell, J.J. Potter, K.R. Juluri, D. Maag, S. Kim, A.S. Huang, M.J. Dailey, M. Saleh, A.M. Snowman, T.H. Moran, E. Mezey, and S.H. Snyder. 2010. Inositol pyrophosphates inhibit Akt signaling, thereby regulating insulin sensitivity and weight gain. *Cell* 143:897–910.

Chattopadhyay, A., and E. London. 1987. Parallax method for direct measurement of membrane penetration depth utilizing fluorescence quenching by spin-labeled phospholipids. *Biochemistry* 26:39–45.

Chilvers, E.R., I.H. Batty, R.A.J. Challis, P.J. Barnes, and S.R. Nahorski. 1991. Determination of mass changes in phosphatidylinositol 4,5-bisphosphate and evidence for agonist-stimulated metabolism of inositol 1,4,5-trisphosphate in airway smooth muscle cells. *Biochem. J.* 275:373–379.

Christy, A.H., A.C. Kim, S.M. Marfatia, M. Lutchman, M. Hanspal, H. Jindal, S.C. Liu, P.S. Low, G.A. Rouleau, N. Mohandas, J.A. Chasis, J.G. Conboy, P. Gascard, Y. Takakuwa, S.C. Huang, E.J. Benz, Jr., A. Bretscher, R.G. Fehon, J.F. Gusella, V. Ramesh, F. Solomon, V.T. Marchesi, S. Tsukita, M. Arpin, D. Louvard, N.K. Tonks, J.M. Anderson, A.S. Fanning, P.J. Bryant, D.F. Woods, and K.B. Hoover. 1998. The FERM domain: A unique module involved in the linkage of cytoplasmic proteins to the membrane. *Trends Biochem. Sci.* 23:281–2.

Clark, J., K.E. Anderson, V. Juvin, T.S. Smith, F. Karpe, M.J. Wakelam, L.R. Stephens, and P.T. Hawkins. 2011. Quantification of PtdInsP_3 molecular species in cells and tissues by mass spectrometry. *Nat. Methods* 8:267–72.

Cohen, G.B., R. Ren, and D. Baltimore. 1995. Modular binding domains in signal transduction proteins. *Cell* 80:237–48.

Cohen, L.A., A. Honda, P. Varnai, F.B. Brown, T. Balla, and J.G. Donaldson. 2007. Active Arf6 recruits ARNO/Cytohesin GEFs to the PM by binding their PH domains. *Mol. Biol. Cell* 18:2244–2253.

Coronas, S., D. Ramel, C. Pendaries, F. Gaits-Iacovoni, H. Tronchere, and B. Payrastre. 2007. PtdIns5*P*: A little phosphoinositide with big functions? *Biochem. Soc. Symp.* 74:117–28.

Cronin, T.C., J.P. DiNitto, M.P. Czech, and D.G. Lambright. 2004. Structural determinants of phosphoinositide selectivity in splice variants of Grp1 family PH domains. *EMBO J.* 23:3711–20.

Csordas, G., P. Varnai, T. Golenar, S. Roy, G. Purkins, T.G. Schneider, T. Balla, and G. Hajnoczky. 2010. Imaging interorganelle contacts and local calcium dynamics at the ER-mitochondrial interface. *Mol. Cell* 39:121–32.

De Matteis, M.A., A. Di Campli, and A. Godi. 2005. The role of the phosphoinositides at the Golgi complex. *Biochim. Biophys. Acta* 1744:396–405.

DeRose, R., C. Pohlmeyer, N. Umeda, T. Ueno, T. Nagano, S. Kuo, and T. Inoue. 2012. Spatio-temporal manipulation of small GTPase activity at subcellular level and on timescale of seconds in living cells. *J. Vis. Exp.* 61:pii:3794.

Di Paolo, G., H.S. Moskowitz, K. Gipson, M.R. Wenk, S. Voronov, M. Obayashi, R. Flavell, R.M. Fitzsimonds, T.A. Ryan, and P. De Camilli. 2004. Impaired PtdIns(4,5)P_2 synthesis in nerve terminals produces defects in synaptic vesicle trafficking. *Nature* 431:415–22.

Dong, X.P., D. Shen, X. Wang, T. Dawson, X. Li, Q. Zhang, X. Cheng, Y. Zhang, L.S. Weisman, M. Delling, and H. Xu. 2010. PI(3,5)P(2) controls membrane trafficking by direct activation of mucolipin Ca($^{2+}$) release channels in the endolysosome. *Nat. Commun.* 1:38.

Dormann, D., G. Weijer, C.A. Parent, P.N. Devreotes, and C.J. Weijer. 2002. Visualizing PI3 kinase-mediated cell-cell signaling during *Dictyostelium* development. *Curr. Biol.* 12:1178–88.

Dove, S.K., R.C. Piper, R.K. McEwen, J.W. Yu, M.C. King, D.C. Hughes, J. Thuring, A.B. Holmes, F.T. Cooke, R.H. Michell, P.J. Parker, and M.A. Lemmon. 2004. Svp1p defines a family of phosphatidylinositol 3,5-bisphosphate effectors. *EMBO J.* 23:1922–33.

Dowler, S., R.A. Currie, D.G. Campbell, M. Deak, G. Kular, C.P. Downes, and D.R. Alessi. 2000. Identification of pleckstrin-homology-domain-containing proteins with novel phosphoinositide-binding specificities. *Biochem. J.* 351:19–31.

Downes, C.P., and R.H. Michell. 1981. The polyphosphoinositide phosphodiesterase of erythrocyte membranes. *Biochem. J.* 198:133–40.

Dries, D.R., L.L. Gallegos, and A.C. Newton. 2007. A single residue in the C1 domain sensitizes novel protein kinase C isoforms to cellular diacylglycerol production. *J. Biol. Chem.* 282:826–30.

Ellson, C.D., S. Gobert-Gosse, K.E. Anderson, K. Davidson, H. Erdjument-Bromage, P. Tempst, J.W. Thuring, M.A. Cooper, Z.Y. Lim, A.B. Holmes, P.R.J. Gaffney, J. Coadwell, E.R. Chilvers, P.T. Hawkins, and L.R. Stephens. 2001. PtdIns(3)*P* regulates the neutrophil oxidase complex by binding to the PX domain of p40phox. *Nat. Cell Biol.* 3:679–82.

Epand, R.M., P. Vuong, C.M. Yip, S. Maekawa, and R.F. Epand. 2004. Cholesterol-dependent partitioning of PtdIns(4,5)P_2 into membrane domains by the N-terminal fragment of NAP-22 (neuronal axonal myristoylated membrane protein of 22 kDa). *Biochem. J.* 379:527–32.

Fairn, G.D., M. Hermansson, P. Somerharju, and S. Grinstein. 2011. Phosphatidylserine is polarized and required for proper Cdc42 localization and for development of cell polarity. *Nat. Cell Biol.* 13:1424–30.

Falkenburger, B.H., J.B. Jensen, and B. Hille. 2010. Kinetics of PIP2 metabolism and KCNQ2/3 channel regulation studied with a voltage-sensitive phosphatase in living cells. *J. Gen. Physiol.* 135:99–114.

Fedele, C.G., L.M. Ooms, M. Ho, J. Vieusseux, S.A. O'Toole, E.K. Millar, E. Lopez-Knowles, A. Sriratana, R. Gurung, L. Baglietto, G.G. Giles, C.G. Bailey, J.E. Rasko, B.J. Shields, J.T. Price, P.W. Majerus, R.L. Sutherland, T. Tiganis, C.A. McLean, and C.A. Mitchell. 2010. Inositol polyphosphate 4-phosphatase II regulates PI3K/Akt signaling and is lost in human basal-like breast cancers. *Proc. Natl. Acad. Sci. U.S.A.* 107:22231–6.

Ferguson, C.G., R.D. James, C.S. Bigman, D.A. Shepard, Y. Abdiche, P.S. Katsamba, D.G. Myszka, and G.D. Prestwich. 2005. Phosphoinositide-containing polymerized liposomes: Stable membrane-mimetic vesicles for protein-lipid binding analysis. *Bioconjug. Chem.* 16:1475–83.

Field, S.J., N. Madson, M.L. Kerr, K.A. Galbraith, C.E. Kennedy, M. Tahiliani, A. Wilkins, and L.C. Cantley. 2005. PtdIns(4,5)P_2 functions at the cleavage furrow during cytokinesis. *Curr. Biol.* 15:1407–12.

Fili, N., V. Calleja, R. Woscholski, P.J. Parker, and B. Larijani. 2006. Compartmental signal modulation: Endosomal phosphatidylinositol 3-phosphate controls endosome morphology and selective cargo sorting. *Proc. Natl. Acad. Sci. U.S.A.* 103:15473–8.

Folch, J. 1949. Complete fractionation of brain cephalin: Isolation from it of phosphatidyl serine, phosphatidyl ethanolamine, and diphosphoinositide. *J. Biol. Chem.* 177:497–504.

Franke, T.F., D.R. Kaplan, L.C. Cantley, and A. Toker. 1997. Direct regulation of the Akt protooncogene product by PI3,4P2. *Science* 275:665–8.

Friant, S., E.I. Pecheur, A. Eugster, F. Michel, Y. Lefkir, D. Nourrisson, and F. Letourneur. 2003. Ent3p Is a PtdIns(3,5)P_2 effector required for protein sorting to the multivesicular body. *Dev. Cell* 5:499–511.

Fruman, D.A., R.E. Meyers, and L.C. Cantley. 1998. Phosphoinositide kinases. *Annu. Rev. Biochem.* 67:481–507.

Gary, J.D., A.E. Wurmser, C.J. Bonangelino, L.S. Weisman, and S.D. Emr. 1998. Fab1p is essential for PtdIns(3)*P* 5-kinase activity and the maintenance of vacuolar size and membrane homeostasis. *J. Cell Biol.* 143:65–79.

Ghosh, S., S. Moore, R.M. Bell, and M. Dush. 2003. Functional analysis of a phosphatidic acid binding domain in human Raf-1 kinase: Mutations in the phosphatidate binding domain lead to tail and trunk abnormalities in developing zebrafish embryos. *J. Biol. Chem.* 278:45690–6.

Giepmans, B.N., S.R. Adams, M.H. Ellisman, and R.Y. Tsien. 2006. The fluorescent toolbox for assessing protein location and function. *Science* 312:217–24.

Gill, D.J., H. Teo, J. Sun, O. Perisic, D.B. Veprintsev, Y. Vallis, S.D. Emr, and R.L. Williams. 2007. Structural studies of phosphoinositide 3-kinase-dependent traffic to multivesicular bodies. *Biochem. Soc. Symp.* 74:47–57.

Gillooly, D.J., I.C. Morrow, M. Lindsay, R. Gould, N.J. Bryant, L.M. Gaullier, G.P. Parton, and H. Stenmark. 2000. Localization of phosphatidylinositol 3-phosphate in yeast and mammalian cells. *EMBO J.* 19:4577–88.

Godi, A., A. Di Campi, A. Konstantakopoulos, G. Di Tullio, D.R. Alessi, G.S. Kular, T. Daniele, P. Marra, J.M. Lucocq, and M.A. De Matteis. 2004. FAPPs control Golgi-to-cell-surface membrane traffic by binding to ARF and PtdIns(4)*P*. *Nat. Cell Biol.* 6:393–404.

Goedhart, J., D. von Stetten, M. Noirclerc-Savoye, M. Lelimousin, L. Joosen, M.A. Hink, L. van Weeren, T.W. Gadella, Jr., and A. Royant. 2012. Structure-guided evolution of cyan fluorescent proteins towards a quantum yield of 93%. *Nat. Commun.* 3:751.

Golebiewska, U., J.G. Kay, T. Masters, S. Grinstein, W. Im, R.W. Pastor, S. Scarlata, and S. McLaughlin. 2011. Evidence for a fence that impedes the diffusion of phosphatidylinositol 4,5-bisphosphate out of the forming phagosomes of macrophages. *Mol. Biol. Cell* 22:3498–507.

Gozani, O., P. Karuman, D.R. Jones, D. Ivanov, J. Cha, A.A. Logovskoy, C.L. Baird, H. Zhu, S.J. Field, S.L. Lessnick, J. Villasenov, B. Mehrotra, J. Chen, V.R. Rao, J.S. Brugge, C.G. Ferguson, B. Payrastre, D.G. Myszka, L.C. Cantley, G. Wagner, N. Divecha, G.D. Prestwich, and J. Yuan. 2003. The PHD finger of the chromatin-associated protein ING2 functions as a nuclear phosphoinositide receptor. *Cell* 114:99–111.

Gray, A., J. Van Der Kaay, and C.P. Downes. 1999. The pleckstrin homology domains of protein kinase B and GRP1 (general receptor for phosphoinositides-1) are sensitive and selective probes for the cellular detection of phosphatidylinositol 3,4-bisphosphate and/or phosphatidylinositol 3,4,5–trisphosphate in vivo. *Biochem. J.* 344:929–36.

Halaszovich, C.R., D.N. Schreiber, and D. Oliver. 2009. Ci-VSP is a depolarization-activated phosphatidylinositol-4,5–bisphosphate and phosphatidylinositol-3,4,5–trisphosphate 5'-phosphatase. *J. Biol. Chem.* 284:2106–13.

Halet, G. 2005. Imaging phosphoinositide dynamics using GFP-tagged protein domains. *Biol. Cell* 97:501–18.

Halstead, J.R., K. Jalink, and N. Divecha. 2005. An emerging role for PtdIns(4,5)P_2-mediated signalling in human disease. *Trends Pharmacol. Sci.* 26:654–60.

Hammond, G.R., S.K. Dove, A. Nicol, J.A. Pinxteren, D. Zicha, and G. Schiavo. 2006. Elimination of plasma membrane phosphatidylinositol (4,5)-bisphosphate is required for exocytosis from mast cells. *J. Cell Sci.* 119:2084–94.

Hammond, G.R., M.J. Fischer, K.E. Anderson, J. Holdich, A. Koteci, T. Balla, and R.F. Irvine. 2012. PI4P and PI(4,5)P2 are essential but independent lipid determinants of membrane identity. *Science* 337:727–30.

Hammond, G.R., M.P. Machner, and T. Balla. 2014. A novel probe for phosphatidylinositol 4-phosphate reveals multiple pools beyond the Golgi, *J. Cell Biol.* 205:113–126.

Hammond, G.R., G. Schiavo, and R.F. Irvine. 2009a. Immunocytochemical techniques reveal multiple, distinct cellular pools of PtdIns4P and PtdIns(4,5)P_2. *Biochem. J.* 422:23–35.

Hammond, G.R., Y. Sim, L. Lagnado, and R.F. Irvine. 2009b. Reversible binding and rapid diffusion of proteins in complex with inositol lipids serves to coordinate free movement with spatial information. *J. Cell Biol.* 184:297–308.

Han, B.K., and S.D. Emr. 2011. Phosphoinositide [PI(3,5)P2] lipid-dependent regulation of the general transcriptional regulator Tup1. *Genes Dev.* 25:984–95.

Haugh, J.M., F. Codazzi, M. Teruel, and T. Meyer. 2000. Spatial sensing in fibroblasts mediated by 3′ phosphoinositides. *J. Cell Biol.* 151:1269–80.

He, H.T., and D. Marguet. 2011. Detecting nanodomains in living cell membrane by fluorescence correlation spectroscopy. *Annu. Rev. Phys. Chem.* 62:417–36.

He, J., J.L. Scott, A. Heroux, S. Roy, M. Lenoir, M. Overduin, R.V. Stahelin, and T.G. Kutateladze. 2011. Molecular basis of phosphatidylinositol 4-phosphate and ARF1 GTPase recognition by the FAPP1 pleckstrin homology (PH) domain. *J. Biol. Chem.* 286:18650–7.

Heinemann, F., V. Betaneli, F.A. Thomas, and P. Schwille. 2012. Quantifying lipid diffusion by fluorescence correlation spectroscopy: A critical treatise. *Langmuir* 28:13395–404.

Heo, W.D., T. Inoue, W.S. Park, M.L. Kim, B.O. Park, T.J. Wandless, and T. Meyer. 2006. PI(3,4,5)P3 and PI(4,5)P2 lipids target proteins with polybasic clusters to the plasma membrane. *Science* 314:1458–61.

Hess, S.T., T.P. Girirajan, and M.D. Mason. 2006. Ultra-high resolution imaging by fluorescence photoactivation localization microscopy. *Biophys. J.* 91:4258–72.

Hilbi, H., S. Weber, and I. Finsel. 2011. Anchors for effectors: Subversion of phosphoinositide lipids by legionella. *Front. Microbiol.* 2:91.

Hirose, K., S. Kadowaki, M. Tanabe, H. Takeshima, and M. Iino. 1999. Spatiotemporal dynamics of inositol 1,4,5-trisphosphate that underlies complex Ca^{2+} mobilization patterns. *Science* 284:1527–30.

Horan, K.A., K. Watanabe, A.M. Kong, C.G. Bailey, J.E. Rasko, T. Sasaki, and C.A. Mitchell. 2007. Regulation of FcgammaR-stimulated phagocytosis by the 72-kDa inositol polyphosphate 5-phosphatase: SHIP1, but not the 72-kDa 5-phosphatase, regulates complement receptor 3 mediated phagocytosis by differential recruitment of these 5-phosphatases to the phagocytic cup. *Blood* 110:4480–91.

Huang, Y.E., M. Iijima, C.A. Parent, S. Funamoto, R.A. Firtel, and P. Devreotes. 2003. Receptor-mediated regulation of PI3Ks confines PI(3,4,5)P3 to the leading edge of chemotaxing cells. *Mol. Biol. Cell* 14:1913–22.

Hunyady, L., A.J. Baukal, Z. Gaborik, J.A. Olivares-Reyes, M. Bor, M. Szaszak, R. Lodge, K.J. Catt, and T. Balla. 2002. Differential PI 3-kinase dependence of early and late phases of recycling of the internalized AT1 angiotensin receptor. *J. Cell Biol.* 157:1211–22.

Idevall-Hagren, O., E.J. Dickson, B. Hille, D.K. Toomre, and P. De Camilli. 2012. Optogenetic control of phosphoinositide metabolism. *Proc. Natl. Acad. Sci. U.S.A.* 109:E2316–23.

Ikonomov, O.C., D. Sbrissa, M. Foti, J.L. Carpentier, and A. Shisheva. 2003. PIKfyve controls fluid phase endocytosis but not recycling/degradation of endocytosed receptors or sorting of procathepsin D by regulating multivesicular body morphogenesis. *Mol. Biol. Cell* 14:4581–91.

Inoue, T., W.D. Heo, J.S. Grimley, T.J. Wandless, and T. Meyer. 2005. An inducible translocation strategy to rapidly activate and inhibit small GTPase signaling pathways. *Nat. Methods* 2:415–8.

Irvine, R.F., E.E. Anggard, A.J. Letcher, and C.P. Downes. 1985. Metabolism of inositol 1,4,5-trisphosphate and inositol 1,3,4-trisphosphate in rat parotid glands. *Biochem. J.* 229:505–11.

Iwasaki, H., Y. Murata, Y. Kim, M.I. Hossain, C.A. Worby, J.E. Dixon, T. McCormack, T. Sasaki, and Y. Okamura. 2008. A voltage-sensing phosphatase, Ci-VSP, which shares sequence identity with PTEN, dephosphorylates phosphatidylinositol 4,5-bisphosphate. *Proc. Natl. Acad. Sci. U.S.A.* 105:7970–5.

Jefferies, H.B., F.T. Cooke, P. Jat, C. Boucheron, T. Koizumi, M. Hayakawa, H. Kaizawa, T. Ohishi, P. Workman, M.D. Waterfield, and P.J. Parker. 2008. A selective PIKfyve inhibitor blocks PtdIns(3,5)P_2 production and disrupts endomembrane transport and retroviral budding. *EMBO Rep.* 9:164–70.

Jones, J.A., R. Rawles, and Y.A. Hannun. 2005. Identification of a novel phosphatidic acid binding domain in protein phosphatase-1. *Biochemistry* 44:13235–45.

Kanai, F., H. Liu, S.J. Field, H. Akbary, T. Matsuo, G.E. Brown, L.C. Cantley, and M.B. Yaffe. 2001. The PX domains of p47phox and p40phox bind to lipid products of PI(3)K. *Nat. Cell Biol.* 3:675–8.

Karathanassis, D., R.V. Stahelin, J. Bravo, O. Perisic, C.M. Pacold, W. Cho, and R.L. Williams. 2002. Binding of the PX domain of p47phox to phosphatidylinositol 3,4-bisphosphate and phosphatidic acid is masked by an intramolecular interaction. *EMBO J.* 21:5057–68.

Kassas, N., P. Tryoen-Toth, M. Corrotte, T. Thahouly, M.F. Bader, N.J. Grant, and N. Vitale. 2012. Genetically encoded probes for phosphatidic acid. *Methods Cell Biol.* 108:445–59.

Kavran, J.M., D.E. Klein, A. Lee, M. Falasca, S.J. Isakoff, E.Y. Skolnik, and M.A. Lemmon. 1998. Specificity and promiscuity in phosphoinositide binding by pleckstrin homology domains. *J. Biol. Chem.* 273:30497–508.

Kay, J.G., M. Koivusalo, X. Ma, T. Wohland, and S. Grinstein. 2012. Phosphatidylserine dynamics in cellular membranes. *Mol. Biol. Cell* 23:2198–212.

Kim, Y.J., M.L. Guzman-Hernandez, and T. Balla. 2011. A highly dynamic ER-derived phosphatidylinositol-synthesizing organelle supplies phosphoinositides to cellular membranes. *Dev. Cell* 21:813–24.

Kimber, W.A., L. Trinkle-Mulcahy, P.C. Cheung, M. Deak, L.J. Marsden, A. Kieloch, S. Watt, R.T. Javier, A. Gray, C.P. Downes, J.M. Lucocq, and D.R. Alessi. 2002. Evidence that the tandem-pleckstrin-homology-domain-containing protein TAPP1 interacts with Ptd(3,4)P_2 and the multi-PDZ-domain-containing protein MUPP1 in vivo. *Biochem. J.* 361:525–36.

Klarlund, J.K., A. Guilherme, J.J. Holik, J.V. Virbasius, A. Chawla, and M.P. Czech. 1997. Signaling by phosphoinositide-3,4,5-trisphosphate through proteins containing plekstrin and Sec7 homology domains. *Science* 275:1927–30.

Klarlund, J.K., W. Tsiaras, J.J. Holik, A. Chawla, and M.P. Czech. 2000. Distinct polyphosphoinositide binding selectivities for pleckstrin homology domains of GRP1-like proteins based on diglycine versus triglycine motifs. *J. Biol. Chem.* 275:32816–21.

Klein, R.M., C.A. Ufret-Vincenty, L. Hua, and S.E. Gordon. 2008. Determinants of molecular specificity in phosphoinositide regulation. Phosphatidylinositol (4,5)-bisphosphate (PI(4,5)P2) is the endogenous lipid regulating TRPV1. *J. Biol. Chem.* 283:26208–16.

Kner, P., B.B. Chhun, E.R. Griffis, L. Winoto, and M.G. Gustafsson. 2009. Super-resolution video microscopy of live cells by structured illumination. *Nat. Methods* 6:339–42.

Knight, Z.A., B. Gonzalez, M.E. Feldman, E.R. Zunder, D.D. Goldenberg, O. Williams, R. Loewith, D. Stokoe, A. Balla, B. Toth, T. Balla, W.A. Weiss, R.L. Williams, and K.M. Shokat. 2006. A pharmacological map of the PI3-K family defines a role for p110alpha in insulin signaling. *Cell* 125:733–47.

Komander, D., A. Fairservice, M. Deak, G.S. Kular, A.R. Prescott, C.P. Downes, S.T. Safrany, D.R. Alessi, and D.M. van Aalten. 2004. Structural insights into the regulation of PDK1 by phosphoinositides and inositol phosphates. *EMBO J.* 23:3918–28.

Komatsu, T., I. Kukelyansky, J.M. McCaffery, T. Ueno, L.C. Varela, and T. Inoue. 2010. Organelle-specific, rapid induction of molecular activities and membrane tethering. *Nat. Methods* 7:206–8.

Kontos, C.D., T.P. Stauffer, W.P. Yang, J.D. York, L. Huang, M.A. Blanar, T. Meyer, and K.G. Peters. 1998. Tyrosine 1101 of Tie2 is the major site of association of p85 and is required for activation of phosphatidylinositol 3-kinase and Akt. *Mol. Cell. Biol.* 18:4131–40.

Koopman, G., C.P. Reutelingsperger, G.A. Kuijten, R.M. Keehnen, S.T. Pals, and M.H. van Oers. 1994. Annexin V for flow cytometric detection of phosphatidylserine expression on B cells undergoing apoptosis. *Blood* 84:1415–20.

Korzeniowski, M.K., M.A. Popovic, Z. Szentpetery, P. Varnai, S.S. Stojilkovic, and T. Balla. 2009. Dependence of STIM1/Orai1-mediated calcium entry on plasma membrane phosphoinositides. *J. Biol. Chem.* 284:21027–35.

Kutateladze, T.G., K.D. Ogburn, W.T. Watson, T. deBeer, S.D. Emr, C.G. Burd, and M. Overduin. 1999. Phosphatidylinositol 3-phosphate recognition by the FYVE domain. *Mol. Cell* 3:805–11.

Lacroix, J., C.R. Halaszovich, D.N. Schreiber, M.G. Leitner, F. Bezanilla, D. Oliver, and C.A. Villalba-Galea. 2011. Controlling the activity of a phosphatase and tensin homolog (PTEN) by membrane potential. *J. Biol. Chem.* 286:17945–53.

Lehel, C., Z. Olah, G. Jakab, and W.B. Anderson. 1995. Protein kinase C epsilon is localized to the Golgi via its zinc-finger domain and modulates Golgi function. *Proc. Natl. Acad. Sci. U.S.A.* 92:1406–10.

Lemmon, M.A. 2008. Membrane recognition by phospholipid-binding domains. *Nat. Rev. Mol. Cell Biol.* 9:99–111.

Lemmon, M.A., K.M. Ferguson, R. O'Brian, P.B. Sigler, and J. Schlessinger. 1995. Specific and high-affinity binding of inositol phosphates to an isolated plekstrin homology domain. *Proc. Natl. Acad. Sci. U.S.A.* 92:10472–6.

Lenoir, M., U. Coskun, M. Grzybek, X. Cao, S.B. Buschhorn, J. James, K. Simons, and M. Overduin. 2010. Structural basis of wedging the Golgi membrane by FAPP pleckstrin homology domains. *EMBO Rep.* 11:279–84.

Leslie, N.R., D. Bennett, Y.E. Lindsay, H. Stewart, A. Gray, and C.P. Downes. 2003. Redox regulation of PI 3-kinase signalling via inactivation of PTEN. *EMBO J.* 22:5501–10.

Levine, T.P., and S. Munro. 1998. The pleckstrin-homology domain of oxysterol-binding protein recognizes a determinant specific to Golgi membranes. *Curr. Biol.* 8:729–39.

Levine, T.P., and S. Munro. 2001. Dual targeting of Osh1p, a yeast homologue of oxysterol-binding protein, to both the Golgi and the nucleus-vacuole junction. *Mol. Biol. Cell* 6:1633–44.

Levine, T.P., and S. Munro. 2002. Targeting of Golgi-specific pleckstrin homology domains involves both PtdIns 4-kinase-dependent and -independent components. *Curr. Biol.* 12:695–704.

Li, X., X. Wang, X. Zhang, M. Zhao, W.L. Tsang, Y. Zhang, R.G. Yau, L.S. Weisman, and H. Xu. 2013. Genetically encoded fluorescent probe to visualize intracellular phosphatidylinositol 3,5-bisphosphate localization and dynamics. *Proc. Natl. Acad. Sci. U.S.A.* 110:21165–70.

Lindner, M., M.G. Leitner, C.R. Halaszovich, G.R. Hammond, and D. Oliver. 2011. Probing the regulation of TASK potassium channels by PI(4,5)P2 with switchable phosphoinositide phosphatases. *J. Physiol.* 589(Pt 13):3149–62.

Lippincott-Schwartz, J., and S. Manley. 2009. Putting super-resolution fluorescence microscopy to work. *Nat. Methods* 6:21–3.

Lippincott-Schwartz, J., J.F. Presley, K.J. Zaal, K. Hirschberg, C.D. Miller, and J. Ellenberg. 1999. Monitoring the dynamics and mobility of membrane proteins tagged with green fluorescent protein. *Methods Cell Biol.* 58:261–81.

Lipsky, N.G., and R.E. Pagano. 1983. Sphingolipid metabolism in cultured fibroblasts: Microscopic and biochemical studies employing a fluorescent ceramide analogue. *Proc. Natl. Acad. Sci. U.S.A.* 80:2608–12.

Lukacs, V., B. Thyagarajan, P. Varnai, A. Balla, T. Balla, and T. Rohacs. 2007. Dual regulation of TRPV1 by phosphoinositides. *J. Neurosci.* 27:7070–80.

Manna, D., A. Albanese, W.S. Park, and W. Cho. 2007. Mechanistic basis of differential cellular responses of phosphatidylinositol 3,4-bisphosphate- and phosphatidylinositol 3,4,5-trisphosphate-binding pleckstrin homology domains. *J. Biol. Chem.* 282:32093–105.

Matsuda, M., H.F. Paterson, R. Rodriguez, A.C. Fensome, M.V. Ellis, K. Swann, and M. Katan. 2001. Real time fluorescence imaging of PLC gamma translocation and its interaction with the epidermal growth factor receptor. *J. Cell Biol.* 153:599–612.

Matsu-ura, T., T. Michikawa, T. Inoue, A. Miyawaki, M. Yoshida, and K. Mikoshiba. 2006. Cytosolic inositol 1,4,5-trisphosphate dynamics during intracellular calcium oscillations in living cells. *J. Cell Biol.* 173:755–65.

Meyer, T., and E. Oancea. 1998. Protein kinase C as a molecular machine for decoding calcium and diacylglycerol signals. *Cell* 95:307–18.

Michell, R.H. 1975. Inositol phospholipids and cell surface receptor function. *Biochim. Biophys. Acta* 415:81–147.

Michell, R.H., and J.N. Hawthorne. 1965. The site of diphosphoinositide synthesis in rat liver. *Biochem. Biophys. Res. Commun.* 21:333–8.

Michell, R.H., V.L. Heath, M.A. Lemmon, and S.K. Dove. 2005. Phosphatidylinositol 3,5-bisphosphate: Metabolism and cellular functions. *Trends Biochem. Sci.* 31:52–63.

Misra, S., and J.H. Hurley. 1999. Crystal structure of a phosphatidylinositol 3-phosphate-specific membrane-targeting motif, the FYVE domain of Vps27p. *Cell* 97:657–66.

Miyamoto, T., R. DeRose, A. Suarez, T. Ueno, M. Chen, T.P. Sun, M.J. Wolfgang, C. Mukherjee, D.J. Meyers, and T. Inoue. 2012. Rapid and orthogonal logic gating with a gibberellin-induced dimerization system. *Nat. Chem. Biol.* 8:465–70.

Miyazawa, A., M. Umeda, T. Horikoshi, K. Yanagisawa, T. Yoshioka, and K. Inoue. 1988. Production and characterization of monoclonal antibodies that bind to phosphatidylinositol 4,5-bisphosphate. *Mol. Immunol.* 25:1025–31.

Murata, Y., H. Iwasaki, M. Sasaki, K. Inaba, and Y. Okamura. 2005. Phosphoinositide phosphatase activity coupled to an intrinsic voltage sensor. *Nature* 435:1239–43.

Murata, Y., and Y. Okamura. 2007. Depolarization activates the phosphoinositide phosphatase Ci-VSP, as detected in *Xenopus* oocytes coexpressing sensors of PIP2. *J. Physiol.* 583:875–89.

Muthuswamy, S.K., M. Gilman, and J.S. Brugge. 1999. Controlled dimerization of ErbB receptors provides evidence for differential signaling by homo- and heterodimers. *Mol. Cell. Biol.* 19:6845–57.

Nagel, W., P. Schilcher, L. Zeitlmann, and W. Kolanus. 1998. The PH domain and the polybasic c domain of cytohesin-1 cooperate specifically in plasma membrane-association and cellular function. *Mol. Biol. Cell* 9:1981–94.

Nakanishi, H., P. de los Santos, and A.M. Neiman. 2004. Positive and negative regulation of a SNARE protein by control of intracellular localization. *Mol. Biol. Cell* 15:1802–15.

Narayan, K., and M.A. Lemmon. 2006. Determining selectivity of phosphoinositide-binding domains. *Methods* 39:122–33.

Nash, M.S., K.W. Young, G.B. Willars, R.A. Challiss, and S.R. Nahorski. 2001. Single-cell imaging of graded Ins(1,4,5)P3 production following G-protein-coupled-receptor activation. *Biochem. J.* 356:137–42.

Nasuhoglu, C., S. Feng, J. Mao, M. Yamamoto, H.L. Yin, S. Earnest, B. Barylko, J.P. Albanesi, and D.W. Hilgemann. 2002. Nonradioactive analysis of phosphatidylinositides and other anionic phospholipids by anion-exchange high-performance liquid chromatography with suppressed conductivity detection. *Anal. Biochem.* 301:243–54.

Nelson, C.P., S.R. Nahorski, and R.A. Challiss. 2008. Temporal profiling of changes in phosphatidylinositol 4,5-bisphosphate, inositol 1,4,5-trisphosphate and diacylglycerol allows comprehensive analysis of phospholipase C-initiated signalling in single neurons. *J. Neurochem.* 107:602–15.

Newton, A.C. 2009. Lipid activation of protein kinases. *J. Lipid Res.* 50(Suppl):S266–71.

Nishio, M., K.I. Watanabe, J. Sasaki, C. Taya, S. Takasuga, R. Iizuka, T. Balla, M. Yamazaki, H. Watanabe, R. Itoh, S. Kuroda, Y. Horie, I. Forster, T.W. Mak, H. Yonekawa, J.M. Penninger, Y. Kanaho, A. Suzuki, and T. Sasaki. 2007. Control of cell polarity and motility by the PtdIns(3,4,5)P_3 phosphatase SHIP1. *Nat. Cell Biol.* 9:36–44.

Nishioka, T., K. Aoki, K. Hikake, H. Yoshizaki, E. Kiyokawa, and M. Matsuda. 2008. Rapid turnover rate of phosphoinositides at the front of migrating MDCK cells. *Mol. Biol. Cell* 19:4213–23.

Nishioka, T., M.A. Frohman, M. Matsuda, and E. Kiyokawa. 2010. Heterogeneity of phosphatidic acid levels and distribution at the plasma membrane in living cells as visualized by a Foster resonance energy transfer (FRET) biosensor. *J. Biol. Chem.* 285:35979–87.

Oancea, E., M.N. Teruel, A.F.G. Quest, and T. Meyer. 1998. Green fluorescent protein (GFP)-tagged cystein-rich domains from protein kinase C as a fluorescent indicators for diacylglycerol signaling in living cells. *J. Cell Biol.* 140:485–98.

Ogasawara, M., S.C. Kim, R. Adamik, A. Togawa, V.J. Ferrans, K. Takeda, M. Kirby, J. Moss, and M. Vaughan. 2000. Similarities in function and gene structure of cytohesin-4 and cytohesin-1, guanine nucleotide-exchange proteins for ADP-ribosylation factors. *J. Biol. Chem.* 275:3221–30.

Ozaki, S., D.B. DeWald, J.C. Shope, J. Chen, and G.D. Prestwich. 2000. Intracellular delivery of phosphoinositides and inositol phosphates using polyamine carriers. *Proc. Natl. Acad. Sci. U.S.A.* 97:11286–91.

Pendaries, C., H. Tronchere, L. Arbibe, J. Mounier, O. Gozani, L. Cantley, M.J. Fry, F. Gaits-Iacovoni, P.J. Sansonetti, and B. Payrastre. 2006. PtdIns5*P* activates the host cell PI3-kinase/Akt pathway during *Shigella flexneri* infection. *EMBO J.* 25:1024–34.

Pirruccello, M., and P. De Camilli. 2012. Inositol 5-phosphatases: Insights from the Lowe syndrome protein OCRL. *Trends Biochem. Sci.* 37:134–43.

Pittet, D., W. Schlegel, D.P. Lew, A. Monod, and G.W. Mayr. 1989. Mass changes in inositol tetrakis- and pentakisphosphate isomers induced by chemotactic peptide stimulation in HL-60 cells. *J. Biol. Chem.* 264:18489–93.

Prasad, A., Y. Jia, A. Chakraborty, Y. Li, S.K. Jain, J. Zhong, S.G. Roy, F. Loison, S. Mondal, J. Sakai, C. Blanchard, S.H. Snyder, and H.R. Luo. 2011. Inositol hexakisphosphate kinase 1 regulates neutrophil function in innate immunity by inhibiting phosphatidylinositol-(3,4,5)-trisphosphate signaling. *Nat. Immunol.* 12:752–60.

Quinn, K.V., P. Behe, and A. Tinker. 2008. Monitoring changes in membrane phosphatidylinositol 4,5-bisphosphate in living cells using a domain from the transcription factor tubby. *J. Physiol.* 586:2855–71.

Ragaz, C., H. Pietsch, S. Urwyler, A. Tiaden, S.S. Weber, and H. Hilbi. 2008. The *Legionella pneumophila* phosphatidylinositol-4 phosphate-binding type IV substrate SidC recruits endoplasmic reticulum vesicles to a replication-permissive vacuole. *Cell. Microbiol.* 10:2416–33.

Rameh, L.E., A. Arvidsson, K.L. Carraway, III, A.D. Couvillon, G. Rathbun, A. Crompton, B. VanRentherghem, M.P. Czech, K.S. Ravichandran, S.J. Burakoff, D.S. Wang, C.S. Chen, and L.C. Cantley. 1997a. A comparative analysis of the phosphoinositide binding specificity of plekstrin homology domains. *J. Biol. Chem.* 272:22059–66.

Rameh, L.E., K.F. Tolias, B.C. Duckworth, and L.C. Cantley. 1997b. A new pathway for synthesis of phosphatidylinositol-4,5-bisphosphate. *Nature* 390:192–6.

Ratzan, W.J., A.V. Evsikov, Y. Okamura, and L.A. Jaffe. 2011. Voltage sensitive phosphoinositide phosphatases of *Xenopus*: Their tissue distribution and voltage dependence. *J. Cell Physiol.* 226:2740–46.

Raucher, D., T. Stauffer, W. Chen, K. Shen, S. Guo, J.D. York, M.P. Sheetz, and T. Meyer. 2000. Phosphatidylinositol 4,5-bisphosphate functions as a second messenger that regulates cytoskeleton-plasma membrane adhesion. *Cell* 100:221–8.

Rizzo, M.A., K. Shome, S.C. Watkins, and G. Romero. 2000. The recruitment of Raf-1 to membranes is mediated by direct interaction with phosphatidic acid and is independent of association with Ras. *J. Biol. Chem.* 275:23911–8.

Roy, A., and T.P. Levine. 2004. Multiple pools of phosphatidylinositol 4-phosphate detected using the pleckstrin homology domain of Osh2p. *J. Biol. Chem.* 279:44683–9.

Salim, K., M.J. Bottomley, E. Querfurth, M.J. Zvelebil, I. Gout, R. Scaife, R.L. Margolis, R. Gigg, C.I.E. Smith, P.C. Driscoll, M.D. Waterfield, and G. Panayotou. 1996. Distinct specificity in the recognition of phosphoinositides by the plekstrin homology domains of dynamin and Bruton's tyrosine kinase. *EMBO J.* 15:6241–50.

Santagata, S., T.J. Boggon, C.L. Baird, C.A. Gomez, J. Zhao, W.S. Shan, D.G. Myszka, and L. Shapiro. 2001. G-protein signaling through tubby proteins. *Science* 292:2041–50.

Sato, M., Y. Ueda, and Y. Umezawa. 2006. Imaging diacylglycerol dynamics at organelle membranes. *Nat. Methods* 3:797–9.

Schneider, I.C., and J.M. Haugh. 2004. Spatial analysis of 3′phosphoinositide signaling in living fibroblasts: II. Parameter estimates for individual cells from experiments. *Biophys. J.* 86:599–608.

Schoebel, S., W. Blankenfeldt, R.S. Goody, and A. Itzen. 2010. High-affinity binding of phosphatidylinositol 4-phosphate by *Legionella pneumophila* DrrA. *EMBO Rep.* 11:598–604.

Schu, P.V., K. Takegawa, M.J. Fry, J.H. Stack, M.D. Waterfield, and S.D. Emr. 1993. Phosphatidylinositol 3-kinase encoded by yeast VPS34 gene essential for protein sorting. *Science* 260:88–91.

Schultz, A., M. Ling, and C. Larsson. 2004. Identification of an amino acid residue in the protein kinase C C1b domain crucial for its localization to the Golgi network. *J. Biol. Chem.* 279:31750–60.

Sekar, R.B., and A. Periasamy. 2003. Fluorescence resonance energy transfer (FRET) microscopy imaging of live cell protein localizations. *J. Cell Biol.* 160:629–33.

Servant, G., O.D. Weiner, P. Herzmark, T. Balla, J.W. Sedat, and H.R. Bourne. 2000. Polarization of chemoattractant receptor signaling during neutrophil chemotaxis. *Science* 287:1037–40.

Shaner, N.C., P.A. Steinbach, and R.Y. Tsien. 2005. A guide to choosing fluorescent proteins. *Nat. Methods* 2:905–9.

Sharkey, N.A., K.L. Leach, and P.M. Blumberg. 1984. Competitive inhibition by diacylglycerol of specific phorbol ester binding. *Proc. Natl. Acad. Sci. U.S.A.* 81:607–10.

Shen, J., W.M. Yu, M. Brotto, J.A. Scherman, C. Guo, C. Stoddard, T.M. Nosek, H.H. Valdivia, and C.K. Qu. 2009. Deficiency of MIP/MTMR14 phosphatase induces a muscle disorder by disrupting Ca($^{2+}$) homeostasis. *Nat. Cell Biol.* 11:769–76.

Simonsen, A., R. Lippe, S. Christoforidis, J.M. Gaullier, A. Brech, J. Callaghan, B.H. Toh, C. Murphy, M. Zerial, and H. Stenmark. 1998. EEA1 links PI(3)K function to Rab5 regulation of endosome fusion. *Nature* 394:494–8.

Stahelin, R.V., A. Burian, K.S. Bruzik, D. Murray, and W. Cho. 2003. Membrane binding mechanisms of the PX domains of NADPH oxidase p40phox and p47phox. *J. Biol. Chem.* 278:14469–79.

Stahelin, R.V., M.A. Digman, M. Medkova, B. Ananthanarayanan, H.R. Melowic, J.D. Rafter, and W. Cho. 2005. Diacylglycerol-induced membrane targeting and activation of protein kinase Cepsilon: Mechanistic differences between protein kinases Cdelta and Cepsilon. *J. Biol. Chem.* 280:19784–93.

Stauffer, T.P., S. Ahn, and T. Meyer. 1998. Receptor-induced transient reduction in plasma membrane PtdIns(4,5)P_2 concentration monitored in living cells. *Curr. Biol.* 8:343–6.

Stepanenko, O.V., D.M. Shcherbakova, I.M. Kuznetsova, K.K. Turoverov, and V.V. Verkhusha. 2011. Modern fluorescent proteins: From chromophore formation to novel intracellular applications. *Biotechniques* 51:313–4, 316, 318 passim.

Suh, B.C., T. Inoue, T. Meyer, and B. Hille. 2006. Rapid chemically induced changes of PtdIns(4,5)P_2 gate KCNQ ion channels. *Science* 314:1454–7.

Szentpetery, Z., A. Balla, Y.J. Kim, M.A. Lemmon, and T. Balla. 2009. Live cell imaging with protein domains capable of recognizing phosphatidylinositol 4,5-bisphosphate: A comparative study. *BMC Cell Biol.* 10:67.

Szentpetery, Z., P. Varnai, and T. Balla. 2010. Acute manipulation of Golgi phosphoinositides to assess their importance in cellular trafficking and signaling. *Proc. Natl. Acad. Sci. U.S.A.* 107:8225–30.

Tengholm, A., M.N. Teruel, and T. Meyer. 2003. Single cell imaging of PI3K activity and glucose transporter insertion into the plasma membrane by dual color evanescent wave microscopy. *Sci. STKE* 2003:PL4.

Terrillon, S., and M. Bouvier. 2004. Receptor activity-independent recruitment of beta-arrestin2 reveals specific signalling modes. *EMBO J.* 23:3950–61.

Thaler, C., S.V. Koushik, P.S. Blank, and S.S. Vogel. 2005. Quantitative multiphoton spectral imaging and its use for measuring resonance energy transfer. *Biophys. J.* 89:2736–49.

Thomas, C.C., S. Dowler, M. Deak, D.R. Alessi, and D.M. van Aalten. 2001. Crystal structure of the phosphatidylinositol 3,4-bisphosphate-binding pleckstrin homology (PH) domain of tandem PH-domain-containing protein 1 (TAPP1): Molecular basis of lipid specificity. *Biochem. J.* 358:287–94.

Toth, D.J., J. Toth, G. Gulyas, A. Balla, T. Balla, L. Hunyady, and P. Varnai. 2012. Acute depletion of plasma membrane phosphatidylinositol 4,5-bisphosphate impairs specific steps in G protein-coupled receptor endocytosis. *J. Cell Sci.* 125(Pt 9): 2185–97.

van Der Wal, J., R. Habets, P. Varnai, T. Balla, and K. Jalink. 2001. Monitoring phospholipase C activation kinetics in live cells by FRET. *J. Biol. Chem.* 276:15337–44.

van Munster, E.B., and T.W. Gadella. 2005. Fluorescence lifetime imaging microscopy (FLIM). *Adv. Biochem. Eng. Biotechnol.* 95:143–75.

van Rheenen, J., E.M. Achame, H. Janssen, J. Calafat, and K. Jalink. 2005. PIP2 signaling in lipid domains: A critical re-evaluation. *EMBO J.* 24:1664–73.

van Rheenen, J., M. Langeslag, and K. Jalink. 2004. Correcting confocal acquisition to optimize imaging of fluorescence resonance energy transfer by sensitized emission. *Biophys. J.* 86:2517–29.

Várnai, P., and T. Balla. 1998. Visualization of phosphoinositides that bind pleckstrin homology domains: Calcium- and agonist-induced dynamic changes and relationship to myo-[^{3}H]inositol-labeled phosphoinositide pools. *J. Cell Biol.* 143:501–10.

Varnai, P., and T. Balla. 2006. Live cell imaging of phosphoinositide dynamics with fluorescent protein domains. *Biochim. Biophys. Acta* 1761:957–67.

Varnai, P., and T. Balla. 2007. Visualization and manipulation of phosphoinositide dynamics in live cells using engineered protein domains. *Pflugers Arch.* 455:69–82.

Varnai, P., T. Bondeva, P. Tamas, B. Toth, L. Buday, L. Hunyady, and T. Balla. 2005. Selective cellular effects of overexpressed pleckstrin-homology domains that recognize PtdIns(3,4,5)P_3 suggest their interaction with protein binding partners. *J. Cell Sci.* 118:4879–88.

Varnai, P., K.I. Rother, and T. Balla. 1999. Phosphatidylinositol 3-kinase-dependent membrane association of the Bruton's tyrosine kinase pleckstrin homology domain visualized in single living cells. *J. Biol. Chem.* 274:10983–9.

Varnai, P., B. Thyagarajan, T. Rohacs, and T. Balla. 2006. Rapidly inducible changes in phosphatidylinositol 4,5-bisphosphate levels influence multiple regulatory functions of the lipid in intact living cells. *J. Cell Biol.* 175:377–82.

Venkateswarlu, K., F. Gunn-Moore, P.B. Oatey, J.M. Tavare, and P.J. Cullen. 1998a. Nerve growth factor- and epidermal growth factor-stimulated translocation of the ADP-ribosylation factor-exchange factor GRP1 to the plasma membrane of PC12 cells requires activation of phosphatidylinositol 3-kinase and the GRP1 pleckstrin homology domain. *Biochem. J.* 335(Pt 1):139–46.

Venkateswarlu, K., F. Gunn-Moore, J.M. Tavare, and P.J. Cullen. 1999. EGF-and NGF-stimulated translocation of cytohesin-1 to the plasma membrane of PC12 cells requires PI 3-kinase activation and a functional cytohesin-1 PH domain. *J. Cell Sci.* 112:1957–65.

Venkateswarlu, K., P.B. Oatey, J.M. Tavare, and P.J. Cullen. 1998b. Insulin-dependent translocation of ARNO to the plasma membrane of adipocytes requires phosphatidylinositol 3-kinase. *Curr. Biol.* 8:463–6.

Wallrabe, H., and A. Periasamy. 2005. Imaging protein molecules using FRET and FLIM microscopy. *Curr. Opin. Biotechnol.* 16:19–27.

Wang, J., A. Gambhir, G. Hangyas-Mihalyne, D. Murray, U. Golebiewska, and S. McLaughlin. 2002. Lateral sequestration of phosphatidylinositol 4,5-bisphosphate by the basic effector domain of myristoylated alanine-rich C kinase substrate is due to non-specific electrostatic interactions. *J. Biol. Chem.* 277:34401–12.

Wang, X., S.P. Devaiah, W. Zhang, and R. Welti. 2006. Signaling functions of phosphatidic acid. *Prog. Lipid Res.* 45:250–78.

Wang, Y.H., A. Collins, L. Guo, K.B. Smith-Dupont, F. Gai, T. Svitkina, and P.A. Janmey. 2012. Divalent cation-induced cluster formation by polyphosphoinositides in model membranes. *J. Am. Chem. Soc.* 134:3387–95.

Watt, S.A., W.A. Kimber, I.N. Fleming, N.R. Leslie, C.P. Downes, and J.M. Lucocq. 2004. Detection of novel intracellular agonist responsive pools of phosphatidylinositol 3,4-bisphosphate using the TAPP1 pleckstrin homology domain in immunoelectron microscopy. *Biochem. J.* 377:653–63.

Watt, S.A., G. Kular, I.N. Fleming, C.P. Downes, and J.M. Lucocq. 2002. Subcellular localization of phosphatidylinositol 4,5-bisphosphate using the pleckstrin homology domain of phospholipase C delta1. *Biochem. J.* 363:657–66.

Watton, J., and J. Downward. 1999. Akt/PKB localisation and 3′ phosphoinositide generation at sites of epithelial cell-matrix and cell-cell interaction. *Curr. Biol.* 9:433–6.

Weber, S.S., C. Ragaz, K. Reus, Y. Nyfeler, and H. Hilbi. 2006. *Legionella pneumophila* exploits PI(4)P to anchor secreted effector proteins to the replicative vacuole. *PLoS Pathog.* 2:e46.

Weixel, K.M., A. Blumental-Perry, S.C. Watkins, M. Aridor, and O.A. Weisz. 2005. Distinct Golgi populations of phosphatidylinositol 4-phosphate regulated by phosphatidylinositol 4-kinases. *J. Biol. Chem.* 280:10501–8.

Williamson, P., and R.A. Schlegel. 2002. Transbilayer phospholipid movement and the clearance of apoptotic cells. *Biochim. Biophys. Acta* 1585:53–63.

Wong, K.K., J.A. Engelman, and L.C. Cantley. 2010. Targeting the PI3K signaling pathway in cancer. *Curr. Opin. Genet. Dev.* 20:87–90.

Xie, J., B. Sun, J. Du, W. Yang, H.C. Chen, J.D. Overton, L.W. Runnels, and L. Yue. 2011. Phosphatidylinositol 4,5-bisphosphate (PIP(2)) controls magnesium gatekeeper TRPM6 activity. *Sci. Rep.* 1:146.

Yaradanakul, A., and D.W. Hilgemann. 2007. Unrestricted diffusion of exogenous and endogenous PIP(2)in baby hamster kidney and Chinese hamster ovary cell plasmalemma. *J. Membr. Biol.* 220:53–67.

Yeung, T., G.E. Gilbert, J. Shi, J. Silvius, A. Kapus, and S. Grinstein. 2008. Membrane phosphatidylserine regulates surface charge and protein localization. *Science* 319:210–3.

Yu, J.W., J.M. Mendrola, A. Audhya, S. Singh, D. Keleti, D.B. DeWald, D. Murray, S.D. Emr, and M.A. Lemmon. 2004. Genome-wide analysis of membrane targeting by S. cerevisiae pleckstrin homology domains. *Mol. Cell.* 13:677–88.

Yuan, T.L., and L.C. Cantley. 2008. PI3K pathway alterations in cancer: Variations on a theme. *Oncogene* 27:5497–510.

Yudin, Y., V. Lukacs, C. Cao, and T. Rohacs. 2011. Decrease in phosphatidylinositol 4,5-bisphosphate levels mediates desensitization of the cold sensor TRPM8 channels. *J. Physiol.* 589:6007–27.

Zeniou-Meyer, M., N. Zabari, U. Ashery, S. Chasserot-Golaz, A.M. Haeberle, V. Demais, Y. Bailly, I. Gottfried, H. Nakanishi, A.M. Neiman, G. Du, M.A. Frohman, M.F. Bader, and N. Vitale. 2007. Phospholipase D1 production of phosphatidic acid at the plasma membrane promotes exocytosis of large dense-core granules at a late stage. *J. Biol. Chem.* 282:21746–57.

Zhan, Y., J.V. Virbasius, X. Song, D.P. Pomerleau, and G.W. Zhou. 2002. The p40phox and p47phox PX domains of NADPH oxydase target cell membranes via direct and indirect recruitment by phosphoinositides. *J. Biol. Chem.* 277:4512–8.

Zhang, F., Z. Wang, M. Lu, Y. Yonekubo, X. Liang, Y. Zhang, P. Wu, Y. Zhou, S. Grinstein, J.F. Hancock, and G. Du. 2014. Temporal production of the signaling lipid phosphatidic acid by phospholipase D2 determines the output of extracellular signal-regulated kinase signaling in cancer cells. *Mol. Cell. Biol.* 34:84–95.

Zhang, G., M.G. Kazainetz, P.M. Blumberg, and J.H. Hurley. 1995. Crystal structure of the cys2 activator-binding domain of protein kinase Cdelta in complex with phorbol ester. *Cell* 81:917–24.

Zhang, L., Y.S. Mao, P.A. Janmey, and H.L. Yin. 2012. Phosphatidylinositol 4,5 bisphosphate and the actin cytoskeleton. *Subcell. Biochem.* 59:177–215.

Zhu, Y., L. Hu, Y. Zhou, Q. Yao, L. Liu, and F. Shao. 2010. Structural mechanism of host Rab1 activation by the bifunctional *Legionella* type IV effector SidM/DrrA. *Proc. Natl. Acad. Sci. U.S.A.* 107:4699–704.

Zoncu, R., R.M. Perera, R. Sebastian, F. Nakatsu, H. Chen, T. Balla, G. Ayala, D. Toomre, and P.V. De Camilli. 2007. Loss of endocytic clathrin-coated pits upon acute depletion of phosphatidylinositol 4,5-bisphosphate. *Proc. Natl. Acad. Sci. U.S.A.* 104:3793–8.

Zou, J., J. Marjanovic, M.V. Kisseleva, M. Wilson, and P.W. Majerus. 2007. Type I phosphatidylinositol-4,5-bisphosphate 4-phosphatase regulates stress-induced apoptosis. *Proc. Natl. Acad. Sci. U.S.A.* 104:16834–9.

Zwaal, R.F., P. Comfurius, and E.M. Bevers. 1998. Lipid-protein interactions in blood coagulation. *Biochim. Biophys. Acta* 1376:433–53.

Chapter 5

Biosensors of Small GTPase Proteins for Use in Living Cells and Animals

Ellen C. O'Shaughnessy, Jason J. Yi, and Klaus M. Hahn

CONTENTS

5.1 SMALL GTPase BIOSENSOR DESIGN

Low-molecular-weight guanosine triphosphatase proteins (small GTPases, a form of G-protein distinct from heterotrimeric G-proteins) were among the first molecules targeted for study with biosensors. Their long and interesting history illustrates the evolution of biosensor design, principles, and approaches. Although there are some exceptions, GTPase proteins almost always exist in one of two conformations—an inactive conformation bound to GDP and an active conformation bound to GTP. It is only in this active conformation that GTPases can interact productively with their downstream effector proteins (Figure 5.1). In general, the activation of GTPases is regulated by three classes of upstream proteins: guanine nucleotide exchange factors (GEFs), which mediate the binding of GTP; GTPase activating proteins (GAPs), which accelerate the hydrolysis of GTP to GDP; and guanine nucleotide dissociation inhibitors (GDI), which bind the GDP-bound form of the proteins (Bar-Sagi and Hall 2000; Reuther and Der 2000; Takai et al. 2001; Jaffe and Hall 2005). The goal of most biosensors is to track the transient localization and formation of the GTP-bound, "activated" conformation of the GTPase in living cells and animals. For example, the cycling between active and inactive nucleotide states is tightly coupled to changes in membrane localization for many GTPases. In other cases, regulatory mechanisms such as phosphorylation and degradation can influence the activity and localization of the protein. In designing a biosensor, one strives for maximum sensitivity by producing the brightest possible sensor, and for some designs, the greatest possible difference between the fluorescence

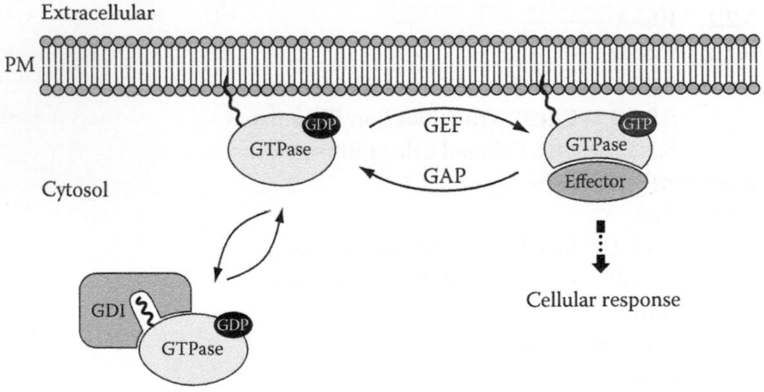

Figure 5.1 The Rho GTPase regulatory cycle. Rho GTPases are molecular switches that cycle between GDP-bound (inactive) and GTP-bound (active) states. Guanine dissociation inhibitors (GDIs) sequester GDP-bound GTPases in the cytoplasm. At the membrane (potentially still associated with GDIs), GTPases are activated by guanine nucleotide exchange factors (GEFs), which promote the exchange of GDP for GTP. Active GTPases interact with multiple effector proteins that govern a wide range of cell responses and behaviors. GEF activity is opposed by GTPase-activating proteins (GAPs), which increase the rate of GTP hydrolysis, thereby returning GTPases to their GDP-bound state.

of the active and inactive conformations. The goal is to perturb the GTPase of interest as little as possible, but it is difficult to modify a GTPase or sense its activity without in some way interfering with biologically important interactions. Most importantly, one must not alter the upstream regulatory interactions that control activation, or the downstream interactions that generate subcellular localization. As choices between different perturbations are often the only viable course, a biosensor may report only a subset of signaling events in a cell, but valuable biological information can be obtained if the limitations of the biosensor are understood. An enduring misconception in the field has been the view that biosensors act simply as "activity stains" akin to antibodies, without regard to the limitations of each design. In this chapter, we discuss the development of small GTPase biosensors, highlighting the utility and limitations of existing sensors.

The design of the first GTPase biosensor, Rac1 FLARE (Kraynov et al. 2000), was based on previously described biochemical GTPase activation assays developed to quantify GTPase activation in cell lysates. In these studies, a fragment of a downstream effector protein that binds only the active form of the GTPase (an "affinity reagent") was used to pull down the GTPases Rac1 or Cdc42 (Benard et al. 1999). Western blotting was then used to gauge the quantity of pulled-down, active GTPase. This affinity-domain paradigm was adapted to produce a biosensor by using Förster/fluorescence resonance energy transfer (FRET) to visualize the binding of the affinity reagent to the GTPase (Kraynov et al. 2000). An effector fragment from p21-activated kinase (Pak) covalently labeled with a fluorescent dye produced FRET when it bound to the active (GTP-loaded) conformation of green fluorescent protein (GFP)-tagged Rac1.

The development of fluorescent proteins (FPs) with different excitation/emission wavelengths capable of undergoing FRET made it possible to generate GTPase biosensors that were genetically encoded, an approach first applied to Ras and Rap1 (Mochizuki et al. 2001). Importantly, the use of genetically encoded components made it possible to link the affinity reagent and the GTPase in a single-chain, simplifying expression and ratio imaging. Genetically encoded versions of both single- and dual-chain designs have since been developed for multiple different GTPases (Figure 5.2). These two design types have important differences that impact the optimization and design of new biosensors, as well as the tailoring of biosensors to specific biological applications. The single-chain designs have predominated because they are simpler to apply using widefield ratio imaging, but the advent of new modes of microscopy (e.g., fluorescence lifetime imaging) for the quantitation of biosensor activity and the greater sensitivity of the dual-chain biosensors are leading to their reemergence (Hinde et al. 2011, 2013). In brief, single-chain biosensors are easier to use for ratiometric imaging, because the GTPase and affinity reagents are

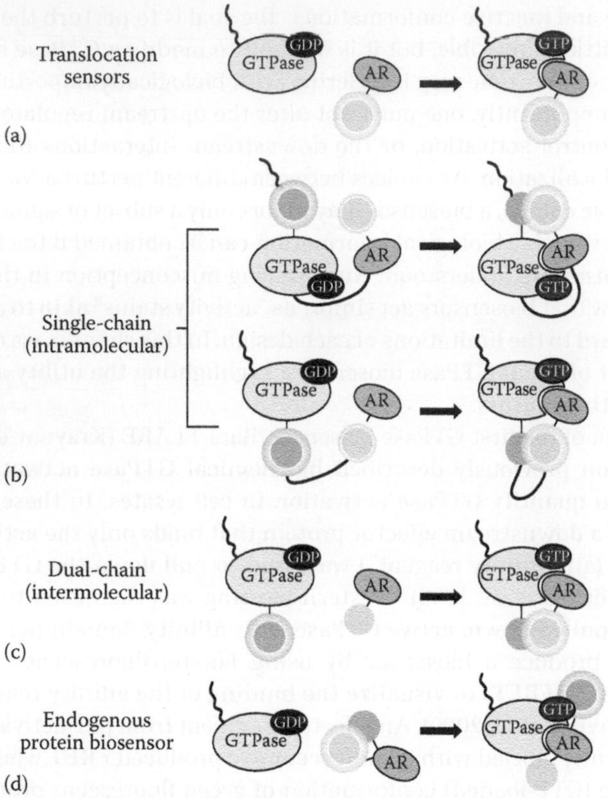

Figure 5.2 Genetically encoded biosensor designs. (a) A protein fragment that selectively binds to the active form of the GTPase (the affinity reagent, AR) is attached to a fluorescent protein. It translocates to the site of GTPase activation, revealing the localization and kinetics of activation. (b) Single-chain intramolecular biosensors. The GTPase, an AR, and fluorescent proteins capable of undergoing FRET are combined in a single chain. When the GTPase is activated, the AR binds the GTPase, affecting the separation and orientation of the fluorescent proteins and thereby affecting FRET. In some cases, the fluorescent protein is attached to the C-terminus (top). Because this destroys the motifs needed for membrane localization, a lipid is attached to the C-terminal fluorescent protein, leading to constitutive membrane localization. In other cases, the fluorescent proteins are inserted in the middle of the chain, leaving the C-terminus of the GTPase unaltered and free for interaction with the membrane and with proteins that regulate reversible membrane localization. (c) Dual-chain intermolecular biosensors. Here, the AR and GTPase are each tagged with a fluorescent protein and are expressed as separate chains. Activation of the GTPase leads to intermolecular AR binding and FRET. (d) Single-chain biosensors directed against endogenous targets. Here, a pair of fluorescent proteins that can undergo FRET is attached to the AR. Their relative positions are affected on AR-GTPase binding, leading to a change in FRET.

distributed identically and therefore do not require correction for spectral bleed-through. However, the dual-chain designs can be more sensitive because there is no residual FRET in the "off state," and bleaching does not affect quantitation in fluorescence lifetime imaging microscopy (FLIM) as it does for single-chain designs. Furthermore, dual-chain designs can be easier to make because they do not require the difficult optimization of multiple linkers and protein orientations (Baird et al. 1999). Unfortunately, dual-chain biosensors have a tendency to produce more heterogeneous data because of variations in the relative expression levels of the two chains. Dual- and single-chain biosensors can introduce distinct biological perturbations, and they differ in their propensity to produce either false negatives or false positives. Single-chain designs have a greater tendency to act as dominant negative protein analogs because, on activation, the intramolecular interaction of the affinity reagent outcompetes interactions with native effector molecules. In dual-chain designs, the affinity reagent can be competed away by endogenous effectors, leading to "false negatives." It is essential to carry out appropriate controls, including titrating the intracellular expression of the biosensor against the perturbation of cellular behavior. Past studies have shown that it is possible to use "tracer amounts" of biosensor that report activation through reversible interactions with upstream regulators while not unduly perturbing the cell behavior under study (Kraynov et al. 2000; Nalbant et al. 2004). For tissue culture cells, cell brightness per unit cell area or volume has proven to be a simple measure that is proportional to biosensor concentration. Fluorescence-activated cell sorting (FACS) can be used to isolate cell populations of the appropriate brightness. One must be careful of the fact that different fluorescent proteins show different levels of misfolding and/or degradation.

It is also important to realize that GTPase biosensor designs can affect the membrane interactions of the GTPase, an important aspect of biological regulation. In some designs, the GTPase is constitutively anchored to the membrane (Mochizuki et al. 2001; Itoh et al. 2002; Yoshizaki 2003), while in other designs the sequences required for regulation of reversible membrane interaction remain intact (Figure 5.2) (Jiang and Sorkin 2002; Pertz et al. 2006; Machacek et al. 2009). Biosensors that are constitutively anchored to the membrane have been said to indicate GEF activity rather than overall GTPase activity. Some biosensors simply eliminate membrane localization altogether, a design that is difficult to interpret biologically (Kardash et al. 2010). Clearly, more systematic studies of these effects would be valuable, as it is important to identify precisely which specific subsets of GTPase interactions are being reported by a given design.

Many GTPase biosensors are not based on FRET. So-called "translocation biosensors" monitor the change in the localization of a fluorescently tagged GTPase, or a fluorescent effector domain, on GTPase activation. GTPases often become concentrated in specific subcellular regions upon activation, such as the cell rim during wound healing (Benink and Bement 2005), or at the bottom membrane of the cell where they can be quantified by total internal reflection fluorescence (TIRF) microscopy (Navarro-Lerita et al. 2011; Sato et al. 2012). The signal from translocation biosensors is typically much brighter than that from FRET biosensors, but changes in localization must be pronounced enough to discern over a background of diffusely localized biosensors that is fluorescing at the same wavelengths. In contrast, the sensitivity of FRET is enhanced by the unique spectral signature produced on activation. Translocation biosensors are easy to use and amenable to high-content screening, but it is difficult to use them to discern or quantify subtle activation events such as gradients of activity.

To minimize cell perturbation, it is advantageous to use designs that report the activity of endogenous proteins. The translocation of fluorescently tagged affinity reagents has been used (Table 5.1), and in other cases, sensitivity has been enhanced by covalently modifying the affinity reagent with bright, environmentally sensitive dyes that change their fluorescence intensity or wavelengths when they bind the activated conformation of an endogenous GTPase (Figure 5.3). This technique has been used successfully to study the activation of endogenous Cdc42 and other non-GTPase proteins (Nalbant et al. 2004; Loving et al. 2010; Gulyani et al. 2011). The amount of biosensor needed for visualization is substantially lower than that needed for FRET, because the dyes can be intrinsically brighter than FPs and are directly excited. However, the dye-labeled proteins must be microinjected, electroporated, and so forth, rather than simply being expressed. A final approach to report the activity of endogenous GTPases has been to fuse two FPs to the affinity reagent. The fluorophores are positioned such that FRET is altered on binding of the affinity reagent to the activated GTPase target.

Although the focus of this chapter is on live-cell imaging, we briefly note that several GTPase biosensors have been developed for *in vitro* applications. Environment-sensing dyes attached directly to GTPases report conformational changes, and fluorescently labeled nucleotides have been used *in vitro* to study the kinetics of GTPase activation/deactivation, as well as interactions with various effectors (Nomanbhoy et al. 1996; Nomanbhoy and Cerione 1999; Goguen et al. 2011).

In the following sections, we have organized the small GTPase biosensors by target molecule, hoping that this will be useful to those interested in finding available biosensors for the target(s) they are studying. We discuss the different design strategies employed and provide what we hope is a comprehensive list of published GTPase biosensors in Table 5.1. We sincerely apologize to our colleagues for the inevitable and unintentional omission of some biosensors.

TABLE 5.1 PUBLISHED GTPASE BIOSENSORS

GTPase	Design	References
Rho-Family		
RhoA	Single-chain	Yoshizaki (2003); Pertz et al. (2006)
	Single-chain, endogenous target	Yoshizaki (2003)
	Dual-chain	Murakoshi et al. (2011)
	Translocation sensor	Benink and Bement (2005)
Rac1	Dual-chain, dye-based	Kraynov et al. (2000); Del Pozo et al. (2002)
	Dual-chain, genetically encoded	Machacek et al. (2009)
	Single-chain	Itoh et al. (2002)
	Single-chain, endogenous target	Graham et al. (2001)
Cdc42	Single-chain	Itoh et al. (2002); Seth et al. (2003); Kamiyama and Chiba (2009)
	Dual-chain	Kamiyama and Chiba (2009); Murakoshi et al. (2011)
	Dye-conjugated affinity reagent	Nalbant et al. (2004); Goguen et al. (2011)
	Single-chain, endogenous target	Seth et al. (2003); Lorenz et al. (2004); Graham et al. (2001)
	Translocation sensor	Benink and Bement (2005); Kim et al. (2000); Kumfer et al. (2010)
	Direct dye labeling	Nomanbhoy et al. (1996); Nomanbhoy and Cerione (1999); Goguen et al. (2011)
RhoC	Single-chain	Bravo-Cordero et al. (2011)
	Dual-chain	Zhong et al. (2007)
TC10	Single-chain	Kawase et al. (2006)
	Dual-chain	Pommereit and Wouters (2007)
Ras-Family		
Ras (H, K, N)	Single-chain	Mochizuki et al. (2001)
	Dual-chain	Jiang and Sorkin (2002); Hibino et al. (2003); Yasuda et al. (2006)
	Translocation sensor	Bondeva et al. (2002); Chiu et al. (2002) Caloca et al. (2003); Augsten et al. (2006)
	Dye-conjugated affinity reagent	Murakoshi et al. (2004)

(*Continued*)

TABLE 5.1 (CONTINUED) PUBLISHED GTPASE BIOSENSORS

GTPase	Design	References
	Ras-Family	
R-Ras	Single-chain	Takaya et al. (2007)
Rap1	Single-chain	Mochizuki et al. (2001)
	Translocation sensor	Bivona et al. (2004)
RalA/B	Single-chain	Takaya et al. (2004)
	Rab/Ran-Family	
Ran	Single-chain, endogenous target	Kaláb et al. (2002, 2006)
	Single-chain	Kaláb et al. (2002)
	Dual-chain	Plafker (2002)
Rab5	Single-chain	Kitano et al. (2008)
	Dual-chain	Galperin (2003)
Rab6	Dual-chain	Thyrock et al. (2010)
Rab35	Single-chain	Ishido et al. (2011)
Rab10	Dual-chain	Chen et al. (2009)
	Sar1/Arf and Rheb-Families	
Arf6	Dual-chain	Hall et al. (2008)
Rheb	Dual-chain	Li et al. (2007)

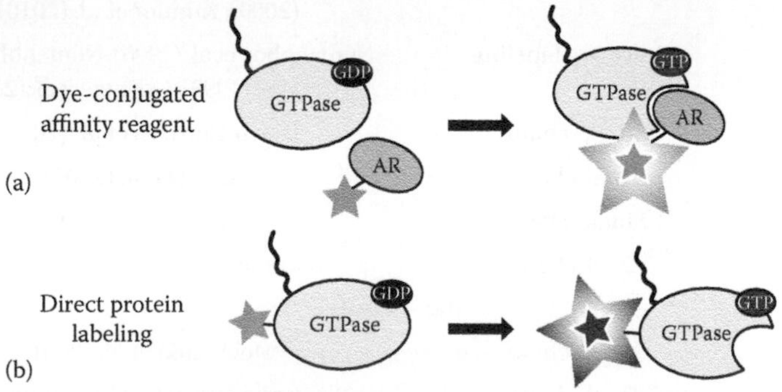

Figure 5.3 GTPase biosensors based on solvatochromic dyes. (a) The AR is covalently labeled with a dye whose fluorescence is affected by the environment. When the AR binds the GTPase, the dye encounters a more hydrophobic environment and/or undergoes interactions with specific amino acids, leading to a change in fluorescence. (b) The GTPase can be directly labeled with an environmentally sensitive fluorescent dye. Changes in conformation affect residues around the dye, altering fluorescence.

5.2 Rho-FAMILY GTPases

The Rho family of GTPases comprises 20 homologous proteins expressed ubiquitously in mammals. They are essential signaling components in a broad range of dynamic cellular events that require precise spatial and temporal control. These processes include cytoskeletal rearrangements, vesicular trafficking, cell migration, and polarization. In this section, we discuss biosensors for the three best characterized members of Rho-family proteins: RhoA, Rac1, and Cdc42.

5.2.1 Rac1

Based on sequence homology, Rac1, Rac2, Rac3, and RhoG form a subgroup within the Rho family of GTPases (Boureux et al. 2006). Rac proteins initiate lamellipodia and membrane ruffling and are also involved in membrane formation during phagocytosis. Like the Rho isoforms, each Rac subfamily member appears to have nonredundant functions. For example, whereas Rac1 and RhoG are widely expressed (Vincent et al. 1992), Rac2 expression is restricted to hematopoietic cells (Didsbury et al. 1989; Shirsat et al. 1990), and Rac3 is abundantly expressed in the brain (Haataja et al. 1997; Bolis et al. 2003; Corbetta et al. 2005). Moreover, whereas Rac1-null mice are embryonic lethal (Sugihara et al. 1998), Rac2-, Rac3-, and RhoG-null mice have no overt developmental abnormalities but present subtle cell-type–specific defects (Vincent et al. 1992; Roberts et al. 1999; Vigorito et al. 2004; Cho et al. 2005; Corbetta et al. 2005). Despite the pleiotropic effects of Rac homologues, all biosensors published in the literature have thus far centered on Rac1, but these designs may well be readily extended to the other homologues.

The first GTPase biosensor for living cells, Rac–FLARE (fluorescence activation indicator for Rho proteins), was based on two components: Rac1 fused to a GFP molecule and the p21-binding domain (PBD) of Pak1, which bound selectively to the activated conformation of the GTPase (Kraynov et al. 2000). The Pak1-binding domain (the affinity reagent) was labeled on a cysteine near its N-terminus with the FRET-acceptor dye Alexa 546. The design preserved normal binding to GDI and regulation of membrane translocation and showed a dynamic range of greater than 400%. Imaging studies with Rac–FLAIR provided our first glimpse of the highly dynamic nature of Rho GTPase signaling inside living cells. Local increases in Rac activity were shown to correlate in time and space with cell protrusion and the production of membrane ruffles, and a broad gradient of Rac activity was demonstrated at the leading edge of motile cells. These previously unattainable observations supported specific models of Rac's role in polarization and were consistent with biochemical and genetic studies indicating a role in protrusion. Genetically encoded versions of this biosensor have since been generated and coupled with novel image analysis approaches for the precise quantitation of Rac dynamics during protrusion and retraction (Machacek et al. 2009).

Rac biosensors took advantage of FPs engineered to undergo FRET, including enhanced blue fluorescent protein (EBFP) and enhanced green fluorescent protein (EGFP). In 2001, a group led by Peter Chalk explored the latitude for preserving FRET in two fused FPs by inserting flexible linkers of increasing length between EGFP and EBFP molecules. Surprisingly, not only were linkers as long as 50 amino acids tolerated, but also the FRET between EBFP and EGFP increased in proportion to linker length (Graham et al. 2001). This suggested that biologically relevant protein sequences could be inserted between two fluorescent molecules to change FRET efficiencies on target protein binding. Such a simple arrangement could be applied as a general design to construct additional biosensors. They tested this hypothesis by inserting the PBD of Pak1 between EGFP and EBFP and employed various *in vitro* assays to assess FRET changes on binding to active Rac1 and Cdc42 (Graham et al. 2001). In experiments with purified proteins, high levels of constitutive FRET in the absence of activated GTPase were reduced by approximately 60% when the PBD bound to active Rac1. The use of the Chalk et al. biosensor in living cells was not reported, likely because of the relatively small amount of FRET produced and because EBFP is relatively dim, photobleaches rapidly, and requires "cytotoxic" excitation wavelengths. Subsequent improvements in FPs greatly aided FRET-based biosensor design by expanding the color palette of FPs and enhancing their spectral properties. Almost all Rho-family biosensors have been based on derivatives of enhanced cyan fluorescent protein (ECFP) and enhanced yellow fluorescent protein (EYFP), including the ECFP variants mCerulean (Rizzo et al. 2004), CyPet (Nguyen and Daugherty 2005), Azurite (Mena et al. 2006), EBFP2 (Pédelacq et al. 2006), and TFP1 (Ai et al. 2006) and the EYFP variants mVenus (Nagai et al. 2002), YPet (Nguyen and Daugherty 2005), and Citrine (Griesbeck et al. 2001).

By harnessing these genetically encoded FRET pairs and incorporating innovations such as a single-chain design, the Matsuda group generated Rho-family biosensors that were substantially more practical than their predecessors. These included iterations of their "Raichu" design scheme, as well as an improved version of the Chalk et al. biosensor. Their first sensor, named Raichu–Rac, included an N-terminal EYFP followed by the PBD of Pak1, a flexible linker, Rac1, CFP, and the C-terminal farnesylation moiety from K-Ras (Itoh et al. 2002). In this arrangement, the biosensor changes conformation when the PBD binds to active Rac1, bringing the two terminal FPs together to produce FRET. This design resulted in an *in vitro* dynamic range of approximately 80%. Their second biosensor, named Raichu–CRIB, employed the design of Chalk et al. but substituted EGFP and EBFP with EYFP and ECFP. Raichu–CRIB could report the activity of both endogenous Rac1 and Cdc42 (Itoh et al. 2002). This biosensor responded to active Cdc42 with a dynamic range of approximately 40% and to Rac1 with approximately 10%. In the Raichu sensor designs, the C-terminus of the GTPase is replaced by ECFP. Because this displaces the C-terminal lipid modifications required for membrane localization, a C-terminal fusion of the farnesyl moiety from K-Ras was added, leading to constitutive membrane localization. The authors reported that these modifications abrogated binding to RhoGDI, eliminating this pathway for downregulating GTPase activity (Itoh et al. 2002). Newer derivatives of these biosensors, in which the terminal FP has been shifted to an internal site, enable native regulation of GTPase localization (Benard et al. 1999; Pertz et al. 2006). Raichu–Rac1 that is constitutively membrane bound likely reports GEF activity at the membrane rather than the GTPase cycle itself. Others have modified the initial Raichu–Rac1 design by removing the K-Ras localization sequence altogether, thereby generating a cytosolic Rac1 sensor (Kardash et al. 2010). Such designs must be approached with caution, as membrane localization is an important determinant of GTPase specificity.

5.2.2 RhoA

RhoA, along with its highly homologous isoforms RhoB and RhoC (85% amino acid sequence identity), is well known for its ability to regulate cell contractility and stress fiber formation when overexpressed in fibroblasts (Wheeler and Ridley 2004). The cellular functions of Rho proteins have been studied extensively with dominant-negative strategies or with clostridial enzyme C3 transferase, which modifies and inhibits all three Rho isoforms. However, recent evidence suggests that individual Rho isoforms have distinct roles in cellular events. For example, dominant-negative RhoA, RhoB, and RhoC each have distinguishable effects on cells (Rondanino et al. 2007), and genetic studies suggest that RhoB-null and RhoC-null mice have distinct functions *in vivo* (Liu et al. 2001; Hakem et al. 2005).

RhoA biosensors have been designed by the Matsuda group based on the Raichu framework (Raichu–RhoA) (Yoshizaki 2003). Several effector domains of RhoA were tested as affinity reagents, along with different modular orientations of RhoA and the affinity reagent to develop a single-chain activity reporter with optimized dynamic range. The best response was achieved using the RhoA-binding domain (RBD) of protein kinase N (PKN), with other components remaining in the order described for Raichu–Rac1 (EYFP–PKN–linker–RhoA–ECFP–farnesyl moiety). This resulted in a constitutively membrane-bound Raichu–RhoA sensor with an *in vitro* dynamic range of 33%. The second sensor was based on the design of Raichu–CRIB, with two modifications: the RBD of Rhotekin was used instead of the CRIB domain to generate a biosensor specific for RhoA (Raichu–RBD), and the K-Ras farnesyl moiety was removed from the C-terminus (Yoshizaki 2003). The resulting Raichu–RBD biosensor produced an *in vitro* dynamic range of approximately 20%. Intriguingly, the response of Raichu–RBD to overexpression of wild-type RhoA was indistinguishable from its response to constitutively active RhoA (RhoA–Q63L).

Our group also created a RhoA biosensor, named RhoA–FLARE (Pertz et al. 2006). This single-chain design consisted of the RBD of Rhotekin, CFP, an unstructured linker of optimized length, YFP, and full-length RhoA (RBD–CFP–linker–YFP–RhoA, with C-terminus intact for interaction with GDI and membrane localization). On RhoA activation, the RBD binds RhoA, bringing the two fluorescent proteins into closer proximity to enhance FRET. The biosensor has a dynamic range of approximately 40%, as assessed in intact HEK293 cells. Similar to Raichu–RBD, wild-type RhoA–FLARE and constitutively active RhoA–FLARE mutants initially yielded similar results. However, the elevated response of wild-type sensors was greatly reduced when the biosensor was coexpressed with RhoGDI, suggesting that overexpression of RhoA biosensors can saturate the ability of endogenous GDI to inactivate the biosensor. This response to GDI concentration has been seen with biosensors that have an intact C-terminus, but is not seen when the C-terminus is altered. These observations illustrate the profound influence that molecular design can impart on biosensor behavior and interpretation. By titrating down the intracellular concentration of the GDI-responsive sensor to a level where reversible membrane localization was observed, we were able to visualize activation events that had previously been obscured, and activity at the membrane was greatly reduced.

A notable difference between biosensors is the composition of their linkers. Detailed studies examining linker length and composition have demonstrated that both factors can have profound effects on the fluorescence properties of FRET-based systems (Graham et al. 2001; Ohashi et al. 2007), but a neglected aspect has been the effect on degradation, which can generate species that affect biology yet produce no fluorescence response. Most designs have used extended polypeptides rich in glycine and serine or threonine residues because they confer enhanced solubility and unrestricted movement of the protein domains. Detailed protein engineering studies examining extended Gly-Ser/Thr linkers in single-chain Fv antibody fragments have found that such linkers are subject to aggregation and proteolytic cleavage (Whitlow et al. 1993). The RhoA–FLARE biosensor employs an optimized linker that has been demonstrated to remain flexible when fused to proteins, resist aggregation, and resist proteolytic cleavage (Whitlow et al. 1993), and may therefore be useful in other biosensors.

5.2.3 Cdc42

Cdc42 is an ancient molecule with a conserved role in regulating cell polarity and the actin cytoskeleton. It is involved in a wide range of eukaryotic cell processes including yeast budding, epithelial polarity, migratory polarity, and cell-fate specification. It was first identified as a cell cycle mutant in *Saccharomyces cerevisiae*, as the loss of Cdc42 prevents budding and mating projection (Adams et al. 1990). It is the most heavily studied Rho-family GTPase in terms of the diversity of biosensors that have been developed to investigate its regulation *in vitro*, in live-cell imaging, and in whole-animal studies.

5.2.3.1 Dye-Based Cdc42 Biosensors

The first biosensors to probe the regulation of Cdc42 and its interactions with effectors *in vitro* were described in a series of papers by Cerione and colleagues in the 1990s. Nomanbhoy et al. (1996) reported that a specific native lysine could be selectively labeled with a fluorescent succinimidyl ester (sNBD) and that the attached dye underwent a change in fluorescence intensity upon GTP loading. A nucleotide-dependent conformational change in Cdc42 likely altered the quenching of dye fluorescence by surrounding residues. The kinetics of GDP/GTP exchange and GTP hydrolysis were not affected by this covalent modification. Although the authors did not pursue live-cell studies, their work paved the way to derivatize Cdc42 with environmentally sensitive dyes designed for *in vivo* imaging. Direct labeling of endogenous proteins with reporter dyes has proven effective for tracing protein activity in vivo (Hahn et al. 1992; Toutchkine et al. 2003; Garrett et al. 2008; Macnevin et al. 2013) and has the advantage of reduced biological perturbation.

In contrast to the direct labeling of Cdc42, our laboratory took an alternate approach to visualize the activation of endogenous, untagged Cdc42. A solvent-sensitive dye with properties optimized for live-cell applications was attached to a fragment of the Cdc42 effector Wiskott Aldrich syndrome protein (WASP) (Nalbant et al. 2004). This fragment bound selectively to the active conformation of Cdc42, leading to a change in the fluorescence of the attached dye that revealed the localization and kinetics of Cdc42 activation. This approach was potentially less perturbing than some others because bright dyes were directly excited, and because there was no need to express exogenous modified Cdc42. This biosensor, termed MeroCBD (merocyanine dye–Cdc42 binding domain) did not interact with homologous Rho-family GTPases such as RhoA and Rac1 but did bind to the very closely related GTPase TC10 (Figure 5.4). The MeroCBD sensor revealed the dynamics of Cdc42 activation at broad protrusions, filopodia, and the Golgi apparatus of motile fibroblasts. Cdc42 was shown to be activated during protrusion in a microtubule-dependent manner, and activity was correlated in time and space with the extension/retraction of the cell edge (Machacek et al. 2009). Although this design enabled the detection of endogenous protein activity with high sensitivity, it had to be introduced into the cell using microinjection.

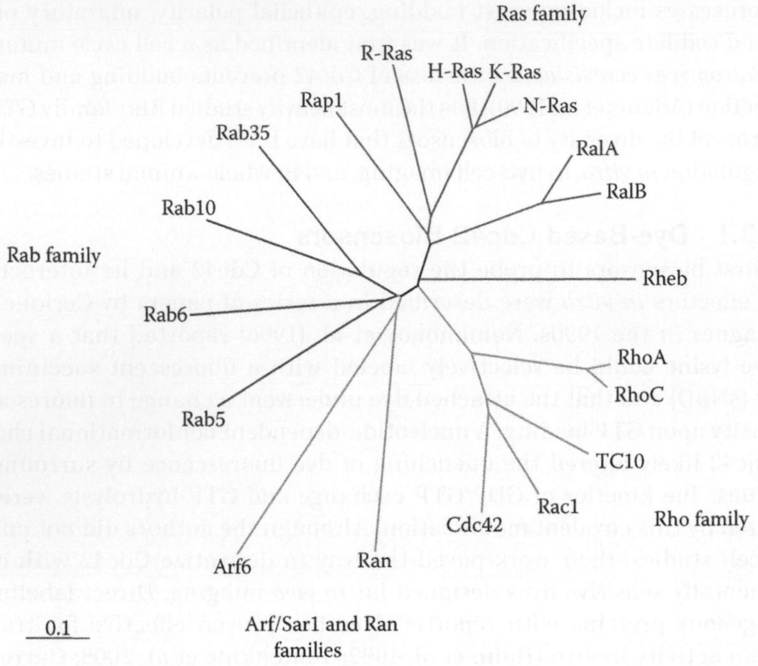

Figure 5.4 Phylogenetic tree of the small monomeric G-protein family. The amino acid sequences of human monomeric GTPases were aligned using the ClustalW program, and an unrooted phylogenetic tree was generated using the TreeView program. The 19 representative members of the family can be classified into four subfamilies. These include the Rab, Ras, Rho, and Arf/Sar1 and Ran subfamilies. Scale bar represents 0.1 amino acid substitutions per site.

5.2.3.2 Cdc42 Translocation Biosensors

In addition to dye-based biosensors, several translocation biosensors have been used to study Cdc42. Kim et al. (2000) studied E-cadherin–dependent Cdc42 activation in MCF-7 epithelial cells by fusing the Cdc42-binding domain of WASP to EGFP and imaging its accumulation at cell-cell junctions. Benink and Bement (2005) used a similar design to co-image activated Cdc42 and RhoA simultaneously in a *Xenopus* wound-healing assay. They fused the RhoA binding domain of Rhotekin to EGFP and the Cdc42 binding domain of N-WASP to monomeric red fluorescent protein (mRFP), thereby demonstrating the formation of spatially discrete rings of RhoA and Cdc42 activity around the wound. These distinct zones of GTPase activity formed rapidly and moved inward as the wounds healed. Through the use of these sensors, they demonstrated that the formation, segregation, and/or movement of these discrete rings were affected by microtubules, the actin cytoskeleton, and crosstalk between RhoA and Cdc42. In a series of elegant experiments, this group observed changes in the formation and segregation of both the Cdc42 and RhoA rings, leading them to hypothesize that RhoA negatively regulates Cdc42. Moreover, RhoA requires a small amount of Cdc42 to become active, but RhoA is inhibited by specific localized regions of Cdc42 activity.

Although highly useful, sensing activation by observing the accumulation of effector-binding domains presents difficulties in quantitation and in observing some activation events. High contrast is required to resolve changing activation in small subcellular structures such as filopodia, and ratio imaging against volume indicators must be used to determine whether changes in fluorophore intensities are due to GTPase dynamics or alterations in cell thickness and/or illumination. Despite these challenges, translocation sensors have been very valuable because of their ease of design and implementation and have provided unambiguous readouts of Cdc42 activation where translocation clearly occurs.

The straightforward readout provided by translocation sensors can simplify studies in complex environments. Kumfer et al. (2010) adapted the Cdc42 translocation sensor to *Caenorhabditis elegans* to study the spatiotemporal dynamics of Cdc42 activation in establishing and maintaining cell polarity in the single-cell worm embryo. They fused the Cdc42-binding domain of the *C. elegans* WASP homolog (WSP-1) to EGPF and imaged GTPase activity throughout the developing embryo. Using this probe, they demonstrated the coordinated movement of Cdc42 activity from the posterior to the anterior embryo at specific stages in development and then used an RNAi library to identify regulators of this movement. Based on sequence analysis, they selected 18 putative GEFs and 22 putative GAPs and determined how knockdown of each protein affects the posterior–anterior migration of Cdc42 activity. They identified a novel GEF and GAP for Cdc42 in *C. elegans* and demonstrated that precise spatiotemporal coordination of these regulators is required to achieve proper cell polarity.

5.2.3.3 FRET-Based Cdc42 Biosensors

Matsuda and co-workers produced a single-chain, intramolecular FRET biosensor for Cdc42 based on the Raichu design, which was made and tested in conjunction with Raichu–Rac1 (Itoh et al. 2002). Both biosensors use a fragment of Pak as the affinity reagent and are based on the following design: CFP–GTPase–PAK fragment–YFP–K-Ras membrane-targeting sequence. In this study, both Rac1 and Cdc42 were shown to localize to the leading edge in motile HT1080 cells. However, Cdc42 was very tightly localized to the edge, whereas Rac1 was more broadly activated.

Seth et al. (2003) developed two biosensors based on different designs, each described and characterized as reporters of Cdc42 GEF activity. The GTPase-binding domain of WASP was inserted between terminal ECFP and EYFP fluorophores, either alone or with a C-terminal VCA domain. The biosensor was characterized carefully both *in vitro* and in live-cell assays, but they found that the *in vivo* sensitivity of these biosensors was too low to detect endogenous protein. They therefore performed experiments in cells by coexpressing the biosensor along with ectopic GTPase. Interestingly, they found that the rate of GEF exchange for this sensor was comparable to that of endogenous Cdc42, but the rate of GAP-mediated hydrolysis was 16-fold slower, leading the authors to classify this as a GEF biosensor. With important implications for GTPase biosensors in general, the authors suggested that single-chain sensors may have greater difficulty in reporting deactivation kinetics, as GAPs and effectors typically bind to overlapping domains on the GTPase. Affinity reagents derived from effector proteins would have similar competitive binding interactions.

Finally, Kamiyama and Chiba (2009) describe a FRET-based biosensor designed for whole-animal studies. They developed both single- and dual-chain activation probes termed A-probe.1 and .2, respectively. A-probe.1 is fully genetically encoded and uses a p21-binding domain from *Drosophila* WASP in the configuration CFP–Cdc42–PBD–YFP. In contrast to A-probe.1, the dual-chain version is not fully genetically encoded but instead consists of EGFP–Cdc42 and PBD conjugated to the dye Alexa546. To use A-probe.2, the embryos were dissected, fixed prior to incubation in a bath of Alexa546-PBD, and washed before imaging. The authors report qualitatively similar results with both probes and performed the majority of studies with A-probe.1. Using a constitutive promoter to drive the expression of A-probe.1 in the whole embryo, Kamiyama and Chiba demonstrated that, although Cdc42 is ubiquitously expressed in the developing embryo in all stages, the GTPase is not active until more than two-thirds of development is complete. This was also observed when the sensor was restricted to a specific tissue such as the trachea or CNS. Interestingly, the deletion of Cdc42 is 100% fatal, but the majority of observable development proceeds normally. Their biosensor data provide a potential explanation for these results. The authors also investigated the spatiotemporal dynamics of Cdc42 activation in specific cells such as aCC motor neurons. Typically, neuronal polarization initiates with the specification and selection of the axon, and dendrogenesis occurs after axon specification. In aCC motor neurons, the authors found no detectable FRET in the axons until 4 h after axon initiation. Active Cdc42 was restricted primarily to the proximal axon just before dendrogenesis, suggesting a spatially restricted concentration of Cdc42 during neuronal morphogenesis.

5.3 Ras-FAMILY GTPases

The Ras subfamily consists of 36 members with three main branches: Ras, Rap, and Ral (Reuther and Der 2000; Takai et al. 2001; Malumbres and Barbacid 2003). Ras is the founding member of a GTPase superfamily that now comprises more than 150 proteins. Ras-family GTPases are responsible for mediating mitogenic processes including cell growth, proliferation, and differentiation. Their role in these essential functions, as well as their contribution to many cancers, has led to more than 30 years of very active research into Ras-family proteins.

5.3.1 Ras

H-Ras, N-Ras, K-Ras-4A, and K-Ras-4B, collectively referred to as Ras, are highly homologous proteins (approximately 80% homology). They were initially identified as viral oncogenes, and mutations in Ras have been implicated in approximately 20–30% of all cancers. As with many homologous GTPases, the various Ras proteins were initially thought to mediate functionally redundant processes, but this idea has been challenged on a number of fronts. Although all Ras isoforms are ubiquitously expressed, the relative amounts of each variant differ across cell types and, typically, mutation of a specific isoform is enriched in a given type of tumor (Almoguera et al. 1988; Grady and Markowitz 2002; Mitsuuchi and Testa 2002). The greatest variability among the isoforms is in the C-terminal hypervariable region (HVR), and this region is the primary determinant of localization and membrane association (Reuther and Der 2000; Castellano and Santos 2011). Ras proteins are regulated in part by differential membrane association mediated by posttranslational lipid modification of the HVR. H-Ras, N-Ras, and K-Ras-4A undergo palmitoylation and are brought to the plasma membrane via the secretory pathway. In contrast, K-Ras-4B does not undergo palmitoylation and is shuttled directly from the endoplasmic reticulum to the plasma membrane, bypassing the Golgi and vesicular trafficking (Reuther and Der 2000). In addition to differences in lipid modification and localization, evidence is mounting that regulators of Ras activity (e.g., GEFs and GAPs) have distinct specificity toward various isoforms, and individual isoforms show preferential specificity toward different effectors (Castellano and Santos 2011). It is well accepted that Ras signals from the plasma membrane, but whether or not signaling occurs from the endoplasmic reticulum, Golgi, and/or mitochondrial membranes has been the subject of vigorous debate. Several researchers have attempted to address questions regarding both the spatial and temporal regulation of Ras signaling through the use of live-cell biosensor imaging.

5.3.1.1 Ras Translocation Biosensors

The majority of Ras biosensors are translocation sensors, typically utilizing various domains of the effector protein Raf1 fused to GFP. The first such sensors, reported in 2002, led to new insight into Ras signaling but also gave rise to apparently conflicting conclusions. Using a GFP-tagged Raf1 RBD in conjunction with ectopic H-Ras expression, Chiu and colleagues observed the accumulation of active Ras in the endoplasmic reticulum and Golgi and in the plasma membrane in response to serum stimulation (Chiu et al. 2002). They also found that when H-Ras was mutated to prohibit palmitoylation, thereby blocking secretory pathway trafficking, the GFP–RBD sensor no longer accumulated in internal membranes. Strikingly, they showed that the kinetics of translocation were different at the endoplasmic reticulum and Golgi versus at the plasma membrane, leading the authors to suggest that Ras participates in distinct signaling pathways at these different membranes. At the same time, Bondeva and colleagues used a similar GFP–RBD probe to assay endogenous Ras activity in normal and virally H-Ras-transformed NIH3T3 cells (Bondeva et al. 2002). These authors found that both the RBD and an additional Raf1 domain, the cysteine-rich domain (CRD), were required to monitor translocation of the probe to detect endogenous Ras. They found significant translocation (and therefore activation) at the plasma membrane but not at the Golgi. Similarly, Augsten et al. (2006) developed a translocation sensor designed to image endogenous Ras. They used the oligomerization of up to three Raf1 RBD domains fused to GFP to increase the affinity of the sensor for Ras. This increase in affinity enabled the characterization of Ras activation in the absence of exogenous expression. They found that in response to growth factor stimulation, endogenous Ras was active on the plasma membrane but not in the Golgi. It has been suggested that the discrepancies between these studies can be explained by ectopic expression of Ras. However, differences in experimental conditions, cell types, growth factors, and coexpressed molecules, and so forth, preclude the drawing of firm conclusions. As discussed in Section 5.1, these translocation probes require a high contrast between probe accumulation and background. Unfortunately, GFP–RBDs are broadly localized and concentrated in the thickest part of the cell, near the nucleus and Golgi, making it difficult to identify activity in this important region (Bondeva et al. 2002; Bivona et al. 2004). Ras-family GTPases such as Rap1 have effector-binding domains identical to Ras and bind many of the same effector proteins. Therefore, Ras translocation probes that have been engineered to have a high affinity for Ras may also interact with endogenous Rap.

5.3.1.2 FRET-Based Ras Biosensors

Several single- and dual-chain Ras biosensors have been developed, potentially addressing some of the ambiguities discussed in Section 5.3.1.1. The first FRET-based Ras biosensor was the founding member of the Raichu sensor family developed by Matsuda and colleagues (Mochizuki et al. 2001), named for the Ras and interacting protein chimeric unit (and/or for the fire-loving Pokémon character). Raichu–Ras consists of H-Ras and the Raf1-binding domain flanked by CFP and YFP and includes a C-terminal K-Ras4B tag to constitutively target the construct to the plasma membrane (YFP–Ras–Raf1 RBD–CFP–KRas-4B tag). This sensor revealed that Ras is activated predominately at the plasma membrane in EGF-stimulated COS-1 cells and that sustained Ras activation in nerve growth factor (NGF)-stimulated PC12 cells is seen only in extended neurites. A dual-chain Ras biosensor developed by Jiang and Sorkin (2002) consists of YFP–RBD and CFP–Ras. This design enabled regulation of the CAAX-box protein domain involved in membrane targeting. Using this sensor as part of a larger study on signaling initiated by the epidermal growth factor receptor (EGFR), the authors found that Ras is activated both at the plasma membrane and in endosomes. Furthermore, they showed that although both H-Ras and K-Ras are ubiquitously expressed, H-Ras is seen to a greater extent in endosomes, whereas K-Ras is preferentially localized to the plasma membrane. Therefore, as with the translocation sensors, different conclusions have been drawn from experiments utilizing these distinct designs, leading to a vigorous debate regarding the regulation of Ras activity in specific cellular compartments. The distinction between biosensors responding to different modes of regulation could inform this discussion.

The majority of Ras biosensors have been developed to determine the membrane from which Ras signals and the kinetics of its activation in distinct subcellular compartments. In contrast, Murakoshi et al. (2004) designed a novel biosensor to enable single-molecule studies of activated Ras in an effort to understand how Ras mediates protein–protein interactions and thereby orchestrates complex signaling networks. These authors stably expressed YFP-tagged Ras in KB cells and microinjected dye-labeled GTP. This design utilizes YFP as a donor and the dye BodipyTR as an acceptor for FRET. Using these two bright fluorophores in TIRF microscopy enabled the detection of single activated Ras molecules. Ras was shown to be activated in the plasma membrane in response to EGF, and activation strongly affected the diffusion rate of Ras molecules. These findings led the authors to hypothesize that Ras becomes part of a large, fixed signaling complex upon activation and does not mediate downstream signaling by random collision with effector molecules. Furthermore, K-Ras and H-Ras had different diffusion rates under certain experimental conditions, providing evidence that these highly homologous GTPases are not simply redundant. Single-molecule studies were also carried out by Hibino et al. (2003), primarily using a full-length GFP-tagged Raf1 translocation sensor. Using this technique, the authors found that activated Ras in the plasma membrane is localized specifically to ruffles in EGF-stimulated HeLa cells and that Raf1 dissociates from the plasma membrane on two different characteristic time scales. Tools developed for quantitative microscopy, such as two-photon fluorescence lifetime imaging (2pFLIM), would facilitate quantitation of signaling on a subcellular scale. In a series of elegant studies, Yasuda and colleagues developed 2pFLIM-optimized dual-chain biosensors of Ras, Rac1, and Cdc42 to enable the study of very fine structures including individual dendrites and spines in CA1 hippocampal neurons (Yasuda et al. 2006; Murakoshi et al. 2011).

5.3.2 Rap

The small GTPases Rap1A and Rap1B are responsible for mediating both mitogenic and adhesion signaling. Rap1A and Rap1B are 95% homologous and mediate overlapping functions, though they show differences in cell-type–specific expression and localization (Wittchen et al. 2011). Rap1 was initially cloned based on homology to Ras and identified as a Ras antagonist capable of reversing Ras-mediated transformation (Kitayama et al. 1989). Two decades of work on Rap proteins have since identified Ras-independent mitogenic and adhesion-based signaling roles. Rap1 has been shown to regulate cell differentiation and proliferation as well as integrin activation and cadherin-mediated cell junction formation (Kitayama et al. 1989; Hattori 2003; Bos 2005; Frische and Zwartkruis 2010). Rap1 is important in adhesion and extravasation of immune cells, owing largely to its role in mediating integrin processes, and is also essential in maintaining the epithelium (Hogg et al. 2011; Wittchen et al. 2011). Furthermore, Rap1 has been shown to be the master regulator of polarization in both neurons and T cells, where it regulates Rho-family GTPases such as Cdc42 and Rac1, as well as the Par polarity complex, to orchestrate cellular asymmetry (Schwamborn and Püschel 2004; Gerard et al. 2007; Schwamborn et al. 2007; Iden and Collard 2008). The ability of Rap1 to regulate such diverse processes and its role in highly localized phenomena such as junction formation and polarization has led to the development of tools to try to address some of these questions in live cells.

A single-chain Raichu–Rap1 biosensor was made using the configuration CFP–Rap1–Raf1 binding domain–YFP (Mochizuki et al. 2001). As with Raichu–Ras, the membrane localization of Raichu–Rap1 was achieved using a K-Ras4B tag that induced constitutive membrane localization. Strikingly, despite the plasma membrane localization tag, the biosensor was activated predominantly in the perinuclear region of COS-1 cells stimulated with EGF. Raichu–Ras and Raichu–Rap were used in the same study to understand how these similar molecules achieve specificity. Mochizuki et al. (2001) found that whereas Rap1 was predominantly activated in perinuclear regions, Ras was activated exclusively at the plasma membrane (see above), leading the authors to suggest that these GTPases mediate distinct processes through disparate cellular localization. In a subsequent study, a different sensor design was used to probe the dynamic regulation of Rap1 localization and trafficking. Bivona et al. (2004) constructed a translocation sensor using the Rap1 binding domain of RalGDS fused to GFP to track the location of GTP-loaded Rap1. With this biosensor, Rap1 was found to be broadly localized in the perinuclear region, endosomes, and cytosol, whereas the activated protein was detected only at the periphery in the plasma membrane. Further study is needed to resolve these discrepancies; the development of biosensors that quantitatively report on the activity of Rap1 under native regulation may prove useful.

5.4 OTHER GTPase FAMILIES

5.4.1 Ran

The small GTPase Ran has long been known to maintain and compartmentalize genomic DNA within the cell (Clarke and Zhang 2008), to regulate spindle assembly (Clarke and Zhang 2001; Kaláb et al. 2006), and to control nuclear transport (Moore 1998). The only known Ran GEF, RCC1, has a high affinity for chromatin and is localized to the nucleus, whereas RanGAP is cytosolic. Indeed, the affinity of RCC1 for chromatin allows it to be used as a global marker for chromosomal organization. This cellular polarization of positive and negative Ran regulators was thought to heighten Ran activity in the nucleus, while diminishing its activity in the cytosol, and establish a gradient of Ran activity. Recent studies have demonstrated important roles for active Ran in the cytosol as well. For example, Ran mediates retrograde flow from the tips of axons toward the cell body in injured nerves, and neurons deficient in Ran show increased axonal branching and blebbing (Yudin and Fainzilber 2009). Moreover, the known Ran binding protein RanBP10 binds to microtubules in the cytoplasm of megakaryocytes and has been shown to have nucleotide exchange activity toward Ran, challenging the assumption that nuclear RCC1 is the only RanGEF (Schulze et al. 2008). Several Ran biosensors have been developed in an effort to study these phenomena *in vivo*.

Although many researchers hypothesized that local gradients of active Ran were required to facilitate Ran's varied functions, Kaláb et al. (2002) developed biosensors that could demonstrate their existence experimentally. They designed two biosensors, each with a binding domain flanked by the fluorophores CFP and YFP. In the first design, the affinity reagent directly bound RanGTP, and the interaction between the probe and Ran resulted in decreased FRET. In the second design, the affinity reagent (IBB) bound importin-β, because the interaction between IBB and importin-β is disrupted only in the presence of activated RanGTP. This latter design reported the activation of Ran through an increase in FRET efficiency. These biosensors demonstrated the existence of a sharp RanGTP gradient during spindle formation and revealed a role for RanGTP as a chromosomal positional marker. Unfortunately, the authors found that the expression of IBB, derived from importin-α, disrupted cell cycle progression and could be toxic to somatic cells. Therefore, they refined their design to incorporate an importin-β–binding domain from snurportin 1 that is not toxic and they used the fluorophores EYFP and Cerulean for fluorescence lifetime microscopy (Kaláb et al. 2006). This new biosensor, termed Rango, is a measure of the RanGTP-mediated binding of importin-β to an affinity reagent and is not a direct measure of RanGTP itself.

5.4.2 Arf6

Members of the Arf family of GTPases were first identified as the cellular factors required for the toxic effect of cholera toxin through ADP ribosylation of the Gs heterotrimeric G protein (Kahn and Gilman 1986; O'Neal 2005). This ADP ribosylation factor (ARF) activity is shared by several homologous proteins, which were subsequently numbered Arf1–Arf6. Like other GTPases, the Arf proteins bind to their downstream effector targets only when bound to GTP. Nonetheless, there are several unique characteristics of Arf proteins that make them divergent from other small GTPases. Foremost among these, Arf proteins are not true GTPases. Like other GTPases, Arf proteins cycle through active and inactive states that are dependent on GTP binding and hydrolysis. However, Arf proteins do not possess intrinsic hydrolase activity and require the activity of a GAP to hydrolyze GTP. Moreover, Arf proteins share a set of structural features that define a larger family that encompasses Sar1 and a set of Arf-like (Arl) proteins (Pasqualato et al. 2002; Kahn et al. 2006). These proteins have a myristoylated N-terminal amphipathic helix rather than the C-terminal lipid modification common to other GTPases. GTP binding displaces the N-terminal amphipathic helix from a hydrophobic pocket within the protein, and this promotes the insertion of the helix into the plasma membrane (Goldberg 1998; Pasqualato et al. 2002). Such structural and mechanistic divergence from classical small GTPases has provided unique challenges in designing biosensors for the Arf family.

Nearly all genetically encoded GTPase biosensors fuse an FP to either the N- or C-terminus of the protein of interest. Unfortunately, neither terminus could be altered in Arf sensors. A group led by Martin Schwartz constructed a dual-chain FRET biosensor for Arf6 using the crystal structure to identify internal insertion sites (Hall et al. 2008). First, they identified an exposed loop composed of amino acid residues 140–148. Whereas direct insertion of a circular permutant of GFP after Ile144 resulted in an unstable protein, the introduction of 6-amino acid spacers at the N- and C- termini of the insertion site and the use of CyPet resulted in a stable chimeric protein. A fragment of the clathrin adaptor protein GGA was used as an affinity reagent. Using structural information from cocrystallized Arf6 and GGA, the second chain was created by fusing YPet to residues 148–303 from GGA3 (Hall et al. 2008). This resulted in a biosensor that showed a 200% difference in FRET between constitutively active (Q67L) and dominant-negative (T27N) Arf6 mutants. This effect was also observed when the biosensor was coexpressed with ACAP1, a GAP for Arf6, or ARNO, an Arf6 GEF, indicating that the biosensor was subject to upstream regulators of GTPase activity. Indeed, the authors found that in fibroblasts stimulated with PDGF, Arf6 activity increased in a rapid but transient manner (approximately 10 min duration), suggesting that endogenous receptors and signaling molecules could function to properly regulate the biosensor's activation/deactivation cycle. However, the two components of the biosensor showed obvious differences in localization; YPet–GGA3 was primarily at the Golgi apparatus, whereas Arf6–CyPet was predominantly cytosolic and often concentrated within vesicles. When ratiometric imaging techniques were used, this required careful bleed-through correction.

5.5 CONCLUSION

The ability to observe dynamic changes in GTPase activity in living cells has provided tremendous insights into the molecular events that govern cell behavior. Unlike classical genetic methods, which do not reveal spatiotemporal information, biosensors shed light on the dynamic nature of protein interactions and structural changes. Clearly, a rich ensemble of experimentally validated biosensor designs now exists for Rho GTPases, and understanding the design elements used to formulate these reagents will allow the creation of additional small GTPase biosensors. For example, the Rab-family GTPases, known to regulate vesicle trafficking, are just beginning to be explored (Kitano et al. 2008; Chen et al. 2009; Thyrock et al. 2010).

Many approaches have been used to construct biosensors, but as highlighted here, no design is perfect. Importantly, each biosensor reports specific subsets of interactions, and each is subject to specific restrictions. This is not necessarily a disadvantage. Carefully characterizing and understanding of the limitations and focus of each biosensor will produce a deeper understanding the spatiotemporal dynamics that only biosensors can reveal. The wealth of biosensor designs with different sensitivities presents a means to compare the roles of different regulatory pathways. For example, negative results with one biosensor and not another, examined in light of their designs, can demonstrate the localized activity of negative regulatory pathways reflected by one biosensor.

Future work will couple new biosensor designs with new modes of microscopy and new computational methods to simplify animal imaging of protein activity, enhance quantitation of signaling dynamics with minimal perturbation, and enable imaging of multiple different molecules in the same cell (Welch et al. 2011). Protein labeling within living cells and protein import will enhance the applicability of dye-labeled biosensors (Griffin et al. 1998; Meyer et al. 2006;

Slavoff et al. 2011), and biosensors based on engineered protein scaffolds will provide access to previously intractable targets (Chen et al. 1994; Brient-Litzler et al. 2010; Gulyani et al. 2011).

REFERENCES

Adams, A.E. et al. 1990. Cdc42 and Cdc43, two additional genes involved in budding and the establishment of cell polarity in the yeast *Saccharomyces cerevisiae*. *Journal of Cell Biology*, 111(1), pp. 131–142.

Ai, H.-W. et al. 2006. Directed evolution of a monomeric, bright and photostable version of Clavularia cyan fluorescent protein: Structural characterization and applications in fluorescence imaging. *Biochemical Journal*, 400(3), p. 531.

Almoguera, C. et al. 1988. Most human carcinomas of the exocrine pancreas contain mutant c-K-*ras* genes. *Cell*, 53(4), pp. 549–554.

Augsten, M. et al. 2006. Live-cell imaging of endogenous Ras-GTP illustrates predominant Ras activation at the plasma membrane. *EMBO Reports*, 7(1), pp. 46–51.

Baird, G.S., Zacharias, D.A. and Tsien, R.Y. 1999. Circular permutation and receptor insertion within green fluorescent proteins. *Proceedings of the National Academy of Sciences of the United States of America*, 96(20), pp. 11241–11246.

Bar-Sagi, D. and Hall, A. 2000. Ras and Rho GTPases: A family reunion. *Cell*, 103(2), pp. 227–238.

Benard, V., Bohl, B.P. and Bokoch, G.M. 1999. Characterization of Rac and Cdc42 activation in chemoattractant-stimulated human neutrophils using a novel assay for active GTPases. *Journal of Biological Chemistry*, 274(19), pp. 13198–13204.

Benink, H.A. and Bement, W.M. 2005. Concentric zones of active RhoA and Cdc42 around single cell wounds. *Journal of Cell Biology*, 168(3), pp. 429–439.

Bivona, T.G. et al. 2004. Rap1 up-regulation and activation on plasma membrane regulates T cell adhesion. *Journal of Cell Biology*, 164(3), pp. 461–470.

Bolis, A. et al. 2003. Differential distribution of Rac1 and Rac3 GTPases in the developing mouse brain: Implications for a role of Rac3 in Purkinje cell differentiation. *The European Journal of Neuroscience*, 18(9), pp. 2417–2424.

Bondeva, T. et al. 2002. Structural determinants of Ras-Raf interaction analyzed in live cells. *Molecular Biology of the Cell*, 13(7), pp. 2323–2333.

Bos, J.L. 2005. Linking Rap to cell adhesion. *Current Opinion in Cell Biology*, 17(2), pp. 123–128.

Boureux, A. et al. 2006. Evolution of the Rho family of Ras-like GTPases in eukaryotes. *Molecular Biology and Evolution*, 24(1), pp. 203–216.

Bravo-Cordero, J.J. et al. 2011. A novel spatiotemporal RhoC activation pathway locally regulates cofilin activity at invadopodia. *Current Biology: CB*, 21(8), pp. 635–644.

Brient-Litzler, E., Plückthun, A. and Bedouelle, H. 2010. Knowledge-based design of reagentless fluorescent biosensors from a designed ankyrin repeat protein. *Protein Engineering Design and Selection*, 23(4), pp. 229–241.

Caloca, M.J., Zugaza, J.L. and Bustelo, X.R. 2003. Exchange factors of the RasGRP family mediate Ras activation in the Golgi. *Journal of Biological Chemistry*, 278(35), pp. 33465–33473.

Castellano, E. and Santos, E. 2011. Functional specificity of Ras isoforms: So similar but so different. *Genes & Cancer*, 2(3), pp. 216–231.

Chen, S.Y., Bagley, J. and Marasco, W.A. 1994. Intracellular antibodies as a new class of therapeutic molecules for gene therapy. *Human Gene Therapy*, 5(5), pp. 595–601.

Chen, Y. et al. 2009. GDI-1 preferably interacts with Rab10 in insulin-stimulated GLUT4 translocation. *Biochemical Journal*, 422(2), pp. 229–235.

Chiu, V.K. et al. 2002. Ras signalling on the endoplasmic reticulum and the Golgi. *Nature Cell Biology*, 4, pp. 343–350.

Cho, Y.J. et al. 2005. Generation of Rac3 null mutant mice: Role of Rac3 in Bcr/Abl-caused lymphoblastic leukemia. *Molecular and Cellular Biology*, 25(13), pp. 5777–5785.

Clarke, P.R. and Zhang, C. 2001. Ran GTPase: A master regulator of nuclear structure and function during the eukaryotic cell division cycle? *Trends in Cell Biology*, 11(9), pp. 366–371.

Clarke, P.R. and Zhang, C. 2008. Spatial and temporal coordination of mitosis by Ran GTPase. *Nature Reviews Molecular Cell Biology*, 9(6), pp. 464–477.

Corbetta, S. et al. 2005. Generation and characterization of Rac3 knockout mice. *Molecular and Cellular Biology*, 25(13), pp. 5763–5776.

Del Pozo, M.A. et al. 2002. Integrins regulate GTP-Rac localized effector interactions through dissociation of Rho-GDI. *Nature Cell Biology*, 4(3), pp. 232–239.

Didsbury, J. et al. 1989. Rac, a novel Ras-related family of proteins that are botulinum toxin substrates. *Journal of Biological Chemistry*, 264(28), pp. 16378–16382.

Frische, E.W. and Zwartkruis, F.J.T. 2010. Rap1, a mercenary among the Ras-like GTPases. *Developmental Biology*, 340(1), pp. 1–9.

Galperin, E. 2003. Visualization of Rab5 activity in living cells by FRET microscopy and influence of plasma-membrane-targeted Rab5 on clathrin-dependent endocytosis. *Journal of Cell Science*, 116(23), pp. 4799–4810.

Garrett, S.C. et al. 2008. A biosensor of S100A4 metastasis factor activation: Inhibitor screening and cellular activation dynamics. *Biochemistry*, 47(3), pp. 986–996.

Gerard, A. et al. 2007. The Par polarity complex regulates Rap1- and chemokine-induced T cell polarization. *Journal of Cell Biology*, 176(6), pp. 863–875.

Goguen, B.N., Loving, G.S. and Imperiali, B. 2011. Development of a fluorogenic sensor for activated Cdc42. *Bioorganic & Medicinal Chemistry Letters*, 21(17), pp. 5058–5061.

Goldberg, J. 1998. Structural basis for activation of ARF GTPase: Mechanisms of guanine nucleotide exchange and GTP-myristoyl switching. *Cell*, 95(2), pp. 237–248.

Grady, W.M. and Markowitz, S.D. 2002. Genetic and epigenetic alterations in colon cancer. *Annual Review of Genomics and Human Genetics*, 3, pp. 101–128.

Graham, D., Lowe, P.N. and Chalk, P.A. 2001. A method to measure the interaction of Rac/Cdc42 with their binding partners using fluorescence resonance energy transfer between mutants of green fluorescent protein. *Analytical Biochemistry*, 296(2), pp. 208–217.

Griesbeck, O. et al. 2001. Reducing the environmental sensitivity of yellow fluorescent protein. Mechanism and applications. *Journal of Biological Chemistry*, 276(31), pp. 29188–29194.

Griffin, B.A., Adams, S.R. and Tsien, R.Y. 1998. Specific covalent labeling of recombinant protein molecules inside live cells. *Science*, 281(5374), pp. 269–272.

Gulyani, A. et al. 2011. A biosensor generated via high-throughput screening quantifies cell edge Src dynamics. *Nature Chemical Biology*, 7(7), pp. 437–444.

Haataja, L., Groffen, J. and Heisterkamp, N. 1997. Characterization of Rac3, a novel member of the Rho family. *Journal of Biological Chemistry*, 272(33), pp. 20384–20388.

Hahn, K., DeBiasio, R. and Taylor, D.L. 1992. Patterns of elevated free calcium and calmodulin activation in living cells. *Nature*, 359(6397), pp. 736–738.

Hakem, A. et al. 2005. RhoC is dispensable for embryogenesis and tumor initiation but essential for metastasis. *Genes & Development*, 19(17), pp. 1974–1979.

Hall, B. et al. 2008. A fluorescence resonance energy transfer activation sensor for Arf6. *Analytical Biochemistry*, 374(2), pp. 243–249.

Hattori, M. 2003. Rap1 GTPase: Functions, regulation, and malignancy. *Journal of Biochemistry*, 134(4), pp. 479–484.

Hibino, K. et al. 2003. Single- and multiple-molecule dynamics of the signaling from H-Ras to cRaf-1 visualized on the plasma membrane of living cells. *ChemPhysChem*, 4(7), pp. 748–753.

Hinde, E. et al. 2011. Biosensor Förster resonance energy transfer detection by the phasor approach to fluorescence lifetime imaging microscopy. *Microscopy Research and Technique*, 75(3), pp. 271–281.

Hinde, E. et al. 2013. Millisecond spatiotemporal dynamics of FRET biosensors by the pair correlation function and the phasor approach to FLIM. *Proceedings of the National Academy of Sciences of the United States of America*, 110(1), pp. 135–140.

Hogg, N., Patzak, I. and Willenbrock, F. 2011. The insider's guide to leukocyte integrin signalling and function. *Nature Reviews Immunology*, 11(6), pp. 416–426.

Iden, S. and Collard, J.G. 2008. Crosstalk between small GTPases and polarity proteins in cell polarization. *Nature Reviews Molecular Cell Biology*, 9(11), pp. 846–859.

Ishido, N. et al. 2011. How to make FRET biosensors for Rab family GTPases. *Biosensors-Emerging Materials and Applications*. Ed., P. A. Serra, Intech (online publisher). pp. 1–18.

Itoh, R.E. et al. 2002. Activation of Rac and Cdc42 video imaged by fluorescent resonance energy transfer-based single-molecule probes in the membrane of living cells. *Molecular and Cellular Biology*, 22(18), pp. 6582–6591.

Jaffe, A.B. and Hall, A. 2005. Rho GTPases: Biochemistry and biology. *Annual Review of Cell and Developmental Biology*, 21(1), pp. 247–269.

Jiang, X. and Sorkin, A. 2002. Coordinated traffic of Grb2 and Ras during epidermal growth factor receptor endocytosis visualized in living cells. *Molecular Biology of the Cell*, 13(5), pp. 1522–1535.

Kahn, R.A. et al. 2006. Nomenclature for the human Arf family of GTP-binding proteins: ARF, ARL, and SAR proteins. *Journal of Cell Biology*, 172(5), pp. 645–650.

Kahn, R.A. and Gilman, A.G. 1986. The protein cofactor necessary for ADP-ribosylation of Gs by cholera toxin is itself a GTP binding protein. *Journal of Biological Chemistry*, 261(17), pp. 7906–7911.

Kaláb, P. et al. 2006. Analysis of a RanGTP-regulated gradient in mitotic somatic cells. *Nature*, 440(7084), pp. 697–701.

Kaláb, P., Weis, K. and Heald, R. 2002. Visualization of a Ran-GTP gradient in interphase and mitotic *Xenopus* egg extracts. *Science*, 295(5564), pp. 2452–2456.

Kamiyama, D. and Chiba, A. 2009. Endogenous activation patterns of Cdc42 GTPase within *Drosophila* embryos. *Science*, 324(5932), pp. 1338–1340.

Kardash, E. et al. 2010. A role for Rho GTPases and cell-cell adhesion in single-cell motility in vivo. *Nature Cell Biology*, 12(1), pp. 47–53; Suppl. pp. 1–11.

Kawase, K. et al. 2006. GTP hydrolysis by the Rho family GTPase TC10 promotes exocytic vesicle fusion. *Developmental Cell*, 11(3), pp. 411–421.

Kim, S.H., Li, Z. and Sacks, D.B. 2000. E-cadherin-mediated cell-cell attachment activates Cdc42. *Journal of Biological Chemistry*, 275(47), pp. 36999–37005.

Kitano, M. et al. 2008. Imaging of Rab5 activity identifies essential regulators for phagosome maturation. *Nature*, 453(7192), pp. 241–245.

Kitayama, H. et al. 1989. A Ras-related gene with transformation suppressor activity. *Cell*, 56(1), pp. 77–84.

Kraynov, V.S. et al. 2000. Localized Rac activation dynamics visualized in living cells. *Science*, 290(5490), pp. 333–337.

Kumfer, K.T. et al. 2010. CGEF-1 and CHIN-1 regulate CDC-42 activity during asymmetric division in the *Caenorhabditis elegans* embryo. *Molecular Biology of the Cell*, 21(2), pp. 266–277.

Li, Y. et al. 2007. Bnip3 mediates the hypoxia-induced inhibition on mammalian target of rapamycin by interacting with Rheb. *Journal of Biological Chemistry*, 282(49), pp. 35803–35813.

Liu, A.X. et al. 2001. RhoB is dispensable for mouse development, but it modifies susceptibility to tumor formation as well as cell adhesion and growth factor signaling in transformed cells. *Molecular and Cellular Biology*, 21(20), pp. 6906–6912.

Lorenz, M. et al. 2004. Imaging sites of N-WASP activity in lamellipodia and invadopodia of carcinoma cells. *Current Biology*, 14(8), pp. 697–703.

Loving, G.S., Sainlos, M. and Imperiali, B. 2010. Monitoring protein interactions and dynamics with solvatochromic fluorophores. *Trends in Biotechnology*, 28(2), pp. 73–83.

Machacek, M. et al. 2009. Coordination of Rho GTPase activities during cell protrusion. *Nature*, 461(7260), pp. 99–103.

Macnevin, C.J. et al. 2013. Environment-sensing merocyanine dyes for live cell imaging applications. *Bioconjugate Chemistry*, 24(2), pp. 215–223.

Malumbres, M. and Barbacid, M. 2003. Ras oncogenes: The first 30 years. *Nature Reviews Cancer*, 3(6), pp. 459–465.

Mena, M.A. et al. 2006. Blue fluorescent proteins with enhanced brightness and photostability from a structurally targeted library. *Nature Biotechnology*, 24(12), pp. 1569–1571.

Meyer, B.H. et al. 2006. Covalent labeling of cell-surface proteins for in-vivo FRET studies. *FEBS Letters*, 580(6), pp. 1654–1658.

Mitsuuchi, Y. and Testa, J.R. 2002. Cytogenetics and molecular genetics of lung cancer. *American Journal of Medical Genetics Part B: Neuropsychiatric Genetics*, 115(3), pp. 183–188.

Mochizuki, N. et al. 2001. Spatio-temporal images of growth-factor-induced activation of Ras and Rap1. *Nature*, 411(6841), pp. 1065–1068.

Moore, M.S. 1998. Ran and nuclear transport. *Journal of Biological Chemistry*, 273(36), pp. 22857–22860.

Murakoshi, H. et al. 2004. Single-molecule imaging analysis of Ras activation in living cells. *Proceedings of the National Academy of Sciences of the United States of America*, 101(19), pp. 7317–7322.

Murakoshi, H. et al. 2011. Local, persistent activation of Rho GTPase during plasticity of single dendritic spines. *Nature*, 472, pp. 100–104.

Nagai, T. et al. 2002. A variant of yellow fluorescent protein with fast and efficient maturation for cell-biological applications. *Nature Biotechnology*, 20(1), pp. 87–90.

Nalbant, P. et al. 2004. Activation of endogenous Cdc42 visualized in living cells. *Science*, 305(5690), pp. 1615–1619.

Navarro-Lerita, I. et al. 2011. A palmitoylation switch mechanism regulates Rac1 function and membrane organization. *EMBO Journal*, 31(3), pp. 534–551.

Nguyen, A.W. and Daugherty, P.S. 2005. Evolutionary optimization of fluorescent proteins for intracellular FRET. *Nature Biotechnology*, 23(3), pp. 355–360.

Nomanbhoy, T. and Cerione, R.A. 1999. Fluorescence assays of Cdc42 interactions with target/effector proteins. *Biochemistry*, 38(48), pp. 15878–15884.

Nomanbhoy, T.K. et al. 1996. Investigation of the GTP-binding/GTPase cycle of Cdc42Hs using extrinsic reporter group fluorescence. *Biochemistry*, 35(14), pp. 4602–4608.

Ohashi, T. et al. 2007. An experimental study of GFP-based FRET, with application to intrinsically unstructured proteins. *Protein Science*, 16(7), pp. 1429–1438.

O'Neal, C.J. 2005. Structural basis for the activation of cholera toxin by human ARF6-GTP. *Science*, 309(5737), pp. 1093–1096.

Pasqualato, S., Renault, L. and Cherfils, J. 2002. Arf, Arl, Arp and Sar proteins: A family of GTP-binding proteins with a structural device for "front-back" communication. *EMBO Reports*, 3(11), pp. 1035–1041.

Pédelacq, J.-D. et al. 2006. Engineering and characterization of a superfolder green fluorescent protein. *Nature Biotechnology*, 24(1), pp. 79–88.

Pertz, O. et al. 2006. Spatiotemporal dynamics of RhoA activity in migrating cells. *Nature*, 440(7087), pp. 1069–1072.

Plafker, K. 2002. Fluorescence resonance energy transfer biosensors that detect Ran conformational changes and a Ranmiddle dotGDP-Importin-beta-RanBP1 complex in vitro and in intact cells. *Journal of Biological Chemistry*, 277(33), pp. 30121–30127.

Pommereit, D. and Wouters, F.S. 2007. An NGF-induced Exo70-TC10 complex locally antagonises Cdc42-mediated activation of N-WASP to modulate neurite outgrowth. *Journal of Cell Science*, 120(15), pp. 2694–2705.

Reuther, G.W. and Der, C.J. 2000. The Ras branch of small GTPases: Ras family members don't fall far from the tree. *Current Opinion in Cell Biology*, 12(2), pp. 157–165.

Rizzo, M.A. et al. 2004. An improved cyan fluorescent protein variant useful for FRET. *Nature Biotechnology*, 22(4), pp. 445–449.

Roberts, A.W. et al. 1999. Deficiency of the hematopoietic cell-specific Rho family GTPase Rac2 is characterized by abnormalities in neutrophil function and host defense. *Immunity*, 10(2), pp. 183–196.

Rondanino, C. et al. 2007. RhoB-dependent modulation of postendocytic traffic in polarized Madin-Darby canine kidney cells. *Traffic*, 8(7), pp. 932–949.

Sato, M. et al. 2012. The small GTPase Cdc42 modulates the number of exocytosis-competent dense-core vesicles in PC12 cells. *Biochemical and Biophysical Research Communications*, 420(2), pp. 417–421.

Schulze, H. et al. 2008. RanBP10 is a cytoplasmic guanine nucleotide exchange factor that modulates noncentrosomal microtubules. *Journal of Biological Chemistry*, 283(20), pp. 14109–14119.

Schwamborn, J.C. and Püschel, A.W. 2004. The sequential activity of the GTPases Rap1B and Cdc42 determines neuronal polarity. *Nature Neuroscience*, 7(9), pp. 923–929.

Schwamborn, J.C. et al. 2007. Ubiquitination of the GTPase Rap1B by the ubiquitin ligase Smurf2 is required for the establishment of neuronal polarity. *EMBO Journal*, 26(5), pp. 1410–1422.

Seth, A. et al. 2003. Rational design of genetically encoded fluorescence resonance energy transfer-based sensors of cellular Cdc42 signaling. *Biochemistry*, 42(14), pp. 3997–4008.

Shirsat, N.V. et al. 1990. A member of the ras gene superfamily is expressed specifically in T, B and myeloid hemopoietic cells. *Oncogene*, 5(5), pp. 769–772.

Slavoff, S.A. et al. 2011. Imaging protein–protein interactions inside living cells via interaction-dependent fluorophore ligation. *Journal of the American Chemical Society*, 133(49), pp. 19769–19776.

Sugihara, K. et al. 1998. Rac1 is required for the formation of three germ layers during gastrulation. *Oncogene*, 17(26), pp. 3427–3433.

Takai, Y., Sasaki, T. and Matozaki, T. 2001. Small GTP-binding proteins. *Physiological Reviews*, 81(1), pp. 153–208.

Takaya, A. et al. 2004. RalA activation at nascent lamellipodia of epidermal growth factor-stimulated Cos7 cells and migrating Madin-Darby canine kidney cells. *Molecular Biology of the Cell*, 15(6), pp. 2549–2557.

Takaya, A. et al. 2007. R-Ras regulates exocytosis by Rgl2/Rlf-mediated activation of RalA on endosomes. *Molecular Biology of the Cell*, 18(5), pp. 1850–1860.

Thyrock, A. et al. 2010. Characterizing the interaction between the Rab6 GTPase and Mint3 via flow cytometry based FRET analysis. *Biochemical and Biophysical Research Communications*, 396(3), pp. 679–683.

Toutchkine, A., Kraynov, V. and Hahn, K. 2003. Solvent-sensitive dyes to report protein conformational changes in living cells. *Journal of the American Chemical Society*, 125(14), pp. 4132–4145.

Vigorito, E. et al. 2004. Immunological function in mice lacking the Rac-related GTPase RhoG. *Molecular and Cellular Biology*, 24(2), pp. 719–729.

Vincent, S., Jeanteur, P. and Fort, P. 1992. Growth-regulated expression of rhoG, a new member of the ras homolog gene family. *Molecular and Cellular Biology*, 12(7), pp. 3138–3148.

Welch, C.M. et al. 2011. Imaging the coordination of multiple signalling activities in living cells. *Nature Reviews Molecular Cell Biology*, 12(11), pp. 749–756.

Wheeler, A.P. and Ridley, A.J. 2004. Why three Rho proteins? RhoA, RhoB, RhoC, and cell motility. *Experimental Cell Research*, 301(1), pp. 43–49.

Whitlow, M. et al. 1993. An improved linker for single-chain Fv with reduced aggregation and enhanced proteolytic stability. *Protein Engineering*, 6(8), pp. 989–995.

Wittchen, E.S., Aghajanian, A. and Burridge, K. 2011. Isoform-specific differences between Rap1A and Rap1B GTPases in the formation of endothelial cell junctions. *Small GTPases*, 2(2), pp. 65–76.

Yasuda, R. et al. 2006. Supersensitive Ras activation in dendrites and spines revealed by two-photon fluorescence lifetime imaging. *Nature Neuroscience*, 9(2), pp. 283–291.

Yoshizaki, H. 2003. Activity of Rho-family GTPases during cell division as visualized with FRET-based probes. *Journal of Cell Biology*, 162(2), pp. 223–232.

Yudin, D. and Fainzilber, M. 2009. Ran on tracks—Cytoplasmic roles for a nuclear regulator. *Journal of Cell Science*, 122(5), pp. 587–593.
Zhong, W. et al. 2007. Picosecond-resolution fluorescence lifetime imaging microscopy: A useful tool for sensing molecular interactions in vivo via FRET. *Optics Express*, 15(26), pp. 18220–18235.

SUPPLEMENTARY READINGS

Fritz, R.D. et al., *Science Signaling* 6, rs12–rs12 (2013).
Moshfegh, Y. et al., *Nature* 16, 574–586 (2014).
Nakamura, T. et al., *Genes to Cells* 18, 1020–1031 (2013).
Oliviera, A.F. and R. Yasuda. *PLoS ONE* 8, e52874 (2013).
Wynne, J.P. et al., *J Cell Biol* 199, 317–329 (2012).
Zawistowski, J.S. et al., *PLos ONE* 8, e79877 (2013).

Chapter 6

Molecular Beacon–Type RNA Imaging

Felix Hövelmann and Oliver Seitz

CONTENTS

6.1 INTRODUCTION

Various fields in biomedical research, drug discovery, and disease diagnostics call for methods allowing the sequence-specific detection of DNA or RNA molecules. Recent developments in sequencing technologies have led to dramatic reductions in cost and sequencing time (Guo et al. 2008; Mardis 2008; Clarke et al. 2009). The widely used sequencing infrastructure has prompted the discovery of an ever-growing number of DNA and RNA targets with hitherto unknown function. These discoveries increase the need for chemical probes that allow the analysis of the expression of newly discovered target genes in real time and within living cells. Conventional methods of DNA/RNA detection such as quantitative polymerase chain reaction (qPCR) (Saiki et al. 1985), northern blotting (Alwine et al. 1979), and DNA-microarray technologies, as well as sequencing, provide a general overview of the ensemble of the collected cells (Schena et al. 1995; Ooi et al. 2001; Swain et al. 2002; Raser and O'Shea 2004), which need to be harvested before analysis. Sample collection, preparation, and analysis often take several hours and are complicated by low sample stability (e.g., affected by nucleases), contamination, and the loss of a considerable amount of sample material. Fluorogenic oligonucleotide probes enable direct measurements within living cells. Imaging experiments allow the visualization of gene expression dynamics and reveal details about the transport and degradation of RNA molecules. Perhaps most importantly, RNA imaging experiments address the complex interplay between the target and the multitude of biomolecules present within the biological matrix and thereby provide insight into the cellular functions of RNA.

Fluorogenic oligonucleotide probes and protein-based RNA imaging methods are particularly useful in the real-time analysis of cellular RNA molecules and RNA-regulated processes (Lipski et al. 2001; Marti et al. 2007). The protein-based, "biological" methods rely on the concomitant use of genetically encoded fluorescent fusion proteins and cognate RNA tags (Robinett et al. 1996; Golding and Cox 2004). However, the levels of RNA expressed from recombinant genomes may differ from those of wild-type genomes. Furthermore, the high molecular weight added to the RNA molecule of interest influences its dynamics and localization. In contrast, chemical probes based on fluorescent oligonucleotides allow the analysis of wild-type organisms. Advances in microscopy make it possible to detect changes in mRNA level/localization in single cells as well as in tissues. Therefore, fluorescent probes may become relevant for mechanistic investigations and safety assessments in the drug-discovery process, and probe-guided methods may someday pave the way to better discrimination between healthy and disease tissue in tumor surgery.

Simple fluorescently labeled probes can be used for fluorescence *in situ* hybridization (FISH) experiments, where excess or unbound probes can be washed easily from the fixed cells. However, for live-cell imaging, the washing of unbound probe is impossible, resulting in a too-high background signal. For this reason, extensive work has been invested in the development of fluorescent oligonucleotide probes that enable the induction of fluorescence only on hybridization, but only a fraction of these probes has yet been successfully applied in the analysis of intracellular targets (Bao et al. 2009; Guo et al. 2012). Rather than compiling the vast amount of DNA/RNA-detection methods based on fluorescent hybridization probes, we focus mainly on those probes that were applied in a biological context within living cells.

6.2 SINGLE-LABELED FLUORESCENT PROBES FOR ISH

6.2.1 Linear Probes (FISH)

Even though many fluorescent compounds such as ethidium, 4′,6-diamidino-2-phenylindole (DAPI), or SYBR® dyes allow for unspecific staining of DNA and RNA, sequence-specific nucleic acid detection requires the covalent attachment of a fluorophore to an oligonucleotide (Figure 6.1). Such probes are readily accessible and commercially available. The synthesis places no restrictions on the choice of chromophore and wavelength. Without additional modifications, such probes typically fluoresce regardless of the presence or absence of the target. Therefore, unbound probes need to be removed by washing steps, which need to be applied in so-called FISH experiments on fixed cells. In live-cell imaging, washing steps cannot be performed. Therefore, the omission of washing steps would automatically lead to an extremely high background signal. These restrictions limit the utility of linear probes against RNA targets that are present in high amounts in cells. In pioneering RNA imaging experiments, fluorescein- and tetramethyl-6-carboxyrhodamine (TAMRA)-labeled poly(T) probes have been used to detect the poly(A) tail sequence of mRNA (Molenaar et al. 2001). The poly(A) tail is a ubiquitous target that allows the adjacent hybridization of multiple probes. The local enrichment of fluorophores helped distinguish between unbound probes (background) and hybridized probes

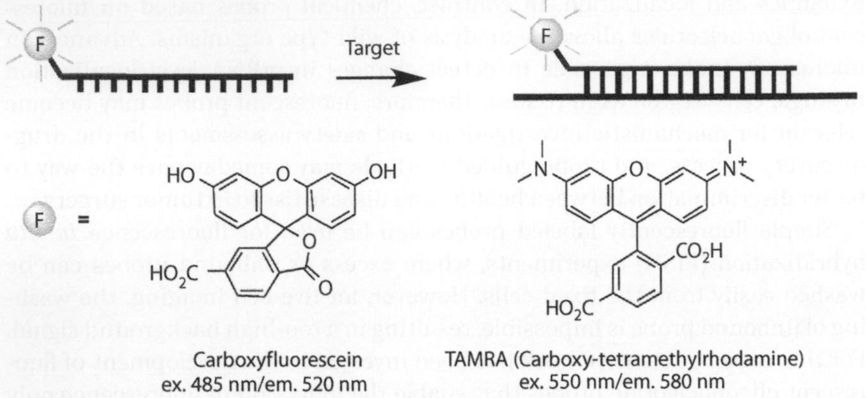

Figure 6.1 Linear fluorescent oligonucleotides.

(signal) and therefore allowed the tracking of mRNA in the nucleus and the cytoplasm (Politz et al. 1998).

A target that occurs in high concentrations is the 28S ribosomal RNA (rRNA). Linear probes labeled with fluorescein or rhodamine allowed for the detection of ribosomal RNA molecules in the nucleus before they reached the cytoplasm (Carmo-Fonseca et al. 1991). The signal-to-background ratio (SBR) can be optimized by photobleaching excess probes (Molenaar et al. 2004). In 1997, RNA imaging experiments also described the monitoring of the cellular delivery of oligonucleotides, which was achieved by using transfection agents or by treatment with streptolysin-O (SLO) (Paillasson et al. 1997).

Linear fluorescently labeled oligonucleotides have been used in single-molecule FISH experiments. Raj et al. designed 48 different Alexa 594-labeled probes for binding to multiple segments within one RNA molecule (Raj et al. 2008). Assisted by computational analysis, three different mRNAs (FKBP5, PTGS2, and FAM105A) could be simultaneously localized in human carcinoma cells (A549) by using sets of Cy5-, Alexa 594-, and *N,N,N′,N′*-tetramethyl-5-carboxyrhodamine (TMR)-labeled probes.

6.2.2 Linear Förster/Fluorescence Resonance Energy Transfer Probes

The adjacent hybridization of two linear probes enables Förster/fluorescence resonance energy transfer (FRET) between a donor and an acceptor chromophore. As a result of the excitation of the donor and the readout of the acceptor, the SBR increases dramatically since FRET does not occur when the target is absent. Tsuji et al. were the first to apply linear FRET probes (Figure 6.2) for the

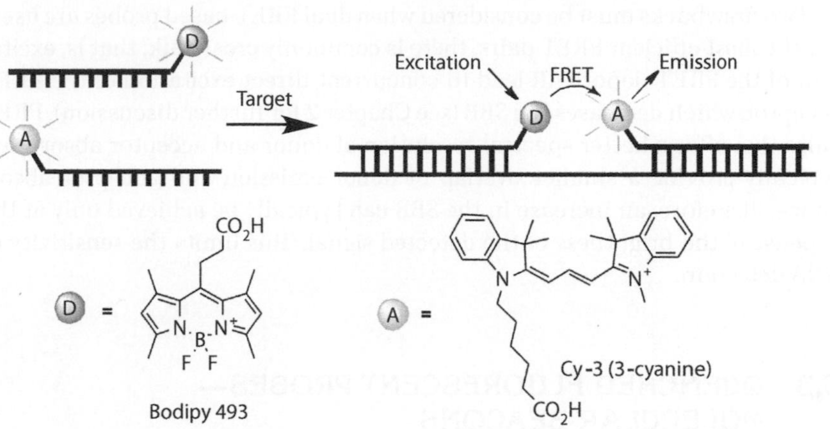

Figure 6.2 Linear FRET probes.

detection of specific mRNA molecules within living cells. They investigated the localization of c-fos mRNA expressed in Cos7-cells by using adjacent hybridizing Bodipy493/503- and Cy-5-labeled probes (Tsuji et al. 2000). To avoid accumulation in the nucleus, the oligonucleotides were bound to streptavidin by means of a covalently attached biotin label. The efficiency of the FRET process and the obtained SBR depend primarily on the spectral overlap between donor emission and acceptor absorption and the distance between both labels (e.g., $1/r^6$; see Chapters 1 and 2) (Förster 1948). To increase the SBR, multiple donor or acceptor chromophores can be accommodated within the same probe (Okamura et al. 2000).

The length of the RNA target region accessed on simultaneous hybridization of two probes is significantly higher than that needed for single-hybridization probes. Therefore, the folding of target mRNA and binding to proteins must be investigated more carefully to predict regions that permit the adjacent hybridization of two probes. This problem can be overcome by fluorescently labeled peptide nucleic acids (PNAs) (Egholm et al. 1993; Nielsen 2004). Such probes proved advantageous for two reasons: On the one hand, the pseudopeptide backbone of PNA is completely stable toward nucleases and does not induce RNase-H–induced degradation of the RNA target. On the other hand, its high binding affinity toward target nucleic acids facilitates access to folded RNA segments. Taylor et al. successfully applied adjacent hybridizing fluorescein (FAM)- and Cy5-labeled PNA probes for the live-cell imaging of inducible nitric oxide synthase (iNOS)–mRNA in mouse macrophages (Wang et al. 2013). The probes were delivered with the help of positively charged cSCK nanoparticles, which were loaded with the corresponding probes after hybridization to complementary DNA.

Two drawbacks must be considered when dual FRET-based probes are used. For the most efficient FRET pairs, there is commonly cross-talk, that is, excitation of the FRET donor will lead to concurrent direct excitation of the FRET acceptor, which decreases the SBR (see Chapter 2 for further discussion). FRET pairs that offer a better spectral resolution of donor and acceptor absorption typically provide a smaller overlap of donor emission and acceptor absorbance. Therefore, an increase in the SBR can typically be achieved only at the expense of the brightness of the detected signal. This limits the sensitivity of RNA detection.

6.3 QUENCHED FLUORESCENT PROBES—MOLECULAR BEACONS

6.3.1 Conventional Molecular Beacons

The most widely used class of fluorescent probes for live-cell experiments are molecular beacons (MBs; Figure 6.3) (Tyagi and Kramer 1996). In their original format, MB probes comprise four components: a loop, a stem, a quencher, and a fluorophore. The loop is spanned by a target-specific oligonucleotide sequence of 15–25 bases. The stem is formed upon intramolecular hybridization of the two complementary arm sequences, which are typically four to six nucleotides long. The fluorophore and the quencher are attached at the 5′- and 3′-terminal ends, respectively. In the absence of target, the self-complementary stem is

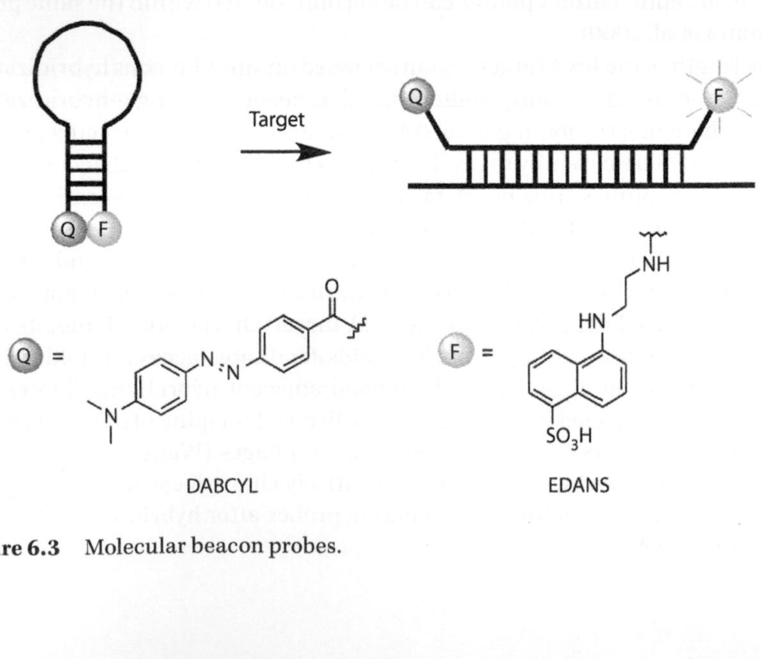

Figure 6.3 Molecular beacon probes.

closed. The fluorophore and quencher are held in close proximity, which very effectively reduces the background signal. The hairpin opens on hybridization because the double-stranded stem region and the loop–target duplex cannot coexist. This separates the fluorophore and the quencher, which leads to a (partial) restoration of donor fluorescence.

Carefully designed probes can discriminate between target sequences with single-base mutations and can provide an up to 200-fold fluorescence enhancement (Root et al. 2004). For the best performance, the length and sequence of the stem and loop regions need to be adjusted to the experimental conditions (e.g., temperature, target sequence) (Bonnet et al. 1998, 1999; Tsourkas et al. 2002). If the thermal stability of the formed probe–target complex is too high, the probe no longer discriminates (single) mismatches from matched targets. This may result in a false-positive signal. A false-negative signal will occur when the affinity of the loop sequence for the target sequence is too low. Under these conditions, the hairpin may not open. The specificity and sensitivity of signaling provided by MB probes can be controlled by adjusting the stem structure. A short stem-region will lead to high responsiveness and thus high sensitivity. In this case, the MB hairpin may open also upon nonspecific binding (e.g., nucleic acid binding proteins). Therefore, long stem regions are required to prevent false-positive signals caused by the undesired opening of the hairpin. This is due to the difference in the energy-penalty for unwinding the closed stem on hybridization between matched or mismatched targets. However, long stem regions will suffer from slow hybridization kinetics and false negatives. Information to aid in the design of MB-probes is available online at www.molecular-beacon.org.

MBs have successfully been used in many imaging studies. A test for the specificity of the signal involves the use of control MBs for which no target is known. Alternatively, target-depleted cells or cells lacking the target may be used. Many of the targets were certainly chosen owing to their high abundance in cells. In a number of studies, RNAs coding for viral proteins, growth factors, or cancer-related proteins were targeted. Matsuo tracked the basic fibroblast growth factor (bFGF) mRNA in human trabecular cells by using a 5-((2-aminoethyl) amino)naphthalene-1-sulfonic acid (EDANS)/4-(4-dimethylaminophenyl) diazenylbenzoic acid (DABCYL)–labeled MB (Matsuo 1998). Yang et al. monitored the expression of the cyclin D1 and survivin mRNAs in different cancer cells (MCF-7, SKBr-3, MDA-MB231) in real time. This work confirmed previous qPCR experiments and survivin-antibody staining (Peng et al. 2005). Cui et al. (2005) used MB probes to study the localization of polio virus plus-strand mRNA and investigated the interaction with microtubule and plasma membrane in Vero cells. In most cases, probes were delivered by transfection agents such as lipofectamine. Another widely used method for probe delivery is the reversible permeabilization of the cell membrane with SLO. Santangelo et al. (2006) analyzed the different infection stages of cells infected by the bovine respiratory syncytial virus (bRSV) and monitored the propagation of the infection to neighboring cells over a period of 7 days. By using dual-color readout experiments, Wang et al. (2008) were able to colocalize viral mRNA and cellular proteins in MDCK cells infected by the influenza A virus.

False-positive signals may arise on nuclease-mediated degradation of the MB. The prevention of MB degradation is strictly required in time-lapse imaging experiments. Modifications based on 2′-*O*-methyl RNA and locked nucleic acid (LNA) nucleotides provide a solution to this problem (Koshkin et al. 1998; Obika et al. 1998; Wang et al. 2005). The 2′-*O*-methyl RNA units are particularly useful because they combine high target affinity with high nuclease stability. In addition, 2′-*O*-methyl oligoribonucleotides prevent the triggering of RNase-H cleavage, which otherwise induces target degradation (Bratu et al. 2003; Yang et al. 2007). Special care is required when LNA modifications are introduced; a completely LNA-modified stem will suffer from extremely slow hybridization kinetics. Wu et al. (2008) extensively investigated chimeric MBs, consisting of DNA and LNA. A fully LNA-modified loop and a chimeric stem proved to be stable and responsive (few minutes until full signal) for at least 24 h in MDA-MB-231 cells. To achieve fast hybridization kinetics, high target affinity, and stability in living cells, so-called "tiny molecular beacons" have been developed (Bratu et al. 2011). These probes comprise 2′-*O*-methyl ribonucleotides and LNA monomers. The latter are positioned within the loop region, at least at every third position. The number and position of the LNA modifications need to be optimized based on the sequence (T_m, GC content, stem region). In the beacons used by Catrina et al. (2012), three to five LNA modifications were placed within an 11-nt loop, which was closed by means of a 4-nt stem. The high target affinity of LNA is expected to facilitate binding to structured mRNA regions. However, it should be considered that the unfolding of RNA may interfere with RNA function and localization.

6.3.2 Dual-FRET MBs

Dual-FRET MBs (Figure 6.4) provide a general solution to the problem of false-positive signals. In analogy to the linear FRET probes, the adjacent hybridization of two probes is required to produce the FRET signal. For more efficient energy transfer from the donor to the acceptor, the stem regions can be designed to bind to the target sequence. Such probes have been termed shared-stem molecular beacons (Bratu et al. 2003). If either or both of the two probes are opened in the absence of target, only the signal of the particular chromophore is observed, whereas FRET remains absent. Santangelo et al. (2004) applied dual-FRET molecular beacons to image survivin and K-Ras mRNA in human dermal fibroblasts (HDFs). By using unmodified MBs and SLO delivery, they

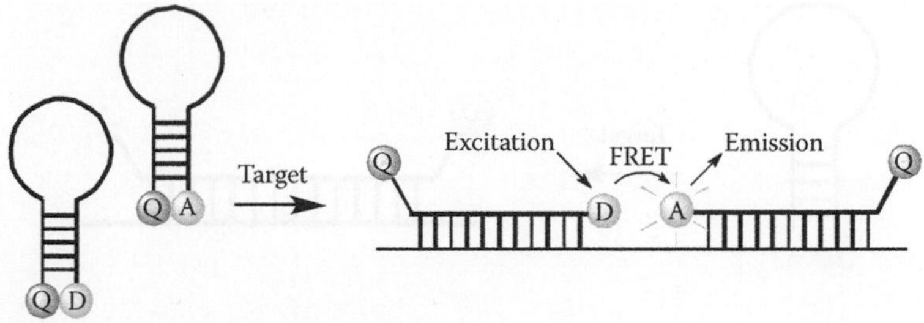

Figure 6.4 Dual-FRET molecular beacons.

could distinguish among false-positive signals, degraded MBs signals (donor only), and correct signals (FRET channel).

The hairpin structure of molecular beacons provides a generic scaffold for the design of nucleic acid conjugates that couple hybridization with changes of measurable observables. This has inspired many researchers to seek new signaling principles. In the following sections, we discuss common strategies for the design of improved MBs that are useful for imaging studies.

6.3.3 Wavelength-Shifting MBs

In dual-fluorophore–labeled MBs, the quencher is replaced by a second fluorophore. In the closed state, the donor and acceptor are again forced into close proximity, resulting in efficient energy transfer. The delivery and localization of closed/unbound probes can be tracked by measuring acceptor emission on donor excitation. Hybridization separates the fluorophores, and energy transfer is no longer efficient. The increase in donor fluorescence reports the target-bound state. Reading out both the donor emission signal and the FRET signal can be achieved by adjusting the emission filter and promises the ratiometric measurement of bound and unbound probe. However, the fluorescence enhancement obtained by separating the two chromophores gives only up to a 10-fold fluorescence enhancement due to the remaining energy transfer in the hybridized state. In addition, this concept is inevitably restricted to readouts at the shorter wavelengths (Figure 6.5).

Alternatively, two fluorochromes can be combined with a quencher (Tyagi et al. 2000). For example, fluorescein was used as a donor and DABCYL as a quencher along with a second (emitter) chromophore that is excited by fluorescein emission. In contrast to the previously described dual-chromophore–labeled MBs, a redshift on hybridization is obtained. As the emitter, 6-carboxyrhodamine 6G (em. 550 nm), tetramethylrhodamine (em. 580 nm), or Texas Red (em. 610 nm) were coupled to the donor. Fluorescein emission is quenched in the closed state. Therefore, FRET between fluorescein and the

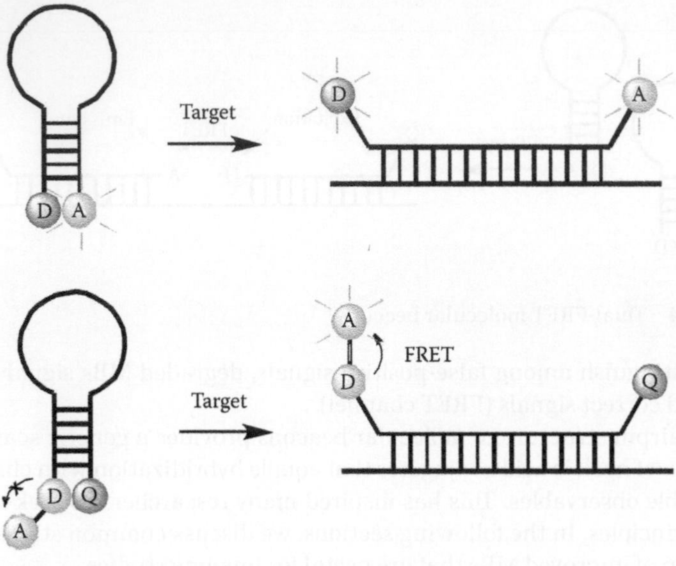

Figure 6.5 Wavelength-shifting molecular beacons.

emitter is prevented. Separation of the chromophore and quencher "activates" the donor. By using only one common excitation wavelength, the FRET readout at three different channels signals the hybridization state of three different MBs. Such probes proved useful for multiplexed PCR genotyping experiments (Tyagi et al. 2000).

6.4 MBs WITH INTERNAL LABELS—IN-STEM MBs

6.4.1 Perylene-Based In-Stem MBs

Fujii et al. (2009) and Kashida et al. (2009) replaced nucleobases within the stem region of MBs with artificial chromophores. Perylene units served as reporter fluorophores that were introduced via a threoninol building block and paired against an abasic site. Quencher groups such as anthraquinone were placed in a zipper-like arrangement at the opposite site. This design leads to an intimate fluorophore–quencher contact that effectively reduces the background. To allow for high fluorescence enhancement on hybridization, the stem region is designed such that the perylenes are part of the newly formed duplex (Figure 6.6). This concept provided up to a 60-fold SBR (Kashida et al. 2009). A linear, quencher-less version of these probes, carrying multiple perylenes or attached via a threoninol linkage, was recently developed (Asanuma

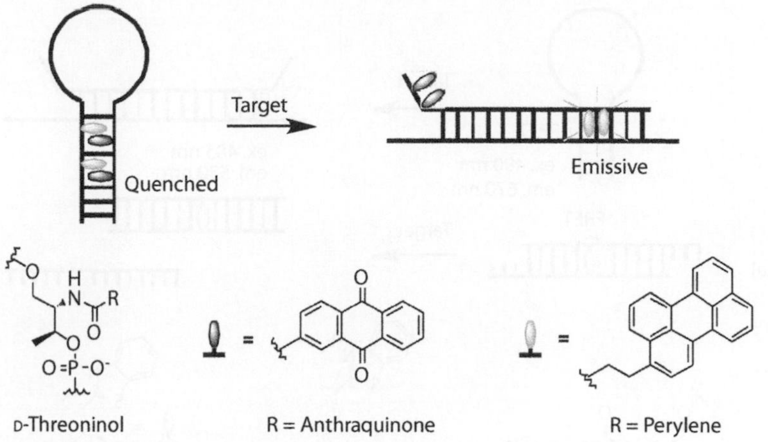

Figure 6.6 Perylene-based in-stem molecular beacons.

et al. 2012). The rather low brightness of fluorescent emission and the spectral overlap with cellular autofluorescence will probably complicate the imaging of intracellular RNA in living cells.

6.4.2 Thiazole Orange–Based ISMBs

Menacher et al. (2008) and Berndl and Wagenknecht (2009) used an amino-propanediol linker for the incorporation of asymmetric cyanine dyes in the stem region. They introduced thiazole orange (TO) and/or thiazole red (TR) cater-cornered to each other and tethered via a flexible linker (Holzhauser et al. 2010). In the closed state, efficient energy transfer from TO to TR is obtained, resulting in a Stokes shift of around 120 nm (530 nm–650 nm). The addition of the target opens the stem, and the disruption of energy transfer results in a wavelength shift via TO emission at 530 nm. Efficient fluorescence signaling was achieved when the chromophores flanked the probe–target duplex (Figure 6.7a) (Holzhauser and Wagenknecht 2011). Aside from being used in conventional MBs this concept has recently been applied to an siRNA system (Figure 6.7b) (Holzhauser and Wagenknecht 2012). The internally TO/TR–labeled siRNA probes were microinjected into CHO-K1 cells. The FRET from TO in one RNA strand to TR in the other RNA strand proved useful for monitoring the integrity of the siRNA duplex in the cell. The dissociation of the siRNA duplex probe was indicated by the wavelength shift toward the emission of TO. It was shown that the doubly modified siRNA probes still maintained the ability to downregulate the expression of mRNA encoding enhanced green fluorescent protein (eGFP) (Holzhauser et al. 2013).

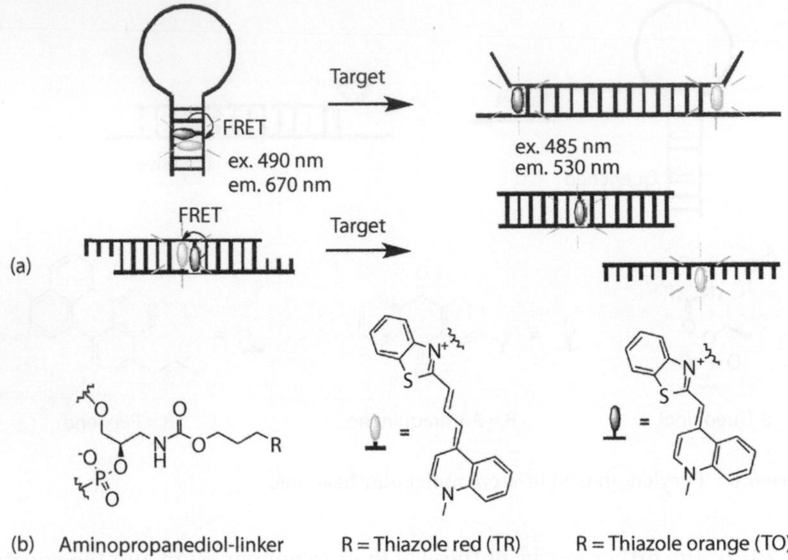

Figure 6.7 (a) Thiazole orange and thiazole red–based in-stem molecular beacon and (b) wavelength-shifting siRNA.

6.5 AVOIDING ACCUMULATION IN THE NUCLEUS

Low-molecular-weight probes easily pass through the nuclear pore, and the high content of DNA-binding proteins within the nucleus offers interaction sites for oligonucleotide-based probes. Therefore, the accumulation of oligonucleotide probes within the nucleus is a frequent problem. Nuclear trapping is associated with the decreased availability of the probes in the cytosol. However, the entry of probes into the nucleus can be prevented by increasing the molecular weight. To this end, streptavidin has been complexed with biotinylated oligonucleotides. Biotin-containing modifiers are commercially available and allow the introduction of 3′- or 5′-terminal or internal biotin. Owing to the strong binding of biotin to streptavidin (which needs to be present in the cells for the experiment), the formed complex is stable under cellular conditions and too large to pass through nuclear pores. Studies of the hybridization kinetics, as well as localization experiments, revealed marginal inference when biotin was attached via a 3′-triethyleneglycol spacer. This concept was applied for the staining of β-actin mRNA in chicken embryonic fibroblasts. Mhlanga et al. (2005) attached a MB to a tRNA molecule. This can be achieved either via a covalent linkage or via the hybridization of sticky ends. After microinjection of such modified probes into the nucleus, their export into the cytoplasm of

CHO cells could be observed via fluorescence microscopy. Depending on the sequence of the MB and therefore the accessibility of target, the attachment of tRNA can reduce the fluorescence signal by up to 50% and the hybridization kinetics might be affected.

6.6 HYBRIDIZATION PROBES THAT DO NOT RELY ON THE CONCEPT OF MBs

6.6.1 Forced Intercalation of Thiazole Orange–Forced Intercalation Probes

Whereas most fluorescent probes are based on the spatial separation of two chromophores, the forced intercalation (FIT) probes (Figure 6.8) introduced by our laboratory require only a single, environmentally sensitive fluorophore, which functions as a base surrogate (Kohler and Seitz 2003). Initially, FIT probes were based on the PNA oligomers. The PNA backbone comprises an aminoethylglycine scaffold, which anchors the nucleobases via peptide bonds. This artificial structure is resistant to degradation by nucleases. PNA is synthesized using known protocols of solid-phase peptide synthesis (Jarikote et al. 2005).

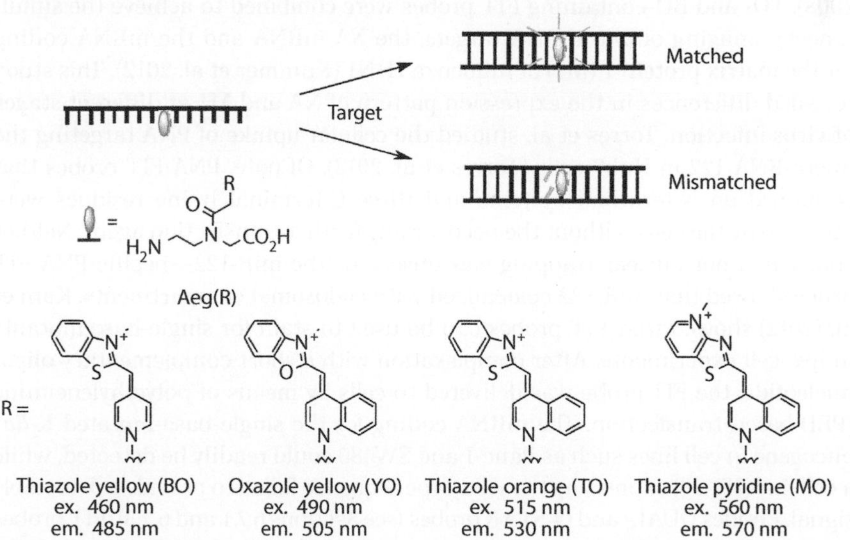

Figure 6.8 Forced intercalation (FIT) probes fluoresce upon hybridization with matched targets.

FIT probes rely on an asymmetric cyanine dye such as TO, which is able to replace any nucleobase within the probe sequence. Owing to the torsional flexibility of the two heterocyclic ring systems, low background fluorescence is observed in the single-stranded state. Hybridization with the target leads to the intercalation of TO within the resulting duplex (Karunakaran et al. 2006). The accompanying restriction of the torsional flexibility results in a strong enhancement of TO fluorescence (SBR > 10). The use of a short linkage (between backbone and fluorophore) confers steric hindrance and hinders interactions of TO with nucleobases in *cis* (Kohler et al. 2005). The unique feature of PNA FIT probes is the responsiveness of TO emission to perturbations of neighboring Watson–Crick base pairing. An attenuation of fluorescence is observed when TO is forced to intercalate next to a mismatched base pair. FIT probes give off for strong positive signals on matched hybridization, while attenuated fluorescence is observed on mismatched hybridization even under nonstringent conditions. The fluorescence properties of TO depend on the sequence context (Kohler et al. 2004; Jarikote et al. 2007), and it is advisable to screen different TO positions to obtain the best performance.

Kummer et al. (2011) used a polyethylene glycol (PEG)ylated FIT probe for the imaging of influenza H1N1-neuramididase (NA) mRNA in virus-infected MDCK cells. The PEGylation not only proved useful for solubility but also prevented the accumulation of the probes in the nucleus. The probes were delivered by SLO treatment. Aside from TO, other fluorochromes have proven useful (Bethge et al. 2008). TO- and BO-containing FIT probes were combined to achieve the simultaneous imaging of two mRNA targets, the NA mRNA and the mRNA coding for the matrix protein-1 (M1) of influenza H1N1 (Kummer et al. 2012). This study revealed differences in the expression pattern of NA and M1 at different stages of virus infection. Torres et al. studied the cellular uptake of PNA targeting the micro-RNA 122 in Huh7 cells (Torres et al. 2012). Of note, PNA FIT probes that contained an N-terminal cysteine and three C-terminal lysine residues were taken up by the cells without the need for any further transfection agent. Neither endosomal nor nuclear trapping was observed. The miR-122–specific PNA FIT probe showed that miR-122 colocalized with endosomal compartments. Kam et al. (2012) showed that FIT probes can be used to stain for single-base mutants in live-cell experiments. After complexation with a short complementary oligonucleotide, the FIT-probe was delivered to cells by means of polyethyleneimine (PEI)-based transfection. The mRNA coding for the single-base-mutated K-*Ras* oncogene in cell lines such as Panc-1 and SW480 could readily be detected, while a cell line (HT-29) expressing the wild-type sequence failed to provide a detectable signal. Besides QUAL- and Q-STAR probes (see Sections 6.7.1 and 6.7.3), FIT probes are probably among the few probes that provide this level of sequence specificity.

The responsiveness of FIT probes can be improved when the TO dye is accompanied by a second, terminally appended fluorophore. Single-stranded

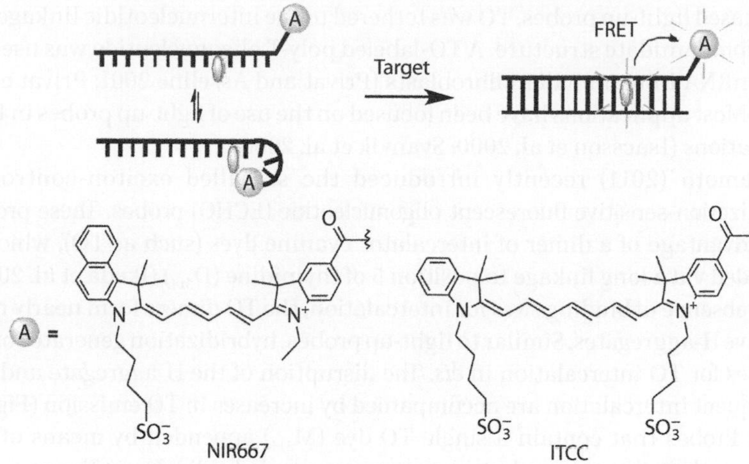

Figure 6.9 FIT–FRET probes.

PNA is highly flexible and readily adopts conformations in which the emission of both dyes is quenched via collisional interactions (Figure 6.9). The tendency of single-stranded PNA to adopt "collapsed" structures in water probably fosters contact quenching. As a result, dual-labeled FIT-probes are extremely dark in the unbound state (Socher et al. 2008). Hybridization with the target leads to rigidification, such that contact or collision between the dyes is prevented. Depending on the choice of the second chromophore, hybridization can be read out by measuring TO emission or FRET. The use of the indotricarbocyanine (ITCC) dye, which is an efficient quencher of TO in the single-stranded state, enabled up to 450-fold enhancement of TO emission on hybridization. The NIR667 dye was used for FRET signaling and provided up to 250-fold SBR (Socher et al. 2012). Recently, DNA-based FIT probes have been developed. Measurements in cell lysates revealed that DNA FIT probes reduce false-positive signals (Bethge et al. 2010; Hövelmann et al. 2012). DNA-FIT-probes have been used to image osk mRNA in eggchambers of *Drosophila melanogaster* (Hövelmann et al. 2013, 2014).

6.6.2 Light-Up Probes and Exciton-Controlled Hybridization-Sensitive Fluorescent Oligonucleotide Probes

Svanvik et al. (2000b) developed the co-called light-up probes, which are PNA-based probes that contain a TO dye appended to one of the terminal ends via a long, flexible linker. TO has little affinity for PNA. The reported single-stranded probes remained dark. Hybridization with DNA fosters backfolding, which can be accompanied by an up to 45-fold enhancement of TO emission. In

DNA-based light-up probes, TO was tethered to the internucleotidic linkage via a phosphoramidate structure. A TO-labeled poly-T oligonucleotide was used to stain mRNA in mammalian fibroblasts (Privat and Asseline 2001; Privat et al. 2002). Most applications have been focused on the use of light-up probes in PCR applications (Isacsson et al. 2000; Svanvik et al. 2000a).

Okamoto (2011) recently introduced the so-called exciton-controlled hybridization-sensitive fluorescent oligonucleotide (ECHO) probes. These probes take advantage of a dimer of intercalator cyanine dyes (such as TO), which is appended via a long linkage to position 5 of thymidine (D_{514}) (Ikeda et al. 2008). In the absence of binding sites for intercalation, the TO dimers form nearly non-emissive H-aggregates. Similar to light-up probes, hybridization generates binding sites for TO intercalation in *cis*. The disruption of the H-aggregate and the subsequent intercalation are accompanied by increases in TO emission (Figure 6.10a). Probes that contain a single TO dye (M_{514}) appended by means of the flexible tether showed modest responsiveness to hybridization (Figure 6.10b)

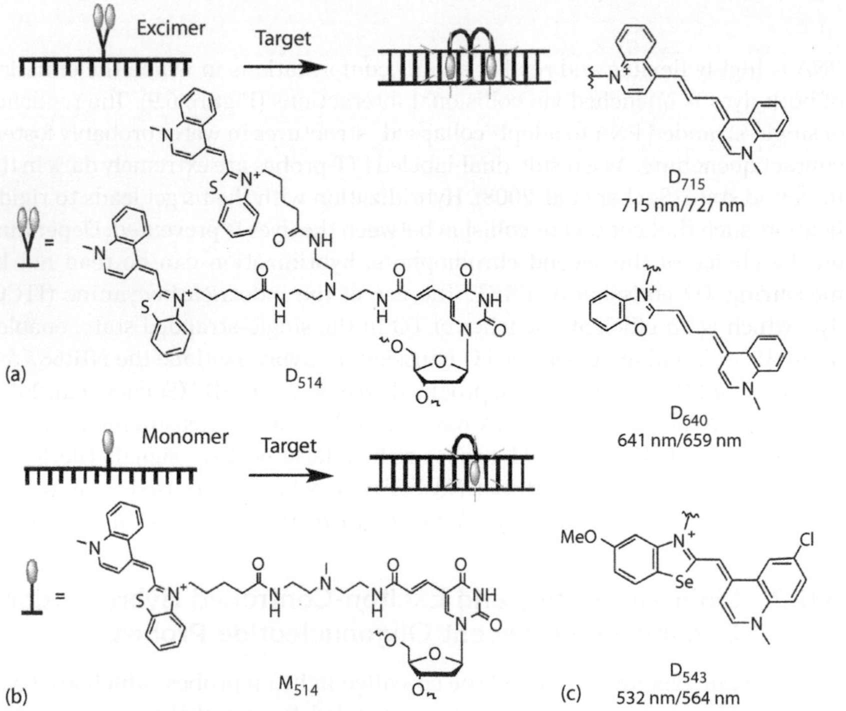

Figure 6.10 (a) ECHO probes show exciton coupling in absence of target and strong fluorescence enhancement upon hybridization; (b) singly labelled probe that allows backfolding of the TO-moiety into the duplex; (c) fluorophores used in ECHO-probes.

(Ikeda and Okamoto 2008). This highlights the importance of excimer formation, which seems to penalize the backbinding of TO in single-stranded probes.

It is essential to avoid (partial) self-dimerization of ECHO probes (Ikeda et al. 2010). Preferably, the fluorophores are appended to A-T–rich regions. The ECHO probes proved useful for FISH and live-cell imaging of total cellular mRNA in HeLa cells (poly(dT) probes) (Kubota et al. 2009). Furthermore, the mRNA coding for the fluorescent protein DsRed2–Mito overexpressed in HeLa cells was imaged by means of a single D_{514}-modified probe. To show the potential for multiplexing, three different ECHO probes (containing D_{514}, D_{543}, and D_{640}) were co-microinjected into the nucleus of HeLa cells along with their corresponding targets, miR-20, miR-17, and miR-30 (Ikeda et al. 2009).

6.7 REACTIVE PROBES

A completely different approach for generating a specific signal on target hybridization is based on a chemical reaction between two (or more) hybridized probes. Even if the requirement for target accessibility can be a limiting factor (binary probes in Section 6.2.2), this strategy combines several advantages. In principle, this binary approach offers outstanding specificity because the reactive probes must be in close proximity to trigger a reaction. In addition, reactive probes are the only reporter system that offers the potential for turnover. Once a signal has been irreversibly generated by a chemical reaction, the target can again anneal two reactive probes and trigger the formation of a second equivalent of signaling molecules. Hence, the target adopts the role of a catalyst, and one target molecule can induce the formation of many signaling molecules. This increases the sensitivity of the detection method.

The design of templated reactions requires special care. The reactions should be sufficiently fast for biological applications, yet the background reaction in the absence of target should be slow. In addition, the generated signal will not remain at its origin but will diffuse throughout the cell, thereby prohibiting localization experiments. Furthermore, temporal changes in target abundance (observed in viral infection or on cellular stress conditions) cannot be monitored, as the generation of the fluorescent signal is irreversible. Although this has not been discussed in the relevant literature, it should be considered that irreversibility leads to a "memory effect," such that even very rare events (including nonspecific responses) will produce a signal that accumulates over time.

6.7.1 Quenched Autoligation Probes

Sando and Kool (2002) and Sando et al. (2004) designed a probe system in which one probe contains a reporter fluorophore (F) appended near a 5′-linked quencher group. The quencher is attached via a sulfonate linkage and can act as a leaving

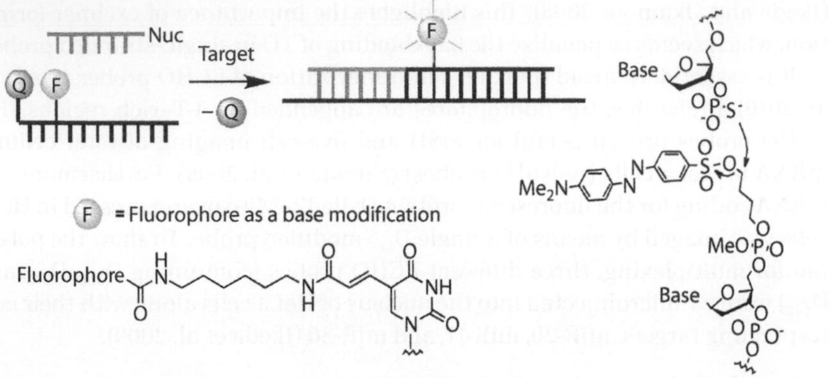

Figure 6.11 Quenched autoligation probes.

group. The second oligonucleotide features a 3′-terminal nucleophile such as a phosphorothioate. Adjacent annealing will enhance the proximity of the two reactive sites and trigger a ligation reaction. The templated nucleophilic substitution results in the release of the quencher, which is accompanied by an irreversible enhancement of the fluorescent signal (Figure 6.11). The combination of DABCYL as the quencher and fluorescein as the reporter provided a nearly 70-fold enhancement of fluorescence. Of note, after the substitution, the two oligonucleotides are covalently linked and therefore will show an increased duplex stability compared to the initial probes.

Short probes provide excellent sequence specificity. For example, a combination of a 7–9-nt long phosphothioate probe with a 13–15-nt long reporter probe was directed against the 16S rRNA in *Escherichia coli* and *Salmonella enterica*, which differ by a single nucleotide. A FAM-labeled probe (match for *E. coli*, and single mismatch for *S. enterica*) and a TMR-labeled probe (match for *S. enterica* and single mismatch for *E. coli*) were applied along with a common phosphothioate probe for the imaging of the two bacterial strains (Sando et al. 2004). Quenched autoligation (QUAL)–FRET probes that carried Cy5 as the acceptor at the thioate probe and FAM as a donor on the quenched probe were used in the live cell imaging of β-actin and glyceraldehyde 3-phosphate dehydrogenase (GAPDH)–mRNA in human cells (Abe and Kool 2006). It was necessary to use nuclease-resistant 2′-*O*-methyl–modified probes.

6.7.2 Probes Based on Fluorophore/Quencher Release on Azide Reduction

Recent developments in target-directed chemistries focus on reactions that do not lead to ligation but rather result in the interconversion of the probe's functional groups. In these cases, the probes after reaction offer the same number of nucleotides for base pairing than before reaction. As a result the reaction products have the same target affinity as the reactive probes. This facilitates turnover in target because the target-directed reactions can be performed under conditions of dynamic strand exchange. The most powerful chemistries so far are the acyl transfer and nitrogen release reactions (Cai et al. 2004; Dose et al. 2006; Grossmann and Seitz 2006). For the latter applications in intracellular RNA detection have been shown.

6.7.3 Quenched Staudinger-Triggered α-Azidoether Release Probes

Franzini and Kool (2009) attached a quencher group via an α-azidoether group to a fluorescently labeled oligonucleotide (Figure 6.12). Triphenylphosphine is attached at the 3′-terminal end of a second oligonucleotide probe. Adjacent hybridization will bring the reducing agent into proximity with the azido group. This triggers a Staudinger reduction and leads to the formation of a hemiaminal linkage, which is subject to hydrolysis and releases the quencher from the fluorescently labeled oligonucleotide probe (Staudinger and Meyer 1919). Many fluorophore–quencher combinations are feasible as long as the quencher can be removed by the chemical reduction. Quenched Staudinger-triggered α-azidoether release (Q-STAR) probes showed up to 200-fold fluorescence enhancements in *ex cellulo* experiments (Franzini and Kool 2011a, b). The background reaction (quencher release in the absence of target) is slow, which helps avoiding false-positive signals. In a proof-of-concept study, the probes were directed against bacterial 16S rRNA and used to distinguish between *E. coli* and *S. enterica* by fluorescence cytometry analysis. Recent improvements involved the introduction of a second quencher molecule or a third probe strand, which resulted in increased sequence specificity and SBR (Kleinbaum and Kool 2010).

In terms of delivery and biological stability, QUAL and Q-STAR probes face the same challenges as MBs. Enhanced stability and decreased cleavage by RNase-H were obtained by 2′-*O*-methyl backbone modification. To address the

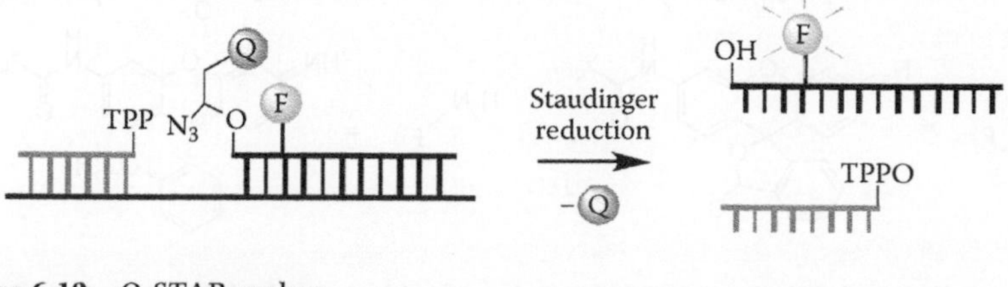

Figure 6.12 Q-STAR probes.

problem of target accessibility to highly structured RNA regions, co-delivery of so-called helper probes, which bind in proximity to the actual probes, were successfully applied (Fuchs et al. 2000).

6.7.4 Staudinger Reduction as a Trigger for Activation of Fluorescence

Many fluorophores are quenched by appended azido groups. Reduction of the azido groups to aniline-like amino groups of fluorochromes often restores fluorescence. Cai et al. (2004), Abe et al. (2008), and Pianowski et al. (2009) developed reporter systems in which one probe was equipped with an azido-coumarin or an azido-rhodamine while the other probe contained the reducing agent (typically a phosphine). Adjacent annealing on the DNA/RNA target will, again, trigger the reduction of the azido group and thus activate fluorescence (Figure 6.13).

Pianowski et al. (2009) fashioned binary reactive probes, which were used to detect intracellular mRNA coding for *O*-6-methylguanine-DNA transferase. To aid the delivery of the probes into cells, positively charged GPNA hybrids were used. In GPNA, the aminoethylglycine backbone in PNA is replaced by aminoethylarginine (Zhou et al. 2003; Dragulescu-Andrasi et al. 2006). These probes are stable against degradation by nucleases and do not induce RNase-H. *Ex cellulo* studies revealed an up to 30-fold SBR. The use of *bis*-azido-rhodamine probes conferred even higher fluorescence enhancements (Gorska et al. 2011). These probes allowed measurements of the miRNA-21 expression levels in five

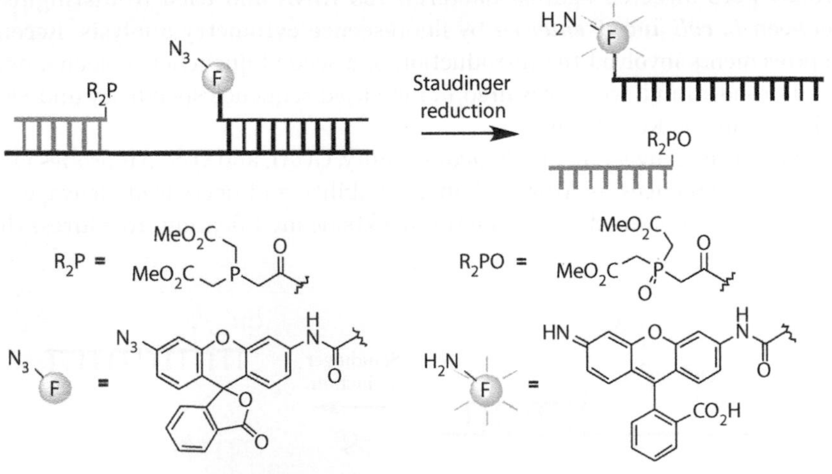

Figure 6.13 Reduction-triggered activation of azide-quenched fluorogenic probes.

different cell lines. Recently, the phosphine in the reducing probe was replaced by a ruthenium (II) complex, which triggers reduction on excitation with visible light (Rothlingshofer et al. 2012; Sadhu et al. 2013).

6.8 DELIVERY

The performance obtained with fluorogenic probes critically depends on the efficiency of delivery. Furthermore, a localization bias should be avoided. Even the best performing probes cannot give good results when they are trapped in the endosome, lysosome, or in the nucleus. There are only a few different methods for delivery that are widely used, and these methods typically require the use of phenol-red–free cell medium to minimize interference with probe fluorescence. In the event that the probes are vulnerable to nuclease digestion, serum-free medium should also be used.

6.8.1 Microinjection

For microinjection, very small low-binding tips are loaded with picoliter amounts of probe solution and injected into the cells by using finely regulated pressure. The injection is monitored by light or fluorescence microscopy. In spite of the availability of high-throughput systems, the number of cells to be microinjected with probes is rather low, and because every injection must be monitored, the method is very elaborate. Microinjection is robust when large cells such as oocytes are used, but the method has been successfully applied to other cell lines, such as PtK2 and MCF-7, as well (Perlette and Tan 2001; Mhlanga et al. 2005; Chen et al. 2007). Once microinjection is completed, imaging can be performed immediately. The amount of probe delivered should be optimized to avoid a high background signal. Most reported protocols involve the use of 0.5–2.0 μM probes in buffer or water and the injection of picoliter amounts (Perlette and Tan 2001).

6.8.2 Microporation/Electroporation

The generation of heat has been a recurring problem for cell viability and delivery efficiency using standard electroporation. However, the recent downsizing to "microporation" has dramatically improved the usefulness of this method. Cell viability is nearly 100%, and optimized protocols give a delivery efficiency of greater than 90%. Usually, 1 μL of a 5–10 μM solution of probe is added to 10 μL of cell suspension (approximately 10^5 cells). On the application of three pulses (each 10 ms) with 1500 V, the resulting change in membrane potential delivers the negatively charged oligonucleotide probes into the cells (Chen et al. 2008).

Microporation has been compared with microinjection. It was found that microporation delivered more probes into the nucleus, where they were rapidly digested by nucleases. To avoid unspecific signaling by MB-based probes, degradation must be prevented as discussed in Section 6.3.1. In addition, the conjugation of probes with PEG or neutravidine provided improved stability after microporation (Chen et al. 2008). Because of the potential for high throughput, the combination of microporation and FACS analysis is often used (Lim et al. 2010).

6.8.3 Transfection

Cellular delivery of probes may be achieved by applying transfection methods that are commonly used to deliver expression vectors. Transfection is the least demanding method in terms of equipment. Many different transfection reagents are available from different commercial suppliers. Typically, positively charged cholesterol or lipid derivatives or polyamines form complexes with the negatively charged oligonucleotides during preincubation (10–15 min) (Dokka and Rojanasakul 2000). Compared to the other methods, delivery using transfection reagents is relatively slow and may take up to 48 h. However, new generation reagents enable efficient delivery within 1–3 h (TransFast, Promega: 65 min; Oligofectamine, Invitrogen: 3 h) (Cui et al. 2005). A drawback of this method is that probes can become trapped in endosomes or lysosomes. The amount (and type) of transfection reagent and probe must be carefully adjusted.

Transfection is in principle restricted to oligonucleotides, but it can be expanded for PNA-based probes by using a partially complementary DNA as a shuttle (Kam et al. 2012).

6.8.4 Pore-Forming Complex—Streptolysin-O

Streptolysin-O (SLO) is a pore-forming bacterial exotoxin that enables the delivery of any kind of macromolecule up to 100 kDa (probes, dextrans, proteins) into any kind of cell. However, the delivery efficiency is dependent on the cell line. Again, the amounts of SLO and probe must be optimized to enable efficient delivery without being detrimental to cell viability (Santangelo et al. 2006). In a typical experiment, a solution containing 0.2–2.0 units of SLO/mL cell medium (activated by reduction with tris (2-carboxyethyl)phosphine [TCEP] or dithiothreitol [DTT]) is preincubated with the probes (2 μM) for 10 min before addition to the cells. After 30 min of incubation, the medium is replaced with growth medium to reseal the cells, which takes up to 30–60 min (Wang et al. 2008). In principle, imaging can be started during resealing.

6.8.5 Cell-Penetrating Peptides

Probes that are conjugated with a so-called cell-penetrating peptide (CPP) can be delivered without the aid of additional reagents or devices. CPPs have been widely used in experiments aimed at delivering toxic compounds. However, reports on the CPP-mediated delivery of RNA imaging probes are less frequent. The best studied example involved the conjugation of a MB oligonucleotide with a peptide derived from the protein transduction domain (PTD) of the HIV TAT Protein, called the TAT peptide (N-TyrGlyArgLysLysArgArgGlnArgArgArg-C) (Ho et al. 2001). Bao et al. described multiple possible attachment modes including (1) covalent linking by addition of a thiol-modified MB to a TAT peptide-maleimide; (2) modification by means of noncovalent interactions between biotinylated oligonucleotides and biotinylated TAT peptides with streptavidin; and (3) a "cleavable" disulfide linkage between a thiolated MB and a thiol-containing TAT peptide (Nitin et al. 2004). Delivery was achieved by incubating the cells with a 0.25–2.0 μM solution of TAT-conjugated MBs for 30–90 min, which is fast compared to standard transfection. Compared with transfection agents, CPPs can avoid the endocytic pathway and therefore may escape trapping in vesicles (Yeh et al. 2008). Torres et al. (2012) added cysteine and lysine residues to PNA-based FIT probes (such as Cys-K-PNA-KKK); these probes were rapidly internalized via both clathrin-mediated and clathrin-independent endocytosis. It should be mentioned here that the use of a nuclear localization sequence (NLS) is not sufficient for delivery. These sequences must be combined with another delivery method to provide nuclear localization (Nitin and Bao 2008).

6.9 CONCLUSION

In light of the recent improvements in ultrafast DNA sequencing, it may appear at first glance that fluorogenic hybridization probes will become redundant. However, the opposite is true. A growing number of RNA targets with hitherto unknown function will likely be identified owing to the increased availability of modern sequencing infrastructure. Once an important function has been discovered, there will—almost inevitably—be demands to screen or trace the relevant molecule *in cellulo* or in tissue. Ultimately, the intracellular role of an RNA molecule can only be uncovered when the dynamics of gene expression, transport, and interactions are known within the biological matrix. This and the increasing importance assigned to regulatory RNA molecules are the most likely reasons for the growing interest in RNA imaging methods.

Impressive efforts in interdisciplinary research comprising physical, chemical, and biological sciences have led to a growing repertoire of imaging methods. An outstanding example for the usefulness of a relatively simple concept is the aforementioned use of 48 linear, fluorescently labeled oligonucleotide probes which enabled the localization of a single RNA molecule. An ever increasing amount of effort is being invested in the development of "smart" probes that provide enhanced fluorescent emission on hybridization with the target. If directed at a suitable target segment, such probes promise RNA imaging without interference with natural RNA localization and function. The advent of the hairpin-shaped MBs provided a generic platform for the design of probes that couple target recognition with changes in a detectable, easy-to-read observable such as fluorescence. This together with commercial availability promoted the acceptance of this technology. MBs have been used to monitor mRNAs coding for viral proteins, cancer-related proteins, growth factors, and micro-RNAs. The hairpin-to-duplex transition provided further inspiration for the improvement of the signal-to-background ratio, which is required to increase the contrast in fluorescence microscopy images. For example, the integration of fluorophores within the stem region allowed for remarkable enhancements in the signal-to-background ratio and provided options for ratiometric imaging.

The majority of the effort invested in the optimization of hairpin-based MBs was focused on reducing background in the absence of the target. However, quenchers also reduce the brightness of fluorescence in the target-bound state and therefore set limits to the sensitivity of RNA imaging. In addition, probes that rely on changes in the distance between two chromophores are prone to false positives, for example, when DNA-binding proteins induce conformational rearrangements. This issue is addressed by quencher-less probe technologies that rely on "smart" fluorophores. Probes such as FIT and ECHO probes use intercalator dyes to signal hybridization, and at least the former completely eliminates false-positive signals. The usefulness of FIT probes has been demonstrated in the simultaneous imaging of two viral mRNA targets.

Recently, reactive probes have been added to the repertoire of imaging probes. A chemical reaction between two adjacently annealed hybridization probes is used to trigger fluorescence. The reactions can be designed to allow turnover of the template. Signal amplification (one target molecule can induce the formation of many signaling molecules) leads to bright fluorescence; however, this comes at the expense of spatial information.

Regardless of the signaling mechanism, all probe technologies face the challenges of degradation of both the probe and the target by nucleases, as well as cellular delivery. Whereas the degradation problem has been successfully addressed by the introduction of modified DNA backbones such as PNA, LNA, or 2′-*O*-ribonucleotides, there is no general solution to the cell delivery problem. The most widely used delivery methods rely on transfection reagents, which typically require long incubation times (12–48 h). Significantly faster delivery is required when viral mRNA is monitored. This can be achieved by using either microinjection or SLO treatment. In summary, there is at present no silver bullet that can be universally applied to any imaging problem. With currently available probes, highly abundant targets such as ribosomal RNA, ubiquitous poly(A) caps, viral targets, and actin can easily be monitored. Even the tracking of less abundant targets such as cellular mRNA, growth factors, and cancer-related targets is possible. However, the investigation of low-copy targets and their intracellular localization remains challenging. The use of multiple hybridization probes (each one directed to a different target segment) provides "single-molecule sensitivity" but comes at the risk of interfering with RNA structure and function. Therefore, beacon technology will have to focus on improving the brightness of fluorescence without degrading responsiveness and spatial information. Given the currently rapid pace of development in this burgeoning field, we are confident that these probes will become available. The future of RNA imaging looks bright.

REFERENCES

Abe, H. and E. T. Kool. 2006. Flow cytometric detection of specific RNAs in native human cells with quenched autoligating FRET probes. *Proc. Natl. Acad. Sci. U.S.A.* 103:263–268.

Abe, H., J. Wang, K. Furukawa et al. 2008. A reduction-triggered fluorescence probe for sensing nucleic acids. *Bioconjug. Chem.* 19:1219–1226.

Alwine, J. C., D. J. Kemp, B. A. Parker et al. 1979. Detection of specific RNAs or specific fragments of DNA by fractionation in gels and transfer to diazobenzyloxymethyl paper. *Methods Enzymol.* 68:220–242.

Asanuma, H., M. Akahane, N. Kondo, T. Osawa, T. Kato and H. Kashida. 2012. Quencher-free linear probe with multiple fluorophores on an acyclic scaffold. *Chem. Sci.* 3:3165–3169.

Bao, G., W. J. Rhee and A. Tsourkas. 2009. Fluorescent probes for live-cell RNA detection. *Annu. Rev. Biomed. Eng.* 11:25–47.

Berndl, S. and H.-A. Wagenknecht. 2009. Fluorescent color readout of DNA hybridization with thiazole orange as an artificial DNA base. *Angew. Chem. Int. Ed.* 48:2418–2421.

Bethge, L., D. V. Jarikote and O. Seitz. 2008. New cyanine dyes as base surrogates in PNA: Forced intercalation probes (FIT-probes) for homogeneous SNP detection. *Bioorg. Med. Chem.* 16:114–125.

Bethge, L., I. Singh and O. Seitz. 2010. Designed thiazole orange nucleotides for the synthesis of single labelled oligonucleotides that fluoresce upon matched hybridization. *Org. Biomol. Chem.* 8:2439–2448.

Bonnet, G., O. Krichevsky and A. Libchaber. 1998. Kinetics of conformational fluctuations in DNA hairpin-loops. *Proc. Natl. Acad. Sci. U.S.A.* 95:8602–8606.

Bonnet, G., S. Tyagi, A. Libchaber and F. R. Kramer. 1999. Thermodynamic basis of the enhanced specificity of structured DNA probes. *Proc. Natl. Acad. Sci. U.S.A.* 96:6171–6176.

Bratu, D. P., I. E. Catrina and S. A. E. Marras. 2011. Tiny molecular beacons for in vivo mRNA detection. *Methods Mol. Biol.* 714:141–157.

Bratu, D. P., B. J. Cha, M. M. Mhlanga, F. R. Kramer and S. Tyagi. 2003. Visualizing the distribution and transport of mRNAs in living cells. *Proc. Natl. Acad. Sci. U.S.A.* 100:13308–13313.

Cai, J. F., X. X. Li, X. Yue and J. S. Taylor. 2004. Nucleic acid-triggered fluorescent probe activation by the Staudinger reaction. *J. Am. Chem. Soc.* 126:16324–16325.

Carmo-Fonseca, M., R. Pepperkok, B. S. Sproat, W. Ansorge, M. S. Swanson and A. I. Lamond. 1991. In vivo detection of snRNP-rich organelles in the nuclei of mammalian cells. *EMBO J.* 10:1863–1873.

Catrina, I. E., S. A. E. Marras and D. P. Bratu. 2012. Tiny molecular beacons: LNA/2′-*O*-methyl RNA chimeric probes for imaging dynamic mRNA processes in living cells. *ACS Chem. Biol.* 7:1586–1595.

Chen, A. K., M. A. Behlke and A. Tsourkas. 2007. Avoiding false-positive signals with nuclease-vulnerable molecular beacons in single living cells. *Nucleic Acids Res.* 35:e105.

Chen, A. K., M. A. Behlke and A. Tsourkas. 2008. Efficient cytosolic delivery of molecular beacon conjugates and flow cytometric analysis of target RNA. *Nucleic Acids Res.* 36:e69.

Clarke, J., H. C. Wu, L. Jayasinghe, A. Patel, S. Reid and H. Bayley. 2009. Continuous base identification for single-molecule nanopore DNA sequencing. *Nat. Nanotechnol.* 4:265–270.

Cui, Z. Q., Z. P. Zhang, X. E. Zhang, J. K. Wen, Y. F. Zhou and W. H. Xie. 2005. Visualizing the dynamic behavior of poliovirus plus-strand RNA in living host cells. *Nucleic Acids Res.* 33:3245–3252.

Dokka, S. and Y. Rojanasakul. 2000. Novel non-endocytic delivery of antisense oligonucleotides. *Adv. Drug Del. Rev.* 44:35–49.

Dose, C., S. Ficht and O. Seitz. 2006. Reducing product inhibition in DNA-template-controlled ligation reactions. *Angew. Chem. Int. Ed.* 45:5369–5373.

Dragulescu-Andrasi, A., S. Rapireddy, B. M. Frezza, C. Gayathri, R. R. Gil and D. H. Ly. 2006. A simple gamma-backbone modification preorganizes peptide nucleic acid into a helical structure. *J. Am. Chem. Soc.* 128:10258–10267.

Egholm, M., O. Buchardt, L. Christensen et al. 1993. PNA hybridizes to complementary oligonucleotides obeying the Watson-Crick hydrogen-bonding rules. *Nature* 365:566–568.

Förster, T. 1948. Zwischenmolekulare Energiewanderung Und Fluoreszenz. *Ann. Phys.* 2:55–75.

Franzini, R. M. and E. T. Kool. 2009. Efficient nucleic acid detection by templated reductive quencher release. *J. Am. Chem. Soc.* 131:16021–16023.

Franzini, R. M. and E. T. Kool. 2011a. Improved templated fluorogenic probes enhance the analysis of closely related pathogenic bacteria by microscopy and flow cytometry. *Bioconjug. Chem.* 22:1869–1877.

Franzini, R. M. and E. T. Kool. 2011b. Two successive reactions on a DNA template: A strategy for improving background fluorescence and specificity in nucleic acid detection. *Chem. Eur. J.* 17:2168–2175.

Fuchs, B. M., F. O. Glockner, J. Wulf and R. Amann. 2000. Unlabeled helper oligonucleotides increase the in situ accessibility to 16S rRNA of fluorescently labeled oligonucleotide probes. *Appl. Environ. Microbiol.* 66:3603–3607.

Fujii, T., H. Kashida and H. Asanuma. 2009. Analysis of coherent heteroclustering of different dyes by use of threoninol nucleotides for comparison with the molecular exciton theory. *Chem. Eur. J.* 15:10092–10102.

Golding, I. and E. C. Cox. 2004. RNA dynamics in live *Escherichia coli* cells. *Proc. Natl. Acad. Sci. U.S.A.* 101:11310–11315.

Gorska, K., I. Keklikoglou, U. Tschulena and N. Winssinger. 2011. Rapid fluorescence imaging of miRNAs in human cells using templated Staudinger reaction. *Chem. Sci.* 2:1969–1975.

Grossmann, T. N. and O. Seitz. 2006. DNA-catalyzed transfer of a reporter group. *J. Am. Chem. Soc.* 128:15596–15597.

Guo, J., J. Y. Ju and N. J. Turro. 2012. Fluorescent hybridization probes for nucleic acid detection. *Anal. Bioanal. Chem.* 402:3115–3125.

Guo, J., N. Xu, Z. Li et al. 2008. Four-color DNA sequencing with 3′-*O*-modified nucleotide reversible terminators and chemically cleavable fluorescent dideoxynucleotides. *Proc. Natl. Acad. Sci. U.S.A.* 105:9145–9150.

Ho, A., S. R. Schwarze, S. J. Mermelstein, G. Waksman and S. F. Dowdy. 2001. Synthetic protein transduction domains: Enhanced transduction potential in vitro and in vivo. *Cancer Res.* 61:474–477.

Holzhauser, C., S. Berndl, F. Menacher, M. Breunig, A. Gopferich and H. A. Wagenknecht. 2010. Synthesis and optical properties of cyanine dyes as fluorescent DNA base substitutions for live cell imaging. *Eur. J. Org. Chem.* 2010:1239–1248.

Holzhauser, C., R. Liebl, A. Goepferich, H.-A. Wagenknecht and M. Breunig. 2013. RNA "traffic lights": An analytical tool to monitor siRNA integrity. *ACS Chem. Biol.* 8:890–894.

Holzhauser, C. and H.-A. Wagenknecht. 2011. In-stem-labeled molecular beacons for distinct fluorescent color readout. *Angew. Chem. Int. Ed.* 50:7268–7272.

Holzhauser, C. and H. A. Wagenknecht. 2012. "DNA traffic lights": Concept of wavelength-shifting DNA probes and application in an Aptasensor. *ChemBioChem* 13:1136–1138.

Hövelmann, F., L. Bethge and O. Seitz. 2012. Single labeled DNA FIT probes for avoiding false-positive signaling in the detection of DNA/RNA in qPCR or cell media. *ChemBioChem* 13:2072–2081.

Hövelmann, F., I. Gaspar, A. Ephrussi and O. Seitz. 2013. Brightness enhanced DNA FIT-probes for wash-free RNA imaging in tissue. *J. Am. Chem. Soc.* 135:19025–19032.

Hövelmann, F., I. Gaspar, I. Loibl, E. A. Ermilov, B. Röder, J. Wengel, A. Ephrussi and O. Seitz. 2014. Brightness through local constraint—LNA-enhanced FIT hybridization probes for in vitro ribonucleotide particle tracking. *Angew. Chem. Int. Ed.* doi: 10.1003/anie.201406022.

Ikeda, S., T. Kubota, K. Kino and A. Okamoto. 2008. Sequence dependence of fluorescence emission and quenching of doubly thiazole orange labeled DNA: Effective design of a hybridization-sensitive probe. *Bioconjug. Chem.* 19:1719–1725.

Ikeda, S., T. Kubota, M. Yuki and A. Okamoto. 2009. Exciton-controlled hybridization-sensitive fluorescent probes: Multicolor detection of nucleic acids. *Angew. Chem. Int. Ed.* 48:6480–6484.

Ikeda, S., T. Kubota, M. Yuki, H. Yanagisawa, S. Tsuruma and A. Okamoto. 2010. Hybridization-sensitive fluorescent DNA probe with self-avoidance ability. *Org. Biomol. Chem.* 8:546–551.

Ikeda, S. and A. Okamoto. 2008. Hybridization-sensitive on-off DNA probe: Application of the exciton coupling effect to effective fluorescence quenching. *Chem. Asian J.* 3:958–968.

Isacsson, J., H. Cao, L. Ohlsson et al.. 2000. Rapid and specific detection of PCR products using light-up probes. *Mol. Cell. Probes* 14:321–328.

Jarikote, D. V., O. Kohler, E. Socher and O. Seitz. 2005. Divergent and linear solid-phase synthesis of PNA containing thiazole orange as artificial base. *Eur. J. Org. Chem.* 2005:3187–3195.

Jarikote, D. V., N. Krebs, S. Tannert, B. Röder and O. Seitz. 2007. Exploring base-pair-specific optical properties of the DNA stain thiazole orange. *Chem. Eur. J.* 13:300–310.

Kam, Y., A. Rubinstein, A. Nissan, D. Halle and E. Yavin. 2012. Detection of endogenous K-ras mRNA in living cells at a single base resolution by a PNA molecular beacon. *Mol. Pharmacol.* 9:685–693.

Karunakaran, V., J. L. Perez Lustres, L. Zhao, N. P. Ernsting and O. Seitz. 2006. Large dynamic Stokes shift of DNA intercalation dye thiazole orange has contribution from a high-frequency mode. *J. Am. Chem. Soc.* 128:2954–2962.

Kashida, H., T. Takatsu, T. Fujii et al. 2009. In-stem molecular beacon containing a pseudo base pair of threoninol nucleotides for the removal of background emission. *Angew. Chem. Int. Ed.* 48:7044–7047.

Kleinbaum, D. J. and E. T. Kool. 2010. Sandwich probes: Two simultaneous reactions for templated nucleic acid detection. *Chem. Commun.* 46:8154–8156.

Kohler, O., D. V. Jarikote and O. Seitz. 2004. Ensemble hybridisation—A new method for exploring sequence dependent fluorescence of dye-nucleic acid conjugates. *Chem. Commun.* 2004:2674–2675.

Kohler, O., D. V. Jarikote and O. Seitz. 2005. Forced intercalation probes (FIT probes): Thiazole orange as a fluorescent base in peptide nucleic acids for homogeneous single-nucleotide-polymorphism detection. *ChemBioChem* 6:69–77.

Kohler, O. and O. Seitz. 2003. Thiazole orange as fluorescent universal base in peptide nucleic acids. *Chem. Commun.* 2003:2938–2939.

Koshkin, A. A., P. Nielsen, M. Meldgaard, V. K. Rajwanshi, S. K. Singh and J. Wengel. 1998. LNA (locked nucleic acid): An RNA mimic forming exceedingly stable LNA: LNA duplexes. *J. Am. Chem. Soc.* 120:13252–13253.

Kubota, T., S. Ikeda and A. Okamoto. 2009. Doubly thiazole orange-labeled DNA for live cell RNA imaging. *Bull. Chem. Soc. Jpn.* 82:110–117.

Kummer, S., A. Knoll, E. Socher, L. Bethge, A. Herrmann and O. Seitz. 2011. Fluorescence imaging of influenza H1N1 mRNA in living infected cells using single-chromophore FIT-PNA. *Angew. Chem. Int. Ed. Engl.* 50:1931–1934.

Kummer, S., A. Knoll, E. Socher, L. Bethge, A. Herrmann and O. Seitz. 2012. PNA FIT-probes for the dual color imaging of two viral mRNA targets in influenza H1N1 infected live cells. *Bioconjug. Chem.* 23:2051–2060.

Lim, J. Y., S. H. Park, C. H. Jeong et al. 2010. Microporation is a valuable transfection method for efficient gene delivery into human umbilical cord blood-derived mesenchymal stem cells. *BMC Biotechnol.* 10:38.

Lipski, A., U. Friedrich and K. Altendorf. 2001. Application of rRNA-targeted oligonucleotide probes in biotechnology. *Appl. Microbiol. Biotechnol.* 56:40–57.

Mardis, E. R. 2008. Next-generation DNA sequencing methods. *Annu. Rev. Genom. Hum. Genet.* 9:387–402.

Marti, A. A., S. Jockusch, N. Stevens, J. Y. Ju and N. J. Turro. 2007. Fluorescent hybridization probes for sensitive and selective DNA and RNA detection. *Acc. Chem. Res.* 40:402–409.

Matsuo, T. 1998. In situ visualization of messenger RNA for basic fibroblast growth factor in living cells. *Biochim. Biophys. Acta Gen. Subj.* 1379:178–184.

Menacher, F., M. Rubner, S. Berndl and H. A. Wagenknecht. 2008. Thiazole orange and Cy3: Improvement of fluorescent DNA probes with use of short range electron transfer. *J. Org. Chem.* 73:4263–4266.

Mhlanga, M. M., D. Y. Vargas, C. W. Fung, F. R. Kramer and S. Tyagi. 2005. tRNA-linked molecular beacons for imaging mRNAs in the cytoplasm of living cells. *Nucleic Acids Res.* 33:1902–1912.

Molenaar, C., A. Abdulle, A. Gena, H. J. Tanke and R. W. Dirks. 2004. Poly(A)(+) RNAs roam the cell nucleus and pass through speckle domains in transcriptionally active and inactive cells. *J. Cell Biol.* 165:191–202.

Molenaar, C., S. A. Marras, J. C. M. Slats et al. 2001. Linear 2′*O*-methyl RNA probes for the visualization of RNA in living cells. *Nucleic Acids Res.* 29:e89.

Nielsen, P. E. 2004. PNA technology. *Mol. Biotechnol.* 26:233–248.

Nitin, N. and G. Bao. 2008. NLS peptide conjugated molecular beacons for visualizing nuclear RNA in living cells. *Bioconjug. Chem.* 19:2205–2211.

Nitin, N., P. J. Santangelo, G. Kim, S. M. Nie and G. Bao. 2004. Peptide-linked molecular beacons for efficient delivery and rapid mRNA detection in living cells. *Nucleic Acids Res.* 32:e58.

Obika, S., D. Nanbu, Y. Hari et al. 1998. Stability and structural features of the duplexes containing nucleoside analogues with a fixed N-type conformation, 2′-*O*,4′-*C*-methyleneribonucleosides. *Tetrahedron Lett.* 39:5401–5404.

Okamoto, A. 2011. ECHO probes: A concept of fluorescence control for practical nucleic acid sensing. *Chem. Soc. Rev.* 40:5815–5828.

Okamura, Y., S. Kondo, I. Sase et al. 2000. Double-labeled donor probe can enhance the signal of fluorescence resonance energy transfer (FRET) in detection of nucleic acid hybridization. *Nucleic Acids Res.* 28:e107.

Ooi, S. L., D. D. Shoemaker and J. D. Boeke. 2001. A DNA microarray-based genetic screen for nonhomologous end-joining mutants in *Saccharomyces cerevisiae*. *Science* 294:2552–2556.

Paillasson, S., M. van de Corput, R. W. Dirks, H. J. Tanke, M. Robert-Nicoud and X. Ronot. 1997. In situ hybridization in living cells: Detection of RNA molecules. *Exp. Cell Res.* 231:226–233.

Peng, X. H., Z. H. Cao, J. T. Xia et al. 2005. Real-time detection of gene expression in cancer cells using molecular beacon imaging: New strategies for cancer research. *Cancer Res.* 65:1909–1917.

Perlette, J. and W. H. Tan. 2001. Real-time monitoring of intracellular mRNA hybridization inside single living cells. *Anal. Chem.* 73:5544–5550.

Pianowski, Z., K. Gorska, L. Oswald, C. A. Merten and N. Winssinger. 2009. Imaging of mRNA in live cells using nucleic acid-templated reduction of azidorhodamine probes. *J. Am. Chem. Soc.* 131:6492–6497.

Politz, J. C., E. S. Browne, D. E. Wolf and T. Pederson. 1998. Intranuclear diffusion and hybridization state of oligonucleotides measured by fluorescence correlation spectroscopy in living cells. *Proc. Natl. Acad. Sci. U.S.A.* 95:6043–6048.

Privat, E. and U. Asseline. 2001. Synthesis and binding properties of oligo-2′-deoxyribonucleotides covalently linked to a thiazole orange derivative. *Bioconjug. Chem.* 12:757–769.

Privat, E., T. Melvin, F. Merola et al. 2002. Fluorescent properties of oligonucleotide-conjugated thiazole orange probes. *Photochem. Photobiol.* 75:201–210.

Raj, A., P. van den Bogaard, S. A. Rifkin, A. van Oudenaarden and S. Tyagi. 2008. Imaging individual mRNA molecules using multiple singly labeled probes. *Nat. Methods* 5:877–879.

Raser, J. M. and E. K. O'Shea. 2004. Control of stochasticity in eukaryotic gene expression. *Science* 304:1811–1814.

Robinett, C. C., A. Straight, G. Li et al. 1996. In vivo localization of DNA sequences and visualization of large-scale chromatin organization using lac operator/repressor recognition. *J. Cell Biol.* 135:1685–1700.

Root, D. D., C. Vaccaro, Z. Zhang and M. Castro. 2004. Detection of single nucleotide variations by a hybridization proximity assay based on molecular beacons and luminescence resonance energy transfer. *Biopolymers* 75:60–70.

Rothlingshofer, M., K. Gorska and N. Winssinger. 2012. Nucleic acid templated uncaging of fluorophores using Ru-catalyzed photoreduction with visible light. *Org. Lett.* 14:482–485.

Sadhu, K. K., M. Rothlingshofer and N. Winssinger. 2013. DNA as a platform to program assemblies with emerging functions in chemical biology. *Isr. J. Chem.* 53:75–86.

Saiki, R. K., S. Scharf, F. Faloona et al. 1985. Enzymatic amplification of beta-globin genomic sequences and restriction site analysis for diagnosis of sickle cell anemia. *Science* 230:1350–1354.

Sando, S., H. Abe and E. T. Kool. 2004. Quenched auto-ligating DNAs: Multicolor identification of nucleic acids at single nucleotide resolution. *J. Am. Chem. Soc.* 126:1081–1087.

Sando, S. and E. T. Kool. 2002. Imaging of RNA in bacteria with self-ligating quenched probes. *J. Am. Chem. Soc.* 124:9686–9687.

Santangelo, P., N. Nitin, L. LaConte, A. Woolums and G. Bao. 2006. Live-cell characterization and analysis of a clinical isolate of bovine respiratory syncytial virus, using molecular beacons. *J. Virol.* 80:682–688.

Santangelo, P. J., B. Nix, A. Tsourkas and G. Bao. 2004. Dual FRET molecular beacons for mRNA detection in living cells. *Nucleic Acids Res.* 32:e57.

Schena, M., D. Shalon, R. W. Davis and P. O. Brown. 1995. Quantitative monitoring of gene expression patterns with a complementary DNA microarray. *Science* 270:467–470.

Socher, E., L. Bethge, A. Knoll, N. Jungnick, A. Herrmann and O. Seitz. 2008. Low-noise stemless PNA beacons for sensitive DNA and RNA detection. *Angew. Chem. Int. Ed.* 47:9555–9559.

Socher, E., A. Knoll and O. Seitz. 2012. Dual fluorophore PNA FIT-probes—Extremely responsive and bright hybridization probes for the sensitive detection of DNA and RNA. *Org. Biomol. Chem.* 10:7363–7371.

Staudinger, H. and J. Meyer. 1919. Über neue organische Phosphorverbindungen III. Phosphinmethylenderivate und Phosphinimine. *Helv. Chim. Acta* 2:635–646.

Svanvik, N., A. Stahlberg, U. Sehlstedt, R. Sjoback and M. Kubista. 2000a. Detection of PCR products in real time using light-up probes. *Anal. Biochem.* 287:179–182.

Svanvik, N., G. Westman, D. Y. Wang and M. Kubista. 2000b. Light-up probes: Thiazole orange-conjugated peptide nucleic acid for detection of target nucleic acid in homogeneous solution. *Anal. Biochem.* 281:26–35.

Swain, P. S., M. B. Elowitz and E. D. Siggia. 2002. Intrinsic and extrinsic contributions to stochasticity in gene expression. *Proc. Natl. Acad. Sci. U.S.A.* 99:12795–12800.

Torres, A. G., M. M. Fabani, E. Vigorito et al. 2012. Chemical structure requirements and cellular targeting of microRNA-122 by peptide nucleic acids anti-miRs. *Nucleic Acids Res.* 40:2152–2167.

Tsourkas, A., M. A. Behlke and G. Bao. 2002. Structure-function relationships of shared-stem and conventional molecular beacons. *Nucleic Acids Res.* 30:4208–4215.

Tsuji, A., H. Koshimoto, Y. Sato et al. 2000. Direct observation of specific messenger RNA in a single living cell under a fluorescence microscope. *Biophys. J.* 78:3260–3274.

Tyagi, S. and F. R. Kramer. 1996. Molecular beacons: Probes that fluoresce upon hybridization. *Nat. Biotechnol.* 14:303–308.

Tyagi, S., S. A. E. Marras and F. R. Kramer. 2000. Wavelength-shifting molecular beacons. *Nat. Biotechnol.* 18:1191–1196.

Wang, L., C. Y. J. Yang, C. D. Medley, S. A. Benner and W. H. Tan. 2005. Locked nucleic acid molecular beacons. *J. Am. Chem. Soc.* 127:15664–15665.

Wang, W., Z. Q. Cui, H. Han et al. 2008. Imaging and characterizing influenza A virus mRNA transport in living cells. *Nucleic Acids Res.* 36:4913–4928.

Wang, Z., K. Zhang, Y. Shen et al. 2013. Imaging mRNA expression levels in living cells with PNA·DNA binary FRET probes delivered by cationic shell-crosslinked nanoparticles. *Org. Biomol. Chem.* 11:3159–3167.

Wu, Y. R., C. J. Yang, L. L. Moroz and W. H. Tan. 2008. Nucleic acid beacons for long-term real-time intracellular monitoring. *Anal. Chem.* 80:3025–3028.

Yang, C. J., L. Wang, Y. R. Wu et al. 2007. Synthesis and investigation of deoxyribonucleic acid/locked nucleic acid chimeric molecular beacons. *Nucleic Acids Res.* 35:4030–4041.

Yeh, H. Y., M. V. Yates, A. Mulchandani and W. Chen. 2008. Visualizing the dynamics of viral replication in living cells via Tat peptide delivery of nuclease-resistant molecular beacons. *Proc. Natl. Acad. Sci. U.S.A.* 105:17522–17525.

Zhou, P., M. Wang, L. Du, G. W. Fisher, A. Waggoner and D. H. Ly. 2003. Novel binding and efficient cellular uptake of guanidine-based peptide nucleic acids (GPNA). *J. Am. Chem. Soc.* 125:6878–6879.

Chapter 7

Optical Probes for Metabolic Signals

Yin Pun Hung and Gary Yellen

CONTENTS

7.1 INTRODUCTION

Cell metabolism involves a collection of biochemical reactions to process carbohydrates, amino acids, and lipids to generate energy and to synthesize essential cellular building blocks. Catalyzed by metabolic enzymes, each reaction transforms a substrate, be it the initial nutrient or an intermediate metabolite, into a product for the subsequent reaction. Together, these reactions constitute elaborate metabolic pathways for particular fuels, such as glycolysis to convert glucose into pyruvate, as well as β-oxidation to break down fatty acids. From a labyrinthine ensemble of metabolic pathways, cells derive energy and macromolecules to perform diverse physiological functions that encompass circadian rhythm (Bass and Takahashi 2010; Asher and Schibler 2011), cell proliferation (Buchakjian and Kornbluth 2010; Lunt and Vander Heiden 2011), cell survival (Vaughn and Deshmukh 2008; Yi et al. 2011), and gene regulation (Zheng et al. 2003; Wellen et al. 2009). Metabolism thus provides crucial housekeeping functions for all cells.

A renaissance of interest in cell metabolism research is currently under way, driven in part by discoveries in cancer metabolism. Many types of tumors exhibit abnormal metabolism; despite the presence of oxygen, these tumors show hallmarks of anaerobic glucose metabolism, with elevated levels of glucose uptake and lactate production (Warburg et al. 1927; Warburg 1956). For decades, this phenomenon, known as the Warburg effect, has been exploited for clinical diagnostic imaging of tumors using [^{18}F]deoxyglucose as a radiotracer for positron emission tomography (FDG-PET; Vander Heiden et al. 2009). Recently, the molecular underpinnings of cancer metabolism have begun to be elucidated: cancer cells have been found to harbor aberrations in metabolic enzymes in various pathways, for instance, the pyruvate kinase M2 isoform (PKM2) in glycolysis (Christofk et al. 2008; Anastasiou et al. 2011) and phosphoglycerate dehydrogenase (PHGDH) in serine metabolism (Locasale et al. 2011; Possemato et al. 2011). Furthermore, known oncogenes and tumor suppressors, such as c-*Myc* and *p53*, were found to regulate metabolic pathways, thus connecting genetic alterations in tumors to their distinctive metabolism (Gao et al. 2009; Vousden and Ryan 2009; Yuneva et al. 2012). Therefore, metabolic pathways have emerged not only as potential contributing factors but also as therapeutic targets in cancer.

Besides oncology, there is a burgeoning interest in how metabolism can be exploited for therapy, for instance, in neurological disorders. Normally, the brain relies on glucose as the major fuel (Sokoloff et al. 1977). However, during a high-fat, low-carbohydrate diet known as the ketogenic diet, the levels of circulating ketone bodies increase, replacing glucose to support brain metabolism (Owen et al. 1967; Haymond et al. 1983). This fuel switch dramatically alters brain activity and has been used to treat refractory epilepsy (Vining 1999; Bailey et al. 2005). The anticonvulsant effect of the ketogenic diet is a clinical example that highlights the intimate connection between metabolism and brain function (Yellen 2008).

7.2 METABOLIC MEASUREMENTS IN SINGLE LIVE CELLS

Concurrent with this renaissance in metabolism research, several optical probes have been developed to report metabolic dynamics in live cells, providing single-cell measurements that could not be obtained using conventional biochemical methods. Biochemical studies from decades ago have already characterized the metabolic enzymes, substrates, and cofactors in major pathways, for instance, glycolysis and purine biosynthesis (Lowry and Passonneau 1964; Krebs 1977). However, because most of these studies require the use of tissue or cell extracts, each measurement reflects a cell population rather than an individual cell. This masks any differences among various cell types within the tissue (e.g., cancer cells and stromal cells in tumors, or neurons and glia in brain tissues), as well as any cell-to-cell heterogeneity within the same cell type. Optical probes circumvent this drawback by enabling single-cell measurements. Although many individual components of metabolic pathways have been defined biochemically, it remains unclear how these components are integrated in the metabolic system as a whole and how metabolite levels are regulated in an intact cell in a physiological setting. By providing measurements in intact cells, optical probes are powerful tools for addressing systems-level questions in cell metabolism.

This chapter reviews the design and use of optical probes for assessing the spatiotemporal dynamics of metabolism in the context of single intact cells. We begin with strategies to engineer these probes, focusing on genetically encoded fluorescent sensors. Next, we consider various aspects of sensor design, such as sensitivity, specificity, and calibration. Finally, we review sensor examples that have been used to detect the levels of metabolites and other metabolic parameters, including glucose, glutathione redox state, reactive oxygen species (ROS), adenosine triphosphate (ATP), adenosine monophosphate–dependent protein kinase (AMPK) activity, and nicotinamide adenine dinucleotide (NADH). Although this review is not comprehensive, we hope to give our readers a sense of how to design, utilize, and evaluate optical probes of cell metabolism.

7.3 DESIGN STRATEGIES FOR FLUORESCENT SENSORS OF METABOLISM

Fluorescent probes are typically based on either small molecules or fluorescent proteins (FPs). Small molecule indicators have been engineered to report metabolic signals such as pH and ROS (Wardman 2007; Han and Burgess 2010).

Nevertheless, a major challenge of using these indicators remains in delivering them into cells, as well as ensuring sufficient retention for imaging. Compared with small-molecule indicators, FP-based probes can be introduced into cells genetically and are more versatile for intracellular applications.

Although genetically encoded fluorescent sensors can be engineered by various approaches (Giepmans et al. 2006; Frommer et al. 2009; Palmer et al. 2011; Tantama et al. 2012), we focus on four of the most common designs: FP translocation reporters, environmentally sensitive FP reporters, fluorescence resonance energy transfer (FRET) sensors, and circularly permuted fluorescent protein (cpFP) sensors. Figure 7.1 illustrates a schematic of these four types of sensor designs, and Table 7.1 lists several published fluorescent biosensors of metabolic signals.

7.3.1 FP Translocation Reporters

On detecting certain metabolic signals and ions, many proteins alter their intracellular locations. By tagging these protein domains with FPs and tracking their translocation, one can monitor these signals. For instance, phosphatidylinositol 4,5-bisphosphate can be detected by using the green fluorescent protein (GFP)-tagged pleckstrin homology domain of phospholipase C, which on ligand binding transiently moves from the cytosol to the plasma membrane (Stauffer et al. 1998; Raucher et al. 2000). Reporters based on FP translocation have also been used to monitor metabolites such as diacylglycerol (Oancea and Meyer 1998; Oancea et al. 1998), inositol 1,4,5-triphosphate (Hirose et al. 1999), and phosphatidylserine (Yeung et al. 2008). Nonetheless, for specific sensing, the scaffold proteins need to be chosen carefully, as many of them respond to multiple signals, the most common of which is intracellular calcium.

7.3.2 Environmentally Sensitive FP Reporters

Although the FP chromophore is enclosed within a polypeptide β-barrel that protects it from quenching by the bulk solvent and most solutes, mutations or circular permutation can render an FP environmentally sensitive. Even though wild-type *Aequorea* GFP is fairly pH resistant, mutations can be engineered to make it sensitive to physiological pH changes, suitable for intracellular imaging (Llopis et al. 1998; Miesenböck et al. 1998). GFP variants have been designed to detect intracellular anions such as chloride (Jayaraman et al. 2000) or to monitor intracellular chloride and pH simultaneously (Arosio et al. 2010). Besides pH and chloride, FPs can be designed to sense thiol-disulfide redox. Cysteine residues can be engineered near the chromophore to yield redox sensitive FPs, such as rxYFP and roGFP (Ostergaard et al. 2001; Dooley et al. 2004); these FPs can be used to assess general thiol-disulfide redox state in different subcellular

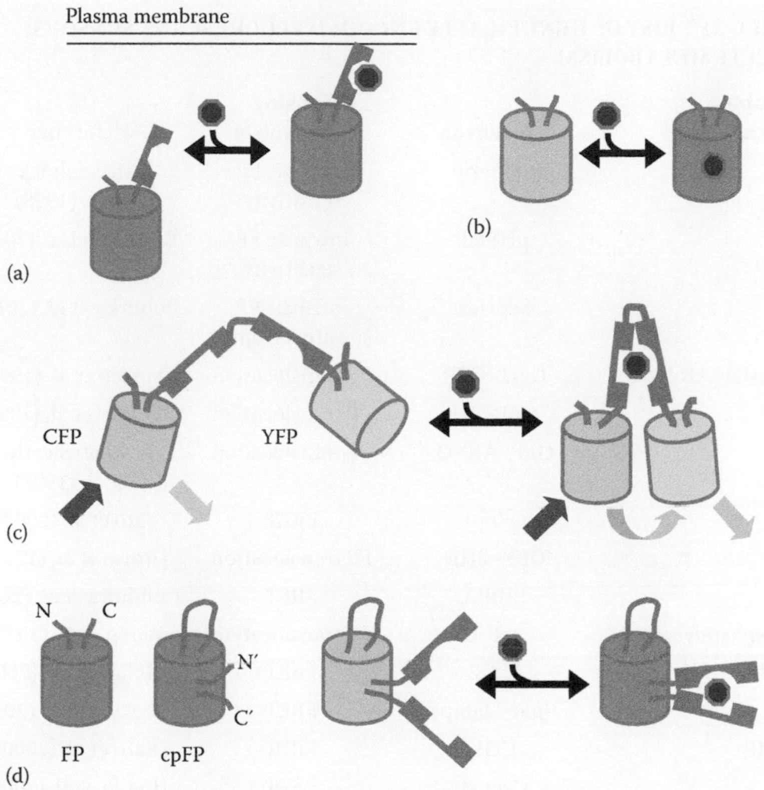

Figure 7.1 Four types of fluorescent biosensor designs. (a) In an FP translocation reporter, an FP is fused to a protein domain that alters its intracellular location (e.g., from the cytosol to the plasma membrane) on sensing the parameter of interest. (b) In an environmentally sensitive FP reporter, the fluorescence responds to a particular environmental signal (e.g., pH or chloride). (c) A fluorescence resonance energy transfer (FRET) sensor comprises protein domain(s) connected to a donor FP (e.g., CFP) and an acceptor FP (e.g., YFP), with spectral overlap between donor emission and acceptor absorption. On sensing the parameter of interest, the sensor undergoes conformational changes that alter the relative distance and/or orientation of the two FPs, thus changing the fluorescence emission of the acceptor via FRET. (d) (Left) Circular permutation of an FP involves linking its original termini (N and C) with a polypeptide linker and introducing new termini (N′ and C′) elsewhere, such as near the chromophore. (Right) A circularly permuted fluorescent protein (cpFP) sensor comprises protein domain(s) connected to a cpFP, creating a conformational coupling between ligand binding and cpFP fluorescence.

TABLE 7.1 LIST OF GENETICALLY ENCODED FLUORESCENT SENSORS OF CELL METABOLISM

Metabolic Parameter	Biosensor	Sensing Mechanism	Reference
pH	pHluorin	Intrinsic FP sensitivity	Miesenböck et al. (1998)
	pHRed	Intrinsic FP sensitivity	Tantama et al. (2011)
	SypHer	Intrinsic FP sensitivity; cpFP	Poburko et al. (2011)
Diacylglycerol	Cys1–GFP	FP translocation	Oancea et al. (1998)
PIP2	GFP–PH	FP translocation	Stauffer et al. (1998)
PIP3	GFP–ARNO	FP translocation	Venkateswarlu et al. (1998)
	Fllip	FRET	Sato et al. (2003)
IP3	GFP–PHD	FP translocation	Hirose et al. (1999)
	LIBRA	FRET	Tanimura et al. (2004)
Phosphatidylserine	Lact–C2	FP translocation	Yeung et al. (2008)
cAMP	ICUE	FRET	DiPilato et al. (2004)
	Epac–camps	FRET	Nikolaev et al. (2004)
cGMP	CGY	FRET	Sato et al. (2000)
	Cygnet	FRET	Honda et al. (2001)
	cGES–DE5	FRET	Nikolaev et al. (2006)
	FlincG	cpFP	Nausch et al. (2008)
Glutamate	FLIPE	FRET	Okumoto et al. (2005)
	SuperGluSnFR	FRET	Hires et al. (2008)
Glucose	FLII^{12}Pglu700µΔ6	FRET	Hou et al. (2011)
General thiol redox	roGFP	Intrinsic FP sensitivity	Dooley et al. (2004)
Glutathione redox	Grx1–roGFP2	Redox relay	Gutscher et al. (2008)
H_2O_2	HyPer	cpFP	Belousov et al. (2006)
	roGFP2–Orp1	Redox relay	Gutscher et al. (2009)
Superoxide	cpYFP	Intrinsic FP sensitivity	Wang et al. (2008)
ATP/ADP ratio	Perceval	cpFP	Berg et al. (2009)

(Continued)

TABLE 7.1 (CONTINUED) LIST OF GENETICALLY ENCODED FLUORESCENT SENSORS OF CELL METABOLISM

Metabolic Parameter	Biosensor	Sensing Mechanism	Reference
ATP	ATeam	FRET	Imamura et al. (2009)
AMPK activity	AMPKAR	FRET	Tsou et al. (2011)
Cytosolic NADH/ NAD^+ ratio	Peredox	cpFP	Hung et al. (2011)
NADH	Frex	cpFP	Zhao et al. (2011b)

compartments (Hanson et al. 2004; Merksamer et al. 2008). In addition, environmentally sensitive FPs can be created by circular permutation (see Section 7.3.4). Circularly permuted yellow fluorescent protein (YFP) has been used to detect superoxide (Wang et al. 2008), though the sensing mechanism remains unclear. Given the few environmentally sensitive FP reporters and their particular designs, it may be difficult to generalize this approach to sense other metabolic ligands or parameters.

7.3.3 Fluorescence Resonance Energy Transfer Sensors

Numerous genetically encoded FP-based sensors have been designed based on fluorescence (or Förster) resonance energy transfer (FRET). A common strategy to establish FRET involves using a pair of fluorophores, a donor and an acceptor, with spectral overlap between donor emission and acceptor absorption. When the two fluorophores are in close proximity to each other, donor excitation can lead to fluorescence emission of the acceptor via FRET. The efficiency of FRET varies with the distance between the fluorophores, their spectral overlap, and their relative orientations (Zhang et al. 2002a; Campbell 2009). For a genetically encoded FRET sensor, the two FP fluorophores, for instance cyan fluorescent protein (CFP) and YFP, can be fused either separately to two protein domains (bimolecular or intermolecular FRET) or both to a single domain (unimolecular or intramolecular FRET). On sensing the parameter of interest, the reporter undergoes conformational changes that alter the relative distance and/or orientation of the two FP fluorophores, creating changes in FRET (Miyawaki 2011). After the first FP-based FRET reporters for calcium (Miyawaki et al. 1997; Romoser et al. 1997), numerous genetically encoded FRET sensors have been engineered to report intracellular parameters, including protein kinase activation, membrane receptor dimerization, small ions and molecules (Frommer et al. 2009; Herbst et al. 2009; Vinkenborg et al. 2010), as well as—pertinent to cell metabolism research—glucose level (Fehr et al.

2003; Takanaga et al. 2008), ATP level (Imamura et al. 2009), and AMPK activity (Tsou et al. 2011).

With limited dynamic range in the fluorescence response, many genetically encoded FP-based FRET sensors can be difficult to use. Typically, the dynamic range in many FRET sensors may be as low as 5%–10%. Compared to such dynamic range, signal variations across cells at baseline as well as noise variations are often substantial, making it difficult to acquire robust and accurate measurements. There have been efforts to improve the dynamic range of FRET sensors by varying the FP orientation via circular permutation (Nagai et al. 2004) or by varying the linker length (Hires et al. 2008; Komatsu et al. 2011).

7.3.4 Circularly Permuted Fluorescent Protein Sensors

Circularly permuted fluorescent proteins (cpFPs) have become part of the standard repertoire in the FP sensor design. Circular permutation of a GFP involves linking its original amino and carboxyl termini with a polypeptide and then introducing new termini elsewhere in the protein, such as near the chromophore (Baird et al. 1999). When a cpFP is fused to a ligand-binding protein domain, this creates a conformational coupling between ligand binding and cpFP fluorescence (Nagai et al. 2001). Based on this strategy, cpFP sensors have been engineered to report calcium, hydrogen peroxide, cyclic 3′,5′-guanosine monophosphate (cGMP), ATP/ADP ratio, and NADH–NAD^+ redox state (Nagai et al. 2001; Belousov et al. 2006; Nausch et al. 2008; Berg et al. 2009; Zhao et al. 2011b; Hung et al. 2011). Typically, cpFP sensors require only single FPs to work and are spectrally ratiometric. For those that are not spectrally ratiometric, a second FP can be attached in tandem for ratiometric imaging (Shimozono et al. 2004; Hung et al. 2011). Compared to most published FRET sensors, cpFP sensors generally show greater dynamic range, in some cases approximately 20-fold increase in fluorescence response, that is, 2000% change (Souslova et al. 2007; Zhao et al. 2011a).

Most cpFP sensors are notably pH sensitive, and this poses a challenge to obtaining accurate measurements for cell metabolism research. The pK_a's in most cpFP sensors are near the physiological pH range (Nagai et al. 2001; Belousov et al. 2006; Nausch et al. 2008; Wang et al. 2008; Berg et al. 2009; Zhao et al. 2011b). Metabolic manipulations often perturb intracellular pH, occasionally by as much as 0.5 unit (Llopis et al. 1998; Tantama et al. 2011), which can produce a substantial response in many cpFP sensors; this response aliases a bona fide change in the nominal analyte. As a somewhat ironic example to highlight this pH artifact, a sensor for mitochondrial pH has been produced by spoiling the sensing of the original target ligand in a highly pH-sensitive cpFP sensor (Poburko et al. 2011). To correct for the pH sensitivity of many cpFP sensors, one can monitor intracellular pH using a red small molecule indicator (Berg et al.

2009) or a red FP (Tantama et al. 2011); both the pH readout and the cpFP signal can then be spectrally compatible and concurrently measured. An alternative, simpler solution is to start with other cpFP variants with lower pK_a to engineer a sensor that is fairly pH resistant (Hung et al. 2011).

The four types of sensor design—FP translocation, environmentally sensitive FP, FRET, and cpFP—are not mutually exclusive, and a single probe can incorporate multiple sensing strategies. For instance, a cpFP sensor for hydrogen peroxide has been fused to a modified PH domain that undergoes translocation on phosphatidylinositol 3,4,5-trisphosphate (PIP3) binding (Mishina et al. 2012). This single probe can be used to monitor both hydrogen peroxide and PIP3 simultaneously using distinct readouts: its fluorescence response reports hydrogen peroxide, whereas its intracellular distribution reports PIP3.

7.4 BASIC PRINCIPLES OF METABOLIC SENSOR DESIGN

Although genetically encoded fluorescent sensors are powerful tools to assess metabolism in single live cells, there are caveats and questions associated with these tools: How can one design or evaluate a fluorescent sensor of metabolism? What kinds of attributes are desirable or undesirable, with regard to affinity, specificity, and kinetics? How can one convert sensor measurements into biologically meaningful parameters? These questions are important, as the design and usage of the sensor underlie the accuracy and meaning of its measurements. Currently, not many fluorescent sensors have been used beyond their laboratories of origin in published studies (Lemke and Schultz 2011), as some of these sensors have not been carefully vetted, and the investment needed to master the imaging techniques might not be trivial. Through the following discussion, we hope to give our readers a sense of how to design or evaluate these sensors, to understand their power and limits, and to utilize them effectively in cell metabolism research.

7.4.1 FP Characteristics and Sensor Engineering

The reliability of a genetically encoded fluorescent sensor depends critically on the component FP(s). Given the availability of numerous FPs with distinct colors and features (Shaner et al. 2005; Day and Davidson 2009), one should consider the following desirable characteristics:

1. *Optimal brightness and color*: The FP should be far brighter than the endogenous autofluorescence. Typically, for live-cell microscopy, the autofluorescence background is higher in the blue wavelength region (Wagnières et al. 1998). Thus, a blue FP would need to be brighter than a red FP for comparable signal detection.

2. *Ratiometric signal with large dynamic range*: This is not an intrinsic property of the FP alone but rather of the FP-based sensor. As calculated from fluorescence measurements at two wavelengths, a ratiometric signal not only normalizes for sensor concentration but also reduces artifacts due to changes in cell volume or cell movement (Nagai et al. 2001; Berg et al. 2009). A large dynamic range also facilitates sensor imaging, particularly given large noise and signal variations.
3. *High photostability*: FPs that are photostable, or resistant to photobleaching, are extremely useful for long-term imaging, though some FPs exhibit complex photobleaching curves (Shaner et al. 2005) and can be difficult to use in quantitative experiments.
4. *Fast maturation*: Slowly maturing FPs can be difficult to express in mammalian cells at physiological temperatures (Tsien 1998). Fortunately, FPs with improved maturation have been engineered (Nagai et al. 2002; Shaner et al. 2004) and are more suitable.
5. *Lack of oligomerization or aggregation*: As some FPs, particularly of the coral family, often oligomerize and aggregate in cell cultures and *in vivo* (Yanushevich et al. 2002; Hirrlinger et al. 2005; Katayama et al. 2008), the resulting fluorescent puncta render imaging experiments difficult. Monomeric FPs are preferable.
6. *Lack of photoconversion*: Many FPs can exhibit photoconversion, in which excitation at certain wavelengths change the FP color, for instance, from yellow to cyan, red to green, or orange to far-red (Valentin et al. 2005; Kirber et al. 2007; Kremers et al. 2009). Depending on the instrumentation and protocols used, photoconversion may be evident only in some cases (Kremers et al. 2009). Ideally, FPs that can undergo photoconversion should be avoided.
7. *Lack of environmental sensitivity*: Except when intended (as in environmentally sensitive FP sensors), the FP should be resistant to fluctuations in the intracellular environment. The main factors to consider include pH and concentrations of anions such as chloride. Some FPs are notoriously pH sensitive or chloride sensitive (Griesbeck et al. 2001; Shaner et al. 2005) and should be avoided. In addition to the FP, the metabolite-sensing protein domain can be environmentally sensitive; in the end, an FP-based sensor should be checked carefully for potential interference before intracellular applications.

Regardless of the strategy (FRET- or cpFP-based), designing a sensor remains an empirical endeavor. Although there are innumerable ways to combine an FP and a metabolite-sensing protein domain, only a small fraction may make useful sensors in the end. Initially, one can vary the lengths or the compositions of the peptide linkers, optimize the metabolite-sensing protein domain(s) by mutation or truncation, or screen iteratively for responsive candidates. Once identified, promising variants can be characterized and improved on to yield the eventual genetically encoded FP-based sensor.

7.4.2 Sensor Tuning and Sensitivity

The sensor should be tuned to the appropriate range for the parameter of interest. For instance, for a pH indicator to work effectively in cells, its pK_a should lie within the range of intracellular pH. In general, for a sensor of metabolic ligand A, its dissociation constant (K_d, defined as the concentration of A needed for a half maximal sensor response) should be near the intracellular concentration of A; otherwise, the sensor response could be undetectable or minimal. Thus, the sensor should be sensitive, recording whenever the parameter of interest changes.

Nonetheless, for sensor design, the intracellular concentrations of metabolites reported by biochemical studies may not be applicable, as the molecules may be concentrated in subcellular compartments or microdomains. Furthermore, a substantial fraction of these molecules can be reversibly bound to cellular proteins and thus not accessible to the biosensor. For a sensor of ligand A, its K_d should thus be tuned not to the total concentration of A, but rather to the free concentration of A. If this free concentration is known or can be estimated, one can design a reporter with its K_d based on that information. Alternatively, from variants with different affinities, one can empirically test for the optimal sensor response in cells.

Most metabolite-sensing protein domains are under complex regulation by several ligands rather than a single one; in the end, such protein domains may elegantly detect the relevant energetic parameters. For instance, a bacterial ATP-binding protein GlnK1 has been found to bind both ATP and adenosine diphosphate (ADP) with extremely high affinity in the submicromolar range, in such a way that it reports a major energetic parameter, the ATP/ADP ratio (Berg et al. 2009). As another example, a bacterial NADH-binding protein Rex has been found to bind both NADH and NAD^+ with remarkable pH sensitivity; Rex turns out to report the parameter $[NADH]/[NAD^+] \times [H^+]$, which reflects the redox potential of NAD(H)-dependent dehydrogenases (Hung et al. 2011). Therefore, when working with protein domains, one should determine if they are already tuned to endogenous metabolic parameters, as well as if these properties can be exploited to engineer fluorescent sensors of cell metabolism.

7.4.3 Sensor Specificity and Kinetics

Besides appropriate tuning, a sensor should be specific, recognizing the target ligand(s) and ignoring all others. The sensor specificity depends on both its affinity for the target as compared with interfering ligands, as well as the abundance of the target as compared with interfering ligands. Accordingly, it can be difficult to create sensors specific for rare metabolites, due to competition from interfering ligands of high abundance. For specificity assessment, one should test if the interfering ligands activate the sensor alone, as well as if they compete with the target to modulate the sensor response.

A sensor should also be fast. Genetically encoded fluorescent sensors—based on FP translocation, environmentally sensitive FP, FRET, or cpFP—are typically so, capable of detecting dynamics changes with a time resolution of seconds or less.

7.4.4 Detection Limit

As a sensor is expressed in cells, its intracellular concentration imposes a detection limit on its metabolic measurements. For FP-based sensors that work by direct ligand binding, when the concentration of sensor exceeds the concentration of ligand, the sensor response can be limited by stoichiometric binding and not by ligand titration. Typically, most FP-based sensors are expressed at 1–10 μM (Niswender et al. 1995; Akerboom et al. 2009). For rare metabolic ligands, such as those that exist at low nanomolar concentrations, at best only approximately 0.1% of the sensors can bind the ligands at a given time even if the sensor has extremely high affinity. In addition, under these circumstances, the sensor can strongly perturb the free ligand concentration. Thus, for FP-based biosensors, the target ligands should be far more abundant, in the submillimolar to millimolar range.

For detection of rare metabolites, this concentration limit can be overcome by an amplification scheme. Using enzymatic catalysis, primary signals from direct ligand binding can be amplified to stronger secondary signals. For instance, detection of nitric oxide (NO) at extremely low, picomolar concentrations can be achieved by expressing in cells both soluble guanylate cyclase and an FP-based cGMP reporter (Sato et al. 2006). On binding of minute amounts of NO, the amplified generation of cGMP by guanylate cyclase leads to detectable signals in the cGMP reporter.

7.4.5 Measurement Calibration

Measurements made with metabolic biosensors can be either qualitative or quantitative. In qualitative use, the sensor reports *relative* changes in the parameter (e.g., an elevation of the $NADH/NAD^+$ ratio), whereas for quantitative measurements, one can determine the *actual* metabolic parameter (e.g., an $NADH/NAD^+$ ratio of 0.1).

Quantitative measurements can be obtained only after careful calibration of the sensor signals in the intracellular milieu. For sensor calibration, cells are maintained in several metabolic conditions where the corresponding values of the parameter of interest are known, and the sensor signals are measured to generate a standard curve; the imaging data can then be converted into biologically meaningful measurements. One can also check if the sensor behaves as expected, by comparing the calibration data in cells with the biochemical data from purified sensor proteins. Although the standard curve should contain several points, a major challenge in sensor calibration is to settle on the metabolic conditions that present well-defined parameters and are experimentally feasible. Occasionally, when one can obtain only a single calibration point in cells, one may incorporate the purified protein data to generate a standard curve by assuming similar sensor behaviors in cells as in purified proteins. How often a sensor should be calibrated depends on the variability of its behavior among individual cells and/or experiments in identical metabolic states. If such variations are small, calibration data from one set can be applicable to other sets. Otherwise, calibration should be conducted on a cell-by-cell or experiment-by-experiment basis, for instance at the end of each imaging experiment. Overall, only through sensor calibration can one derive from the fluorescence data the actual metabolic parameters of interest.

Several FP-based sensors of metabolism can be routinely calibrated to yield quantitative measurements; examples include pH sensors and redox sensors. To calibrate pH sensors, one can use a combination of ionophores to clamp intracellular pH at various values (Llopis et al. 1998; Tantama et al. 2011). For calibration of general thiol redox sensors of the roGFP family (Dooley et al. 2004; Hanson et al. 2004) or the glutathione sensor Grx1–roGFP2 (Gutscher et al. 2008), one can utilize oxidants such as hydrogen peroxide and reducing agents such as dithiothreitol (DTT) to generate maximal and minimal sensor responses. Finally, for calibration of the cytosolic NADH–NAD^+ redox sensor Peredox (Hung et al. 2011), one can supply cells with a mixture of lactate and pyruvate, but without glucose. By using different lactate/pyruvate ratios and exploiting the endogenous lactate dehydrogenase reaction, one can poise the cytosolic NADH–NAD^+ redox state in cells, and the sensor can be shown to exhibit similar behaviors in cells and in purified proteins.

Even when calibration conditions for a sensor are well defined, it may still fall to each laboratory to conduct its own calibration, owing to the substantial variations in illuminator intensity and detector efficiency between instruments. An exception may be in fluorescence lifetime measurements (FLIM): Some biosensors exhibit substantial changes in fluorescence lifetime (e.g., Haj et al. 2002; Lee et al. 2009; Tantama et al. 2011), which can be measured reliably by time domain measurements in multiphoton microscopy. FLIM offers the possibility of calibrated fluorescence measurements that can be compared among laboratories and microscopy setups.

7.4.6 Potential Toxicity and Perturbations

The expression of any protein, including a fluorescent biosensor, may adversely affect cell physiology. In addition to potential FP misfolding and aggregation, the biosensor may interact with endogenous proteins, although the possibility of such signaling cross-talk could be minimized by using protein scaffolds derived from distantly related organisms. Also, because most metabolic sensors detect their ligands by direct binding or via enzymatic reactions, such buffering or catalysis may drastically alter metabolite levels and perturb cell physiology. To avoid these adverse effects, sensor expression ideally should be maintained at the lowest level that is sufficient for imaging; also, the imaging data should be checked for any dependence on the level of sensor expression.

7.5 EXAMPLES OF GENETICALLY ENCODED METABOLIC BIOSENSORS

After considering the basic principles of sensor design, we focus on a few examples of sensors to monitor the following parameters: glucose, general and glutathione redox state, ROS, ATP, AMPK activity, and NADH.

7.5.1 Glucose

Glucose is one of the major carbon sources for energetics and biosynthesis. After entering cells via specialized transporters (Carruthers 1990), glucose is phosphorylated and metabolized by diverse pathways: glycolysis, glycogen synthesis, the pentose phosphate pathway, and the glycosylation pathway (Vander Heiden et al. 2009; Wellen et al. 2010). Intracellular levels of glucose can thus exert profound effects on cell metabolism and signaling.

Fluorescent analogs of glucose have been used to estimate glucose uptake in single live cells. The intracellular accumulation of these glucose derivatives, such as 2-[*N*-(7-nitrobenz-2-oxa-1,3-diazol-4-yl)amino]-2-deoxy-D-glucose (2-NBDG), can be optically monitored to estimate glucose uptake (Yoshioka et al. 1996; Yamada et al. 2000; Loaiza et al. 2003). However, these compounds may exhibit unconventional uptake behaviors by the glucose transporters (Barros et al. 2009). In addition, for accurate measurements of intracellular 2-NBDG, careful control of perfusion is needed to reduce background fluorescence from extracellular 2-NBDG (Yamada et al. 2007).

Genetically encoded fluorescent sensors of glucose have also been used to monitor intracellular glucose levels and glucose flux. Each of these sensors is constructed by flanking a bacterial periplasmic glucose/galactose binding protein with two FPs of a FRET pair, such that the binding of glucose alters the FRET signal (Fehr et al. 2003; Hou et al. 2011). Among the many published sensor variants, FLII^{12}Pglu700μΔ6 (also known as FLII^{12}Pglu600μΔ6) may be the most useful in mammalian cells; its K_d for glucose is approximately 600–700 μM (Takanaga et al. 2008). Using this fluorescent glucose sensor, one can assess the intracellular glucose level as well as the rate of glucose utilization (Bittner et al. 2010, 2011). Glycolytic flux, or more precisely the rate of glucose elimination by metabolism, has been estimated by abruptly lowering extracellular glucose or inhibiting glucose transport, then monitoring changes in the intracellular glucose level using the biosensor. Nevertheless, for an accurate determination of the glucose level and its elimination rate, the proper calibration of the sensor response is essential. In this case, it may not suffice to rely on data from purified proteins, as the glucose sensor appears to behave differently as a purified protein compared with in cells (Takanaga et al. 2008).

7.5.2 General Thiol-Disulfide and Glutathione Redox State

Reduction and oxidation (redox) reactions abound in the intracellular environment. Cells utilize numerous sets of redox couples, such as NADH–NAD$^+$, NADPH–NADP$^+$, and the glutathione system, to transfer reducing equivalents in biochemical reactions. Each of these redox systems performs distinct metabolic functions, and their spatiotemporal dynamics are tightly regulated (Meyer and Dick 2010).

Redox-sensitive FPs and their variants have been used to monitor the cellular thiol redox states. In the rxYFP and roGFP family of proteins, the oxidation of two cysteine thiols placed near the FP chromophore can alter its fluorescence (Ostergaard et al. 2001; Dooley et al. 2004). This redox sensing is reversible because of the regeneration of free thiols by endogenous antioxidants. These redox-sensitive FPs can thus report on the dynamics of the general thiol-disulfide redox in live cells. Because of their ratiometric spectral responses, roGFP variants are more widely used; they have been targeted to various subcellular compartments for redox monitoring (Hanson et al. 2004; Merksamer et al. 2008; Guzman et al. 2010). Although detection by roGFP alone is not specific for any redox couple, a fluorescent sensor specific for the glutathione redox state has been engineered using a roGFP variant (Gutscher et al. 2008). Connecting roGFP2 via a peptide linker to the human glutaredoxin Grx1 facilitates rapid equilibration between the thiol redox state of the FP and the glutathione redox state in the intracellular environment. The resulting sensor, Grx1–roGFP2, can report on glutathione redox dynamics associated with various physiological stimuli. Furthermore, this sensor has been expressed as a transgene in *Drosophila* to assess the glutathione redox state in various tissues during development and aging (Albrecht et al. 2011). Overall, fluorescent probes based on roGFP variants have become convenient and versatile tools to study redox signaling.

7.5.3 ROS

Continuously generated as byproducts of oxidative metabolism, ROS are key mediators of diverse physiological processes and pathologies (Balaban et al. 2005; Winterbourn 2008; Murphy et al. 2011). ROS can be detected using various small molecule indicators (Wardman 2007). In addition, genetically encoded fluorescent sensors have been used to study the spatiotemporal dynamics of certain ROS metabolites, such as hydrogen peroxide and superoxide. Currently, there are two fluorescent sensors with distinct designs for the detection of hydrogen peroxide: On the one hand, the sensor HyPer was constructed by combining a cpFP with the bacterial peroxide-sensing domain OxyR (Belousov et al. 2006); the oxidation of two cysteine thiols in OxyR alters the sensor conformation and the cpFP fluorescence. On the other hand, the sensor roGFP2–Orp1 was constructed by linking roGFP2 to the yeast peroxiredoxin Orp1 (Gutscher et al. 2009). Here, roGFP2–Orp1 senses hydrogen peroxide via a redox relay mechanism similar to that of the glutathione sensor Grx1–roGFP2. Both HyPer and roGFP2–Orp1 respond specifically to hydrogen peroxide and provide a ratiometric fluorescence response. Interestingly, the detection of superoxide can also be achieved using the cpFP variant cpYFP alone, without conjugation to any ROS recognition domain (Wang et al. 2008). cpYFP has been used to visualize superoxide flashes in mitochondria in mammalian cells, yet the mechanism through which superoxide interacts with the cpYFP chromophore, as well as whether this interaction is specific, remains unclear. Finally, similar to many cpFP reporters, cpYFP and HyPer are sensitive to changes in physiological pH (Belousov et al. 2006; Wang et al. 2008); therefore, pH should be monitored to rule out artifacts.

7.5.4 ATP

ATP is the well-known currency for bioenergetics: Energy for cellular reactions is supplied by the conversion of ATP into related cofactors, ADP and AMP. Because the free ATP/ADP ratio is kept far from the chemical equilibrium in intracellular environments (Nicholls and Ferguson 2002), it can be utilized to do work and drive biosynthetic reactions. For bioenergetics, what matters is not the total amount of ATP alone but rather the free ATP/ADP or ATP/AMP ratio (Atkinson 1968; Pradet and Raymond 1983). Although the total ATP/ADP ratio can be chemically determined, this may underestimate the free ATP/ADP ratios, as relatively more ADP is bound to cellular proteins (Mörikofer-Zwez and Walter 1989).

Both indirect and direct methods have been devised to assess ATP levels in live cells. ATP can be monitored indirectly via Mg^{2+} measurements. Owing to the stronger binding of Mg^{2+} by ATP than by ADP, changes in ATP levels can be accompanied by reciprocal changes in Mg^{2+} levels, which can be detected with a small molecule indicator (Harman et al. 1990; Jung et al. 1990). However, this signal can be aliased by ATP-independent changes in Mg^{2+} levels. In addition, ATP can be measured directly using a luciferase reporter system to generate ATP-dependent luminescence. This approach has been used to detect cytosolic and mitochondrial ATP in single cells (Jouaville et al. 1999; Kennedy et al. 1999). However, the luminescence signal is typically extremely low and requires a sensitive photon-counting charge coupled device (CCD) camera for detection, and the consumption of ATP in this method may perturb cell metabolism.

Genetically encoded biosensors of ATP and the ATP/ADP ratio have been engineered by fusing FPs with ATP-binding protein domains. Unlike the luciferase method, these sensors do not consume ATP, and their signals are far brighter. Interestingly (and somewhat disturbingly), physiological amounts of ATP have been shown to alter the fluorescence of a simple tandem CFP–YFP FRET pair or the fluorescence lifetime of CFP alone (Willemse et al. 2007; Borst et al. 2010), though this phenomenon has not yet been used to measure ATP in live cells. Instead, ATP-binding protein domains have been used in conjunction with FPs to create reporters of ATP and related energetic parameters.

A fluorescent sensor of the cellular ATP/ADP ratio has been engineered by linking a cpFP to the bacterial protein GlnK1 (Berg et al. 2009). As GlnK1 binds both ATP and ADP with high affinity but with distinct conformations, the resulting sensor Perceval effectively reports the free ATP/ADP ratio. Perceval has been used to monitor dynamic changes in the ATP/ADP ratio in mammalian cells upon energetic challenges. Currently, this sensor is tuned to detect fairly low ATP/ADP ratios of <1. Although it is suitable for use in sensing energy deficits or in certain cell types, Perceval may not be responsive in most mammalian cells, which typically have higher resting ATP/ADP ratios. Also, like many cpFP sensors, Perceval can be pH sensitive and require correction; this necessitates concurrent measurements of cellular pH using a spectrally compatible dye indicator (Berg et al. 2009) or a red pH biosensor, pHRed (Tantama et al. 2011).

Aside from the ATP/ADP ratio, intracellular ATP can be measured using genetically encoded biosensors for ATP (Imamura et al. 2009). Known as the ATeams, these sensors have been created using a FRET strategy, with the ε-subunit of a bacterial F_0F_1-ATP synthase flanked by two FPs that form a FRET pair; the binding of ATP alters the protein conformation and results in a FRET change. Together, the ATeams cover a broad range of affinities for ATP and can report ATP levels in distinct subcellular compartments and on energetic challenges. Through the use of different FPs, an ATeam variant has been designed to be spectrally compatible with fura-2, allowing simultaneous imaging of both ATP and calcium (Nakano et al. 2011). As these FP-based sensors show notable pH sensitivity in their dynamic range and ATP binding affinity, pH should be measured concurrently or in parallel with sensor measurements to rule out pH artifacts.

7.5.5 AMPK Activity

AMPK is a master regulator of metabolism, and its dysregulation has been implicated in various metabolic and neoplastic diseases (Sarbassov et al. 2005; Shaw and Cantley 2006). On energy challenges, the AMP level increases and leads to the allosteric activation of AMPK (Hardie 2004). With multiple downstream targets, AMPK maintains cellular energy balance by promoting catabolic processes such as fatty acid oxidation, inhibiting macromolecule biosynthesis, and coordinating diverse physiological processes including autophagy, mitochondrial biogenesis, and cell division (Hardie 2004; Reznick and Shulman 2006; Banko et al. 2011; Mihaylova and Shaw 2011).

The spatiotemporal dynamics of AMPK activity can be monitored in single cells using a genetically encoded fluorescent sensor. Similar to other biosensors of kinase activity (Allen and Zhang 2006; Gallegos et al. 2006), this sensor of AMPK activity is based on FRET, with two FPs of a FRET pair linked to an optimized substrate peptide together with the phosphothreonine binding domain FHA1 (Tsou et al. 2011). On activation of AMPK, the phosphorylated peptide binds to the FHA1 domain, altering the sensor conformation and thus the FRET signal. This fluorescent sensor AMPKAR yields reversible FRET signal changes because the phosphorylated peptide can be acted on by endogenous phosphatases; thus, AMPKAR technically detects the balance of activity between AMPK and endogenous phosphatases. To check for sensor specificity, the phosphorylation site within the AMPK substrate peptide can be spoiled as a control. Using AMPKAR, one can examine the dynamics and the heterogeneity of AMPK activity among individual cells. Depending on the stimuli, AMPK activation occurs in distinct subcellular compartments. Although calcium elevation leads to AMPK activation both in the cytosol and the nucleus, the application of 2-deoxyglucose induces AMPK activation that is confined to the cytosol only. In addition, AMPKAR has been used to track AMPK activity during cell division, showing that AMPK coordinates energetic status with mitotic regulation (Banko et al. 2011).

7.5.6 NADH

Nicotinamide adenine dinucleotide (reduced: NADH; oxidized NAD^+) is a key cofactor in redox reactions. In mammalian cells, NADH–NAD^+ redox environments vastly differ between the cytosol and the mitochondrial matrix (Nicholls and Ferguson 2002), involving distinct metabolic pathways, for instance, glycolysis in the cytosol and the tricarboxylic acid (TCA) cycle in mitochondria. In addition, there are dedicated shuttle systems that transfer reducing equivalents of NADH from the cytosol into mitochondria (Schoolwerth and LaNoue 1985).

Although NADH can be measured using biochemical methods or autofluorescence, these traditional approaches have drawbacks. With biochemical determination, one can indirectly measure the free NADH/NAD^+ ratio, the pertinent parameter in redox reactions. Though NADH and NAD^+ can be directly extracted, such measurements of the total NADH/NAD^+ ratio greatly overestimate the actual free NADH/NAD^+ ratio, as a greater fraction of NADH than NAD^+ is protein bound (Williamson et al. 1967; Zhang et al. 2002b). Instead, by assuming a constant, known pH and exploiting a dehydrogenase reaction presumably at equilibrium (such as the lactate dehydrogenase), one can use the ratio of that redox couple (e.g., the lactate/pyruvate ratio) to infer the free NADH/NAD^+ ratio (Williamson et al. 1967). However, this is not a single-cell measurement but rather an average from millions of cells. Alternatively, autofluorescence imaging can be used to assess NADH in single cells, but this method has several limitations: First, it cannot distinguish between NADH and the related cofactor NADPH (Rocheleau et al. 2004); distinguishing between these two cofactors is crucial, as each governs different redox reactions (Klingenberg and Bücher 1960). Second, because autofluorescence comes mostly from protein-bound forms of the cofactors, intracellular changes such as pH can alter the signal by affecting the NAD(P)H–protein interactions (Ogikubo et al. 2011). Third, this method cannot yield the free NADH/NAD^+ ratio. Finally, mitochondrial signals often dominate the measurements, whereas cytosolic autofluorescence is notably weak (Patterson et al. 2000).

To overcome these drawbacks, two groups have independently created FP-based sensors of NADH (Hung et al. 2011; Zhao et al. 2011b). Although both groups adopted a cpFP strategy using the bacterial NADH-binding protein Rex, the sensors differ in their properties and in the redox parameters they detect. In the first report, while a cpFP inserted into Rex reports the parameter $[NADH]/[NAD^+] \times [H^+]$, the optimized sensor Peredox measures the cytosolic free NADH/NAD^+ ratio in mammalian cells (Hung et al. 2011). Based on a cpFP variant of T-Sapphire with a low pK_a (Zapata-Hommer and Griesbeck 2003), the Peredox response is notably pH resistant, unlike many other cpFP sensors. Using high-content image analysis, Peredox can report on the dynamics of glycolysis as indicated by the cytosolic free NADH/NAD^+ ratios in hundreds of live cells. With its current affinity, Peredox is tuned to sensing NADH/NAD^+ ratios in the cytosol but not in the mitochondrial matrix, where free NADH/NAD^+ ratios can be 100- to 1000-fold higher (Williamson et al. 1967).

In contrast, the fluorescent sensor Frex detects the NADH level rather than the NADH/NAD^+ ratio (Zhao et al. 2011b). This sensor was constructed by combining a cpYFP with part of the Rex protein. After being targeted to the mitochondrial matrix, the sensor reveals mitochondrial NADH changes during energetic or peroxide challenges. To account for pH interference on Frex signals, cpYFP alone has been used as a control in parallel experiments. However, this may not be adequate to correct for the effect of pH on NADH binding (Hung et al. 2011), as well as the different pH sensitivities often seen in the bound and unbound states of cpYFP-based sensors (Belousov et al. 2006; Berg et al. 2009).

Overall, as genetically encoded fluorescent sensors, Peredox and Frex may be used to reveal cytosolic and mitochondrial NADH–NAD^+ redox dynamics in single cells with unprecedented detail. When used in conjunction with other fluorescent reporters, they may allow us to obtain an integrated view of the dynamics and regulation of metabolic pathways.

7.6 CONCLUSION

In this chapter, we have reviewed the basic design principles for creating optical probes of cell metabolism, and we have elaborated on examples of several fluorescent biosensors. Indeed, the field of metabolic sensor design is still in its infancy. Looking ahead, we anticipate the following challenges and developments: First, many existing fluorescent biosensors could be greatly improved, as most FRET and cpFP sensors suffer from limited dynamic range or notable pH sensitivity, rendering them difficult to use. Efficient methods to generate and screen for improved variants would be valuable (Zhao et al. 2011a); desirable properties to look for include greater dynamic range, pH resistance, distinct colors for multiplex imaging, or larger lifetime signals for two-photon FLIM in deep tissues. Furthermore, given the vastness of the metabolome, our existing toolbox remains extremely limited and could be further expanded. In designing sensors for novel metabolic targets, the focus should be on key molecules that are involved in the regulation of core metabolic pathways, and potential sensor scaffolds may be found amidst the abundance of metabolite-binding protein domains from diverse organisms. In addition to FPs and metabolite-binding protein domains, fluorescent biosensors can also be engineered using RNA aptamers combined with small-molecule fluorophores to image diverse intracellular metabolites (Paige et al. 2011, 2012). Lastly, the current and upcoming probes may help us to elucidate metabolic regulation in diverse areas, such as microbiology, metabolic engineering, cancer metabolism, and neurobiology (Gerosa and Sauer 2011; Lunt and Vander Heiden 2011; Reaves and Rabinowitz 2011; Tantama et al. 2012). Although fluorescent sensors cannot monitor the fluxes of metabolic pathways per se, their measurements of metabolite levels or metabolic parameters in individual intact cells provide valuable information to constrain and evaluate computational models of metabolism. In summary, genetically encoded optical probes should open the door to unraveling the elaborate organization and complexity of cell metabolism, as well as the consequences of such metabolic regulation in health and disease.

ACKNOWLEDGMENTS

We are grateful to Mathew Tantama for his comments. This work was supported by the US National Institutes of Health (R01 NS055031 to G. Y.), as well as an Albert J. Ryan Fellowship and a Stuart H. Q. and Victoria Quan Predoctoral Fellowship in Neurobiology (both to Y. P. H.).

REFERENCES

Akerboom, J., Rivera, J. D. V., Guilbe, M. M. R., Malavé, E. C. A., Hernandez, H. H., Tian, L., Hires, S. A., Marvin, J. S., Looger, L. L. and Schreiter, E. R. (2009). Crystal structures of the GCaMP calcium sensor reveal the mechanism of fluorescence signal change and aid rational design. *J Biol Chem 284*, 6455–6464.

Albrecht, S. C., Barata, A. G., Grosshans, J., Teleman, A. A. and Dick, T. P. (2011). In vivo mapping of hydrogen peroxide and oxidized glutathione reveals chemical and regional specificity of redox homeostasis. *Cell Metab 14*, 819–829.

Allen, M. D. and Zhang, J. (2006). Subcellular dynamics of protein kinase A activity visualized by FRET-based reporters. *Biochem Biophys Res Commun 348*, 716–721.

Anastasiou, D., Poulogiannis, G., Asara, J. M., Boxer, M. B., Jiang, J.-K., Shen, M., Bellinger, G., Sasaki, A. T., Locasale, J. W., Auld, D. S., Thomas, C. J., Vander Heiden, M. G. and Cantley, L. C. (2011). Inhibition of pyruvate kinase M2 by reactive oxygen species contributes to cellular antioxidant responses. *Science 334*, 1278–1283.

Arosio, D., Ricci, F., Marchetti, L., Gualdani, R., Albertazzi, L. and Beltram, F. (2010). Simultaneous intracellular chloride and pH measurements using a GFP-based sensor. *Nat Methods 7*, 516–518.

Asher, G. and Schibler, U. (2011). Crosstalk between components of circadian and metabolic cycles in mammals. *Cell Metab 13*, 125–137.

Atkinson, D. E. (1968). The energy charge of the adenylate pool as a regulatory parameter. Interaction with feedback modifiers. *Biochemistry 7*, 4030–4034.

Bailey, E. E., Pfeifer, H. H. and Thiele, E. A. (2005). The use of diet in the treatment of epilepsy. *Epilepsy Behav 6*, 4–8.

Baird, G. S., Zacharias, D. A. and Tsien, R. Y. (1999). Circular permutation and receptor insertion within green fluorescent proteins. *Proc Natl Acad Sci U S A 96*, 11241–11246.

Balaban, R. S., Nemoto, S. and Finkel, T. (2005). Mitochondria, oxidants, and aging. *Cell 120*, 483–495.

Banko, M. R., Allen, J. J., Schaffer, B. E., Wilker, E. W., Tsou, P., White, J. L., Villén, J., Wang, B., Kim, S. R., Sakamoto, K., Gygi, S. P., Cantley, L. C., Yaffe, M. B., Shokat, K. M. and Brunet, A. (2011). Chemical genetic screen for AMPKα2 substrates uncovers a network of proteins involved in mitosis. *Mol Cell 44*, 878–892.

Barros, L. F., Bittner, C. X., Loaiza, A., Ruminot, I., Larenas, V., Moldenhauer, H., Oyarzún, C. and Alvarez, M. (2009). Kinetic validation of 6-NBDG as a probe for the glucose transporter GLUT1 in astrocytes. *J Neurochem 109 Suppl 1*, 94–100.

Bass, J. and Takahashi, J. S. (2010). Circadian integration of metabolism and energetics. *Science 330*, 1349–1354.

Belousov, V. V., Fradkov, A. F., Lukyanov, K. A., Staroverov, D. B., Shakhbazov, K. S., Terskikh, A. V. and Lukyanov, S. (2006). Genetically encoded fluorescent indicator for intracellular hydrogen peroxide. *Nat Methods 3*, 281–286.

Berg, J., Hung, Y. P. and Yellen, G. (2009). A genetically encoded fluorescent reporter of ATP:ADP ratio. *Nat Methods 6*, 161–166.

Bittner, C. X., Loaiza, A., Ruminot, I., Larenas, V., Sotelo-Hitschfeld, T., Gutiérrez, R., Córdova, A., Valdebenito, R., Frommer, W. B. and Barros, L. F. (2010). High resolution measurement of the glycolytic rate. *Front Neuroenergetics 2*, pii: 26.

Bittner, C. X., Valdebenito, R., Ruminot, I., Loaiza, A., Larenas, V., Sotelo-Hitschfeld, T., Moldenhauer, H., San Martín, A., Gutiérrez, R., Zambrano, M. and Barros, L. F. (2011). Fast and reversible stimulation of astrocytic glycolysis by K^+ and a delayed and persistent effect of glutamate. *J Neurosci 31*, 4709–4713.

Borst, J. W., Willemse, M., Slijkhuis, R., van der Krogt, G., Laptenok, S. P., Jalink, K., Wieringa, B. and Fransen, J. A. M. (2010). ATP changes the fluorescence lifetime of cyan fluorescent protein via an interaction with His148. *PLoS One 5*, e13862.

Buchakjian, M. R. and Kornbluth, S. (2010). The engine driving the ship: metabolic steering of cell proliferation and death. *Nat Rev Mol Cell Biol 11*, 715–727.

Campbell, R. E. (2009). Fluorescent-protein-based biosensors: Modulation of energy transfer as a design principle. *Anal Chem 81*, 5972–5979.

Carruthers, A. (1990). Facilitated diffusion of glucose. *Physiol Rev 70*, 1135–1176.

Christofk, H. R., Vander Heiden, M. G., Harris, M. H., Ramanathan, A., Gerszten, R. E., Wei, R., Fleming, M. D., Schreiber, S. L. and Cantley, L. C. (2008). The M2 splice isoform of pyruvate kinase is important for cancer metabolism and tumour growth. *Nature 452*, 230–233.

Day, R. N. and Davidson, M. W. (2009). The fluorescent protein palette: Tools for cellular imaging. *Chem Soc Rev 38*, 2887–2921.

DiPilato, L. M., Cheng, X. and Zhang, J. (2004). Fluorescent indicators of cAMP and Epac activation reveal differential dynamics of cAMP signaling with discrete subcellular compartments. *Proc Natl Acad Sci U S A 101*, 16513–16518.

Dooley, C. T., Dore, T. M., Hanson, G. T., Jackson, W. C., Remington, S. J. and Tsien, R. Y. (2004). Imaging dynamic redox changes in mammalian cells with green fluorescent protein indicators. *J Biol Chem 279*, 22284–22293.

Fehr, M., Lalonde, S., Lager, I., Wolff, M. W. and Frommer, W. B. (2003). In vivo imaging of the dynamics of glucose uptake in the cytosol of COS-7 cells by fluorescent nanosensors. *J Biol Chem 278*, 19127–19133.

Frommer, W. B., Davidson, M. W. and Campbell, R. E. (2009). Genetically encoded biosensors based on engineered fluorescent proteins. *Chem Soc Rev 38*, 2833–2841.

Gallegos, L. L., Kunkel, M. T. and Newton, A. C. (2006). Targeting protein kinase C activity reporter to discrete intracellular regions reveals spatiotemporal differences in agonist-dependent signaling. *J Biol Chem 281*, 30947–30956.

Gao, P., Tchernyshyov, I., Chang, T.-C., Lee, Y.-S., Kita, K., Ochi, T., Zeller, K. I., De Marzo, A. M., Van Eyk, J. E., Mendell, J. T. and Dang, C. V. (2009). c-Myc suppression of miR-23a/b enhances mitochondrial glutaminase expression and glutamine metabolism. *Nature 458*, 762–765.

Gerosa, L. and Sauer, U. (2011). Regulation and control of metabolic fluxes in microbes. *Curr Opin Biotechnol 22*, 566–575.

Giepmans, B. N. G., Adams, S. R., Ellisman, M. H. and Tsien, R. Y. (2006). The fluorescent toolbox for assessing protein location and function. *Science 312*, 217–224.

Griesbeck, O., Baird, G. S., Campbell, R. E., Zacharias, D. A. and Tsien, R. Y. (2001). Reducing the environmental sensitivity of yellow fluorescent protein. Mechanism and applications. *J Biol Chem 276*, 29188–29194.

Gutscher, M., Pauleau, A.-L., Marty, L., Brach, T., Wabnitz, G. H., Samstag, Y., Meyer, A. J. and Dick, T. P. (2008). Real-time imaging of the intracellular glutathione redox potential. *Nat Methods 5*, 553–559.

Gutscher, M., Sobotta, M. C., Wabnitz, G. H., Ballikaya, S., Meyer, A. J., Samstag, Y. and Dick, T. P. (2009). Proximity-based protein thiol oxidation by H_2O_2-scavenging peroxidases. *J Biol Chem 284*, 31532–31540.

Guzman, J. N., Sanchez-Padilla, J., Wokosin, D., Kondapalli, J., Ilijic, E., Schumacker, P. T. and Surmeier, D. J. (2010). Oxidant stress evoked by pacemaking in dopaminergic neurons is attenuated by DJ-1. *Nature 468*, 696–700.

Haj, F. G., Verveer, P. J., Squire, A., Neel, B. G. and Bastiaens, P. I. H. (2002). Imaging sites of receptor dephosphorylation by PTP1B on the surface of the endoplasmic reticulum. *Science 295*, 1708–1711.

Han, J. and Burgess, K. (2010). Fluorescent indicators for intracellular pH. *Chem Rev 110*, 2709–2728.

Hanson, G. T., Aggeler, R., Oglesbee, D., Cannon, M., Capaldi, R. A., Tsien, R. Y. and Remington, S. J. (2004). Investigating mitochondrial redox potential with redox-sensitive green fluorescent protein indicators. *J Biol Chem 279*, 13044–13053.

Hardie, D. G. (2004). The AMP-activated protein kinase pathway—new players upstream and downstream. *J Cell Sci 117*, 5479–5487.

Harman, A. W., Nieminen, A. L., Lemasters, J. J. and Herman, B. (1990). Cytosolic free magnesium, ATP and blebbing during chemical hypoxia in cultured rat hepatocytes. *Biochem Biophys Res Commun 170*, 477–483.

Haymond, M. W., Howard, C., Ben-Galim, E. and DeVivo, D. C. (1983). Effects of ketosis on glucose flux in children and adults. *Am J Physiol 245*, E373–E378.

Herbst, K. J., Ni, Q. and Zhang, J. (2009). Dynamic visualization of signal transduction in living cells: from second messengers to kinases. *IUBMB Life 61*, 902–908.

Hires, S. A., Zhu, Y. and Tsien, R. Y. (2008). Optical measurement of synaptic glutamate spillover and reuptake by linker optimized glutamate-sensitive fluorescent reporters. *Proc Natl Acad Sci U S A 105*, 4411–4416.

Hirose, K., Kadowaki, S., Tanabe, M., Takeshima, H. and Iino, M. (1999). Spatiotemporal dynamics of inositol 1,4,5-trisphosphate that underlies complex Ca^{2+} mobilization patterns. *Science 284*, 1527–1530.

Hirrlinger, P. G., Scheller, A., Braun, C., Quintela-Schneider, M., Fuss, B., Hirrlinger, J. and Kirchhoff, F. (2005). Expression of reef coral fluorescent proteins in the central nervous system of transgenic mice. *Mol Cell Neurosci 30*, 291–303.

Honda, A., Adams, S. R., Sawyer, C. L., Lev-Ram, V., Tsien, R. Y. and Dostmann, W. R. (2001). Spatiotemporal dynamics of guanosine 3′,5′-cyclic monophosphate revealed by a genetically encoded, fluorescent indicator. *Proc Natl Acad Sci U S A 98*, 2437–2442.

Hou, B.-H., Takanaga, H., Grossmann, G., Chen, L.-Q., Qu, X.-Q., Jones, A. M., Lalonde, S., Schweissgut, O., Wiechert, W. and Frommer, W. B. (2011). Optical sensors for monitoring dynamic changes of intracellular metabolite levels in mammalian cells. *Nat Protoc 6*, 1818–1833.

Hung, Y. P., Albeck, J. G., Tantama, M. and Yellen, G. (2011). Imaging cytosolic NADH-NAD^+ redox state with a genetically encoded fluorescent biosensor. *Cell Metab 14*, 545–554.

Imamura, H., Nhat, K. P. H., Togawa, H., Saito, K., Iino, R., Kato-Yamada, Y., Nagai, T. and Noji, H. (2009). Visualization of ATP levels inside single living cells with fluorescence resonance energy transfer-based genetically encoded indicators. *Proc Natl Acad Sci U S A 106*, 15651–15656.

Jayaraman, S., Haggie, P., Wachter, R. M., Remington, S. J. and Verkman, A. S. (2000). Mechanism and cellular applications of a green fluorescent protein-based halide sensor. *J Biol Chem 275*, 6047–6050.

Jouaville, L. S., Pinton, P., Bastianutto, C., Rutter, G. A. and Rizzuto, R. (1999). Regulation of mitochondrial ATP synthesis by calcium: evidence for a long-term metabolic priming. *Proc Natl Acad Sci U S A 96*, 13807–13812.

Jung, D. W., Apel, L. and Brierley, G. P. (1990). Matrix free Mg^{2+} changes with metabolic state in isolated heart mitochondria. *Biochemistry 29*, 4121–4128.

Katayama, H., Yamamoto, A., Mizushima, N., Yoshimori, T. and Miyawaki, A. (2008). GFP-like proteins stably accumulate in lysosomes. *Cell Struct Funct 33*, 1–12.

Kennedy, H. J., Pouli, A. E., Ainscow, E. K., Jouaville, L. S., Rizzuto, R. and Rutter, G. A. (1999). Glucose generates sub-plasma membrane ATP microdomains in single islet beta-cells. Potential role for strategically located mitochondria. *J Biol Chem 274*, 13281–13291.

Kirber, M. T., Chen, K. and Keaney, J. F., Jr. (2007). YFP photoconversion revisited: Confirmation of the CFP-like species. *Nat Methods 4*, 767–768.

Klingenberg, M. and Bücher, T. (1960). Biological oxidations. *Annu Rev Biochem 29*, 669–708.

Komatsu, N., Aoki, K., Yamada, M., Yukinaga, H., Fujita, Y., Kamioka, Y. and Matsuda, M. (2011). Development of an optimized backbone of FRET biosensors for kinases and GTPases. *Mol Biol Cell 22*, 4647–4656.

Krebs, H. A. (1977). Regulatory mechanisms in purine biosynthesis. *Adv Enzyme Regul 16*, 409–422.

Kremers, G.-J., Hazelwood, K. L., Murphy, C. S., Davidson, M. W. and Piston, D. W. (2009). Photoconversion in orange and red fluorescent proteins. *Nat Methods 6*, 355–358.

Lee, S.-J. R., Escobedo-Lozoya, Y., Szatmari, E. M. and Yasuda, R. (2009). Activation of CaMKII in single dendritic spines during long-term potentiation. *Nature 458*, 299–304.

Lemke, E. A. and Schultz, C. (2011). Principles for designing fluorescent sensors and reporters. *Nat Chem Biol 7*, 480–483.

Llopis, J., McCaffery, J. M., Miyawaki, A., Farquhar, M. G. and Tsien, R. Y. (1998). Measurement of cytosolic, mitochondrial, and Golgi pH in single living cells with green fluorescent proteins. *Proc Natl Acad Sci U S A 95*, 6803–6808.

Loaiza, A., Porras, O. H. and Barros, L. F. (2003). Glutamate triggers rapid glucose transport stimulation in astrocytes as evidenced by real-time confocal microscopy. *J Neurosci 23*, 7337–7342.

Locasale, J. W., Grassian, A. R., Melman, T., Lyssiotis, C. A., Mattaini, K. R., Bass, A. J., Heffron, G., Metallo, C. M., Muranen, T., Sharfi, H., Sasaki, A. T., Anastasiou, D., Mullarky, E., Vokes, N. I., Sasaki, M., Beroukhim, R., Stephanopoulos, G., Ligon, A. H., Meyerson, M., Richardson, A. L., Chin, L., Wagner, G., Asara, J. M., Brugge, J. S., Cantley, L. C. and Vander Heiden, M. G. (2011). Phosphoglycerate dehydrogenase diverts glycolytic flux and contributes to oncogenesis. *Nat Genet 43*, 869–874.

Lowry, O. H. and Passonneau, J. V. (1964). The relationships between substrates and enzymes of glycolysis in brain. *J Biol Chem 239*, 31–42.

Lunt, S. Y. and Vander Heiden, M. G. (2011). Aerobic glycolysis: Meeting the metabolic requirements of cell proliferation. *Annu Rev Cell Dev Biol 27*, 441–464.

Merksamer, P. I., Trusina, A. and Papa, F. R. (2008). Real-time redox measurements during endoplasmic reticulum stress reveal interlinked protein folding functions. *Cell 135*, 933–947.

Meyer, A. J. and Dick, T. P. (2010). Fluorescent protein-based redox probes. *Antioxid Redox Signal 13*, 621–650.

Miesenböck, G., De Angelis, D. A. and Rothman, J. E. (1998). Visualizing secretion and synaptic transmission with pH-sensitive green fluorescent proteins. *Nature 394*, 192–195.

Mihaylova, M. M. and Shaw, R. J. (2011). The AMPK signalling pathway coordinates cell growth, autophagy and metabolism. *Nat Cell Biol 13*, 1016–1023.

Mishina, N., Bogeski, I., Bolotin, D. A., Hoth, M., Niemeyer, B. A., Schultz, C., Zagaynova, E. V., Lukyanov, S. and Belousov, V. (2012). Can we see PIP3 and hydrogen peroxide with a single probe? *Antioxid Redox Signal 17*, 505–512.

Miyawaki, A. (2011). Development of probes for cellular functions using fluorescent proteins and fluorescence resonance energy transfer. *Annu Rev Biochem 80*, 357–373.

Miyawaki, A., Llopis, J., Heim, R., McCaffery, J. M., Adams, J. A., Ikura, M. and Tsien, R. Y. (1997). Fluorescent indicators for Ca^{2+} based on green fluorescent proteins and calmodulin. *Nature 388*, 882–887.

Mörikofer-Zwez, S. and Walter, P. (1989). Binding of ADP to rat liver cytosolic proteins and its influence on the ratio of free ATP/free ADP. *Biochem J 259*, 117–124.

Murphy, M. P., Holmgren, A., Larsson, N.-G., Halliwell, B., Chang, C. J., Kalyanaraman, B., Rhee, S. G., Thornalley, P. J., Partridge, L., Gems, D., Nyström, T., Belousov, V., Schumacker, P. T. and Winterbourn, C. C. (2011). Unraveling the biological roles of reactive oxygen species. *Cell Metab 13*, 361–366.

Nagai, T., Ibata, K., Park, E. S., Kubota, M., Mikoshiba, K. and Miyawaki, A. (2002). A variant of yellow fluorescent protein with fast and efficient maturation for cell-biological applications. *Nat Biotechnol 20*, 87–90.

Nagai, T., Sawano, A., Park, E. S. and Miyawaki, A. (2001). Circularly permuted green fluorescent proteins engineered to sense Ca^{2+}. *Proc Natl Acad Sci U S A 98*, 3197–3202.

Nagai, T., Yamada, S., Tominaga, T., Ichikawa, M. and Miyawaki, A. (2004). Expanded dynamic range of fluorescent indicators for Ca^{2+} by circularly permuted yellow fluorescent proteins. *Proc Natl Acad Sci U S A 101*, 10554–10559.

Nakano, M., Imamura, H., Nagai, T. and Noji, H. (2011). Ca^{2+} regulation of mitochondrial ATP synthesis visualized at the single cell level. *ACS Chem Biol 6*, 709–715.

Nausch, L. W. M., Ledoux, J., Bonev, A. D., Nelson, M. T. and Dostmann, W. R. (2008). Differential patterning of cGMP in vascular smooth muscle cells revealed by single GFP-linked biosensors. *Proc Natl Acad Sci U S A 105*, 365–370.

Nicholls, D. G. and Ferguson, S. J. (2002). *Bioenergetics*, 3rd ed. San Diego, CA: Academic Press.

Nikolaev, V. O., Bünemann, M., Hein, L., Hannawacker, A. and Lohse, M. J. (2004). Novel single chain cAMP sensors for receptor-induced signal propagation. *J Biol Chem 279*, 37215–37218.

Nikolaev, V. O., Gambaryan, S. and Lohse, M. J. (2006). Fluorescent sensors for rapid monitoring of intracellular cGMP. *Nat Methods 3*, 23–25.

Niswender, K. D., Blackman, S. M., Rohde, L., Magnuson, M. A. and Piston, D. W. (1995). Quantitative imaging of green fluorescent protein in cultured cells: Comparison of microscopic techniques, use in fusion proteins and detection limits. *J Microsc 180*, 109–116.

Oancea, E. and Meyer, T. (1998). Protein kinase C as a molecular machine for decoding calcium and diacylglycerol signals. *Cell 95*, 307–318.

Oancea, E., Teruel, M. N., Quest, A. F. and Meyer, T. (1998). Green fluorescent protein (GFP)-tagged cysteine-rich domains from protein kinase C as fluorescent indicators for diacylglycerol signaling in living cells. *J Cell Biol 140*, 485–498.

Ogikubo, S., Nakabayashi, T., Adachi, T., Islam, M. S., Yoshizawa, T., Kinjo, M. and Ohta, N. (2011). Intracellular pH sensing using autofluorescence lifetime microscopy. *J Phys Chem B 115*, 10385–10390.

Okumoto, S., Looger, L. L., Micheva, K. D., Reimer, R. J., Smith, S. J. and Frommer, W. B. (2005). Detection of glutamate release from neurons by genetically encoded surface-displayed FRET nanosensors. *Proc Natl Acad Sci U S A 102*, 8740–8745.

Ostergaard, H., Henriksen, A., Hansen, F. G. and Winther, J. R. (2001). Shedding light on disulfide bond formation: Engineering a redox switch in green fluorescent protein. *EMBO J 20*, 5853–5862.

Owen, O. E., Morgan, A. P., Kemp, H. G., Sullivan, J. M., Herrera, M. G. and Cahill, G., Jr. (1967). Brain metabolism during fasting. *J Clin Invest 46*, 1589–1595.

Paige, J. S., Nguyen-Duc, T., Song, W. and Jaffrey, S. R. (2012). Fluorescence imaging of cellular metabolites with RNA. *Science 335*, 1194.

Paige, J. S., Wu, K. Y. and Jaffrey, S. R. (2011). RNA mimics of green fluorescent protein. *Science 333*, 642–646.

Palmer, A. E., Qin, Y., Park, J. G. and McCombs, J. E. (2011). Design and application of genetically encoded biosensors. *Trends Biotechnol 29*, 144–152.

Patterson, G. H., Knobel, S. M., Arkhammar, P., Thastrup, O. and Piston, D. W. (2000). Separation of the glucose-stimulated cytoplasmic and mitochondrial NAD(P)H responses in pancreatic islet beta cells. *Proc Natl Acad Sci U S A 97*, 5203–5207.

Poburko, D., Santo-Domingo, J. and Demaurex, N. (2011). Dynamic regulation of the mitochondrial proton gradient during cytosolic calcium elevations. *J Biol Chem 286*, 11672–11684.

Possemato, R., Marks, K. M., Shaul, Y. D., Pacold, M. E., Kim, D., Birsoy, K., Sethumadhavan, S., Woo, H.-K., Jang, H. G., Jha, A. K., Chen, W. W., Barrett, F. G., Stransky, N., Tsun, Z.-Y., Cowley, G. S., Barretina, J., Kalaany, N. Y., Hsu, P. P., Ottina, K., Chan, A. M.,

Yuan, B., Garraway, L. A., Root, D. E., Mino-Kenudson, M., Brachtel, E. F., Driggers, E. M. and Sabatini, D. M. (2011). Functional genomics reveal that the serine synthesis pathway is essential in breast cancer. *Nature 476*, 346–350.

Pradet, A. and Raymond, P. (1983). Adenine nucleotide ratios and adenylate energy charge in energy metabolism. *Annu Rev Plant Physiol 34*, 199–224.

Raucher, D., Stauffer, T., Chen, W., Shen, K., Guo, S., York, J. D., Sheetz, M. P. and Meyer, T. (2000). Phosphatidylinositol 4,5-bisphosphate functions as a second messenger that regulates cytoskeleton-plasma membrane adhesion. *Cell 100*, 221–228.

Reaves, M. L. and Rabinowitz, J. D. (2011). Metabolomics in systems microbiology. *Curr Opin Biotechnol 22*, 17–25.

Reznick, R. M. and Shulman, G. I. (2006). The role of AMP-activated protein kinase in mitochondrial biogenesis. *J Physiol 574*, 33–39.

Rocheleau, J. V., Head, W. S. and Piston, D. W. (2004). Quantitative NAD(P)H/flavoprotein autofluorescence imaging reveals metabolic mechanisms of pancreatic islet pyruvate response. *J Biol Chem 279*, 31780–31787.

Romoser, V. A., Hinkle, P. M. and Persechini, A. (1997). Detection in living cells of Ca^{2+}-dependent changes in the fluorescence emission of an indicator composed of two green fluorescent protein variants linked by a calmodulin-binding sequence. A new class of fluorescent indicators. *J Biol Chem 272*, 13270–13274.

Sarbassov, D. D., Ali, S. M. and Sabatini, D. M. (2005). Growing roles for the mTOR pathway. *Curr Opin Cell Biol 17*, 596–603.

Sato, M., Hida, N., Ozawa, T. and Umezawa, Y. (2000). Fluorescent indicators for cyclic GMP based on cyclic GMP-dependent protein kinase Ialpha and green fluorescent proteins. *Anal Chem 72*, 5918–5924.

Sato, M., Nakajima, T., Goto, M. and Umezawa, Y. (2006). Cell-based indicator to visualize picomolar dynamics of nitric oxide release from living cells. *Anal Chem 78*, 8175–8182.

Sato, M., Ueda, Y., Takagi, T. and Umezawa, Y. (2003). Production of PtdInsP3 at endomembranes is triggered by receptor endocytosis. *Nat Cell Biol 5*, 1016–1022.

Schoolwerth, A. C. and LaNoue, K. F. (1985). Transport of metabolic substrates in renal mitochondria. *Annu Rev Physiol 47*, 143–171.

Shaner, N. C., Campbell, R. E., Steinbach, P. A., Giepmans, B. N. G., Palmer, A. E. and Tsien, R. Y. (2004). Improved monomeric red, orange and yellow fluorescent proteins derived from *Discosoma* sp. red fluorescent protein. *Nat Biotechnol 22*, 1567–1572.

Shaner, N. C., Steinbach, P. A. and Tsien, R. Y. (2005). A guide to choosing fluorescent proteins. *Nat Methods 2*, 905–909.

Shaw, R. J. and Cantley, L. C. (2006). Ras, PI(3)K and mTOR signalling controls tumour cell growth. *Nature 441*, 424–430.

Shimozono, S., Fukano, T., Kimura, K. D., Mori, I., Kirino, Y. and Miyawaki, A. (2004). Slow Ca^{2+} dynamics in pharyngeal muscles in *Caenorhabditis elegans* during fast pumping. *EMBO Rep 5*, 521–526.

Sokoloff, L., Reivich, M., Kennedy, C., Des Rosiers, M. H., Patlak, C. S., Pettigrew, K. D., Sakurada, O. and Shinohara, M. (1977). The [^{14}C]deoxyglucose method for the measurement of local cerebral glucose utilization: Theory, procedure, and normal values in the conscious and anesthetized albino rat. *J Neurochem 28*, 897–916.

Souslova, E. A., Belousov, V. V., Lock, J. G., Strömblad, S., Kasparov, S., Bolshakov, A. P., Pinelis, V. G., Labas, Y. A., Lukyanov, S., Mayr, L. M. and Chudakov, D. M. (2007). Single fluorescent protein-based Ca^{2+} sensors with increased dynamic range. *BMC Biotechnol 7*, 37.

Stauffer, T. P., Ahn, S. and Meyer, T. (1998). Receptor-induced transient reduction in plasma membrane PtdIns(4,5)P2 concentration monitored in living cells. *Curr Biol 8*, 343–346.

Takanaga, H., Chaudhuri, B. and Frommer, W. B. (2008). GLUT1 and GLUT9 as major contributors to glucose influx in HepG2 cells identified by a high sensitivity intramolecular FRET glucose sensor. *Biochim Biophys Acta 1778*, 1091–1099.

Tanimura, A., Nezu, A., Morita, T., Turner, R. J. and Tojyo, Y. (2004). Fluorescent biosensor for quantitative real-time measurements of inositol 1,4,5-trisphosphate in single living cells. *J Biol Chem 279*, 38095–38098.

Tantama, M., Hung, Y. P. and Yellen, G. (2011). Imaging intracellular pH in live cells with a genetically encoded red fluorescent protein sensor. *J Am Chem Soc 133*, 10034–10037.

Tantama, M., Hung, Y. P. and Yellen, G. (2012). Optogenetic reporters: fluorescent protein-based genetically encoded indicators of signaling and metabolism in the brain. *Prog Brain Res 196*, 235–263.

Tsien, R. Y. (1998). The green fluorescent protein. *Annu Rev Biochem 67*, 509–544.

Tsou, P., Zheng, B., Hsu, C.-H., Sasaki, A. T. and Cantley, L. C. (2011). A fluorescent reporter of AMPK activity and cellular energy stress. *Cell Metab 13*, 476–486.

Valentin, G., Verheggen, C., Piolot, T., Neel, H., Coppey-Moisan, M. and Bertrand, E. (2005). Photoconversion of YFP into a CFP-like species during acceptor photobleaching FRET experiments. *Nat Methods 2*, 801.

Vander Heiden, M. G., Cantley, L. C. and Thompson, C. B. (2009). Understanding the Warburg effect: The metabolic requirements of cell proliferation. *Science 324*, 1029–1033.

Vaughn, A. E. and Deshmukh, M. (2008). Glucose metabolism inhibits apoptosis in neurons and cancer cells by redox inactivation of cytochrome *c*. *Nat Cell Biol 10*, 1477–1483.

Venkateswarlu, K., Oatey, P. B., Tavaré, J. M. and Cullen, P. J. (1998). Insulin-dependent translocation of ARNO to the plasma membrane of adipocytes requires phosphatidylinositol 3-kinase. *Curr Biol 8*, 463–466.

Vining, E. P. (1999). Clinical efficacy of the ketogenic diet. *Epilepsy Res 37*, 181–190.

Vinkenborg, J. L., Koay, M. S. and Merkx, M. (2010). Fluorescent imaging of transition metal homeostasis using genetically encoded sensors. *Curr Opin Chem Biol 14*, 231–237.

Vousden, K. H. and Ryan, K. M. (2009). p53 and metabolism. *Nat Rev Cancer 9*, 691–700.

Wagnières, G. A., Star, W. M. and Wilson, B. C. (1998). In vivo fluorescence spectroscopy and imaging for oncological applications. *Photochem Photobiol 68*, 603–632.

Wang, W., Fang, H., Groom, L., Cheng, A., Zhang, W., Liu, J., Wang, X., Li, K., Han, P., Zheng, M., Yin, J., Wang, W., Mattson, M. P., Kao, J. P. Y., Lakatta, E. G., Sheu, S.-S., Ouyang, K., Chen, J., Dirksen, R. T. and Cheng, H. (2008). Superoxide flashes in single mitochondria. *Cell 134*, 279–290.

Warburg, O. (1956). On the origin of cancer cells. *Science 123*, 309–314.

Warburg, O., Wind, F. and Negelein, E. (1927). The metabolism of tumors in the body. *J Gen Physiol 8*, 519–530.

Wardman, P. (2007). Fluorescent and luminescent probes for measurement of oxidative and nitrosative species in cells and tissues: progress, pitfalls, and prospects. *Free Radic Biol Med 43*, 995–1022.

Wellen, K. E., Hatzivassiliou, G., Sachdeva, U. M., Bui, T. V., Cross, J. R. and Thompson, C. B. (2009). ATP-citrate lyase links cellular metabolism to histone acetylation. *Science 324*, 1076–1080.

Wellen, K. E., Lu, C., Mancuso, A., Lemons, J. M. S., Ryczko, M., Dennis, J. W., Rabinowitz, J. D., Coller, H. A. and Thompson, C. B. (2010). The hexosamine biosynthetic pathway couples growth factor-induced glutamine uptake to glucose metabolism. *Genes Dev 24*, 2784–2799.

Willemse, M., Janssen, E., de Lange, F., Wieringa, B. and Fransen, J. (2007). ATP and FRET—a cautionary note. *Nat Biotechnol 25*, 170–172.

Williamson, D. H., Lund, P. and Krebs, H. A. (1967). The redox state of free nicotinamide-adenine dinucleotide in the cytoplasm and mitochondria of rat liver. *Biochem J 103*, 514–527.

Winterbourn, C. C. (2008). Reconciling the chemistry and biology of reactive oxygen species. *Nat Chem Biol 4*, 278–286.

Yamada, K., Nakata, M., Horimoto, N., Saito, M., Matsuoka, H. and Inagaki, N. (2000). Measurement of glucose uptake and intracellular calcium concentration in single, living pancreatic beta-cells. *J Biol Chem 275*, 22278–22283.

Yamada, K., Saito, M., Matsuoka, H. and Inagaki, N. (2007). A real-time method of imaging glucose uptake in single, living mammalian cells. *Nat Protoc 2*, 753–762.

Yanushevich, Y. G., Staroverov, D. B., Savitsky, A. P., Fradkov, A. F., Gurskaya, N. G., Bulina, M. E., Lukyanov, K. A. and Lukyanov, S. A. (2002). A strategy for the generation of non-aggregating mutants of Anthozoa fluorescent proteins. *FEBS Lett 511*, 11–14.

Yellen, G. (2008). Ketone bodies, glycolysis, and KATP channels in the mechanism of the ketogenic diet. *Epilepsia 49 Suppl 8*, 80–82.

Yeung, T., Gilbert, G. E., Shi, J., Silvius, J., Kapus, A. and Grinstein, S. (2008). Membrane phosphatidylserine regulates surface charge and protein localization. *Science 319*, 210–213.

Yi, C. H., Pan, H., Seebacher, J., Jang, I.-H., Hyberts, S. G., Heffron, G. J., Vander Heiden, M. G., Yang, R., Li, F., Locasale, J. W., Sharfi, H., Zhai, B., Rodriguez-Mias, R., Luithardt, H., Cantley, L. C., Daley, G. Q., Asara, J. M., Gygi, S. P., Wagner, G., Liu, C.-F. and Yuan, J. (2011). Metabolic regulation of protein *N*-alpha-acetylation by Bcl-xL promotes cell survival. *Cell 146*, 607–620.

Yoshioka, K., Takahashi, H., Homma, T., Saito, M., Oh, K. B., Nemoto, Y. and Matsuoka, H. (1996). A novel fluorescent derivative of glucose applicable to the assessment of glucose uptake activity of *Escherichia coli*. *Biochim Biophys Acta 1289*, 5–9.

Yuneva, M. O., Fan, T. W. M., Allen, T. D., Higashi, R. M., Ferraris, D. V., Tsukamoto, T., Matés, J. M., Alonso, F. J., Wang, C., Seo, Y., Chen, X. and Bishop, J. M. (2012). The metabolic profile of tumors depends on both the responsible genetic lesion and tissue type. *Cell Metab 15*, 157–170.

Zapata-Hommer, O. and Griesbeck, O. (2003). Efficiently folding and circularly permuted variants of the Sapphire mutant of GFP. *BMC Biotechnol 3*, 5.

Zhang, J., Campbell, R. E., Ting, A. Y. and Tsien, R. Y. (2002a). Creating new fluorescent probes for cell biology. *Nat Rev Mol Cell Biol 3*, 906–918.

Zhang, Q., Piston, D. W. and Goodman, R. H. (2002b). Regulation of corepressor function by nuclear NADH. *Science 295*, 1895–1897.

Zhao, Y., Araki, S., Wu, J., Teramoto, T., Chang, Y.-F., Nakano, M., Abdelfattah, A. S., Fujiwara, M., Ishihara, T., Nagai, T. and Campbell, R. E. (2011a). An expanded palette of genetically encoded Ca^{2+} indicators. *Science 333*, 1888–1891.
Zhao, Y., Jin, J., Hu, Q., Zhou, H.-M., Yi, J., Yu, Z., Xu, L., Wang, X., Yang, Y. and Loscalzo, J. (2011b). Genetically encoded fluorescent sensors for intracellular NADH detection. *Cell Metab 14*, 555–566.
Zheng, L., Roeder, R. G. and Luo, Y. (2003). S phase activation of the histone H2B promoter by OCA-S, a coactivator complex that contains GAPDH as a key component. *Cell 114*, 255–266.

Chapter 8

Genetically Encoded Voltage Indicators

Hiroki Mutoh, Walther Akemann, and Thomas Knöpfel

CONTENTS

8.1 INTRODUCTION

Lipid membranes that form the surface of cells and organelles typically separate fluids of different ion compositions. Differences in ion concentrations and selective ion permeability create a voltage across the membrane. Neurons, muscles, and some secretory cells use fast regenerative changes in this membrane voltage (i.e., action potentials) as signals of inter- and intracellular communication or to trigger muscle contraction or release of biochemical substances. In other, "nonexcitable" cells and organelles, membrane voltage drives important biological processes such as the transport of ions and metabolites across the membrane and the production of ATP in mitochondria.

Membrane voltage and related processes in larger cells can be efficiently investigated using microelectrode techniques such as patch-clamp. However, microelectrode techniques are limited in their spatial resolution (number of measurement points), and they are always to some extent invasive (disturbance of intracellular milieu, mechanical tissue damage). Optical voltage imaging based on dyes that transduce membrane voltage into a fluorescent readout is, at least in principle, less invasive and provides higher spatial resolution (Grinvald et al. 1988; Canepari and Zecevic 2010; Herron et al. 2012). Voltage-sensitive dyes comprise a large class of low-molecular-weight organic compounds that were developed over the last 40 years, with persistent efforts leading to significantly better performing dyes even very recently (Miller et al. 2012; Yan et al. 2012). In this chapter, we focus on the newer class of genetically encoded voltage indicators (GEVIs) that allow the targeting of specific cell classes (Siegel and Isacoff 1997; Sakai et al. 2001; Ataka and Pieribone 2002; Knöpfel et al. 2006; Dimitrov et al. 2007; Baker et al. 2008; Knöpfel 2012; Mutoh et al. 2012). This feature enables the optogenetic monitoring of defined cell populations, facilitating a large variety of studies, particularly in neurophysiology (Canepari and Zecevic 2010) and cardiology (Herron et al. 2012) but also potentially in other fields such as endocrine physiology. Moreover, protein-based indicators can be targeted to specific cellular compartments and organelles, enabling a variety of cell biological applications.

GEVIs complement the genetically engineered light-activated ion channels and pumps that are the principal tools for optogenetic control of neuronal circuits (Deisseroth 2011; Mattis et al. 2012), many of which are discussed in Chapter 17. This pairing of detection and control tools forms the basis for a comprehensive optogenetic approach to electrophysiology, or as it also has been termed, "electrophysiology in the age of light" (Scanziani and Hausser 2009). The field of optogenetic electrophysiology is still a work in progress, but recent innovations have given considerable cause for optimism. In this chapter, we (1) focus on the four most promising current approaches for optogenetic monitoring of membrane voltage transients; (2) discuss how the performance of different GEVIs is specified and compared; and (3) comment on the application-specific tuning of GEVIs.

8.2 DESIGN PRINCIPLES OF GEVIs

Genetically encoded optical indicators are typically fusions of two protein domains—a sensing domain and a reporting domain (Knöpfel et al. 2006; Knöpfel 2012). In the case of GEVIs, the sensing domain is typically a membrane protein with at least two conformational states whose occupancy is a function of membrane voltage. The reporting domain consists of one or two fluorescent proteins (FPs) (or a low-molecular-weight chromophore in the case of opsin-based

GEVIs). The primary challenge in the engineering of GEVIs is to fuse the two protein components such that the fluorescence of the reporter domain is dependent on the conformational state of the sensing domain (and hence on membrane voltage). This general strategy is illustrated in Figure 8.1 for the case of *Ciona intestinalis* voltage sensor-containing phosphatase (Ci-VSP)-based VSFPs.

The most prominent current approaches to engineer GEVIs are schematically depicted in Figure 8.2. Initial GEVI design concepts were based on

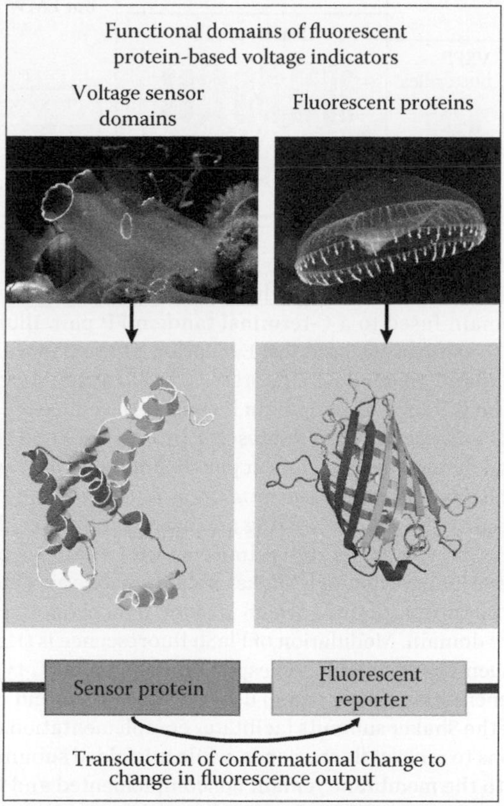

Figure 8.1 Engineering GEVIs by combining naturally occurring protein domains, as illustrated for the case of a voltage-sensitive fluorescent protein (VSFP). Naturally occurring proteins from the sea squirt *Ciona intestinalis* (left) and the jellyfish *Aequorea victoria* (right) are used as the voltage-sensing protein and the fluorescent protein, respectively. These two proteins are assembled to form the VSFP, where the voltage-sensing mechanism is exploited to modulate the reporter fluorescence. (Images from Gittenberger, A., Dutch Ascidians homepage, http://www.ascidians.com; Blakely, S., Aequorea victoria, Wikimedia Commons, 2008.)

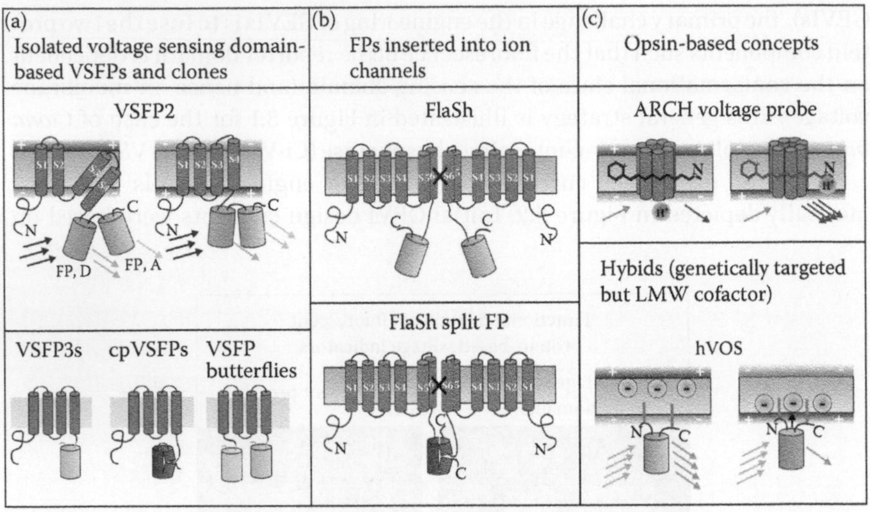

Figure 8.2 Schematic depiction of the most prominent GEVI classes. (a, upper panel) Conceptualization of VSFP2s, which consist of a four-transmembrane-segment voltage-sensing domain fused to a C-terminal tandem FP pair. Illustrated are donor (D) and acceptor (D) fluorescent proteins. Excitation of the donor CFP with 440-nm light results in cyan and yellow FRET fluorescence. A voltage-dependent structural rearrangement of the voltage sensor domain increases the efficiency of FRET between the FRET donor (D) and acceptor (A) fluorescent protein (FP) and thus increases the ratio of acceptor and donor emission. (Lower panel) Single-FP (left) and cpFP (middle)-based VSFPs exhibit fluorescence quenching upon on membrane depolarization. In VSFP-Butterflies (right), the voltage-sensing domain is sandwiched between two FPs. (b) GEVIs of the FlaSh type. In its initial implementation (upper panel), an FP was fused into the C-terminal portion of a Shaker potassium channel subunit. Tetramers of subunits form a channel structure, which is made nonconductive by a point mutation within the pore domain. Modulation of FlaSh fluorescence is triggered by voltage-dependent rearrangements, probably corresponding to channel C-type inactivation. A more recent implementation (lower panel) uses fluorescent protein complementation. Tetramerization of the Shaker subunits facilitates complementation of the two fluorescent protein portions to recover fluorescence, while misfolded subunits and monomers that do not traffic to the membrane remain noncomplemented and hence nonfluorescent. (c, upper panel) Cartoon of microbial rhodopsin-based voltage indicators such as Arch. A change in membrane potential induces increased fluorescence of the retinal molecule. (Lower panel) Schematic depiction of the hybrid voltage sensor hVOS that consists of a combination of a fluorescent protein (genetically encoded component) with dipicrylamine (DPA) (exogenous component). The fluorescent protein is anchored to the intracellular side of the plasma membrane by a prenylation motif. Positively charged DPA is partitioned into the membrane as a function of the membrane potential. With membrane depolarization, DPA moves within Förster distance of the fluorescent protein and quenches its fluorescence.

voltage-dependent structural rearrangements of voltage-gated ion channels or the voltage-sensor domains isolated from those proteins (Siegel and Isacoff 1997; Sakai et al. 2001; Ataka and Pieribone 2002). The first reported fluorescent voltage reporter protein was FlaSh (Figure 8.2b, upper panel), a construct that uses a nonconducting mutant of a voltage-gated potassium channel as the voltage sensor and an FP inserted into the C-terminal region of the channel protein as a reporter (Siegel and Isacoff 1997). A conceptually closely related prototype, SPARC, is based on the insertion of GFP into a skeletal muscle sodium channel (Ataka and Pieribone 2002). A different design principle, for which the acronym VSFP (voltage-sensitive fluorescent protein [Sakai et al. 2001]) was introduced, exploits the voltage-dependent conformational changes of a voltage-sensing domain coupled to a pair of FPs. The basic idea behind the design of VSFPs is that the movement of the voltage-sensing domain will modulate the efficiency of Förster resonance energy transfer (FRET) between the two FPs by shifting their position and orientation relative to each other. The first generation of VSFPs, VSFP1, used the voltage-sensing domain of the Kv2.1 potassium channel subunit (Sakai et al. 2001). Initial versions of FlaSh, SPARC, and VSFP1 provided a proof of principle for the GEVI concept but failed in their intended application in brain tissue because of small (or practically absent) optical signals (Baker et al. 2007). However, subsequent refinement of the molecular design and the discovery of new voltage-reporting mechanisms, described in Section 8.3, overcame this initial setback.

8.2.1 Isolated Voltage-Sensor Domain-Based Voltage Indicators

Voltage-dependent potassium channel (Kv channel) subunits consist of a four-transmembrane-segment voltage-sensor domain (S1–S4) and two transmembrane segments (S5–S6) that form the ion-channel pore in subunit tetramers (Bezanilla 2000). The structure–function relationship of potassium channels has been studied for decades, and therefore the potassium channel voltage-sensor domain (S1–S4) was an obvious choice as a naturally evolved candidate scaffold to actuate the reporter FPs in GEVIs (Sakai et al. 2001). However, because the fourth transmembrane segment (S4) of the voltage-sensor domain contains positively charged amino acids that are presumably exposed to the lipid environment and electric field of the plasma membrane unless shielded by the overall ion channel structure, it has been questioned whether an isolated voltage-sensor domain can exist as a self-contained and functional protein in plasma membranes. This issue was resolved with the description of *Ciona intestinalis* voltage sensor-containing phosphatase (Ci-VSP) (Murata et al. 2005), which consists of a single voltage-sensor domain attached to an intracellular enzyme and does not appear to be dependent on interactions with other membrane proteins. Because Ci-VSP naturally occurs in a monomeric configuration, whereas the obligate tetramerization of Kv channel subunits likely affects the membrane trafficking of isolated Kv voltage-sensor domains, the Ci-VSP voltage-sensor domain has become a promising alternative scaffold for the development of GEVIs such as the VSFPs. During the past several years, several Ci-VSP-based VSFP designs have been developed (Perron et al. 2012). The first design involves a pair of tandem FPs fused to the end of S4 (Figure 8.2a, upper panel). Because this design corresponds to that of the potassium channel-based VSFP1, this family of GEVIs was named VSFP2s. The variant VSFP2.3 (Lundby et al. 2008; Akemann et al. 2010) was the first FRET-based GEVI to enable the optical imaging of spontaneous action and synaptic potentials in neurons.

An alternative VSFP design, the VSFP3s (Figure 8.2a, lower panel), uses a single FP instead of an FP pair (Figure 8.2a, lower panel) (Perron et al. 2009b). VSFP3s offer the advantages of broad coverage of the color spectrum and relatively fast overall kinetics, but with smaller signal amplitudes than the VSFP2s (Perron et al. 2009b). VSFP3 variants have also been designed to incorporate circularly permuted fluorescent proteins (cpFPs; Figure 8.2a, lower panel) (Gautam et al. 2009; Barnett et al. 2012). Although calcium indicator proteins containing cpFPs have proven successful (Nakai et al. 2001; Tian et al. 2009), cpFP-based VSFPs have yet to match this success. A recently reported VSFP3 variant uses a mutated ecliptic pHluorin pH indicator, originally derived from GFP (Jin et al. 2012). VSFP3 variants based on Ci-VSP homologues from other species (e.g., zebrafish) have been explored as well (Baker et al. 2008, 2012).

In a more recent VSFP design, the voltage-sensor domain is sandwiched between two FPs (Figure 8.2a, lower panel). This series of GEVIs, termed VSFP-Butterflies, currently represents the best-performing probes for monitoring subthreshold membrane oscillations of neurons *in vivo* (Akemann et al. 2012).

The latest advances in VSFP engineering have achieved dramatic improvements in voltage-response kinetics. The voltage-sensing mechanism of Ci-VSP is relatively slow (>1–2 ms), as estimated from sensing current measurements (Lundby et al. 2008). Delayed rectifier potassium channels that contribute to the repolarization of fast action potentials have much faster kinetics, as they need to open at the peak of an action potential. Fast action potentials, such as those of cerebellar Purkinje cells, rely on fast-gating Kv3 potassium channels. We therefore transplanted homologous amino acid motifs from the tetrameric voltage-activated potassium channel Kv3.1 into the monomeric voltage sensing domain of Ci-VSP in the VSFP2.3 scaffold (Mishina et al. 2012). We found that motifs extending from 10 to roughly 100 amino acids can be readily transferred from Kv3.1 into Ci-VSP to form engineered VSDs that efficiently incorporate into the plasma membrane and sense voltage. VSFPs based on these engineered chimeric VSDs exhibit very fast activation and deactivation kinetics that are superior to Ci-VSP-based VSFPs (Mishina et al. 2012).

8.2.2 FlaSh-Type Voltage Indicators

Initial versions of FlaSh-type GEVIs (Siegel and Isacoff 1997; Guerrero et al. 2002) (Figure 8.2b, upper panel) suffered from poor membrane targeting in mammalian cells (Baker et al. 2007). A more recent implementation of the FlaSh concept (Figure 8.2b, lower panel) uses FP complementation. To improve membrane localization, a consortium of researches led by Lawrence Cohen (Baker et al. 2008) split the FP into two nonfluorescent halves and attached each half to a Kv channel subunit (Jin et al. 2011). They then screened 56 fluorescent probes (generated by coexpression of Kv subunits containing either half of the FP), of which 30 were expressed at the plasma membrane and were capable of optically reporting changes in membrane potential (Figure 8.2b, lower panel). The largest signal from these novel FlaSh-derived sensors was –1.4% in $\Delta F/F$ for a 100 mV depolarization, with on-time constants of approximately 15 ms and off-time constants of approximately 200 ms. Unfortunately, this "split-can" approach did not yield probes with better performance than previously available GEVIs.

8.2.3 Microbial Opsin-Based Voltage Indicators

The newest concept for GEVI design is based on the use of microbial opsins (Kralj et al. 2011, 2012) (Figure 8.2c, upper panel). These proteins bind retinal (a vitamin A–related organic chromophore) and evolved naturally to function as transducers of light into cellular signals, including changes in membrane voltage. Adam Cohen's group at Harvard found that the natural relationship between light and voltage can be reversed, so that membrane voltage changes are reported as an optical signal. The proof of principle was first demonstrated with a proteorhodopsin-based optical proton sensor (PROPS) from green light-absorbing bacteria (Kralj et al. 2011). PROPS produced signals that appeared to represent voltage fluctuations in *Escherichia coli* but did not target well to plasma membranes of eukaryotic cells. Subsequently, the researchers determined that archaerhodopsin-3 (Arch), a previously established optogenetic control tool, produces a fluorescent signal that correlates with changes in membrane voltage (Kralj et al. 2012). Their findings were the cause of considerable excitement, as the change in Arch fluorescence is very fast and linear, two desirable features for a GEVI. However, because the natural function of Arch is to drive a proton current with the absorbed light energy, voltage sensing also changes voltage. This undesirable effect was fixed by a point mutation in the

Arch protein, Arch(D95N), that abolished its capacity to elicit light-driven currents. However, this mutation also dramatically slowed down the optical signal in response to membrane-potential changes (Kralj et al. 2012). At mammalian body temperature, the voltage response of Arch(D95N) has fast and slow components of roughly equal magnitude (A. Cohen, personal communication). Though additional protein engineering might strengthen the fast response while avoiding photocurrent, the most serious limitation of the Arch class of voltage probes is their very low quantum efficiency (0.001) (Kralj et al. 2012). This results in very low brightness for this prototypic GEVI, which however might eventually be improved by introducing mutations into the Arch scaffold that favor the fluorescent state in its natural photo cycle. Another strategy could be to use Arch as a FRET acceptor together with a bright FP donor. This might lead to voltage-dependent quenching of bright donor fluorescence.

8.2.4 Hybrid Voltage Indicators

A fourth design concept for voltage indicators is a two-component FRET-based system, as originally developed without genetic components (Gonzalez and Tsien 1995, 1997) but subsequently adapted to a genetically targetable probe by Chanda and colleagues (Chanda et al. 2005; Sjulson and Miesenbock 2008; Wang et al. 2010). The first component of their hybrid voltage sensor (hVOS, Figure 8.2c, lower panel) is an FP with attached farnesylated and palmitoylated motifs that anchor the protein to the plasma membrane (Figure 8.2c, lower panel). The second component is the nonfluorescent synthetic compound dipicrylamine (DPA), which serves as a voltage-sensing FRET acceptor (quencher). Because DPA is lipophilic but negatively charged, it distributes in the membrane in a voltage-dependent fashion. When the membrane is depolarized, DPA translocates to the inner layer of the membrane, within Förster distance of the FP, quenching its fluorescence. Unfortunately, DPA increases the membrane capacitance, so care must be taken to ensure that the concentrations used do not disrupt the native physiological responses (Sjulson and Miesenbock 2008; Akemann et al. 2009). Despite this principal concern, the group of Meyer B. Jackson at University of Wisconsin has successfully generated transgenic mice that express the hVOS probes under the neuron-specific thy-1 promoter. Probe expression was quite strong in hippocampal slices, and electrical stimulation elicited robust optical responses similar to those observed with some voltage-sensitive dyes, while adverse effects of DPA were not prominent at the concentrations used for the imaging experiments (Wang et al. 2010).

Another notable "hybrid" strategy, proposed by Peter Fromherz and colleagues, entails activating an organic voltage-sensitive dye via an enzyme that may be genetically targeted to specific cell populations (Hinner et al. 2006;

Ng and Fromherz 2011). This strategy is analogous to the widely used acetyl ester-modified calcium indicators (which are activated by an endogenous enzyme) and could be generalized and applied to new improved organic dyes (Miller et al. 2012).

The general concerns regarding these hybrid strategies relate to the practical difficulty to apply the exogenous lipophilic compound to neuronal membranes with good control of membrane concentration and spatial homogeneity in intact tissue.

8.3 GENERAL PERFORMANCE SPECIFICATIONS OF VOLTAGE REPORTERS

While offering high spatial resolution and (at least in principle) low invasiveness, voltage imaging is typically noisier than microelectrode measurements. The most important photophysical factors that determine the signal-to-noise ratio (SNR) of GEVIs are the sensitivity and the photostability of the indicator. The sensitivity is usually expressed as the change in light intensity normalized to the baseline intensity (i.e., fluorescence, F, in the case of a fluorescent dye) in response to changes in transmembrane potential (e.g., % $\Delta F/F$ per millivolt). The photostability determines the rate at which the dye bleaches under standard illumination conditions. The rate of bleaching in turn determines the maximal illumination intensity that can be employed during a fixed measurement interval. These photophysical factors lead to an SNR performance that is proportional to the product of indicator sensitivity and the square root of the bleaching time constant of the fluorescence (Mutoh et al. 2012).

In addition to the basic photophysical performance of GEVIs, their applications in biological systems can be affected by GEVI-expression-induced alterations in membrane properties. Probably the most severe concern in this context is related to the increase in dynamic membrane capacitance that results from the introduction of mobile charges into the membrane (Akemann et al. 2009). Because voltage-dependent conformational changes in membrane proteins are generally driven by the forces imposed by the membrane's electric field on charged amino acids, all protein-based GEVIs are likely to increase membrane capacitance. The net charge transfer during voltage sensing differs, however, between different GEVIs ranging from 12–14 e^- for FlaSh-type GEVIs to 1.2 e^- for VSFP-type GEVIs (Zagotta et al. 1994; Lundby et al. 2010). Experimental data and detailed computer simulations (Akemann et al. 2009) have revealed that the adverse effects of increased capacitance occur only at VSFP expression levels that exceed those of practical interest. Capacitance increases are a more serious concern with GEVIs that involve a larger number of mobile charges such as FlaSh and DPA-based hybrid probes (Sjulson and Miesenbock 2008). There is also a measurable charge transfer (increase in membrane capacitance) associated with voltage sensing by Arch; this effect might be small and of no concern for biological experiments, but a quantification of this effect has not been reported yet (A. Cohen, personal communication).

The effective performance of any GEVI in a given biological application finally depends on how efficiently the protein traffics to the membrane. The first-generation GEVIs and some VSFP derivatives such as Mermaid (Tsutsui et al. 2008) failed, for instance, when applied to intact brain tissue owing to inefficient membrane trafficking, leading to fluorescence that was mostly localized to organelles or protein aggregates (Baker et al. 2007; Perron et al. 2009a; Akemann et al. 2010).

8.4 TUNING GEVIs FOR SPECIFIC APPLICATIONS

Like voltage-gated ion channels, GEVIs based on ion channel voltage-sensing domains or homologous scaffolds activate with a sigmoidal voltage dependency (Akemann et al. 2009). As a consequence, the sensitivity of VSFPs to probe small voltage changes is maximal at the voltage of half activation. For the very same reason that nonlinear activation is essential for the functional specialization and diversity of ion channels, nonlinearity can be exploited to tune VSFPs to customized probe profiles. For instance, VSFP2.3 and VSFP2.42 exhibit half activation at −30 mV in neurons (Akemann et al. 2010), shifted from +80 mV in VSFP2D by a R217Q point mutation in S4 of the voltage sensor of Ci-VSP (Dimitrov et al. 2007). On the other hand, compared to VSFP2.42, VSFP-Butterfly 1.2 activates at more negative potentials, with a half-activation voltage of −69 ± 3 mV (Akemann et al. 2012). Whereas GEVIs with fast response time constants are required for tracking very fast voltage fluctuations (i.e., neuronal action potentials), slower responding variants are more suitable to track slower voltage signals that are also imaged at slower frame rates. Indeed, a GEVI with a slow response time constant implements a low-pass filter that helps to prevent aliasing of the monitored signal under conditions of temporal undersampling (i.e., action potential of 1 ms duration sampled at 1 kHz or less).

Another parameter that can be adapted to experimental needs is the color of FP-based GEVIs. GEVI variants with excitation and emission at long light wavelengths are desirable in preparations with green autofluorescence, light absorption by hemoglobin, and/or strong scattering of short-wavelength light. VSFP2-type GEVIs have been generated with several different FRET pairs (Tsutsui et al. 2008; Mutoh et al. 2009; Lam et al. 2012), and VSFP3-type GEVIs cover a large portion of the visual spectrum (Perron et al. 2009b; Jin et al. 2012). Color variants of GEVIs are also needed in experimental designs where different probes or optogenetic control tools are used in combination. Ideally, monitoring different parameters, such as the membrane voltage of different cell classes or membrane voltage and intracellular calcium, or combined monitoring and control using optogenetic actuator proteins, would be done by combining suitable color variants such that the spectra of FP excitation, the spectral bands of FP emission detection, and the action spectra of the optogenetic control tools are set up in a way that minimizes cross-talk and functional interference.

8.5 CHOOSING THE MOST APPROPRIATE GEVI

At present, the most advanced GEVIs are those from the VSFP class using the combination of an isolated voltage-sensing domain (either derived from Ci-VSD or engineered as chimeras between fast gating ion channels and Ci-VSD). Microbial opsin-based voltage indicators are still in the early stages of development, and their potential for biological applications is just beginning to be explored. For experiments where the optical signals are confounded by hemodynamic effects and motion artifacts (time-lapse *in vivo* recordings), ratiometric (e.g., FRET-based) GEVIs are the best choice. For other types of experiments, monochromatic sensors offer the advantage of simpler instrumentation. Tables 8.1 and 8.2 provide an overview of published GEVIs to aid in the selection of the most appropriate GEVI variant for a given application, taking into consideration the general framework provided in the previous sections. These tables are taken from http://openoptogenetics.org, and readers are encouraged to consult this resource for updates. For 2P excitation, VSFPs based on the citrine/mKate2 FRET pair fluorescent proteins are preferable (Knöpfel Laboratory, unpublished observations). All GEVIs published to date have response time constants greater than 1 ms (with the exception of unmodified Arch, which generates a photocurrent) and can therefore report only fast action potentials with attenuated SNR (Kralj et al. 2012). Sufficiently bright and side-effect-free GEVIs that respond to both depolarization and repolarization with effective time constants faster than 1 ms remain to be identified. It should be emphasized, however, that a reasonable SNR at high temporal resolution (as is required to resolve action potentials) can be achieved only if the indicator can deliver both large photon fluxes (i.e., is bright and photostable) and high sensitivity (Knöpfel et al. 2006). In addition to satisfying these challenging specifications, multisite optical action potential recordings in intact tissue (either living animals or brain slices) will require optical instrumentation that is far beyond what is currently available "off the shelf."

TABLE 8.1 MONOCHROMATIC (SINGLE FP) GEVIs

Year	Name	Voltage-Sensing Domain	Expression System for Functional Characterization	Chromophore(s) Peak Emission Wavelength(s)	Sensitivity (% $\Delta F/F$ per 100 mV) at xx mV[a]	Response Time Constant τ (On)[b]	Response Time Constant τ (Off)[b]	Reference
1997	FlaSh (wtGFP)	Shaker potassium channel	Xenopus oocytes	GFP/505 nm	5.1%	23 ms	105 ms	Siegel and Isacoff (1997)
2002	FlaSh	Shaker potassium channel	Xenopus oocytes	GFPuv, Ecliptic GFP, eGFP, YFP				Guerrero et al. (2002)
2002	SPARC	Voltage-gated sodium channel	Xenopus oocytes	GFP/505 nm	0.5%	~1 ms	~1 ms	Ataka and Pieribone (2002)
2008	VSFP3.1	ci-VSP	PC12 cells	Cerulean/475 nm	2.2% at –43 mV	1.8 ms	105 ms	Lundby et al. (2008); Perron et al. (2009b)
2009	VSFP3.x	ci-VSP	Cultured neurons	Citrine/530 nm mOrange/562 nm TagRFP/584 nm mKate2/633 nm	2.5–5% at –43 mV	2–100 ms	~100 ms	Perron et al. (2009b)
2011	Split-Venus FlaSh variants	Shaker potassium channel	HEK293 or NIE115 neuroblastoma cells	Split-Venus/528 nm	1.4%	~15 ms	~200 ms	Jin et al. (2011)
2011	Arch(D95N)	Archeorhodopsin (microbial opsin)	Cultured neurons	Retinal/687 nm	60%	41 ms	41 ms	Kralj et al. (2011)
2012	ArcLight A242	Ci-VSP	HEK293 cells and cultured neurons	Super ecliptic pHluorin A227D	35%	~10 ms	—	Jin et al. (2012)
2012	ElectricPk	Ci-VSP	HEK293 cells and cultured neurons	Circularly permutated eGFP	1.5%	2 ms	2 ms	Barnett et al. (2012)

Source: OpenOptogenetics, Genetically-encoded voltage indicators. Available at http://www.openoptogenetics.org/.

[a] The sensitivity is given as change of fluorescence relative to baseline ($\Delta F/F$) measured in cultured cells. The sensitivity of most voltage indicators is voltage-dependent and, if data are available, is given at the membrane voltage of half m-aximal response. If available, data are from experiments where different indicators were compared side by side.

[b] Measured with short lasting and large amplitude depolarizing voltage steps (<10 ms); for kinetics described best with two-exponential functions, the two time constants are given in order of weight.

TABLE 8.2 FRET-BASED GEVIs

Year	Name	Voltage-Sensing Domain	Expression System for Functional Characterization	Chromophore(s) Peak Emission Wavelength(s)	Sensitivity (% $\Delta R/R$ per 100 mV) at xx mV[a]	Response Time Constant τ (On)[b]	Response Time Constant τ (Off)[b]	Reference
2001	VSFP1	Kv2.1	HEK 293 cells	CFP/477 nm YFP/529 nm	1.8%	<1 ms	<1 ms	Sakai et al. (2001)
2007	VSFP2.1	ci-VSP	PC12 cells and neurons	Cerulean/475 nm Citrine/529 nm	6.8% at −70 mV	15 ms	75 ms	Dimitrov et al. (2007)
2008	VSFP2.3	ci-VSP	PC12 cells, cultured neurons, mouse in vivo	Cerulean/480 nm Citrine/530 nm	22% at −50 mV	30 ms ~2 ms	~80 ms	Mutoh et al. (2009)
2008	VSFP2-Mermaid	ci-VSP	Xenopus oocyte, NT cells, primary cultures of brain cells	mUKG /490 nm mKOκ/560 nm	21% at −43 mV	30 ms ~2 ms	~80 ms	Mutoh et al. (2009); Tsutsui et al. (2008)
2009	VSFP2.4	ci-VSP	PC12 cells	Citrine/530 nm mKate2/633 nm	20.5% at −54 mV	30 ms ~2 ms	~80 ms	Mutoh et al. (2009)
2010	VSFP2.42	ci-VSP	Neurons in culture, brain slices and in vivo	Citrine/529 nm mKate2/633 nm	20.5% at −50 mV	80 ms 2 ms	~80 ms	Akemann et al. (2010)
2012	Zahra and Zahra 2	Voltage-sensitive phosphatases of Nematostella and Danio	HEK293 cells	Cerulean/480 nm Citrine/530 nm	0.2%	~5 ms	~5 ms	Baker et al. (2012)
2012	VSFP2-Buterfly 1.2	ci-VSP	Neurons in culture, brain slices and in vivo	Citrine/530 nm mKate2/633 nm	22.2% at −79 mV	10 ms 2 ms	~80 ms	Akemann et al. (2012)

Source: OpenOptogenetics, Genetically-encoded voltage indicators. Available at http://www.openoptogenetics.org/.

[a] The sensitivity is given as change of fluorescence ratio ($\Delta R/R$) measured in cultured cells. The sensitivity of most voltage indicators is voltage-dependent and, if data are available, is given at the membrane voltage of half maximal response. If available, data are from experiments where different indicators were compared side by side.

[b] Measured with short lasting and large amplitude depolarizing voltage steps (<10 ms); for kinetics described best with two-exponential functions, the two time constants are given in order of weight.

8.6 CONCLUDING REMARKS

Since the conceptualization of GEVIs more than 15 years ago, the persistent development and discovery of novel scaffolds and design principles led to probes that found their first applications in neuroscience and cardiac physiology. There is, however, clear space for further improvement, and the next generation of GEVIs will undoubtedly outperform currently available probes. At present, the most convincingly demonstrated advantage of genetically encoded indicators relates to the feasibility of optical imaging experiments over repeated sessions spanning days or even months. The second inherent and unique advantage of GEVIs is the ability to target specific cell populations. We expect that optogenetic voltage imaging approaches will also be extended to multimodal strategies using indicators with fluorescence in different bands of the color spectrum. The combination of genetically encoded tools to monitor neuronal activities with matching tools to control neuronal activities (Deisseroth 2011) will enable a complete optogenetic approach that is complementary to traditional electrophysiology.

ACKNOWLEDGMENTS

We thank all present and past members of the Knöpfel laboratory for their contributions, support, and discussions. Work in the laboratory has received funding from RIKEN; the Japanese Society for Promotion of Science (JSPS); the Human Frontiers Science Program (HFSP); US National Institutes of Health (NIH); and Ministry of Education, Culture, Sport, Science and Technology of Japan (MEXT).

REFERENCES

Akemann, W., Lundby, A., Mutoh, H., and Knöpfel, T. 2009. Effect of voltage sensitive fluorescent proteins on neuronal excitability. *Biophys. J.*, vol. 96, no. 10, pp. 3959–3976.

Akemann, W., Mutoh, H., Perron, A., Kyung, P. Y., Iwamoto, Y., and Knöpfel, T. 2012. Imaging neural circuit dynamics with a voltage-sensitive fluorescent protein. *J. Neurophysiol.*, vol. 108, no. 8, pp. 2323–2337.

Akemann, W., Mutoh, H., Perron, A., Rossier, J., and Knöpfel, T. 2010. Imaging brain electric signals with genetically targeted voltage-sensitive fluorescent proteins. *Nat. Methods*, vol. 7, no. 8, pp. 643–649.

Ataka, K., and Pieribone, V. A. 2002. A genetically targetable fluorescent probe of channel gating with rapid kinetics. *Biophys. J.*, vol. 82, no. 1, Pt 1, pp. 509–516.

Baker, B. J., Jin, L., Han, Z., Cohen, L. B., Popovic, M., Platisa, J., and Pieribone, V. 2012. Genetically encoded fluorescent voltage sensors using the voltage-sensing domain of Nematostella and Danio phosphatases exhibit fast kinetics. *J. Neurosci. Methods*, vol. 208, no. 2, pp. 190–196.

Baker, B. J., Lee, H., Pieribone, V. A., Cohen, L. B., Isacoff, E. Y., Knöpfel, T., and Kosmidis, E. K. 2007. Three fluorescent protein voltage sensors exhibit low plasma membrane expression in mammalian cells. *J. Neurosci. Methods*, vol. 161, no. 1, pp. 32–38.

Baker, B. J., Mutoh, H., Dimitrov, D., Akemann, W., Perron, A., Iwamoto, Y., Jin, L., Cohen, L. B., Isacoff, E. Y., Pieribone, V. A., Hughes, T., and Knöpfel, T. 2008. Genetically encoded fluorescent sensors of membrane potential. *Brain Cell Biol.*, vol. 36, nos. 1–4, pp. 53–67.

Barnett, L., Platisa, J., Popovic, M., Pieribone, V. A., and Hughes, T. 2012. A fluorescent, genetically-encoded voltage probe capable of resolving action potentials. *PLoS One*, vol. 7, no. 9, p. e43454.

Bezanilla, F. 2000. The voltage sensor in voltage-dependent ion channels. *Physiol. Rev.*, vol. 80, no. 2, pp. 555–592.

Canepari, M., and Zecevic, D., editors. 2010. *Membrane Potential Imaging in the Nervous System: Methods and Applications*. Springer, New York.

Chanda, B., Blunck, R., Faria, L. C., Schweizer, F. E., Mody, I., and Bezanilla, F. 2005. A hybrid approach to measuring electrical activity in genetically specified neurons. *Nat. Neurosci.*, vol. 8, no. 11, pp. 1619–1626.

Deisseroth, K. 2011. Optogenetics. *Nat. Methods*, vol. 8, no. 1, pp. 26–29.

Dimitrov, D., He, Y., Mutoh, H., Baker, B. J., Cohen, L., Akemann, W., and Knöpfel, T. 2007. Engineering and characterization of an enhanced fluorescent protein voltage sensor. *PLoS One*, vol. 2, no. 5, p. e440.

Gautam, S. G., Perron, A., Mutoh, H., and Knöpfel, T. 2009. Exploration of fluorescent protein voltage probes based on circularly permuted fluorescent proteins. *Front. Neuroeng.*, vol. 2, p. 14.

Gonzalez, J. E., and Tsien, R. Y. 1995. Voltage sensing by fluorescence resonance energy transfer in single cells. *Biophys. J.*, vol. 69, no. 4, pp. 1272–1280.

Gonzalez, J. E., and Tsien, R. Y. 1997. Improved indicators of cell membrane potential that use fluorescence resonance energy transfer. *Chem. Biol.*, vol. 4, no. 4, pp. 269–277.

Grinvald, A., Frostig, R. D., Lieke, E., and Hildesheim, R. 1988. Optical imaging of neuronal-activity. *Physiol. Rev.*, vol. 68, no. 4, pp. 1285–1366.

Guerrero, G., Siegel, M. S., Roska, B., Loots, E., and Isacoff, E. Y. 2002. Tuning FlaSh: Redesign of the dynamics, voltage range, and color of the genetically encoded optical sensor of membrane potential. *Biophys. J.*, vol. 83, no. 6, pp. 3607–3618.

Herron, T. J., Lee, P., and Jalife, J. 2012. Optical imaging of voltage and calcium in cardiac cells and tissues. *Circ. Res.*, vol. 110, no. 4, pp. 609–623.

Hinner, M. J., Hubener, G., and Fromherz, P. 2006. Genetic targeting of individual cells with a voltage-sensitive dye through enzymatic activation of membrane binding. *ChemBioChem*, vol. 7, no. 3, pp. 495–505.

Jin, L., Baker, B., Mealer, R., Cohen, L., Pieribone, V., Pralle, A., and Hughes, T. 2011. Random insertion of split-cans of the fluorescent protein venus into Shaker channels yields voltage sensitive probes with improved membrane localization in mammalian cells. *J. Neurosci. Methods*, vol. 199, no. 1, pp. 1–9.

Jin, L., Han, Z., Platisa, J., Wooltorton, J. R., Cohen, L. B., and Pieribone, V. A. 2012. Single action potentials and subthreshold electrical events imaged in neurons with a fluorescent protein voltage probe. *Neuron*, vol. 75, no. 5, pp. 779–785.

Knöpfel, T. 2012. Genetically encoded optical indicators for the analysis of neuronal circuits. *Nat. Rev. Neurosci.*, vol. 13, no. 10, pp. 687–700.

Knöpfel, T., Diez-Garcia, J., and Akemann, W. 2006. Optical probing of neuronal circuit dynamics: Genetically encoded versus classical fluorescent sensors. *Trends Neurosci.*, vol. 29, no. 3, pp. 160–166.

Kralj, J. M., Douglass, A. D., Hochbaum, D. R., Maclaurin, D., and Cohen, A. E. 2012. Optical recording of action potentials in mammalian neurons using a microbial rhodopsin. *Nat. Methods*, vol. 9, no. 1, pp. 90–95.

Kralj, J. M., Hochbaum, D. R., Douglass, A. D., and Cohen, A. E. 2011. Electrical spiking in Escherichia coli probed with a fluorescent voltage-indicating protein. *Science*, vol. 333, no. 6040, pp. 345–348.

Lam, A. J., St Pierre, F., Gong, Y., Marshall, J. D., Cranfill, P. J., Baird, M. A., McKeown, M. R., Wiedenmann, J., Davidson, M. W., Schnitzer, M. J., Tsien, R. Y., and Lin, M. Z. 2012. Improving FRET dynamic range with bright green and red fluorescent proteins. *Nat. Methods*, vol. 9, no. 10, pp. 1005–1012.

Lundby, A., Akemann, W., and Knöpfel, T. 2010. Biophysical characterization of the fluorescent protein voltage probe VSFP2.3 based on the voltage-sensing domain of Ci-VSP. *Eur. Biophys. J.*, vol. 39, no. 12, pp. 1625–1635.

Lundby, A., Mutoh, H., Dimitrov, D., Akemann, W., and Knöpfel, T. 2008. Engineering of a genetically encodable fluorescent voltage sensor exploiting fast Ci-VSP voltage-sensing movements. *PLoS One*, vol. 3, no. 6, p. e2514.

Mattis, J., Tye, K. M., Ferenczi, E. A., Ramakrishnan, C., O'Shea, D. J., Prakash, R., Gunaydin, L. A., Hyun, M., Fenno, L. E., Gradinaru, V., Yizhar, O., and Deisseroth, K. 2012. Principles for applying optogenetic tools derived from direct comparative analysis of microbial opsins. *Nat. Methods*, vol. 9, no. 2, pp. 159–172.

Miller, E. W., Lin, J. Y., Frady, E. P., Steinbach, P. A., Kristan, W. B., Jr., and Tsien, R. Y. 2012. Optically monitoring voltage in neurons by photo-induced electron transfer through molecular wires. *Proc. Natl. Acad. Sci. U.S.A.*, vol. 109, no. 6, pp. 2114–2119.

Mishina, Y., Mutoh, H., and Knöpfel, T. 2012. Transfer of Kv3.1 voltage sensor features to the isolated Ci-VSP voltage-sensing domain. *Biophys. J.*, vol. 103, no. 4, pp. 669–676.

Murata, Y., Iwasaki, H., Sasaki, M., Inaba, K., and Okamura, Y. 2005. Phosphoinositide phosphatase activity coupled to an intrinsic voltage sensor. *Nature*, vol. 435, no. 7046, pp. 1239–1243.

Mutoh, H., Akemann, W., and Knöpfel, T. 2012. Genetically engineered fluorescent voltage reporters. *ACS Chem. Neurosci.*, vol. 3, no. 8, pp. 585–592.

Mutoh, H., Perron, A., Dimitrov, D., Iwamoto, Y., Akemann, W., Chudakov, D. M., and Knöpfel, T. 2009. Spectrally-resolved response properties of the three most advanced FRET based fluorescent protein voltage probes. *PLoS One*, vol. 4, no. 2, p. e4555.

Nakai, J., Ohkura, M., and Imoto, K. 2001. A high signal-to-noise Ca(2+) probe composed of a single green fluorescent protein. *Nat. Biotechnol.*, vol. 19, no. 2, pp. 137–141.

Ng, D. N., and Fromherz, P. 2011. Genetic targeting of a voltage-sensitive dye by enzymatic activation of phosphonooxymethyl-ammonium derivative. *ACS Chem. Biol.*, vol. 6 no. 5, pp. 444–451.

OpenOptogenetics. Genetically-encoded voltage indicators. Available at http://www.openoptogenetics.org/index.php?title=Genetically-Encoded_Voltage_Indicators#tab=Probes.

Perron, A., Akemann, W., Mutoh, H., and Knöpfel, T. 2012. Genetically encoded probes for optical imaging of brain electrical activity. *Prog. Brain Res.*, vol. 196, pp. 63–77.

Perron, A., Mutoh, H., Akemann, W., Gautam, S. G., Dimitrov, D., Iwamoto, Y., and Knöpfel, T. (2009a). Second and third generation voltage-sensitive fluorescent proteins for monitoring membrane potential. *Front. Mol. Neurosci.*, vol. 2, p. 5.

Perron, A., Mutoh, H., Launey, T., and Knöpfel, T. (2009b). Red-shifted voltage-sensitive fluorescent proteins. *Chem. Biol.*, vol. 16, no. 12, pp. 1268–1277.

Sakai, R., Repunte-Canonigo, V., Raj, C. D., and Knöpfel, T. 2001. Design and characterization of a DNA-encoded, voltage-sensitive fluorescent protein. *Eur. J. Neurosci.*, vol. 13, no. 12, pp. 2314–2318.

Scanziani, M., and Hausser, M. 2009. Electrophysiology in the age of light. *Nature*, vol. 461, no. 7266, pp. 930–939.

Siegel, M. S., and Isacoff, E. Y. 1997. A genetically encoded optical probe of membrane voltage. *Neuron*, vol. 19, no. 4, pp. 735–741.

Sjulson, L., and Miesenbock, G. 2008. Rational optimization and imaging *in vivo* of a genetically encoded optical voltage reporter. *J. Neurosci.*, vol. 28, no. 21, pp. 5582–5593.

Tian, L., Hires, S. A., Mao, T., Huber, D., Chiappe, M. E., Chalasani, S. H., Petreanu, L., Akerboom, J., McKinney, S. A., Schreiter, E. R., Bargmann, C. I., Jayaraman, V., Svoboda, K., and Looger, L. L. 2009. Imaging neural activity in worms, flies and mice with improved GCaMP calcium indicators. *Nat. Methods*, vol. 6, no. 12, pp. 875–881.

Tsutsui, H., Karasawa, S., Okamura, Y., and Miyawaki, A. 2008. Improving membrane voltage measurements using FRET with new fluorescent proteins. *Nat. Methods*, vol. 5, no. 8, pp. 683–685.

Wang, D., Zhang, Z., Chanda, B., and Jackson, M. B. 2010. Improved probes for hybrid voltage sensor imaging. *Biophys. J.*, vol. 99, no. 7, pp. 2355–2365.

Yan, P., Acker, C. D., Zhou, W. L., Lee, P., Bollensdorff, C., Negrean, A., Lotti, J., Sacconi, L., Antic, S. D., Kohl, P., Mansvelder, H. D., Pavone, F. S., and Loew, L. M. 2012. Palette of fluorinated voltage-sensitive hemicyanine dyes. *Proc. Natl. Acad. Sci. U.S.A.*, vol. 109, no. 50, pp. 20443–20448.

Zagotta, W. N., Hoshi, T., Dittman, J., and Aldrich, R. W. 1994. Shaker potassium channel gating. II: Transitions in the activation pathway. *J. Gen. Physiol.*, vol. 103, no. 2, pp. 279–319.

Chapter 9

Prototypical Kinase Sensor Design Motifs for *In Vitro* and *In Vivo* Imaging

Gary C. H. Mo, Ambhighainath Ganesan, and Jin Zhang

CONTENTS

9.1 INTRODUCTION

Phosphorylation is a versatile and reversible posttranslational modification that affects the subcellular location, catalytic activity, and interaction of proteins. The importance of phosphorylation is evident not only in its pervasive influence over approximately one third of the human proteome but also in the nodal functions served by many phosphorylated proteins. The protein kinases that confer phosphorylation are thus pivotal components in many signaling cascades. Commensurate with this central role, kinase signaling is often finely regulated through mechanisms such as compartmentalization and feedback interactions. All of these processes are important pieces in the kinase-signaling puzzle, yet it is often difficult to assess their impact in a live cell context using traditional biochemical approaches. However, the recent development of fluorescent reporters, along with advances in optical microscopy, has enabled us to monitor the spatiotemporal dynamics of signaling molecules such as kinases in real time in living cells, thus greatly complementing traditional biochemistry. The simplest optical readout that can be elicited from a fluorescence-based kinase reporter is the subcellular location of the kinase of interest; however, when the signal from a biosensor changes in parallel with the activity of a kinase, microscopy can further reveal the spatiotemporal dynamics of the kinase and allow a more meaningful examination of signaling. Such signaling dynamics are best captured by sensors with high dynamic ranges, high specificity, and rapid reversibility. The purpose of this chapter is to showcase prototypical kinase sensor designs and review the considerations that go into making optimal sensors that satisfy the aforementioned criteria. We introduce the general principles that underlie the design of these biosensors and highlight how such biosensors have aided the dissection of signaling pathways.

The core function of an optical kinase biosensor is to translate phosphorylation events into detectable, optical signals. Consequently, sensors consist of two components: a sensing unit and a reporting unit. The sensing unit is a molecular switch that changes its conformation on the detection of a kinase-dependent signal. This change is then registered by the reporting unit and transmitted to the observer. The reporting unit can be a small molecular dye, an enzyme, or a fluorescent protein (FP), whose changing photophysical properties manifest as an optical readout. The simplest readout is a change in fluorescence intensity (intensiometric measure). In this respect, small molecule fluorescent dyes have long been utilized as indicators for cations (Valeur and Leray 2000; Jiang and Guo 2004; Nolan and Lippard 2008; Hyman and Franz 2012) and oxygen (Acker et al. 2006), because processes such as hydrogen bonding and chelation affect their fluorescent intensity. Transitioning between the hydrophilic solvent and hydrophobic environments (such as between protein complexes or within a membrane bilayer) can also modulate the fluorescence intensity of small molecules. An alternate intensiometric strategy involves generating photons as the products of biochemical reactions, as in the case of bioluminescence. This strategy uses an enzyme called luciferase that metabolizes a substrate called luciferin. The enzymatic oxidation of luciferin generates an unstable, excited-state product, which quickly returns to the ground state and contributes to an increase in luminescence intensity through the emission of photons (Navizet et al. 2011).

Whereas a single fluorophore can exhibit changes in intensity, a pair of fluorophores may interact with one another through a photophysical process known as Förster resonance energy transfer (FRET). In this process, the donor fluorophore can transfer its excited-state energy to a proximal acceptor fluorophore, leading to an overall increase in acceptor fluorescence and a reciprocal decrease in donor fluorescence. FRET is general to all fluorophores, including small molecule dyes and FPs. Given the sensitivity of FRET to the distance between and the orientation of the two fluorophores, FRET is a suitable strategy for reporting on conformational changes. Changes in FRET therefore serve as an effective readout for interactions between proteins or protein domains (see Kalab and Soderholm [2010] and Zhou et al. [2012a] for comprehensive reviews on FRET and its biological applications). FRET imaging can be performed by determining the ratio of the relative changes in donor versus acceptor intensity (ratiometric measure). Unlike an intensiometric measurement, a ratiometric measurement is less sensitive to changes in absolute fluorophore concentrations. In live-cell imaging, this translates into less sensitivity toward variations in the probe level and cell thickness. Alternatively, FRET efficiency can be measured directly through the decrease in donor fluorescence lifetime on the nanosecond timescale using specialized equipment.

Traditionally, the sensors used to monitor signaling activities have been based mostly on small-molecule fluorescent dyes, owing to the availability of a wealth of different kinds of dyes. Although these types of sensors generally provide large dynamic ranges, the need to attach fluorescent dyes to kinases or their substrates complicates the production of these biosensors, as well as their introduction into living cells. In contrast, genetically encoded, FP-based biosensors are expressed directly within live cells, thereby eliminating the need for further manipulation. These genetically encoded probes can also be subcellularly targeted using native sequences, which can be used to reveal signaling details in different cellular locales. In the following sections, we highlight the design principles behind both chemical and genetically encoded biosensors for kinase signaling.

9.2 KINASE ACTIVATION SENSORS

Kinase signaling is often characterized by multiple levels of regulation. Upstream molecules, including other kinases, can activate a given kinase, which then undergoes a conformational change on activation by these upstream signaling components. Therefore, the design of kinase *activation* reporters often utilizes the kinase of interest as the sensing unit sandwiched between a FRET reporting pair (either FPs or dyes), thus linking a change in the FRET response with these conformational changes and hence kinase activation (Figure 9.1a). A truly optimal design for this type of sensor can be achieved if crystallographic information on the sensing unit is available, as this structural information can provide insight into the conformational changes that may occur in the sensing unit upon activation. Based on this information, the reporting units can be appropriately placed so as to efficiently transduce the molecular movements into optical changes. This can be straightforwardly performed using purified kinases. In a very early example, Cotton et al. used expressed protein ligation to conjugate a FRET pair consisting of fluorescein and tetramethylrhodamine to the N- and C-termini of purified, full-length CrkII (Cotton and Muir 2000). This *in vitro* Abl kinase sensor generated a low (3%) response. Alternatively, routine

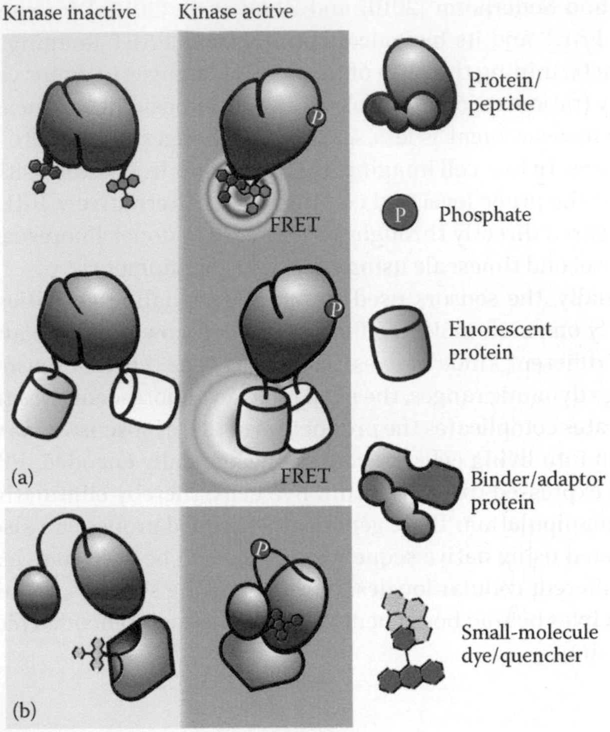

Figure 9.1 Biosensors for sensing kinase activation. Light-blue and red regions designate the nonphosphorylated and phosphorylated state of the sensor, respectively. Dashed outlines indicate a nonfluorescent component. (a) Phosphorylation of a purified, fluorescently labeled protein by a kinase causes a conformational change, leading to a change in FRET efficiency between either dye labels or fluorescent proteins. (b) A dye-labeled adaptor protein specifically binds an activated kinase and gains fluorescence by preventing solvent-quenching of the dye.

molecular biology techniques can be employed to generate genetically encoded FRET-based kinase activation sensors comprising two fluorescent proteins.

Genetically encoded kinase activation biosensors are of great use in dissecting signaling cascades, such as the phosphoinositide 3-kinase (PI3K)/Akt pathway, that consist of a hierarchy of kinases. Typically, receptor tyrosine kinases are activated in response to different growth factors, resulting in the activation of PI3K. Activated PI3K then generates the lipid second messenger phosphoinositol-(3,4,5)-phosphate (PIP_3). This increase in PIP_3 levels then mobilizes the protein kinase Akt, also known as protein kinase B (PKB), to effect a vast number of different cellular outcomes including growth, cell-cycle arrest, glucose metabolism, or death/apoptosis. A long-standing question in systems with similar "bow-tie" signaling network architectures is, How do the convergent signals maintain the specificity required to effect different functions? Using genetically encoded kinase activation sensors, researchers have recently been able to assemble crucial clues to answer this question by studying the spatiotemporal regulation of kinase activation in the PI3K/Akt pathway.

One of the earliest sensors of Akt activation, simply called GFP-Akt-YFP (Calleja et al. 2003), utilizes the full-length Akt kinase as the sensing unit and green fluorescent protein (GFP)/yellow fluorescent protein (YFP) as a FRET FP pair. These two FPs possess strongly overlapping spectra and are therefore typically avoided in a FRET sensor due to spectral cross-contamination. However, these authors took advantage of the strong overlap to detect the significant reduction in GFP donor lifetime on Akt activation. The resulting Akt activity maps were consistent with earlier studies (Andjelkovic et al. 1997; Stokoe et al. 1997; Zhang and Vik 1997) and showed that activated Akt translocated to the membrane. Later, a ratiometric Akt activation reporter called ReAktion (Ananthanarayanan et al. 2007), which comprises full-length Akt and the FRET pair enhanced cyan fluorescent protein (ECFP)/citrine, was developed and utilized in a study of Akt regulation. This study emphasized the advantage of using biosensors that incorporate full-length kinase, as these probes can recapitulate the behaviors of their endogenous counterparts. Thus, ReAktion is able to translocate to the plasma membrane on activation just like endogenous Akt, making it useful in studying the relationship between the activation and translocation of Akt at the molecular level. Mutant ReAktion sensors were therefore generated in which two key phosphorylatable residues within Akt were mutated. These mutants displayed a variety of behaviors, from a lack of translocation to a lack of conformational change (i.e., activation). Notably, a phosphorylatable T308 residue is required to facilitate the conformational change that leads to Akt activation. T308 phosphorylation not only enhances Akt activity but may also reduce the membrane binding affinity of Akt, thereby facilitating its detachment from the membrane. This residue, which is present in the activation loop of Akt, is known to be phosphorylated by the upstream enzyme phosphoinositide-dependent kinase-1 (PDK1). These findings led us to examine the spatiotemporal dynamics of PDK1.

To visualize the activation of PDK1, we generated the PDK activation reporter (PARE) using full-length PDK1 as the sensing unit and ECFP/citrine as the reporting FRET FP pair (Gao et al. 2011). Using PARE, we were able to demonstrate that PDK1 is not constitutively active as previously thought but is actually activated in membrane raft microdomains, suggesting that the spatial activity/activation patterns of Akt might be controlled by the spatial dynamics of the upstream kinase PDK1. This agrees with previous observations that Akt is also preferentially activated in these microdomains (Gao and Zhang 2008).

Using a genetically targeted lipid phosphatase, as well as small molecule perturbations, we further demonstrated that specifically disrupting this membrane compartmentalization could abolish the activity of the PI3K/Akt pathway, thereby leading to functional disabilities such as insulin resistance.

In many situations, the activation of a kinase involves interactions with other signaling molecules. The reporters used to monitor kinase activation in these cases can therefore comprise bimolecular FRET constructs. Such an approach was used by Offterdinger et al. to observe epidermal growth factor receptor (EGFR) activation. This group utilized the association of a phosphotyrosine binding (PTB) domain from Shc with phosphorylated EGFR to report on EGFR autophosphorylation, which precedes activation (Offterdinger et al. 2004). They first constructed two fusions, EGFR–ECFP and PTB–EYFP, which are simply ECFP- and EYFP-tagged EGFR and PTB, respectively. In cells coexpressing these two constructs, the authors found that PTB translocated to the membrane on EGF stimulation and returned to the cytosol after addition of the EGFR inhibitor AG1478. This translocation event brings the two FPs together, increasing the intermolecular FRET efficiency and enabling the quantitative measurement of EGFR activation. One caveat of this bimolecular approach is the diffusion-limited collision of the PTB with activated EGFR. As a result, EGFR dynamics were not faithfully reported using this bimolecular method. The authors therefore connected the two fusion proteins with a flexible linker to generate a unimolecular construct. Because of the linker, the recruitment of the PTB by activated EGFR is no longer diffusion limited. The authors also showed that intermolecular FRET was not significant, that is, the binding of the PTB from one reporter molecule with the EGFR from another reporter molecule did not contribute significantly to the observed FRET signal. Although the dynamic range of the response is reduced (from 50% to 10%) compared to the bimolecular design, the authors claim that the unimolecular design is advantageous owing to its improved response kinetics. This illustrates a common compromise that should be considered when deciding between uni- or bimolecular sensor designs.

An alternative strategy for detecting endogenous, active kinases involves the use of specific affinity reagents. Like antibodies, these affinity reagents are polypeptides that can bind specifically to a protein of interest. To detect active kinases, this design requires a genetically encodable affinity reagent that is highly specific for an active kinase to serve as the sensing unit and a fluorescent reporting tag to highlight the localization of the sensor (Figure 9.1b). Jha and colleagues recently illustrated such an approach by developing a kinase sensor for p21-activated kinase (PAK1; Jha et al. 2011). They first developed PAcKer, an active-PAK1 binding domain derived from the autoinhibitory switch domain that natively binds inactive PAK1. Using cysteine/thiol chemistry, the authors conjugated an environmentally sensitive fluorophore to the predicted binding interface of the PAK1-binding domain to act as the main reporting unit.

Fusion to CFP provided improved stability, as well as a ratiometric readout. These efforts led to an engineered affinity binder that reports PAK1 activation *in vitro* with a 40% dynamic range. Even though it has not been utilized in a live-cell context, this study demonstrates a different approach to kinase activation sensing, which can target endogenous enzymes.

9.3 KINASE ACTIVITY BIOSENSORS

On activation, a kinase catalyzes the hydrolysis and transfer of the γ-phosphate of ATP to a hydroxyl-containing residue within the substrate. In the design of kinase activity biosensors, a specific substrate of the kinase of interest therefore serves as the sensing unit. This substrate can be based on sequences from an endogenous substrate, a portion of an endogenous substrate, or a surrogate substrate designed based on the knowledge of consensus substrate sequences for a given kinase.

The sensing unit in small-molecule–based kinase biosensors often comprises a synthetic peptide substrate that mimics an endogenous substrate. Following the phosphorylation of the peptide by the kinase of interest, the phosphate group can directly interact with the reporting unit to cause changes in fluorescence intensity. It is nontrivial, however, to incorporate a reporting unit without interfering with the specificity of the chosen peptide sequence. Lawrence and coworkers pioneered two different approaches that maintain substrate specificity while utilizing environmentally sensitive small-molecule dye reporting units to generate protein kinase C (PKC) sensors. The first design relied on a phosphothreonine residue to alter the photophysics of dye moieties such as nitro-2-1,3-benzoxadiazol-4-yl (NBD), dansyl, and acridine (Yeh et al. 2002; Figure 9.2a). A library of potential PKC substrates was synthesized as sensing units and subsequently labeled with one of five reporting dyes at the N-terminal. After screening 417 distinct sensors using cell extracts, the authors obtained a peptide-based PKC sensor that is sensitive to all three PKC isoforms *in vitro* and has a dynamic range of greater than 200% in live cells. In a second effort, this group took advantage of the propensity of phosphate groups to chelate divalent cations. Carboxylated dyes were synthesized and conjugated to the N-terminal of a known PKC-sensing peptide substrate; the carboxylated dye and the incipient phospho-residue form a bidentate Ca^{2+} chelator. Phosphorylation of the peptide by active PKC therefore results in successful chelation, which polarizes the dye and thus increases its fluorescence yield (Figure 9.2b). Because this design required precise conformational changes to support chelation, several peptide linkers were tested between the reporting and sensing units. Through this effort, the authors constructed a reporter that displayed a 2.6-fold increase in fluorescence on PKC treatment *in vitro* (Chen et al. 2002).

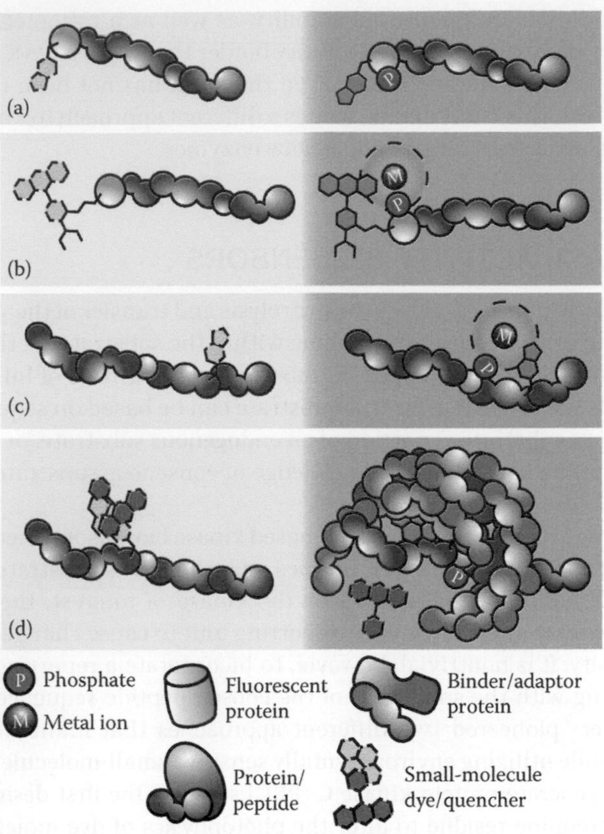

Figure 9.2 Synthetic peptide-based kinase biosensor designs. Light-blue and red regions designate the nonphosphorylated and phosphorylated state of the sensor, respectively. Dashed outlines indicate a nonfluorescent component. (a) Phosphorylation of a peptide changes the microenvironment around a C-terminal dye label and hence its fluorescence intensity. (b) Phosphorylation of the C-terminal residue allows the peptide to chelate metal ions in a bidentate format through its C-terminal dye label, modulating dye fluorescent intensity. (c) Same as (b), except that the phosphorylated residue can be anywhere along the peptide. (d) Phosphorylation of a peptide attracts binder proteins, which prevents solvent-quenching of a dye label.

Although the aforementioned design yielded several PKC probes, the conjugation chemistry limits its application to kinases that phosphorylate N-/C-terminal residues. This approach was later expanded by Imperiali and colleagues using unnatural amino acids (Shults and Imperiali 2003; Shults et al. 2006) (Figure 9.2c). In this method, called chelation-enhanced fluorescence (CHEF), one can introduce the unnatural amino acid Sox at any location within a peptide sensing unit using solid-phase synthesis. The Mg^{2+}-chelating Sox moiety replaces the carboxylated dye in earlier approaches as the reporting unit. The use of unnatural amino acid at positions within the polypeptide vastly increased the possible sensor candidates. These authors were able to screen an extensive library for sensors that optimized both kinase specificity and chelate formation. Sensors were successfully tested for a multitude of kinases including protein kinase A (PKA), PKC, Cdk2, Akt, MK2, and Pim2. Many of these chemosensors exhibited a 4- to 6-fold fluorescence increase on phosphorylation *in vitro* (Shults et al. 2006). An improved set of CHEF sensors has recently been utilized to reveal the kinetic parameters of kinases in a 96-well plate format (Lukovic et al. 2008). These probes are therefore suitable for the high-throughput screening of pharmacological agents such as kinase inhibitors. The dynamic range of a sensor is crucial for high-throughput screening using peptide kinase sensors. Dynamic range can be improved either by increasing the signal maximum or by decreasing the minimum. Protease biosensors, for example, can often achieve 25-fold response increases owing to their low basal signal. On this basis, Lawrence and co-workers suppressed the basal fluorescence of their labeled peptide kinase reporter by using a variety of diffusible quenchers *in vitro* (Sharma et al. 2007) (Figure 9.2d). They synthesized peptides bearing the PKA consensus sequence and attached a pyrene reporting dye at different residues. The recombinant phospho-amino acid–binding domain (PAABD) 14-3-3τ served as the diffusible binding portion of the sensing unit. Before PKA phosphorylation, the quencher is able to access the dye and suppress fluorescence. On substrate phosphorylation, the PAABD binds to the phosphorylated substrate and in the process conveniently shields the dye from the quencher, making the pyrene dye highly fluorescent. An *in vitro* screen identified one sensor with a 64-fold increase in fluorescence on PKA phosphorylation. The success of this strategy is evident from the fact that before this methodology, it was difficult to achieve even a 10-fold response increase in a kinase biosensor (Rothman et al. 2005). These authors speculated on the potential for using this deep-quench methodology in living cells, but to our knowledge it has not yet been applied *in vivo*.

In addition to peptide substrates, early kinase activity sensors also exploited the phosphorylation-dependent conformational changes of certain native substrates and detected such changes using an environmentally sensitive fluorophore or a FRET pair, similar to the scheme depicted in Figure 9.1a. Post et al. (1994) used thiol/iodoacetamide chemistry to label purified myosin regulatory light chain with acrylodan at a single mutated cysteine residue. Acrylodan can report solvent polarity by becoming dim in a hydrophilic environment; both the phosphorylation and the regulatory function of the light chain tolerated the labeling. On phosphorylation by myosin light chain kinase (MLCK) *in vitro*, the fluorescence intensity of this sensor decreased by 60% owing to a conformational change induced by phosphorylation. Unfortunately, when exchanged into myosin II or microinjected into live cells (whereupon the biosensor was incorporated into stress fibers), the dynamic range of the labeled sensor decreased and autofluorescence became a severe problem. The reduced dynamic range is a result of the inability of acrylodan to maintain similar solvent exposure when the light chain is complexed to myosin II. The authors ultimately used another technique to monitor MLCK activity successfully: They separately labeled the purified regulatory light chain with fluorescein and myosin II with rhodamine, then microinjected the two sensors into Swiss 3T3 cells (Post et al. 1995). Using fluorescein and rhodamine as a FRET pair, the authors were able to detect the phosphorylation of the light chain by MLCK *in vivo*.

Although dye-based kinase activity sensors typically have a large dynamic range, the introduction of these sensors into living cells is not straightforward. Genetically encoded sensors are better suited to monitor the activity of signaling molecules in their native milieu. These sensors typically utilize FPs as the reporting unit. Employing a native protein substrate as the sensing unit, Kurokawa et al. and Ting et al. independently developed c-Abl tyrosine kinase activity reporters named Picchu and Abl indicator, respectively. In both cases, the Abl kinase substrate CrkII was utilized as the sensing unit, with the CFP–YFP FRET pair used as the reporting unit. CrkII contains an Abl substrate motif as well as an SH2 domain. Tyrosine phosphorylation of CrkII in the sensor by Abl kinase causes intramolecular complex formation and elicits a response from the FRET reporting pair. Kurokawa et al. (2001) constructed the ratiometric Picchu using C-terminally truncated CrkII, whereas Ting at el. (2001) fused a truncated donor CFP to full-length CrkII. Compared to the earlier attempt using conjugated small-molecule dyes, these authors obtained a higher dynamic range *in vitro* (50% vs. 3%). Both Abl indicator and Picchu exhibit a 10–20% change in emission ratio *in vivo*, indicating room for improving these sensors.

Many current genetically encoded kinase activity reporters (KARs) utilize an artificial molecular switch to transduce the substrate phosphorylation event. Such molecular switches often consist of a peptide substrate based on the consensus sequence of a particular kinase and a PAABD. The binding of the PAABD to the phosphorylated substrate induces a change in FRET between the fluorophore pair employed as the reporting unit (Figure 9.3a). A prototypical kinase activity reporter is the PKA activity sensor AKAR (A kinase activity reporter) (Zhang et al. 2001). PKA is a cAMP-dependent, cGMP-dependent and protein kinase C (AGC) kinase that mediates signals from a variety of G protein–coupled receptors. In the first-generation AKAR (AKAR1), the phosphothreonine binding domain 14-3-3τ was employed to detect the phosphorylation status of a consensus PKA substrate sequence. A FRET FP pair sandwiches this sensing unit at either terminus to report the PKA-induced

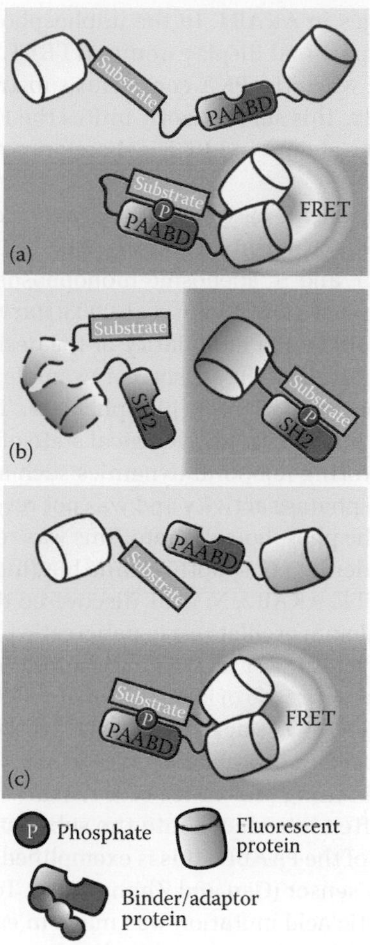

Figure 9.3 Kinase biosensors based on engineered, genetically encodable molecular switches. Light-blue and red regions designate the nonphosphorylated and phosphorylated state of the sensor, respectively. Dashed outlines indicate nonfluorescent components. (a) Phosphorylation of a peptide substrate sequence attracts an intramolecular binding domain, leading to a change in FRET between two fluorescent proteins. (b) Phosphorylation of a substrate sequence attracts an intramolecular binding domain and prevents the chromophore of a fluorescent protein from undergoing solvent-quenching, thereby increasing fluorescence intensity. (c) Bimolecular version of (a). Phosphorylation of a peptide substrate attracts an intermolecular binding domain, which increases FRET between two fluorescent proteins.

phosphorylation changes in AKAR1. In the unphosphorylated state, the two FPs are relatively far apart and display minimal FRET. However, the sensing unit becomes compact when the PKA consensus substrate is phosphorylated and captured by 14-3-3τ. This subsequently unites the FRET pair, resulting in an increase in FRET. The change in FRET is therefore a measure of the change in kinase activity.

This generic domain scheme has proven effective in sensing the activities of many other kinases—c-Jun N-terminal kinase (JNK), PKC, extracellular signal–regulated kinase (ERK), and 5′ adenosine monophosphate-activated protein kinase (AMPK), to name just a few (Violin et al. 2003; Harvey et al. 2008; Fosbrink et al. 2010; Tsou et al. 2011). The modularity of the design facilitates rational improvement on each portion of the sensor. For example, the sensing unit must be available for dephosphorylation by phosphatases. This reversibility is an important feature that reflects the physiological state of the system, especially in the context of monitoring temporal dynamics such as oscillations. AKAR1 did not respond to phosphatase activity and was not reversible *in vivo*, because 14-3-3 binds avidly to the phospho-substrate. This was resolved by substituting 14-3-3 with FHA1, a different phosphothreonine binding domain (Zhang et al. 2005). Using the reversible AKAR2, Ni et al. discovered that PKA activity oscillates in tandem with calcium oscillations in pancreatic β cells, forming a tightly integrated signaling circuit (Ni et al. 2011), with different frequencies of PKA activity oscillations characteristic of local or global cellular signaling.

Replacement of the PAABD as described in the preceding text is nontrivial. The substrate phosphorylation preference of a kinase of interest often does not coincide with the binding preference of a PAABD. To improve the agreement, mutations are often introduced into the substrate sequence to accommodate the preference of the PAABD. This is exemplified by the development of AktAR, an Akt activity sensor (Gao and Zhang 2008). To accommodate FHA1 binding, a single aspartic acid mutation was made in each of the three candidate sequences at the +3 location relative to the phosphothreonine. As a consequence, only the sequence derived from the phosphorylation site of FOXO1 could be phosphorylated by Akt and used to sense Akt activity.

Apart from the kinase substrate and the PAABD, kinase activity sensors may also include docking domains derived from a binding partner or anchoring protein for the kinase of interest. The inclusion of such binding/docking domains in the sensing unit stabilizes kinase binding, thereby increasing the specificity and efficiency of sensor phosphorylation. For example, the ERK activity reporter EKAR uses a substrate peptide that can be recognized not just by ERK but also by other similar MAPKs such as JNK and p38. Harvey et al. therefore used a four-amino-acid docking sequence to specifically enforce ERK binding. Furthermore, they found that candidate sensors incorporating a docking sequence with lower ERK affinity displayed significantly lower

FRET responses, indicating that the docking domain could also be beneficial in terms of efficiency (Harvey et al. 2008). Similarly, Fosbrink et al. (2010) successfully utilized a C-terminal binding motif (from the JNK substrate JDP2) in the sensing unit of JNKAR to ensure JNK specificity.

A sensor with a high dynamic range can register small activity amplitudes, such as those from basal activity, as well as fine details in kinetics. However, it can be difficult to fully optimize the dynamic range of a KAR in the first iteration (Kurokawa et al. 2001; Violin et al. 2003; Gao and Zhang 2008; Harvey et al. 2008). On the other hand, the intensity-based fluorescent calcium sensor GCaMP was initially able to achieve a dynamic range of at least a 400% change in fluorescence intensity (Nakai et al. 2001). Subsequent efforts to engineer the recombinant molecule have successfully led to the improvement of the dynamic range, with a recent version boasting a dynamic range of up to 11,000% (Akerboom et al. 2012). GCaMP works via the reversible closure of a solvent-accessible hole in the barrel of GFP, which is mediated by the binding interaction between calmodulin (CaM) and the M13 peptide. Fused to the N- and C-termini of a circularly permuted variant of GFP (see Chapter 7), CaM and M13 bind tightly upon increases in the calcium concentration, thereby eliminating solvent access to the GFP fluorophore and increasing fluorescence. Inspired by this mechanism, Kawai et al. (2004) produced a series of sensors to monitor insulin-mediated phosphorylation events using circularly permuted FPs. These authors compromised the FP barrel on the strand that passes directly over the phenolate fluorophore (N144/Y145) and attempted to rescue its fluorescence by using a molecular switch consisting of an SH2 domain and a phosphorylation domain derived from the insulin receptor substrate-1 (IRS-1) consensus substrate at the N- and C-termini, respectively (Figure 9.3b). On stimulation with insulin, the IRS-1–derived domain is phosphorylated by the insulin receptor. Subsequently, the SH2 domain binds to the phosphorylated substrate and closes the FP barrel, which is analogous to the action of M13/CaM in GCaMP. This intensity-based method can be used to monitor two or more kinases simultaneously and thus decipher their interdependency. Unfortunately, this intensity-based kinase probe currently does not possess a sufficiently large dynamic range compared with FRET-based kinase reporters.

Other approaches have also been employed to realize better KAR dynamic range, with each successive effort introducing modest improvements. Even though the results are difficult to predict *a priori*, new FP combinations can sometimes lead to better FRET FP reporting units. For example, when different acceptor FPs and orientations were tested in the reversible AKAR2 sensor, a variety of responses were observed (Allen and Zhang 2006). Allen and Zhang tested six different FP pairs, and only the version containing the circularly permuted Venus cpVE172 was found to improve the dynamic range (by 2-fold). Unimolecular AKAR3 was further improved when the donor CFP was substituted with Cerulean (Rizzo et al. 2004), a brighter cyan FP variant (Depry et al. 2011). An alternate strategy is to reduce the basal FRET signal of the sensor, similar to the idea behind deep quench. In the bimolecular approach (Figure 9.3c), two constructs, each comprising one module of the sensing unit tagged to an FP, were introduced into cells to reduce intermolecular contact and the basal FRET level (Herbst et al. 2011). This bimolecular approach has also been echoed by the inclusion of designed long linkers, such as EV and 72G, between the FRET pair in unimolecular sensors to reduce basal FRET (Sato et al. 2002; Harvey et al. 2008). Apart from the obvious benefits, an increase in dynamic range also allows the sensor to register subtle changes such as basal activity. AKAR4, for example, was sufficiently sensitive to reveal locally heightened basal PKA activity when targeted to membrane rafts (Depry et al. 2011).

Finally, in some cases, a substrate-based probe can reveal complex regulation by several different kinases. This is perhaps best exemplified by the use of the reporter for cAMP response element–binding protein (CREB) activation to monitor convergent kinase events. CREB is a transcription factor that can be activated by PKA, PKC, casein kinases, and calmodulin kinases, among others. To monitor this convergent CREB activation, Friedrich et al. (2010) constructed the sensor Indicator of CREB Activation due to Phosphorilation (ICAP) from fragments of both the native CREB kinase-inducible domain and the CREB-binding protein. The use of native elements ensures that the probe can be used to dissect the multiple signaling pathways that converge on CREB. In this study, the authors specifically demonstrated the role of anchoring proteins in modulating PKA-dependent CREB phosphorylation.

9.4 OUTLOOK AND PERSPECTIVE

The reporters reviewed here have greatly aided our passive observation of the responses of signaling molecules to physiological stimuli. But a greater understanding of how the "black box" that is a signaling cascade works can be obtained by combining observation with more active perturbation of various elements within the cascade. The simplest approaches have involved the use of pharmacological agents that are designed to activate or inhibit components of a cascade. Nevertheless, the nonspecificity of such agents has always been a concern and cannot be entirely eliminated in practice. Alternatively, gene knock-in/-out or overexpression can be used to achieve specific perturbations. Yet these approaches can also falter because of the longer time frame required to apply, along with the redundancy and compensation that are built into signaling pathways. Synthetic photoactivatable species overcome this problem by acutely modulating the concentration of a signaling component, thereby overwhelming the natural redundancy. Typically, caged species contain a photolabile protecting moiety that sterically shields the main functional site. Function is restored on the removal of the protection via photo-activation. As an example related to kinase sensing, Nguyen et al. have demonstrated the solid-phase synthesis of a series of caged phosphoserine peptides (Nguyen et al. 2004). By incorporating (nitrophenyl)ethyl caging groups into the consensus peptide sequence recognized by the 14-3-3 group of PAABDs, these authors were able to effectively release a 14-3-3-binding phosphoserine peptide via ultraviolet (UV) illumination. In live cell cultures, it was found that such caged peptides could compete with endogenous 14-3-3 substrates on release and create a phenotype with premature M-phase entry.

Another strategy for acute biochemical perturbations is based on the FKBP–(rapamycin)–FRB dimerization system (Inoue et al. 2005). Using this system, Akt could be recruited to the plasma membrane and could activate the GSK3/NF-κB pathway to a sufficient extent that cells were able to survive an apoptotic challenge (Li et al. 2002). Aside from being useful for recruiting a kinase/phosphatase to the desired subcellular location, the inducible interaction between FKBP and FRB can also be used to activate kinases. Karginov et al. (2010) were able to generate a rapamycin-regulated focal adhesion kinase (RapR-FAK) using a truncated form of FKBP. The secondary structure around the FKBP insertion point became stabilized only on rapamycin/FRB interaction, allowing the FAK fusion to retain many of its physiological features, and this method has been generalized to similarly modify the kinases Src and p38. Irreversibility, however, is the primary disadvantage of rapamycin-based actuation. Optogenetic techniques, on the other hand, employ light to reversibly actuate protein domains and thus retain the advantage of being entirely genetically encoded. One such tool utilizes the LOV (light oxygen voltage) domain, a light-sensitive domain derived from phototropins. This domain consists of a light-absorbing, flavin-binding motif and a Jα helix that unfolds and dissociates in the presence of blue light. Wu et al. showed that the fusion of a LOV domain and Jα helix to Rac1 produced a photoactivatable Rac1, which could then be used to induce cell protrusions and ruffles at will (Wu et al. 2009). Recently, a new, fluorescent-protein–based dimerization system has also been demonstrated. A mutant photoswitchable fluorescent protein was found to dimerize more avidly in its ON (fluorescent) state than its OFF (nonfluorescent) state, and this optical control over the dimerization state was utilized to sterically block the catalytic activities of both the GTPase Cdc42 and a hepatitis C viral protease (Zhou et al. 2012b). In conjunction with kinase sensors, all of these techniques have the potential to enrich the study of kinase signaling in the native biological context.

The outputs of signaling cascades are very often shaped by coordinated inputs from multiple signaling molecules. Cross-regulation between kinases and other signaling molecules is an important feedback mechanism. Therefore, a holistic picture of a signaling process can be obtained by simultaneously observing the coordinated activity dynamics of multiple kinases and signaling molecules. The ongoing development of single-fluorophore reporters, as mentioned previously, can greatly aid in these efforts. Multiple signaling molecules can be monitored simultaneously using single-fluorophore reporters of different colors. Spectral separation can be also achieved by using a FRET-based reporter in combination with a single-fluorophore reporter of a different color. Alternatively, a few approaches have been developed for using two FRET-based reporters for coimaging, including the use of two spectrally distinct fluorophores as donors with a shared acceptor (see Depry et al. [2013]) for a comprehensive review of these approaches).

Computational tools can also be used to simultaneously track multiple signaling activities within the same cell. Recently, a systems biology method was adopted to illustrate the interdependencies between multiple GTPases. Assuming that the relationships between signaling components of the same pathway are invariant across different cells, a spatiotemporal reference point can be established to synchronize different experiments despite cell–cell variations. Such computational multiplexing can integrate data obtained in separate experiments using different biosensors and reveal the order of events. Machacek et al. (2009) compiled live-cell experiments with Rac1, RhoA, and Cdc42, which are all known to be important in cytoskeletal rearrangement during cell protrusion. By setting their reference point at the protrusion of the cell leading edge, they were able to correlate the activity of all three GTPases along the same cell protrusion trajectory. On average, they found that Rac1 and Cdc42 were activated 2 μm behind the leading edge, 40 s after RhoA activation. Applying a similar analysis using a microprinted reference point, another study proposed that PI3K kinase activity follows edge extension but precedes Rac1 activation (Lu et al. 2011). This computational multiplexing approach could be useful in examining the coordination between different kinases.

Finally, it has become increasingly clear that signaling is spatially compartmentalized. However, the study of signaling microdomains that have small dimensions is challenging. Over the years, a number of optical superresolution techniques, such as photoactivation localization microscopy (PALM), stimulated emission depletion (STED), and structured illumination (SI), have allowed a more detailed examination of protein localization below the resolution of standard microscopy (see Chapter 16). Recently, we have developed the pcSOFI technique, which enables live-cell super-resolution imaging using photoswitchable FPs and conventional fluorescence microscopes (Dedecker et al. 2012). It is our hope that some of these methods can be combined with fluorescent biosensors to reveal not only the localization but also the activities of signaling molecules such as protein kinases with improved spatial resolution. Such high-resolution activity maps will help elucidate the finer details underlying the compartmentalization of signal transduction and directly address long-standing questions regarding the function and regulation of minute signaling microdomains.

REFERENCES

Acker, T., Fandrey, J., and Acker, H. (2006). The good, the bad and the ugly in oxygen-sensing: ROS, cytochromes and prolyl-hydroxylases. *Cardiovascular Research 71*, 195–207.

Akerboom, J., Chen, T.-W., Wardill, T. J., Tian, L., Marvin, J. S., Mutlu, S., Calderon, N. C., Esposti, F., Borghuis, B. G., Sun, X. R. et al. (2012). Optimization of a GCaMP calcium indicator for neural activity imaging. *Journal of Neuroscience 32*, 13819–13840.

Allen, M. D., and Zhang, J. (2006). Subcellular dynamics of protein kinase A activity visualized by FRET-based reporters. *Biochemical and Biophysical Research Communications 348*, 716–721.

Ananthanarayanan, B., Fosbrink, M., Rahdar, M., and Zhang, J. (2007). Live-cell molecular analysis of Akt activation reveals roles for activation loop phosphorylation. *Journal of Biological Chemistry 282*, 36634–36641.

Andjelkovic, M., Alessi, D. R., Meier, R., Fernandez, A., Lamb, N. J. C., Frech, M., Cron, P., Cohen, P., Lucocq, J. M., and Hemmings, B. A. (1997). Role of translocation in the activation and function of protein kinase B. *Journal of Biological Chemistry 272*, 31515–31524.

Calleja, V., Ameer-Beg, S. M., Vojnovic, B., Woscholski, R., Downward, J., and Larijani, B. (2003). Monitoring conformational changes of proteins in cells by fluorescence lifetime imaging microscopy. *Biochemical Journal 372*, 33–40.

Chen, C. A., Yeh, R. H., and Lawrence, D. S. (2002). Design and synthesis of a fluorescent reporter of protein kinase activity. *Journal of the American Chemical Society 124*, 3840–3841.

Cotton, G. J., and Muir, T. W. (2000). Generation of a dual-labeled fluorescence biosensor for Crk-II phosphorylation using solid-phase expressed protein ligation. *Chemistry & Biology 7*, 253–261.

Dedecker, P., Mo, G. C. H., Dertinger, T., and Zhang, J. (2012). Widely accessible method for superresolution fluorescence imaging of living systems. *Proceedings of the National Academy of Sciences of the United States of America 109*, 10909–10914.

Depry, C., Allen, M. D., and Zhang, J. (2011). Visualization of PKA activity in plasma membrane microdomains. *Molecular Biosystems 7*, 52–58.

Depry, C., Mehta, S., and Zhang, J. (2013). Multiplexed visualization of dynamic signaling networks using genetically encoded fluorescent protein-based biosensors. *Pflugers Archiv-European Journal of Physiology 465*, 373–381.

Fosbrink, M., Aye-Han, N.-N., Cheong, R., Levchenko, A., and Zhang, J. (2010). Visualization of JNK activity dynamics with a genetically encoded fluorescent biosensor. *Proceedings of the National Academy of Sciences of the United States of America 107*, 5459–5464.

Friedrich, M. W., Aramuni, G., Mank, M., Mackinnon, J. A. G., and Griesbeck, O. (2010). Imaging CREB activation in living cells. *Journal of Biological Chemistry 285*, 23283–23293.

Gao, X., Lowry, P. R., Zhou, X., Depry, C., Wei, Z., Wong, G. W., and Zhang, J. (2011). PI3K/Akt signaling requires spatial compartmentalization in plasma membrane microdomains. *Proceedings of the National Academy of Sciences of the United States of America 108*, 14509–14514.

Gao, X., and Zhang, J. (2008). Spatiotemporal analysis of differential Akt regulation in plasma membrane microdomains. *Molecular Biology of the Cell 19*, 4366–4373.

Harvey, C. D., Ehrhardt, A. G., Cellurale, C., Zhong, H., Yasuda, R., Davis, R. J., and Svoboda, K. (2008). A genetically encoded fluorescent sensor of ERK activity. *Proceedings of the National Academy of Sciences of the United States of America 105*, 19264–19269.

Herbst, K. J., Allen, M. D., and Zhang, J. (2011). Luminescent kinase activity biosensors based on a versatile bimolecular switch. *Journal of the American Chemical Society 133*, 5676–5679.

Hyman, L. M., and Franz, K. J. (2012). Probing oxidative stress: Small molecule fluorescent sensors of metal ions, reactive oxygen species, and thiols. *Coordination Chemistry Reviews 256*, 2333–2356.

Inoue, T., Do Heo, W., Grimley, J. S., Wandless, T. J., and Meyer, T. (2005). An inducible translocation strategy to rapidly activate and inhibit small GTPase signaling pathways. *Nature Methods 2*, 415–418.

Jha, R. K., Wu, Y. I., Zawistowski, J. S., MacNevin, C., Hahn, K. M., and Kuhlman, B. (2011). Redesign of the PAK1 autoinhibitory domain for enhanced stability and affinity in biosensor applications. *Journal of Molecular Biology 413*, 513–522.

Jiang, P. J., and Guo, Z. J. (2004). Fluorescent detection of zinc in biological systems: Recent development on the design of chemosensors and biosensors. *Coordination Chemistry Reviews 248*, 205–229.

Kalab, P., and Soderholm, J. (2010). The design of Forster (fluorescence) resonance energy transfer (FRET)-based molecular sensors for Ran GTPase. *Methods 51*, 220–232.

Karginov, A. V., Ding, F., Kota, P., Dokholyan, N. V., and Hahn, K. M. (2010). Engineered allosteric activation of kinases in living cells. *Nature Biotechnology 28*, 743–747.

Kawai, Y., Sato, M., and Umezawa, Y. (2004). Single color fluorescent indicators of protein phosphorylation for multicolor imaging of intracellular signal flow dynamics. *Analytical Chemistry 76*, 6144–6149.

Kurokawa, K., Mochizuki, N., Ohba, Y., Mizuno, H., Miyawaki, A., and Matsuda, M. (2001). A pair of fluorescent resonance energy transfer-based probes for tyrosine phosphorylation of the CrkII adaptor protein in vivo. *Journal of Biological Chemistry 276*, 31305–31310.

Li, B., Desai, S. A., MacCorkle-Chosnek, R. A., Fan, L., and Spencer, D. M. (2002). A novel conditional Akt "survival switch" reversibly protects cells from apoptosis. *Gene Therapy 9*, 233–244.

Lu, S. Y., Kim, T. J., Chen, C. E., Ouyang, M., Seong, J., Liao, X. L., and Wang, Y. X. (2011). Computational analysis of the spatiotemporal coordination of polarized PI3K and Rac1 activities in micro-patterned live cells. *PLoS One 6*, e21293.

Lukovic, E., Gonzalez-Vera, J. A., and Imperiali, B. (2008). Recognition-domain focused chemosensors: Versatile and efficient reporters of protein kinase activity. *Journal of the American Chemical Society 130*, 12821–12827.

Machacek, M., Hodgson, L., Welch, C., Elliott, H., Pertz, O., Nalbant, P., Abell, A., Johnson, G. L., Hahn, K. M., and Danuser, G. (2009). Coordination of Rho GTPase activities during cell protrusion. *Nature 461*, 99–103.

Nakai, J., Ohkura, M., and Imoto, K. (2001). A high signal-to-noise Ca^{2+} probe composed of a single green fluorescent protein. *Nature Biotechnology 19*, 137–141.

Navizet, I., Liu, Y. J., Ferre, N., Roca-Sanjuan, D., and Lindh, R. (2011). The chemistry of bioluminescence: An analysis of chemical functionalities. *ChemPhysChem 12*, 3064–3076.

Nguyen, A., Rothman, D. M., Stehn, J., Imperiali, B., and Yaffe, M. B. (2004). Caged phosphopeptides reveal a temporal role for 14-3-3 in G1 arrest and S-phase checkpoint function. *Nature Biotechnology 22*, 993–1000.

Ni, Q. A., Ganesan, A., Aye-Han, N. N., Gao, X. X., Allen, M. D., Levchenko, A., and Zhang, J. (2011). Signaling diversity of PKA achieved via a Ca^{2+}-cAMP-PKA oscillatory circuit. *Nature Chemical Biology 7*, 34–40.

Nolan, E. M., and Lippard, S. J. (2008). Tools and tactics for the optical detection of mercuric ion. *Chemical Reviews 108*, 3443–3480.

Offterdinger, M., Georget, V., Girod, A., and Bastiaens, P. I. H. (2004). Imaging phosphorylation dynamics of the epidermal growth factor receptor. *Journal of Biological Chemistry 279*, 36972–36981.

Post, P. L., Debiasio, R. L., and Taylor, D. L. (1995). A fluorescent protein biosensor of myosin-II regulatory light-chain phosphorylation reports a gradient of phosphorylated myosin-II in migrating cells. *Molecular Biology of the Cell 6*, 1755–1768.

Post, P. L., Trybus, K. M., and Taylor, D. L. (1994). A genetically-engineered, protein-based optical biosensor of myosin-II regulatory light-chain phosphorylation. *Journal of Biological Chemistry 269*, 12880–12887.

Rizzo, M. A., Springer, G. H., Granada, B., and Piston, D. W. (2004). An improved cyan fluorescent protein variant useful for FRET. *Nature Biotechnology 22*, 445–449.

Rothman, D. M., Shults, M. D., and Imperiali, B. (2005). Chemical approaches for investigating phosphorylation in signal transduction networks. *Trends in Cell Biology 15*, 502–510.

Sato, M., Ozawa, T., Inukai, K., Asano, T., and Umezawa, Y. (2002). Fluorescent indicators for imaging protein phosphorylation in single living cells. *Nature Biotechnology 20*, 287–294.

Sharma, V., Agnes, R. S., and Lawrence, D. S. (2007). Deep quench: An expanded dynamic range for protein kinase sensors. *Journal of the American Chemical Society 129*, 2742–2743.

Shults, M. D., Carrico-Moniz, D., and Imperiali, B. (2006). Optimal Sox-based fluorescent chemosensor design for serine/threonine protein kinases. *Analytical Biochemistry 352*, 198–207.

Shults, M. D., and Imperiali, B. (2003). Versatile fluorescence probes of protein kinase activity. *Journal of the American Chemical Society 125*, 14248–14249.

Stokoe, D., Stephens, L. R., Copeland, T., Gaffney, P. R. J., Reese, C. B., Painter, G. F., Holmes, A. B., McCormick, F., and Hawkins, P. T. (1997). Dual role of phosphatidylinositol-3,4,5-trisphosphate in the activation of protein kinase B. *Science 277*, 567–570.

Ting, A. Y., Kain, K. H., Klemke, R. L., and Tsien, R. Y. (2001). Genetically encoded fluorescent reporters of protein tyrosine kinase activities in living cells. *Proceedings of the National Academy of Sciences of the United States of America 98*, 15003–15008.

Tsou, P., Zheng, B., Hsu, C.-H., Sasaki, A. T., and Cantley, L. C. (2011). A fluorescent reporter of AMPK activity and cellular energy stress. *Cell Metabolism 13*, 476–486.

Valeur, B., and Leray, I. (2000). Design principles of fluorescent molecular sensors for cation recognition. *Coordination Chemistry Reviews 205*, 3–40.

Violin, J. D., Zhang, J., Tsien, R. Y., and Newton, A. C. (2003). A genetically encoded fluorescent reporter reveals oscillatory phosphorylation by protein kinase C. *Journal of Cell Biology 161*, 899–909.

Wu, Y. I., Frey, D., Lungu, O. I., Jaehrig, A., Schlichting, I., Kuhlman, B., and Hahn, K. M. (2009). A genetically encoded photoactivatable Rac controls the motility of living cells. *Nature 461*, 104–108.

Yeh, R. H., Yan, X. W., Cammer, M., Bresnick, A. R., and Lawrence, D. S. (2002). Real time visualization of protein kinase activity in living cells. *Journal of Biological Chemistry 277*, 11527–11532.

Zhang, J., Hupfeld, C. J., Taylor, S. S., Olefsky, J. M., and Tsien, R. Y. (2005). Insulin disrupts beta-adrenergic signalling to protein kinase A in adipocytes. *Nature 437*, 569–573.

Zhang, J., Ma, Y. L., Taylor, S. S., and Tsien, R. Y. (2001). Genetically encoded reporters of protein kinase A activity reveal impact of substrate tethering. *Proceedings of the National Academy of Sciences of the United States of America 98*, 14997–15002.

Zhang, X., and Vik, T. A. (1997). Growth factor stimulation of hematopoietic cells leads to membrane translocation of AKT1 protein kinase. *Leukemia Research 21*, 849–856.

Zhou, X., Herbst-Robinson, K. J., and Zhang, J. (2012a). Visualizing dynamic activities of signaling enzymes using genetically encodable FRET-based biosensors: From designs to applications. *Meth. Enz. 504*, 317–340.

Zhou, X. X., Chung, H. K., Lam, A. J., and Lin, M. Z. (2012b). Optical control of protein activity by fluorescent protein domains. *Science 338*, 810–814.

Chapter 10

RNA

Samie R. Jaffrey

CONTENTS

10.1 RNA IS SPATIALLY AND TEMPORALLY REGULATED IN CELLS

An emerging concept over the past few years is that transcription and other cell signaling pathways are regulated by a diverse array of noncoding RNAs, such as microRNAs, long intergenic noncoding RNAs, Piwi-interacting RNAs, termini-associated RNAs, and other classes of noncoding RNAs (Aravin et al. 2006; Han et al. 2007; Derrien et al. 2012). Our understanding of mRNA has also evolved. mRNA is no longer viewed as a simple intermediate between DNA and protein, but instead is now known to be subjected to a wide range of posttranscriptional processing events, including splicing, nonsense-mediated decay, RNA editing, exo- and endonucleolytic degradation, adenosine methylation, polyadenylation, and deadenylation (Parker and Song 2004; Meyer et al. 2012). Another intriguing aspect of RNA biology is the finding that trinucleotide repeat–containing mRNAs exert specific gain-of-function toxicities associated with their accumulation in the nucleus and other intracellular sites (Ranum and Day 2004). In addition to these different regulatory pathways, recent studies indicate that RNAs transit through different parts of the cell as they undergo maturation to the final form needed in the cell. For example, nascent mRNA transcripts appear to be trafficked to specific intranuclear sites for processing events, such as splicing, nonsense-mediated decay, or for packaging into transport granules (Kiebler and Bassell 2006; Spector and Lamond 2011). After nuclear export, some RNAs have been localized to RNA-enriched intracellular structures including RNA granules, stress granules, and processing bodies (P-bodies) (Anderson and Kedersha 2002; Kiebler and Bassell 2006). The diversity of these RNA regulatory mechanisms makes it clear that RNA is regulated by a spatially complex and intricate network of regulatory mechanisms that have a critical role in gene expression.

RNA regulatory pathways are particularly prominent in neurons. For example, RNA splicing is more highly regulated and is more complex in neurons than in any other cell type (Dredge et al. 2001). Similarly, RNA editing and diseases caused by trinucleotide repeat expansions in mRNA are especially prominent in neurons despite the widespread expression of these transcripts (Keegan et al. 2004). A recent analysis of 1328 noncoding RNAs (>200 nt) revealed that 64% were expressed in brain, many of which had strikingly specific patterns of expression in discrete brain structures (Mercer et al. 2008).

One form of RNA regulation that has received considerable attention is "local" RNA translation. A compelling argument for a fundamental role for RNA localization in cells was presented in a recent landmark study in which high-throughput gene-specific *in situ* hybridization in *Drosophila* embryos showed that 71% of cellular RNAs exhibit specific and often striking intracellular localizations (Lecuyer et al. 2007). This phenomenon is also prominent in neurons, where mRNAs are enriched in axons and dendrites (Hengst and Jaffrey 2007; Deglincerti and Jaffrey 2012). Local translation of these mRNAs may have evolved to accommodate the highly spatially polarized nature of neuronal morphology, which typically involves axons and dendrites that can extend distances of tens to thousands of micrometers from the cell body. The process of local translation involves mRNAs that are translated directly within dendrites and axons, often within small domains such as a 1–2 μm long dendritic spine or small domains within an axonal growth cone (Steward and Schuman 2003; Ji and Jaffrey 2012). RNAs are dynamically transported between RNA granule structures and P-bodies during synaptic stimulation, which likely regulates mRNA translation (Zeitelhofer et al. 2008). Only a small subset of the total mRNA population is trafficked to axons or dendrites (Deglincerti and Jaffrey 2012), and it appears that discrete 3′ UTR sequences are required for signal-dependent translation of distinct pools of RNAs (Cox et al. 2008; Hengst et al. 2009). Local translation bypasses the time-consuming process of propagating a signal to the nucleus, followed by protein synthesis and subsequent transport of the newly synthesized protein to the specific site of receptor activation (Steward and Schuman 2003). Thus, neurons display complex, time-dependent, and spatially specific regulation of local mRNA translation to accommodate their functional and morphologic demands.

10.2 THE NEED FOR RNA IMAGING TECHNOLOGIES

These diverse studies point to the increasing interest in understanding spatial aspects of RNA biology in cells. In many cases, it is not just the localization of RNA in cells, but also how the localization changes in living cells that has the potential to elucidate the RNA-dependent cellular processes. Questions regarding the specific real-time localization of mRNAs, for example, during their processing in the nucleus and nucleolus, and after export to the cytosol and trafficking to neuronal growth cones, spines, axons, nuclei, organelles, and so forth, are particularly important, especially in terms of specific spliced forms of mRNAs, differentially edited mRNAs, and trinucleotide repeat–containing mRNAs. Furthermore, the timing of mRNA trafficking in response to extracellular signals is of particular interest, such as the role of mRNA trafficking to dendritic spines during synaptic plasticity or in growth cones during axon turning (Zhang et al. 2001; Wu et al. 2005). The role of regulated mRNA degradation in dendrites and axons is also of considerable interest in the neuroscience community (Hengst et al. 2006).

RNA visualization technologies are important not only for understanding mRNA. Imaging technologies are particularly important to shed light on the function of the diverse population of noncoding (nc)RNAs that currently have unknown functions in cells. ncRNAs such as Piwi-interacting RNAs, promoter-associated small RNAs (PASRs), termini-associated small RNAs (TASRs) (Han et al. 2007), as well as a plethora of other small ncRNAs (Hannon et al. 2006) whose function is mysterious, appear likely to have the same influence on molecular biology as the discovery of microRNAs. Many ncRNAs are in the nucleus, while others are in the cytosol (Sharma et al. 2011; Batista and Chang 2013), which points to the importance of assessing their RNA localization to identify potential functions of these RNAs. RNA localization in living cells is particularly important, as it can be influenced by signaling pathways, cell division, or disease processes (Batista and Chang 2013). Furthermore, defects in these localizations could potentially account for abnormal functions of these

RNAs. As a result, methods to image RNA in living cells have become particularly important in molecular biology.

10.3 STRATEGIES TO DETECT RNA IN CELLS

The most commonly used technique to study mRNA localization is *in situ* hybridization (Levsky and Singer 2003). This is a well-established technique, but is not a homogeneous assay and does not allow RNA to be monitored in the same cell at different time points. To achieve real-time, single-cell *in vivo* RNA visualization, one technique has been to synthesize RNA *in vitro* using fluorescent nucleotides and then microinject it into cells (Tyagi 2009). The fate of the RNA can then be monitored by fluorescence microscopy. This approach is limited by the technical difficulty of microinjection, as well as by the low throughput. Another approach is to use molecular beacons, which are oligonucleotides that are typically dual labeled with a fluorophore and a quencher (Tyagi and Alsmadi 2004). The beacon adopts a stem-loop structure that is nonfluorescent because of the proximity of the fluorophore and quencher at the base of the stem. When a target mRNA that exhibits complementarity to the loop hybridizes to the beacon, the stem is disrupted, resulting in separation of the fluorophore and quencher and subsequent fluorescence. However, transfected beacons exhibit nonspecific nuclear sequestration (Tyagi and Alsmadi 2004; Mhlanga and Tyagi 2006) and each mRNA requires a custom-designed beacon for visualization. This makes the procedure technically complex and relatively difficult to adopt by most scientific laboratories.

Because of the inherent difficulties of these synthetic approaches, numerous groups have attempted to develop genetically encoded reporters of RNA localization in cells. We here briefly discuss some of the major genetically encoded reporters, as well as the limitations that have prevented their widespread use.

The most widely used genetically encoded RNA imaging approach is the green fluorescent protein (GFP)–MS2 system (Bertrand et al. 1998). This approach uses two components: MS2, a viral protein, fused to GFP; and MS2-binding elements, which are RNA sequences, inserted into the 3′ UTR of RNAs of interest. GFP–MS2 and MS2-element–containing RNAs, or "fusion RNAs," are expressed in cells from transfected DNA. GFP–MS2 binds to the MS2 element-tagged RNA in cells, and fluorescence signals in these cells should represent RNA–GFP complexes. Because unbound GFP–MS2 molecules diffuse throughout the cytosol there would be, in principle, a high fluorescence background. To alleviate this problem, a nuclear localization signal (NLS) is incorporated in the GFP–MS2 fusion protein so that most of the GFP–MS2 moves into the nucleus (Bertrand et al. 1998). Variations of this protocol using different fluorescent proteins and different RNA binding elements have also been described (Hocine et al. 2013).

Concerns have been raised about nonphysiological localizations of RNAs when they are bound to GFP–MS2 (Tyagi 2009). In particular, RNAs bound to GFP–MS2 are subjected to two trafficking signals: one encoded within the RNA and another being the NLS within GFP–MS2. The presence of two trafficking signals confounds interpretation of the intracellular movements of the tagged mRNA. This issue is typically more prominent in practice, because heterologously expressed RNAs are typically designed so that they bind 24–48 GFP–MS2 sequences. Thus, each RNA is subjected to the effects of 24–48 NLS targeting elements (Tyagi 2009). An additional drawback is that the NLS causes the GFP–MS2 to accumulate in the nucleus, resulting in intense nuclear fluorescence signals and thereby complicating the analysis of nuclear-localized RNAs. Because much RNA biology occurs in the nucleus, such as nonsense-mediated decay, RNA editing, nuclear export of RNA, splicing, pioneer RNA translation, and microRNA processing, nuclear accumulation of GFP–MS2 can complicate the analysis of these events. Thus, even though GFP–MS2 is highly useful, it has various drawbacks that prevented widespread adoption by the research community.

Other related approaches have been described that have important limitations. One recent strategy involves the expression of GFP as two separate halves, each fused to half of the protein eIF4A, an RNA-binding protein (Tyagi 2007). RNAs containing the eIF4A-binding site nucleate the binding of the eIF4A halves, which results in the juxtaposition of each GFP half and the subsequent formation of a stable GFP complex. Because the GFP complex requires approximately 30 min to mature into a fluorescent species (Merzlyak et al. 2007), this method might not allow for visualization of nascent RNAs. In addition, once formed, the fluorescent complex can dissociate spontaneously or after RNA degradation, which can result in high levels of background cytoplasmic fluorescence. Thus, although this approach is promising, there are potential drawbacks that could complicate analysis.

Although these approaches, as well as others not reviewed here because of space limitations (Tyagi 2009), are valuable for imaging RNA in living cells, they are also complicated by the need to introduce many different components into cells. In principle, a much simpler approach would be simply to express an RNA that by itself confers the fluorescence that is needed to image it. This would make the tagging approach analogous to GFP tagging, which is particularly valuable for understanding the functions of proteins in living cells.

10.4 A STRATEGY FOR GENETICALLY ENCODING FLUORESCENT RNA IN CELLS

To monitor RNAs in cells, we were particularly attracted to the idea of tagging RNAs of interest with an RNA sequence that would confer fluorescence exclusively to the tagged RNA. Short RNA sequences that bind other molecules have been termed "aptamers." Using the SELEX (described in Section 10.6) approach, RNA aptamers that bind fluorescent dyes such as fluorescein have been described (Holeman et al. 1998; Sando et al. 2008). However, these RNAs have not found use in live-cell experiments because both bound and unbound dyes are fluorescent and have nearly identical emission spectrum properties. Thus, fluorescent signals from unbound fluorescein would overwhelm the signal from the fluorescein bound to RNA.

Our strategy was to choose fluorophores whose fluorescence would be "switched on" only when bound to the aptamer (Paige et al. 2011). In this strategy, the unbound fluorophore is not fluorescent, so it would not contribute to background fluorescence.

There are numerous molecules that exhibit fluorescence on binding nucleotides. Ethidium bromide and Hoechst dyes are probably the best known, but these molecules bind oligonucleotides relatively nonspecifically. Thus, it is critical to identify "bioorthogonal" molecules that are not switched on by any cellular constituent but could become fluorescent on binding a specific RNA sequence.

To address this challenge, we used a strategy that took advantage of some unique aspects of the chemistry of GFP. After GFP is synthesized, it undergoes a posttranslational, autocatalytic, intramolecular cyclization of an internal Ser-Tyr-Gly tripeptide. The cyclized product is then oxidized to the final 4-hydroxybenzylidene-imidazolinone fluorophore (Figure 10.1). The oxidation results in substantial delocalization of the π-electron system. The intramolecular cyclization and oxidation that occur are catalyzed by the folded GFP protein. The phenolic moiety in the fluorophore is derived from a tyrosine in the protein.

Surprisingly, the chemically synthesized GFP fluorophore is nonfluorescent (Niwa et al. 1996). Similarly, when GFP is denatured, it loses its fluorescence, and the fluorescence returns if the protein is renatured (Bokman and Ward 1981). The explanation for this conditional fluorescence lies in the chemical structure of the fluorophore. After photoexcitation, a molecule can dissipate its energy through either a radiative (i.e., fluorescence) pathway or a nonradiative pathway, which usually involves vibrational or other intramolecular movements. When GFP is folded, the excited fluorophore dissipates its energy by the radiative decay pathway (fluorescence). However, when the protein is unfolded, the fluorophore dissipates its energy by various bond rotations, which likely involves a twisting motion about the ethylenic bridge referred to as a "hula twist" (Baffour-Awuah and Zimmer 2004). The folded GFP protein somehow prevents these movements from occurring, therefore making the radiative decay pathway the sole pathway available to dissipate the energy of the excited state fluorophore (Baffour-Awuah and Zimmer 2004). Unfolding of the protein removes these steric constraints. Consistent with the role of fluorophore

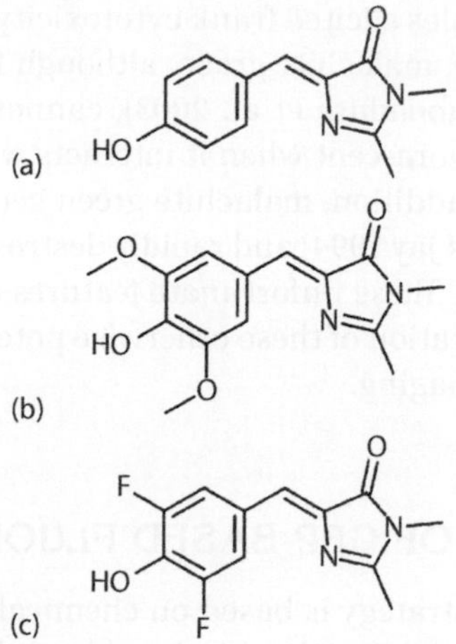

Figure 10.1 Structure of GFP fluorophore and fluorophores switched on by RNA aptamers. (a) Shown is 4-hydroxy-benzylidene imidazolinone, the fluorophore found in GFP. (b) Shown is 3,5-dimethyoxy-4-hydroxy-benzylidene imidazolinone, a fluorophore that is activated by specific RNA aptamers. (c) Shown is 3,5-difluoro-4-hydroxy-benzylidene imidazolinone, the fluorophore activated by the Spinach aptamer.

rigidification in GFP fluorescence, GFP proteins that contain amino acid mutations that permit conformational freedom of the fluorophore have normal fluorophore maturation, but are nonfluorescent (Kummer et al. 1998). A key test of this idea was the finding that artificial restriction of bond movement by immobilizing the fluorophore in ethanol glass, that is, ethanol at 77 K, transformed the fluorophore into an intensely fluorescent species, while the fluorophore in ethanol solution at 25°C is nonfluorescent (Niwa et al. 1996). Thus, the GFP fluorophore is conditionally fluorescent, with fluorescence arising when the GFP protein immobilizes the fluorophore.

Based on these properties, we speculated that an RNA molecule could substitute for the GFP protein and serve to immobilize the GFP fluorophore in a fluorescent conformation. Our idea can be summarized as an attempt to make an RNA-based version of GFP. We also considered other small-molecule dyes that appeared to have the structural features that would make them prone to exhibit increased fluorescence on rigidification, such as triphenylmethane dyes (e.g., malachite green and pararosaniline) (De et al. 2002), stilbenes (Likhtenshtein et al. 1996), cyanine dyes, and numerous other structural classes. However, in our experiments, each of these dyes exhibited fluorescence on application to cells, consistent with the known induction of fluorescence of malachite green and stilbenes by cell membranes (Likhtenshtein et al. 1996; Guidry 1999). In some cases, these molecules elicited frank cytotoxicity at low micromolar concentrations. For example, malachite green, although fluorescent when bound to cognate aptamers (Babendure et al. 2003), cannot be used in living cells because it is intensely fluorescent when it interacts with cell membranes and nuclei (Guidry 1999). In addition, malachite green generates radicals that are cytotoxic (Beermann and Jay 1994) and rapidly destroy the RNA aptamer itself (Grate and Wilson 1999). These unfortunate features of malachite green have prevented the implementation of these otherwise potentially useful malachite green aptamers in cell imaging.

10.5 SYNTHESIS OF GFP-BASED FLUOROPHORES

Our RNA visualization strategy is based on chemically synthesized GFP-like fluorophores and RNA aptamers that bind and immobilize them in a manner that induces its fluorescence. We synthesized GFP-like fluorophores, including novel ones with potentially useful spectral properties. Several groups (Niwa et al. 1996; Kojima et al. 1998) have described a synthesis for the GFP fluorophore. Our best results were obtained with a protocol by Kojima et al. (1998) involving an aldol condensation of *N*-acetyl glycine and substituted benzaldehydes resulting in the lactone intermediate shown (Figure 10.2). The resulting lactone is then subjected to aminolysis and then recyclization to form the final GFP–fluorophore-like final product. Our goal was not to make the exact fluorophore found in GFP, that is, 4-hydoxy-benzylidenene imidazolinone. Instead, we made several derivates by using various substituted benzaldehydes as starting materials. In addition, we used various acylated forms of glycine to make further derivatives of the GFP fluorophore. We reasoned that some of the substituents on these compounds might serve as "handles" that would help the RNA to hold on to the fluorophore, thereby facilitating rigidification. In addition, we wanted to have several RNA–fluorophore complexes, each with a different fluorescent color. We predicted that we could achieve by generating substituted benzylidene imidazolinones that exhibit a range of spectral properties.

Fluorophores need to be synthesized in two forms: (1) a form that can be used for cell-based experiments and (2) a form that can be immobilized on beads for use in SELEX experiments to obtain RNA aptamers (Figure 10.2). Agarose beads that are amine reactive are commercially available. Thus, we developed a synthesis for generating fluorophores containing an aminohexyl linker (Figure 10.2). We used *N*-*t*-butyloxycarbonyl (BOC)-1,6-hexanediamine for aminolysis. The BOC moiety is cleaved with trifluoroacetic acid after purification of the protected compound on silica. This results in fluorophores with pendant linkers that can be readily coupled to amine-reactive agarose beads.

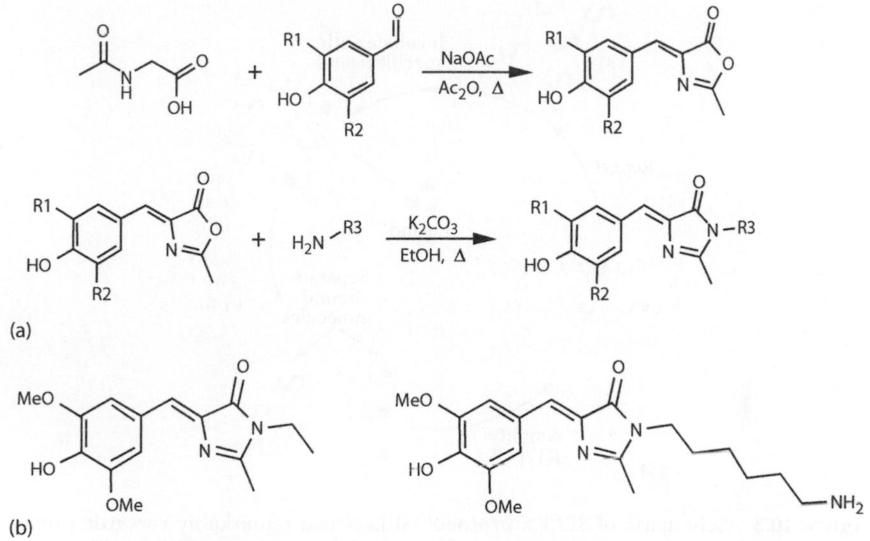

Figure 10.2 Synthesis of substituted 4-hydroxybenzylidene imidazolinones. (a) The synthesis involves two steps; the first generates a lactone intermediate from substituted benzaldehydes (above). This is purified by silica chromatography and then reacted with an alkyl amine to form the substituted imidazolinone. All compounds are synthesized by silica chromatography or HPLC, and characterized by MS and NMR. (b) Fluorophores are synthesized as 1-ethyl (left) or 1-aminohexyl imidazolinones (right). The 1-ethyl derivative is used during SELEX for eluting RNA binders off of fluorophore–agarose beads. This derivative is also the form applied to cells. The 1-aminohexyl derivative is coupled to N-hydroxy-succinimide-activated agarose to form the affinity matrix used in SELEX. DMHBI, which binds certain RNA aptamers, is shown (left), although other fluorophores were synthesized using a similar approach.

10.6 IDENTIFICATION OF RNAs THAT REGULATE THE FLUORESCENCE OF A GFP-BASED FLUOROPHORE

To generate RNAs that bind to GFP-based fluorophores we used the SELEX procedure (Figure 10.3) developed more than 20 years ago independently by Jack Szostak and Larry Gold (Famulok et al. 2000). SELEX (*S*elective *E*nrichment of *L*igands by *EX*ponential enrichment), is a highly effective procedure that can generate RNAs that bind small molecules (Famulok et al. 2000). In this procedure, 10^{14}–10^{15} different RNAs of random sequence are synthesized, and they are allowed to bind to agarose beads that contain the target molecule of interest, which was a fluorophore in this case. After extensive washing, RNAs that are specifically bound are eluted with a buffer that contains millimolar concentrations of fluorophore. The eluted RNAs are precipitated and amplified by reverse

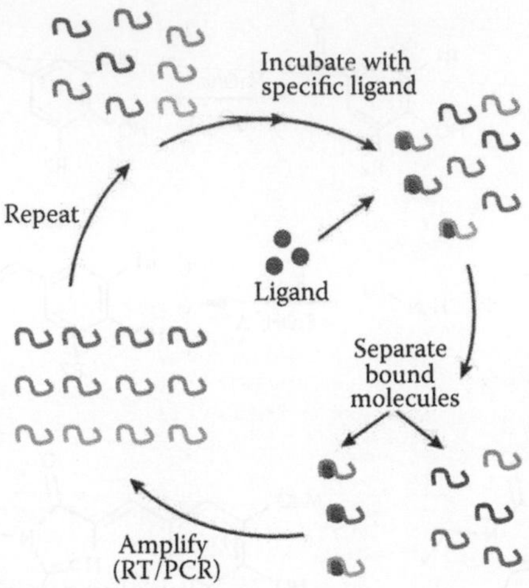

Figure 10.3 Schematic of SELEX protocol. SELEX is a remarkably versatile procedure that is used to identify RNAs that bind to small molecules. In this procedure, a DNA oligonucleotide is used that contains a T7 RNA polymerase priming site and PCR primer sites. A string of N's is required in the starting DNA oligonucleotide and is prepared using phosphoramidite bottles that contain approximately equimolar mixtures of dATP, dGTP, dCTP, and dTTP. Although the starting oligonucleotide is depicted with N's, each individual DNA molecule has a specific random sequence within this region. Transcription products prepared from these DNA templates contain >10^{14} different RNA sequences. RNAs are applied to beads that contain covalently bound fluorophore (Ligand), and bound RNAs are eluted with 2 mM fluorophore. Eluted RNAs are reverse transcribed and PCRed with primers that reintroduce the T7 RNA priming site. These DNAs are then used as a template for an additional round of SELEX. Each round of SELEX results in enrichment of fluorophore-binding RNAs in the RNA pool.

transcriptase-polymerase chain reaction (RT-PCR). Typically the amplified products are used to generate more RNA, and the binding and RT-PCR steps are repeated several times to select for the highest affinity ligands (Figure 10.3). Because of the enormous combinatorial diversity of these mRNAs, and the ability of RNAs to form diverse tertiary structures, this procedure can yield 10–100 different RNAs that bind to any given small molecule (Gold et al. 1995). Failures to identify an aptamer are rare (Gold et al. 1995). Frequently, these ligands bind the aptamers with low micromolar to subnanomolar affinities (Gold et al. 1995).

We next tested our idea that an RNA can mimic GFP and induce the fluorescence of a GFP-based fluorophore. In our initial studies, we focused on 3,5-dimethoxy-4-hydroxybenzylidene imidazolinone (DMHBI; Figure 10.1) because it exhibited intense yellow-green fluorescence in the liquid nitrogen assay. We synthesized DMHBI as well as DMHBI containing an aminohexyl linker (as described in Section 10.5). DMHBI was coupled to agarose beads, and SELEX was performed using a library comprising a stretch of 70 nucleotides of completely randomized nucleotides. After 7–10 rounds of SELEX, we tested if any of the RNA aptamers could switch on the fluorescence of DMHBI, by incubating individual *in vitro* synthesized RNAs with DMHBI. Of several hundred different RNA aptamers tested, a few exhibited activation of DMHBI fluorescence (Paige et al. 2011). In addition, RNA aptamers were generated against other fluorophores. Again, a small subset of the fluorophore-binding aptamers were capable of switching on their fluorescence. Presumably, binding is not sufficient to induce fluorescence. The interaction must occur in a certain manner that leads to fluorescence activation. This interaction may be similar to interactions utilized by fluorescent proteins to activate fluorophore fluorescence. Taken together, these data show that RNA aptamers can be generated that switch on the fluorescence of this class of small-molecule fluorophores.

10.7 AN RNA–FLUOROPHORE COMPLEX THAT MIMICS ENHANCED GFP

Although we found diverse RNA aptamers that can activate the fluorescence of GFP-like fluorophores, in most cases the fluorescence was significantly below that seen with fluorescent proteins (Paige et al. 2011). This occurred because the RNA–fluorophore complex resembled GFP, not enhanced GFP (EGFP). The fluorophore in GFP can exist in a deprotonated phenolate form or a protonated phenol form (Figure 10.4). Each of these forms has its own absorbance

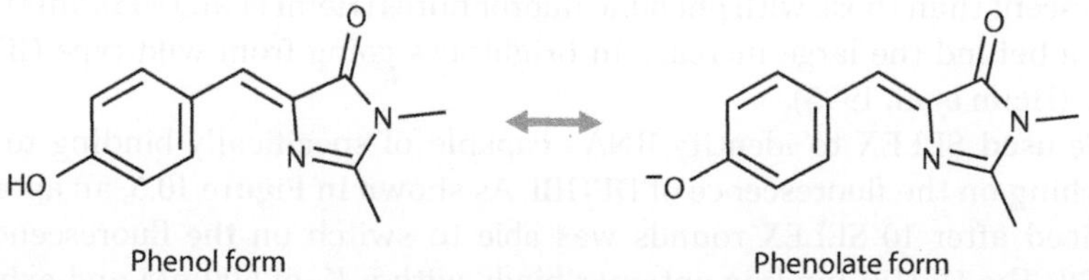

Figure 10.4 Different protonation states of the GFP fluorophore. The GFP fluorophore can exist in a protonated state (left), referred to as the phenolic state, or a deprotonated (right) state, referred to as the phenolate state. The deprotonated form has a higher extinction coefficient and is the predominant form found in EGFP. The protonated form is the major form in GFP. DFHBI is deprotonated under normal buffer conditions and therefore mimics the deprotonated fluorophore found in EGFP.

peak, with the phenol form having an absorbance maximum at approximately 390 nm, and the phenolate having an absorbance at approximately 475 nm. The original GFP cloned from jellyfish has both peaks, indicating that the protein is a heterogeneous mixture of the two forms, with the 390 nm phenol form being the most prominent (Tsien 1998). When the phenol form of the fluorophore is excited by light, the fluorophore uses part of the energy to eject the proton, resulting in the formation of the phenolate form. The remaining energy is dissipated as a green fluorescence emission at approximately 510 nm. Importantly, the phenolate form of the fluorophore has a significantly higher extinction coefficient than the phenol form. Because the degree of fluorescence is dependent on the extinction coefficient and the quantum yield, a major improvement was the generation of EGFP variants in which the fluorophore was predominantly in the higher absorbing phenolate form, which has higher electron density in the conjugated π-electron system (Tsien 1998). The higher absorbance contributed to the overall brighter protein. The absorbance spectra of our RNA–fluorophore aptamer indicated that all the fluorophore was in the phenol form, and none was in the phenolate form, indicating that our RNA–fluorophore complex was more analogous to the original GFP than EGFP (Paige et al. 2011).

We therefore used a biomimetic strategy to create an RNA mimic of EGFP. To do this, we synthesized a new fluorophore that resembles the fluorophore normally found in EGFP. The key issue is that the fluorophore needs to be in the phenolate form during the SELEX procedure and in the cell. We therefore synthesized 3,5-difluoro-4-hydroxybenzylidene imidazolinone (DFHBI; Figure 10.1). We chose DFHBI instead of HBI because the fluorine atoms lower the pK_a of the phenolic moiety to 5.5 compared to 8.1 in HBI (Paige et al. 2011). This ensures that the fluorophore will be present in the phenolate state (deprotonated state) rather than the phenolic state (protonated state) at physiological pH. GFP derivatives containing phenolate fluorophores are significantly more fluorescent than those with phenolic fluorophores (Heim et al. 1995). This is the reason behind the large increase in brightness going from wild-type GFP to EGFP (Heim et al. 1995).

We used SELEX to identify RNAs capable of specifically binding to and switching on the fluorescence of DFHBI. As shown in Figure 10.5, an aptamer obtained after 10 SELEX rounds was able to switch on the fluorescence of DFHBI. The DFHBI-binding aptamer binds with a K_d of 500 nM and exhibits bright fluorescence with a quantum yield of 0.72 and a total molar brightness equal to half that of EGFP. Control RNAs, for example, tRNA, cellular RNA, or scrambled aptamer sequences elicit no fluorescence on incubation with DFHBI. Because of the similarity of the RNA–fluorophore complex to GFP-related proteins, we named the RNA after the vegetable spinach, in analogy to the fruits that have been used to name fluorescent proteins (Paige et al. 2011).

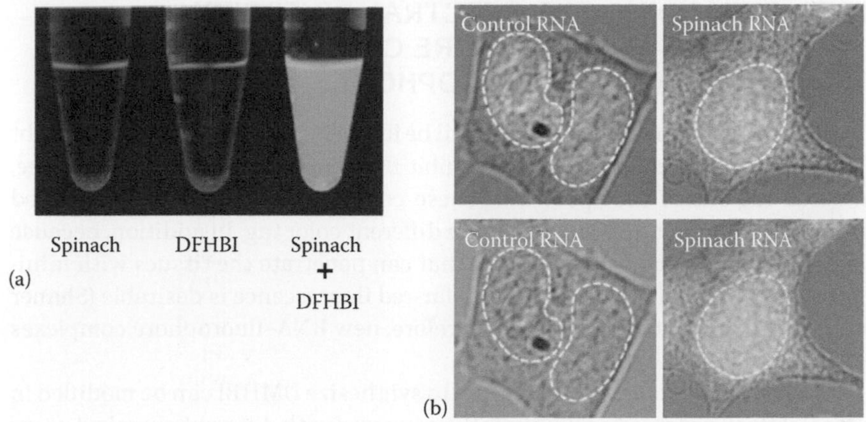

Figure 10.5 RNA mimics of EGFP for live cell imaging. (a) Spinach RNA robustly activates the fluorescence of DFHBI. Each tube contained the indicated solution and was irradiated at 365 nm. (b) HEK 293 cells were transfected with either 5S–control aptamer or 5S–Spinach expressing plasmids and incubated with DFHBI. Nuclei are outlined by dashed white lines. Top images are phase overlay with Hoechst-stained nuclei (blue) and Spinach fluorescence (green). Bottom images show phase and Spinach labeling alone to clearly show labeling of 5S in the nucleus. Control RNA experiments clearly demonstrate that DFHBI fluorescence is not nonspecifically activated by other biological molecules.

10.8 IMAGING RNAs USING SPINACH

We next sought to determine if Spinach could be used to tag RNAs for live cell imaging in a manner analogous to GFP-fusion proteins. To do this, we chose a small noncoding RNA because RNAs of this class have traditionally been excluded from any live-cell imaging technology because of their small size. 5S is a small noncoding RNA transcribed by RNA polymerase III that associates with the large ribosomal subunit (Paige et al. 2011). A 5S–Spinach fusion RNA was expressed from a plasmid containing the endogenous human 5S promoter. After transfection and brief incubation with DFHBI, fluorescence was not detectable in cells expressing a control RNA, while fluorescence was readily detected in 5S–Spinach-expressing cells (Figure 10.5). 5S–Spinach RNA fluorescence was detected throughout cells, with prominent fluorescence signals appearing as diffuse nuclear and cytosolic puncta (Paige et al. 2011). This fluorescence distribution is similar to the patterns seen for endogenous 5S RNA in the same cell type (Paul et al. 2003). These data indicate that Spinach–RNA fusions can be imaged in cells, including in the nucleus, and exhibit localizations consistent with endogenous RNA.

10.9 EXPANDING THE SPECTRAL DIVERSITY OF RNA–FLUOROPHORE COMPLEXES USING NOVEL FLUOROPHORES

In addition to green fluorescence, it will be important to generate highly bright RNA–fluorophore complexes that exhibit fluorescence in the yellow, orange, and red regions of the spectrum. These complexes will enable multiplexed imaging of different RNAs, each with a different color tag. In addition, because imaging in tissue requires radiation that can penetrate the tissues with minimal scatter and background, red and far-red fluorescence is desirable (Shaner et al. 2004; Chudakov et al. 2010). Therefore, new RNA–fluorophore complexes will need to be generated.

The synthetic strategy that we used to synthesize DMHBI can be modified in a straightforward way to allow new fluorophores with different chemical properties to be generated. As described earlier, the synthesis is a two-step reaction (Figure 10.2). In the first step, a benzaldehyde is condensed with an acylated glycine via an aldol condensation, resulting in intermediate that cyclizes to a lactone intermediate. In the second step, the lactone is converted to the final imidazolinone fluorophore via an aminolysis and recyclization reaction. By altering the starting materials, that is, the benzaldehyde and the acylglycine, diverse GFP–fluorophore congeners can be synthesized. Alterations in the substituents in these molecules are expected to exert alterations in either the excitation or emission, potentially creating fluorophores that exhibit desirable spectral properties.

Once a fluorophore is synthesized, it is fairly straightforward to predict the color that it will fluoresce. One strategy is to simply dissolve the fluorophore in ethanol, freeze the sample with liquid nitrogen, and look at its color after ultraviolet (UV) irradiation (Niwa et al. 1996). Although this will provide a reasonable estimate of the likely emission wavelength, the emission could be slightly affected by solvatochromatic effects of ethanol. Another approach is to look at the absorbance spectrum. The absorbance spectrum is typically nearly identical to the fluorescence excitation spectra. Fluorophores that have absorbance peaks that are blue-shifted relative to DMHBI will likely fluoresce either green or blue, depending on how shifted they are. Fluorophores that have absorbance peaks that are red-shifted relative to DMHBI will likely fluoresce yellow, orange, or red. It should be noted that the absorbance of the fluorophore is affected by the specific contacts made by the RNA aptamer, and spectral tuning of the fluorophore is likely. However, in the absence of a cognate RNA, the absorbance spectra of fluorophores are a reasonable starting point.

In addition to these fluorophores, it may be possible to use a biomimetically inspired strategy to create fluorophores that exhibit red fluorescence. In the various red fluorescent proteins, the fluorophore exhibits extended π-bond delocalization (Chudakov et al. 2010). To imitate these red-shifted fluorophores the synthetic strategy could be altered to include either benzoyl or unsaturated acyl substituted glycines that would result in extended π-bond delocalization that mimics the fluorophores seen in red fluorescent protein (RFP). Indeed, the synthesis of these, or highly related fluorophores, has been described (Follenius-Wund et al. 2003).

When fluorophores are synthesized, it is important to characterize them for their suitability for use in RNA–fluorophore complexes. Several parameters should be assessed. The extinction coefficient is critical because the fluorescence is proportional to the extinction coefficient and the quantum yield. Thus, fluorophores that have the highest extinction coefficients are desirable. The absorbance spectra are also important because, as described above, the ultimate emission is likely to be 15–35 nm red-shifted relative to the absorbance maximum of the phenolate peak. In addition to the predicted fluorescence properties, it is essential that the fluorophores exhibit minimal intrinsic fluorescence. Thus, the fluorophores should show negligible fluorescence at room temperature, but substantial fluorescence when frozen in ethanol glass. If there is baseline fluorescence at room temperature, then the fluorophore probably cannot be used unless the RNA–fluorophore complex exhibits a large emission shift relative to the fluorescence of the unbound fluorophore. However, in addition to minimal fluorescence in solution, the fluorophores must exhibit minimal fluorescence when incubated with cellular components, such as cell membranes or endogenous nucleic acid. This is a problem with other dyes, for instance, malachite green and stilbenes (Likhtenshtein et al. 1996; Guidry 1999). Thus, if the fluorophore does not exhibit bioorthogonality, it cannot be used. Lastly, the fluorophores must exhibit minimal cytotoxicity. Many fluorescent dyes exhibit free radical generation on fluorescence excitation, or can form covalent cross-links with endogenous molecules. This is of particular concern with numerous dyes, some of which are exploited for their ability to produce free radicals (Beermann and Jay 1994). Measurements of cytotoxicity should be performed in the presence of illumination compatible with the types that are used during cellular imaging experiments.

Although the preceding discussion involves fluorophores that are related to those found in GFP or RFP, it should be noted that there is no explicit reason why other fluorophores could not be used. For example, derivatives of malachite green that exhibit reduced cytotoxicity and cell-induced fluorescence would be highly desirable because of the high extinction coefficient of these molecules. Similarly, other fluorophores with high extinction coefficients potentially could be switched on by RNA aptamers, and therefore could be valuable in RNA imaging applications in the future.

10.10 SPECTRAL TUNING OF RNA–FLUOROPHORE COMPLEXES

In addition to synthesizing novel fluorophores, another approach to obtain spectral diversity is to create RNA molecules that are capable of inducing significant spectral shifts due to spectral tuning. Indeed, it is not unreasonable that RNA–DFHBI complexes could have a range of fluorescence emissions—a precedent for this is the phenomenon of "spectral tuning" of opsins, in which the same chromophore, that is, retinal, can absorb red, green, or blue light, based on the chromophore microenvironment dictated by the specific opsin protein to which it is covalently coupled.

One approach to identify novel RNAs with altered spectral properties is screen libraries of Spinach mutant RNAs. DNA Libraries encoding Spinach mutants can be prepared using a "doping" approach (Soukup et al. 2001). In this approach, rather than synthesizing the DNA with the specific nucleotide that is found in Spinach at any specific position, the DNA is synthesized using a mixture of phosphoramidites. This mixture is composed predominantly of the nucleotide found in Spinach, but contains a small amount of the other three nucleotides. As a result, individual library members will largely resemble Spinach, but will have a few mutations. These mutations can be screened for fluorescence properties using fluorescence-activated cell sorting (FACS)-based screening approaches in *Escherichia coli* or in mammalian cells. An iterative approach in which proteins were selected that exhibited red-shifted fluorescence emissions, and then remutagenized for further selection, was described recently (Wang et al. 2004). This type of directed evolution approach could conceivably be used to create new Spinach constructs with optimized fluorescence properties.

10.11 CONSIDERATIONS FOR USING SPINACH TO IMAGE RNA IN LIVE CELLS

Several factors need to be considered when imaging RNA using Spinach in living cells. Unlike EGFP, fluorescence is readily obtained after the Spinach RNA is synthesized, which can ensure that there are minimal temporal delays between RNA expression and RNA detection in living cells. However, issues such as the imaging temperature and ionic composition could potentially influence Spinach fluorescence in living cells. RNA is highly sensitive to temperature, and its folding could be inhibited by high temperatures (Brion and Westhof 1997). Importantly, RNA folding could also be influenced by low temperatures that could stabilize misfolded intermediates (Brion and Westhof 1997). Currently, Spinach is optimized for 37°C; however, it will be important to develop mutant forms that can be used at other temperatures. In addition to temperature, ionic considerations are important as well. Most RNA molecules are highly dependent on magnesium structure to achieve a folded structure (Brion and Westhof 1997). Spinach was selected because of its relatively low magnesium dependence. Indeed, the aptamer exhibits maximal fluorescence at approximately 100 μM magnesium. This magnesium dependence is compatible with intracellular magnesium concentrations. However, cells that exhibit low magnesium concentrations, or that exhibit fluctuations in magnesium concentrations, could potentially influence the overall brightness of Spinach. Similarly, low magnesium conditions could potentially reduce the amount of folded Spinach in cells. Therefore, care should be taken to factor potential variations in magnesium concentration into the experimental design.

Another important consideration is the folding of Spinach. Unlike proteins, which exhibit robust context-independent folding, RNA aptamers are highly sensitive to adjacent sequences. Small amounts of hybridization between Spinach and adjacent RNA sequences in the target RNA can impair Spinach folding. Indeed, this has previously been examined in the context of the folding of the group I intron, which exhibits markedly different function when inserted in different places within a target RNA (Roman and Woodson 1998). Thus, it is important to ensure that Spinach can fold properly when inserted into the target RNA. Therefore, a Spinach-tagged RNA should first be synthesized *in vitro*, and the molar fluorescence should be compared with a solution of Spinach at the same concentration. These experiments should be performed in buffers that contain cytosolic magnesium, sodium, and potassium concentrations. We typically express RNA in living cells only if the tagged RNA exhibits the expected fluorescence in solution.

Although biochemical experiments can be performed to ensure that Spinach can fold within the context of a desired target RNA, it will be important to generate Spinach mutants that may be less susceptible to misfolding as a result of diverse RNA contexts. Screening of Spinach mutants that exhibit rapid folding or context-independent folding may help to alleviate this problem. In addition, "insulator" sequences that provide a space between the target RNA and Spinach could also potentially reduce effects of the host RNA on Spinach folding.

Even if the RNA tolerates a Spinach aptamer *in vitro*, it is possible that the RNA is not stable in cells. The insertion of aptamers can affect mRNA stability. Conceivably, insertion of an aptamer could also affect polyadenylation signals, or the function of microRNAs, which are typically located near the 3′ end of the transcript (Hafner et al. 2010). Insertion of an aptamer at the end of the transcript could significantly alter the function of these regulatory elements by displacing them from the 3′ end of transcripts. Therefore, insertion of the aptamer elsewhere may be more desirable. One possibility is to insert the aptamer downstream of the stop codon, but near the 5′ end of the 3′ UTR. Indeed, we have found that this location tends to preserve RNA stability. Nevertheless, direct measures of RNA stability of a heterologous transcript with and without the aptamer should be performed. Interestingly, we typically find minimal effects of Spinach on ncRNAs, while certain mRNAs are affected by insertion of Spinach tags.

10.12 IMAGING RNA IN MULTICELLULAR ORGANISMS

An important direction in RNA imaging is to develop imaging approaches that are compatible with living organisms. To adapt Spinach, and other RNA–fluorophore complexes for living animals, it will be important to ensure that the fluorophores do not exhibit toxicity to the organism. In addition, depending on the route of administration, the fluorophores may need to be altered to enable absorption or penetration into the blood–brain barrier. For other animals, such as *Caenorhabditis elegans*, or plants, ensuring that the fluorophore can penetrate the organism's barriers to drug permeability, such as the cell wall, are important steps that need to be optimized. In the case of DFHBI, it is possible that ester modifications of the phenolic oxygen, or other protecting groups that are hydrolyzed in the cytosol, could be important for ensuring that the fluorophore enters the cell and can bind RNA aptamers.

10.13 STRATEGIES FOR INCREASING SENSITIVITY OF RNA IMAGING

An important goal is imaging RNAs that are low abundance in cells. Many noncoding RNAs appear to function as a single copy and bind to specific promoter regions in the nucleus (Batista and Chang 2013). Similarly, monitoring the fate of RNAs as they are transcribed and trafficked in cells from their "birth" to "death" requires imaging single RNA molecules. Thus, increasing the sensitivity of RNA imaging to a single RNA molecule is of major interest in the research community.

However, in addition to imaging single molecules, it is important to image moderate and low-abundance RNAs in cells. In our original publication, we focused on the 5S RNA, which is expressed at high levels in cells. However, other RNAs of interest are expressed at lower copy numbers in cells. Thus, strategies are needed to enhance the sensitivity of RNA imaging techniques in living cells.

One strategy to increase the sensitivity involves using red fluorescent RNA–fluorophore complexes. Because there is less tissue autofluorescence in the red channel, imaging using red fluorescent complexes is likely to have greater sensitivity than imaging using green fluorescent species. Additional approaches that involve using brighter RNA–fluorophore complexes will also be valuable.

These will involve using fluorophores that have higher extinction coefficients and quantum yields, thereby resulting in higher brightness.

An additional approach to increase the sensitivity of imaging is to tag RNA molecules with cassettes containing multiple Spinach tags. In this approach, tandem Spinach tags are expected to linearly increase the fluorescence of any individual tagged RNA. For example, an RNA could be tagged with 4, 8, 16, or even more Spinach tags. An important concern with this approach is to ensure that individual Spinach aptamers do not hybridize with other Spinach aptamers in the same RNA. This could potentially occur because Spinach contains stem structures that comprise complementary strands (Paige et al. 2011, 2012). Conceivably strands from one aptamer could interact with the complementary strand in another aptamer. Judicious insertion of mutations in each aptamer could help prevent inter-aptamer hybridization.

10.14 FUTURE DIRECTIONS

RNA mimics of GFP comprise an important new approach for imaging RNA living cells. However, further development of this technology will be critical for making this approach widely useful for studies of RNA. Expansion of the spectral versatility of these RNA–fluorophore complexes will enable live imaging and increase the sensitivity. Newer fluorophores could potentially be useful for increasing the overall brightness. Strategies to increase and improve the folding of Spinach could also increase the molar brightness of individual RNA species. In addition, the creation of cassettes that contain multiple Spinach aptamer tags could be useful for increasing the brightness of individual RNA species, potentially enabling the tracking of individual RNA molecules as they are being transcribed to their different processing sites in cells. These approaches will be important for imaging diverse RNAs and uncovering the function of noncoding RNAs in cells.

REFERENCES

Anderson, P., and N. Kedersha. 2002. Visibly stressed: The role of eIF2, TIA-1, and stress granules in protein translation. *Cell Stress Chap* 7:213–221.

Aravin, A., D. Gaidatzis, S. Pfeffer, M. Lagos-Quintana, P. Landgraf, N. Iovino, P. Morris, M. J. Brownstein, S. Kuramochi-Miyagawa, T. Nakano, M. Chien, J. J. Russo, J. Ju, R. Sheridan, C. Sander, M. Zavolan, and T. Tuschl. 2006. A novel class of small RNAs bind to MILI protein in mouse testes. *Nature* 442:203–207.

Babendure, J. R., S. R. Adams, and R. Y. Tsien. 2003. Aptamers switch on fluorescence of triphenylmethane dyes. *J Am Chem Soc* 125:14716–14717.

Baffour-Awuah, N. A., and M. Zimmer. 2004. Hula-twisting in green fluorescent protein. *Chem Phys* 303:7–11.

Batista, P. J., and H. Y. Chang. 2013. Long noncoding RNAs: Cellular address codes in development and disease. *Cell* 152:1298–1307.

Beermann, A. E., and D. G. Jay. 1994. Chromophore-assisted laser inactivation of cellular proteins. *Methods Cell Biol* 44:715–732.

Bertrand, E., P. Chartrand, M. Schaefer, S. M. Shenoy, R. H. Singer, and R. M. Long. 1998. Localization of ASH1 mRNA particles in living yeast. *Mol Cell* 2:437–445.

Bokman, S. H., and W. W. Ward. 1981. Renaturation of Aequorea green-fluorescent protein. *Biochem Biophys Res Commun* 101:1372–1380.

Brion, P., and E. Westhof. 1997. Hierarchy and dynamics of RNA folding. *Annu Rev Biophys Biomol Struct* 26:113–137.

Chudakov, D. M., M. V. Matz, S. Lukyanov, and K. A. Lukyanov. 2010. Fluorescent proteins and their applications in imaging living cells and tissues. *Physiol Rev* 90: 1103–1163.

Cox, L. J., U. Hengst, N. G. Gurskaya, K. A. Lukyanov, and S. R. Jaffrey. 2008. Intra-axonal translation and retrograde trafficking of CREB promotes neuronal survival. *Nat Cell Biol* 10:149–159.

De, S., A. Girigoswami, and S. Mandal. 2002. Enhanced fluorescence of triphenylmethane dyes in aqueous surfactant solutions at supramicellar concentrations—Effect of added electrolyte. *Spectrochim Acta A Mol Biomol Spectrosc* 58:2547–2555.

Deglincerti, A., and S. R. Jaffrey. 2012. Insights into the roles of local translation from the axonal transcriptome. *Open Biol* 2:120079.

Derrien, T., R. Johnson, G. Bussotti, A. Tanzer, S. Djebali, H. Tilgner, G. Guernec, D. Martin, A. Merkel, D. G. Knowles, J. Lagarde, L. Veeravalli, X. Ruan, Y. Ruan, T. Lassmann, P. Carninci, J. B. Brown, L. Lipovich, J. M. Gonzalez, M. Thomas, C. A. Davis, R. Shiekhattar, T. R. Gingeras, T. J. Hubbard, C. Notredame, J. Harrow, and R. Guigo. 2012. The GENCODE v7 catalog of human long noncoding RNAs: Analysis of their gene structure, evolution, and expression. *Genome Res* 22:1775–1789.

Dredge, B. K., A. D. Polydorides, and R. B. Darnell. 2001. The splice of life: Alternative splicing and neurological disease. *Nat Rev* 2:43.

Famulok, M., G. Mayer, and M. Blind. 2000. Nucleic acid aptamers-from selection in vitro to applications in vivo. *Acc Chem Res* 33:591–599.

Follenius-Wund, A., M. Bourotte, M. Schmitt, F. Iyice, H. Lami, J. J. Bourguignon, J. Haiech, and C. Pigault. 2003. Fluorescent derivatives of the GFP chromophore give a new insight into the GFP fluorescence process. *Biophys J* 85:1839–1850.

Gold, L., B. Polisky, O. Uhlenbeck, and M. Yarus. 1995. Diversity of oligonucleotide functions. *Annu Rev Biochem* 64:763–797.

Grate, D., and C. Wilson. 1999. Laser-mediated, site-specific inactivation of RNA transcripts. *Proc Natl Acad Sci U S A* 96:6131–6136.

Guidry, G. 1999. A method for counterstaining tissues in conjunction with the glyoxylic acid condensation reaction for detection of biogenic amines. *J Histochem Cytochem* 47:261–264.

Hafner, M., M. Landthaler, L. Burger, M. Khorshid, J. Hausser, P. Berninger, A. Rothballer, M. Ascano, Jr., A. C. Jungkamp, M. Munschauer, A. Ulrich, G. S. Wardle, S. Dewell, M. Zavolan, and T. Tuschl. 2010. Transcriptome-wide identification of RNA-binding protein and microRNA target sites by PAR-CLIP. *Cell* 141:129–141.

Han, J., D. Kim, and K. V. Morris. 2007. Promoter-associated RNA is required for RNA-directed transcriptional gene silencing in human cells. *Proc Natl Acad Sci U S A* 104:12422–12427.

Hannon, G. J., F. V. Rivas, E. P. Murchison, and J. A. Steitz. 2006. The expanding universe of noncoding RNAs. *Cold Spring Harb Symp Quant Biol* 71:551–564.

Heim, R., A. B. Cubitt, and R. Y. Tsien. 1995. Improved green fluorescence. *Nature* 373:663–664.

Hengst, U., L. J. Cox, E. Z. Macosko, and S. R. Jaffrey. 2006. Functional and selective RNA interference in developing axons and growth cones. *J Neurosci* 26:5727–5732.

Hengst, U., A. Deglincerti, H. J. Kim, N. L. Jeon, and S. R. Jaffrey. 2009. Axonal elongation triggered by stimulus-induced local translation of a polarity complex protein. *Nat Cell Biol* 11:1024–1030.

Hengst, U., and S. R. Jaffrey. 2007. Function and translational regulation of mRNA in developing axons. *Semin Cell Dev Biol* 18:209–215.

Hocine, S., P. Raymond, D. Zenklusen, J. A. Chao, and R. H. Singer. 2013. Single-molecule analysis of gene expression using two-color RNA labeling in live yeast. *Nat Methods* 10:119–121.

Holeman, L. A., S. L. Robinson, J. W. Szostak, and C. Wilson. 1998. Isolation and characterization of fluorophore-binding RNA aptamers. *Fold Des* 3:423–431.

Ji, S. J., and S. R. Jaffrey. 2012. Intra-axonal translation of SMAD1/5/8 mediates retrograde regulation of trigeminal ganglia subtype specification. *Neuron* 74:95–107.

Keegan, L. P., A. Leroy, D. Sproul, and M. A. O'Connell. 2004. Adenosine deaminases acting on RNA (ADARs): RNA-editing enzymes. *Genome Biol* 5:209.

Kiebler, M. A., and G. J. Bassell. 2006. Neuronal RNA granules: Movers and makers. *Neuron* 51:685–690.

Kojima, S., H. Ohkawa, T. Hirano, S. Maki, H. Niwa, M. Ohashi, S. Inouye, and F. I. Tsuji. 1998. Fluorescent properties of model chromophores of tyrosine-66 substituted mutants of Aequorea green fluorescent protein (GFP). *Tetrahedron Lett* 39:5239–5242.

Kummer, A. D., C. Kompa, H. Lossau, F. Pollinger-Dammer, M. E. Michel-Beyerle, C. M. Silva, E. J. Bylina, W. J. Coleman, M. M. Yang, and D. C. Youvan. 1998. Dramatic reduction in fluorescence quantum yield in mutants of green fluorescent protein due to fast internal conversion. *Chem Phys* 237:183–193.

Lecuyer, E., H. Yoshida, N. Parthasarathy, C. Alm, T. Babak, T. Cerovina, T. R. Hughes, P. Tomancak, and H. M. Krause. 2007. Global analysis of mRNA localization reveals a prominent role in organizing cellular architecture and function. *Cell* 131:174–187.

Levsky, J. M., and R. H. Singer. 2003. Fluorescence in situ hybridization: Past, present and future. *J Cell Sci* 116:2833–2838.

Likhtenshtein, G. I., R. Bishara, V. Papper, B. Uzan, I. Fishov, D. Gill, and A. H. Parola. 1996. Novel fluorescence-photochrome labeling method in the study of biomembrane dynamics. *J Biochem Biophys Methods* 33:117–133.

Mercer, T. R., M. E. Dinger, S. M. Sunkin, M. F. Mehler, and J. S. Mattick. 2008. Specific expression of long noncoding RNAs in the mouse brain. *Proc Natl Acad Sci U S A* 105:716–721.

Merzlyak, E. M., J. Goedhart, D. Shcherbo, M. E. Bulina, A. S. Shcheglov, A. F. Fradkov, A. Gaintzeva, K. A. Lukyanov, S. Lukyanov, T. W. Gadella, and D. M. Chudakov. 2007. Bright monomeric red fluorescent protein with an extended fluorescence lifetime. *Nat Methods* 4:555–557.

Meyer, K. D., Y. Saletore, P. Zumbo, O. Elemento, C. E. Mason, and S. R. Jaffrey. 2012. Comprehensive analysis of mRNA methylation reveals enrichment in 3′ UTRs and near stop codons. *Cell* 149:1635–1646.

Mhlanga, M. M., and S. Tyagi. 2006. Using tRNA-linked molecular beacons to image cytoplasmic mRNAs in live cells. *Nat Protoc* 1:1392–1398.

Niwa, H., S. Inouye, T. Hirano, T. Matsuno, S. Kojima, M. Kubota, M. Ohashi, and F. I. Tsuji. 1996. Chemical nature of the light emitter of the Aequorea green fluorescent protein. *Proc Natl Acad Sci U S A* 93:13617–13622.

Paige, J. S., T. Nguyen-Duc, W. Song, and S. R. Jaffrey. 2012. Fluorescence imaging of cellular metabolites with RNA. *Science* 335:1194.

Paige, J. S., K. Y. Wu, and S. R. Jaffrey. 2011. RNA mimics of green fluorescent protein. *Science* 333:642–646.

Parker, R., and H. Song. 2004. The enzymes and control of eukaryotic mRNA turnover. *Nat Struct Mol Biol* 11:121–127.

Paul, C. P., P. D. Good, S. X. Li, A. Kleihauer, J. J. Rossi, and D. R. Engelke. 2003. Localized expression of small RNA inhibitors in human cells. *Mol Ther* 7:237–247.

Ranum, L. P., and J. W. Day. 2004. Myotonic dystrophy: RNA pathogenesis comes into focus. *Am J Hum Genet* 74:793.

Roman, J., and S. A. Woodson. 1998. Integration of the Tetrahymena group I intron into bacterial rRNA by reverse splicing in vivo. *Proc Natl Acad Sci U S A* 95:2134–2139.

Sando, S., A. Narita, M. Hayami, and Y. Aoyama. 2008. Transcription monitoring using fused RNA with a dye-binding light-up aptamer as a tag: A blue fluorescent RNA. *Chem Commun (Camb)* 33:3858–3860.

Shaner, N. C., R. E. Campbell, P. A. Steinbach, B. N. Giepmans, A. E. Palmer, and R. Y. Tsien. 2004. Improved monomeric red, orange and yellow fluorescent proteins derived from *Discosoma* sp. red fluorescent protein. *Nat Biotechnol* 22:1567–1572.

Sharma, S., G. M. Findlay, H. S. Bandukwala, S. Oberdoerffer, B. Baust, Z. Li, V. Schmidt, P. G. Hogan, D. B. Sacks, and A. Rao. 2011. Dephosphorylation of the nuclear factor of activated T cells (NFAT) transcription factor is regulated by an RNA-protein scaffold complex. *Proc Natl Acad Sci U S A* 108:11381–11386.

Soukup, G. A., E. C. DeRose, M. Koizumi, and R. R. Breaker. 2001. Generating new ligand-binding RNAs by affinity maturation and disintegration of allosteric ribozymes. *RNA* 7:524–536.

Spector, D. L., and A. I. Lamond. 2011. Nuclear speckles. *Cold Spring Harb Perspect Biol* 3:1–12.

Steward, O., and E. M. Schuman. 2003. Compartmentalized synthesis and degradation of proteins in neurons. *Neuron* 40:347–359.

Tsien, R. Y. 1998. The green fluorescent protein. *Annu Rev Biochem* 67:509–544.

Tyagi, S. 2007. Splitting or stacking fluorescent proteins to visualize mRNA in living cells. *Nat Methods* 4:391–392.

Tyagi, S. 2009. Imaging intracellular RNA distribution and dynamics in living cells. *Nat Methods* 6:331–338.

Tyagi, S., and O. Alsmadi. 2004. Imaging native beta-actin mRNA in motile fibroblasts. *Biophys J* 87:4153–4162.

Wang, L., W. C. Jackson, P. A. Steinbach, and R. Y. Tsien. 2004. Evolution of new non-antibody proteins via iterative somatic hypermutation. *Proc Natl Acad Sci U S A* 101:16745–16749.

Wu, K. Y., U. Hengst, L. J. Cox, E. Z. Macosko, A. Jeromin, E. R. Urquhart, and S. R. Jaffrey. 2005. Local translation of RhoA regulates growth cone collapse. *Nature* 436:1020–1024.

Zeitelhofer, M., D. Karra, P. Macchi, M. Tolino, S. Thomas, M. Schwarz, M. Kiebler, and R. Dahm. 2008. Dynamic interaction between P-bodies and transport ribonucleoprotein particles in dendrites of mature hippocampal neurons. *J Neurosci* 28:7555–7562.

Zhang, H. L., T. Eom, Y. Oleynikov, S. M. Shenoy, D. A. Liebelt, J. B. Dictenberg, R. H. Singer, and G. J. Bassell. 2001. Neurotrophin-induced transport of a beta-actin mRNP complex increases beta-actin levels and stimulates growth cone motility. *Neuron* 31:261–275.

Chapter 11

Fluorescent Sensors for Imaging Zinc Dynamics in Biological Fluids

Wen-hong Li

CONTENTS

11.1 WHY IMAGING Zn^{2+}?

Zinc ion (Zn^{2+}) is the second most abundant transition metal in living organisms after iron. In cells, Zn^{2+} serves as an important cofactor of numerous proteins to maintain their structures or support their enzymatic activities. It has been estimated that roughly 2800 human proteins potentially bind Zn^{2+} *in vivo*, corresponding to about 10% of the human proteome (Andreini et al. 2006). A well-known example is the zinc finger transcription factors, a large family of Zn^{2+}-binding proteins involved in the regulation of gene expression (Klug 2010). Because of the essential roles Zn^{2+} plays in diverse biochemical, biological, and physiological processes, as well as its involvement in a number of human diseases (Rungby 2010; Fukada et al. 2011; Sensi et al. 2011), increasing interest and effort have been devoted to study the regulation and function of Zn^{2+} homeostasis in different biological systems.

Concentrations of labile or weakly bound Zn^{2+} (free Zn^{2+}) are tightly regulated in both intracellular and extracellular environments. In eukaryotes, two families of Zn^{2+} transporter proteins, the Zn^{2+}-importing proteins (ZIP proteins for Zrt-, Irt-like proteins; or solute-linked carrier 39, SLC39) and Zn^{2+} transporters (ZnT proteins; or solute-linked carrier 30, SLC30A), together with metal-buffering proteins such as metallothioneins, maintain cellular Zn^{2+} homeostasis (Colvin et al. 2010). Whereas ZnTs move Zn^{2+} out of the cell or from the cytosol to the lumen of intracellular organelles, ZIPs do just the opposite. There are at least 10 members of the mammalian ZnT family, ZnT1–ZnT10 (Cousins et al. 2006; Lichten and Cousins 2009). The majority of ZnTs are broadly expressed in different tissues, and are found at intracellular organelles such as the Golgi and the secretory vesicles, with the exception of ZnT1, which is found primarily on the plasma membrane. The mammalian ZIP family has at least 14 members, ZIP1–ZIP14. ZIPs mediate Zn^{2+} influx into the cytosol from the extracellular medium and from intracellular organelles.

In addition to Zn^{2+} transporters, Zn^{2+}-buffering proteins also play a major role in maintaining the steady state of cytosolic free Zn^{2+} concentrations ($[Zn^{2+}]_i$) within a narrow range. The human metallothionein family has more than 18 members expressed in different tissues (Laukens et al. 2009). Despite variations in their amino acid sequences, these low-molecular-weight proteins have in common a high cysteine content, which is responsible for metal binding. Because the thiol group of cysteine is susceptible to oxidation, and because alterations in the thiol/disulfide ratio can change the Zn^{2+}-buffering capacity of metallothioneins, the cellular redox state may affect Zn^{2+} distribution and the level of $[Zn^{2+}]_i$ (Jacob et al. 1998).

Given that so many proteins bind Zn^{2+}, and because Zn^{2+} binding can significantly alter protein functions or activities, it is anticipated that changes in Zn^{2+} activity regulate numerous biochemical and biological processes. Furthermore, a number of mammalian cells contain high levels of Zn^{2+} in their secretory granules (Frederickson et al. 2005). During secretion, Zn^{2+} is co-released with other granular contents into the extracellular space. It has been speculated that the released Zn^{2+} may modulate the activity of cell surface receptors and cell–cell communication. To study the signaling roles of Zn^{2+} in cells and to uncover the mechanisms regulating Zn^{2+} activity, it would be necessary to monitor the dynamics of free Zn^{2+} fluctuations in different cellular compartments, both within and outside of cells. Fluorescence microscopy offers superior sensitivity of detection and very high spatiotemporal resolution; therefore, it serves as an ideal imaging modality for monitoring dynamic Zn^{2+} fluctuations at cellular and subcellular levels, providing that fluorescent sensors with robust and selective Zn^{2+} responsivity are available. Moreover, there has been tremendous progress in recent years in engineering a myriad of fluorescent indicators to image a variety of cellular biochemical events (Newman et al. 2011). This has greatly enhanced our capability to apply multicolor imaging to study cell signaling pathways and their interactions. Combining Zn^{2+} sensors with other biochemical indicators would allow us to examine moment-to-moment correlations in these events that may help to unveil mechanistic insights on the regulation and/or function of Zn^{2+} signaling. Finally, abnormal Zn^{2+} handling has been implicated in a number of human diseases including diabetes, Alzheimer's disease, cancer, and so forth (Chasapis et al. 2012). It is therefore of interest to both researchers and clinicians to monitor Zn^{2+} levels in different tissues or biopsies from patients at various stages of a disease. The high sensitivity and resolution of fluorescence imaging, combined with its ease of quantification and calibration, makes this technique an excellent choice for studying Zn^{2+} homeostasis in physiological preparations and for examining changes between normal and diseased states.

This chapter reviews fluorescent sensors for imaging Zn^{2+} fluctuations in biological fluids, as well as their applications to different biological systems. I first introduce some general principles underlying Zn^{2+} coordination chemistry and sensor design before discussing commonly used fluorescent Zn^{2+} sensors made of small synthetic dyes or of fluorescent proteins. Next, I highlight the utility of these sensors by presenting examples of measuring Zn^{2+} activity inside living cells, in the extracellular environment, and during Zn^{2+} secretion.

11.2 FLUORESCENT SENSORS FOR IMAGING Zn^{2+} DYNAMICS IN BIOLOGICAL FLUIDS

11.2.1 Design Principles—Zinc Coordination Chemistry and Commonly Used Ligands

Zinc has the outer electron configuration of $3d^{10}4s^2$. Because the zinc ion (Zn^{2+}) has a filled *d*-orbital, it shows no stereochemical preference arising from the ligand field stabilization effect (Cotton et al. 1999). Because of its unique combination of polarizability ($3d^{10}$ shell) and high charge-to-radius ratio, which is nearly identical to that of Mg^{2+}, Zn^{2+} is considered to be a borderline Lewis acid that can coordinate with both "hard" and "soft" Lewis bases by accepting their electrons. In fact, Zn^{2+} metalloproteins frequently employ histidine, cysteine, glutamate, and aspartate to provide, respectively, N, S, and O donor atoms to coordinate with Zn^{2+}. Among the many Zn^{2+} metalloproteins examined to date, the binding geometry observed most often is a slightly distorted tetrahedral one with Zn^{2+} coordinated by three or four protein side chains (McCall et al. 2000).

TABLE 11.1 COMMONLY USED Zn^{2+} LIGANDS AND THEIR METAL STABILITY CONSTANTS

Ligand	logK (Zn^{2+})	logK (Ca^{2+})	logK (Mg^{2+})
2,2′-Dipicolylamine (DPA) 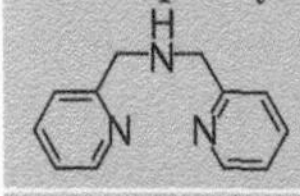	7.63	< 2	< 2
Iminodiacetic acid (IDA) 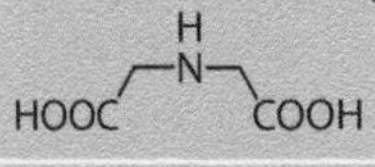	7.15	2.60	2.98
2-Aminoethyl iminodiacetic acid 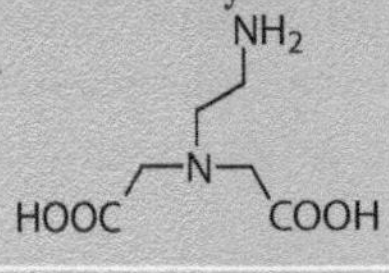	11.93	4.63	4.53

Source: Martell, A. E. and Smith, R. M. (2003). NIST critically selected stability constants of metal complexes database, National Institute of Standards and Technology, Standard Reference Database 46.

Two small ligands, 2,2′-dipicolylamine (DPA) and iminodiacetic acid (IDA, Table 11.1), are frequently employed as the Zn^{2+}-binding motif when designing fluorescent Zn^{2+} sensors. These two ligands bind Zn^{2+} at least four orders of magnitude more selectivity than for competing divalent cations present in biological fluids, including Ca^{2+} and Mg^{2+}. Furthermore, the secondary amine of these two ligands provides a convenient site for chemical derivatization, facilitating conjugation with fluorophores and the tuning of their metal binding affinity. For example, adding a 2-aminoethyl group to IDA boosts its Zn^{2+} affinity by more than four orders of magnitude (Table 11.1).

11.2.2 Small Synthetic Zn^{2+} Indicators

Over the past two decades, numerous fluorescent Zn^{2+} sensors based on DPA or IDA Zn^{2+}-binding motifs have been developed (Jiang and Guo 2004; Kikuchi et al. 2004; Carol et al. 2007; Tomat and Lippard 2010; Koide et al. 2011). A number of these Zn^{2+} sensors possess desirable fluorescent properties for cellular imaging, with excitation and emission wavelengths spanning across the visible spectrum and even extending into the near-infrared region. These newly developed Zn^{2+} sensors quickly replaced the earlier generation of zinc sensors employing quinoline as the sensing moiety. Zn^{2+} sensors based on quinolone, 6-methoxy-8-*p*-toluenesulfonamido-quinoline (TSQ; Frederickson et al. 1987) and ethyl[2-methyl-8-*p*-toluenesulfonamido-6-quinolyloxy]acetate (Zinquin; Zalewski et al. 1993), for example, require ultraviolet (UV) light excitation and show only a modest fluorescence signal. Their utility for live-cell Zn^{2+} imaging is thus limited by the potential phototoxicity of UV light illumination, the low sensitivity of detection, and interference from the cellular background fluorescence.

Table 11.2 lists examples of some commonly used Zn^{2+} sensors exhibiting a robust fluorescence enhancement on Zn^{2+} binding. The majority of these sensors employ 5(6)-amino fluorescein as the fluorescent reporter, exploiting the phenomenon of fluorescence quenching by photo-induced electron transfer (PET) to sense Zn^{2+}: in the absence of the metal, the lone pair electron of the 5(6)-amino group quenches the fluorescence through PET; when $[Zn^{2+}]$ increases, Zn^{2+} coordinates with the 5(6)-amino group and other atoms in the metal binding pocket, effectively reducing or preventing PET and leading to fluorescence dequenching (Figure 11.1).

The Zn^{2+}-binding affinity of these Zn^{2+} sensors can be tuned systematically by modifying the structure of the metal chelator, by varying the steric hindrance of the Zn^{2+}-coordination site, and/or by adjusting the electron density of the atoms responsible for Zn^{2+} chelation. For example, ZnAF2 (Table 11.2), a member of the ZnAF family of Zn^{2+} sensors (Hirano et al. 2000; Komatsu et al. 2005), chelates Zn^{2+} with a dissociation constant (K_d) of 2.7 nM. Changing one of the binding arms of DPA in ZnAF2 from a 2-pyridylmethyl group to a 2-pyridylethyl group leads to ZnAF3, a Zn^{2+} sensor with a K_d (Zn^{2+}) of 0.79 μM—a nearly 300-fold reduction in Zn^{2+} affinity compared to ZnAF2.

To determine the Zn^{2+} affinity of a Zn^{2+} sensor, Zn^{2+} titration is performed in appropriately buffered Zn^{2+} solutions. Depending on the range of Zn^{2+} concentration ($[Zn^{2+}]$) one wishes to clamp, different Zn^{2+} buffers can be used (Aslamkhan et al. 2002). Commonly used Zn^{2+} buffers include tetrakis-(2-pyridylmethyl)ethylenediamine (TPEN), (2-hydroxyethyl)ethylenediaminetriacetic acid (HEDTA), and nitrilotriacetic acid (NTA), which are effective for clamping $[Zn^{2+}]$ near the femtomolar, picomolar, and nanomolar range, respectively (Figure 11.2). To determine the K_d (Zn^{2+}) of fluorescent Zn^{2+} sensors with micromolar or submicromolar Zn^{2+} affinities, it is less critical to use a Zn^{2+} buffer. However, because saline solutions (ionic strength ≥ 0.1 N) prepared from commercially available salts (NaCl, KCl, HEPES, etc.) usually contain trace amounts of Zn^{2+} contamination, it is necessary to either remove residual Zn^{2+} using a Chelex resin (Aslamkhan et al. 2002) or to complex the contaminating Zn^{2+} with a minimal amount of a Zn^{2+} chelator (Li et al. 2011). The subsequent addition of known amounts of $ZnCl_2$ will then yield the desired $[Zn^{2+}]$ in the micromolar range or above.

In addition to possessing the appropriate Zn^{2+} affinity, fluorescent Zn^{2+} sensors also need to discriminate against other cations to ensure specific reporting

TABLE 11.2 EXAMPLES OF SMALL SYNTHETIC FLUORESCENT Zn^{2+} SENSORS

Zn^{2+} Sensor	K_d	k_{on} ($M^{-1}\,s^{-1}$)	k_{off} (s^{-1})	F_{max}/F_{min}	Reference
ZnAF-2 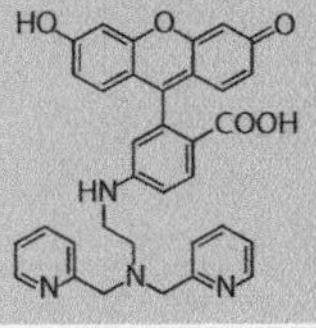	2.7 nM	3.1×10^6	8.4×10^{-3}	51	Hirano et al. (2000)
ZnAF-2F 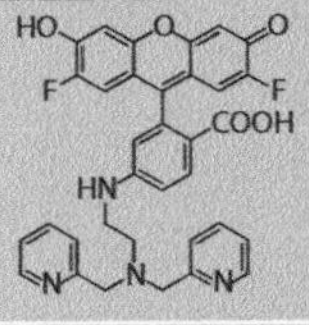	5.5 nM	3.2×10^6	1.8×10^{-2}	60	Hirano et al. (2002)
ZnAF-3 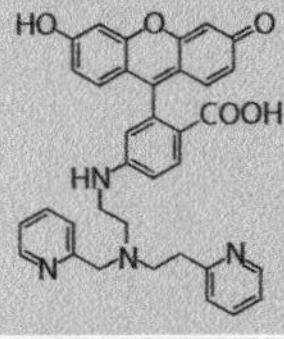	0.79 μM	1.4×10^6	1.1	13	Komatsu et al. (2005)
FluoZin-3 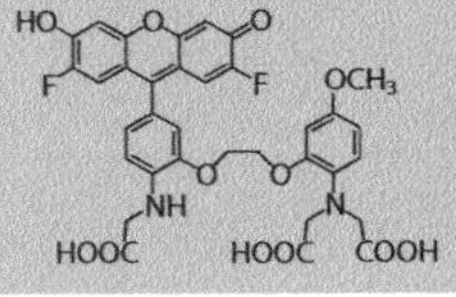	15 nM	3.2×10^6	0.05	> 100	Gee et al. (2002); Zhao et al. (2008)

RhodZin-3 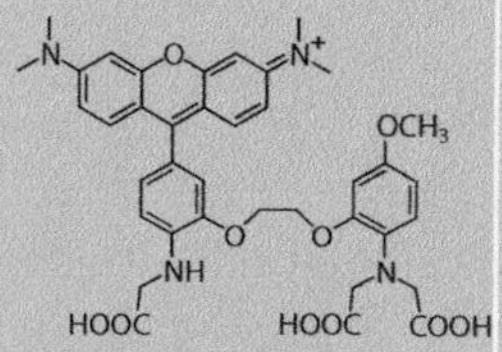	65 nM	NR	NR	75	Sensi et al. (2003b)
Newport Green DCF 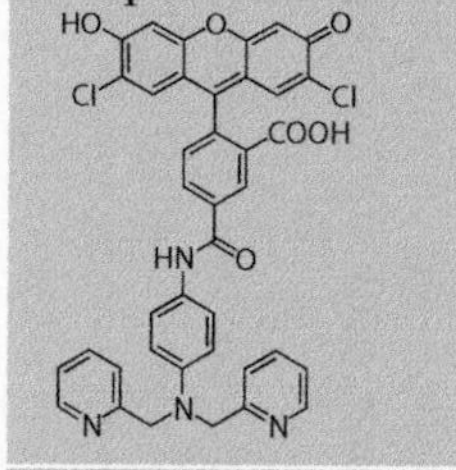	1.5 μM	NR	NR	1.5–6	Thompson et al. (2002); Qian and Noebels (2005)
ZP4 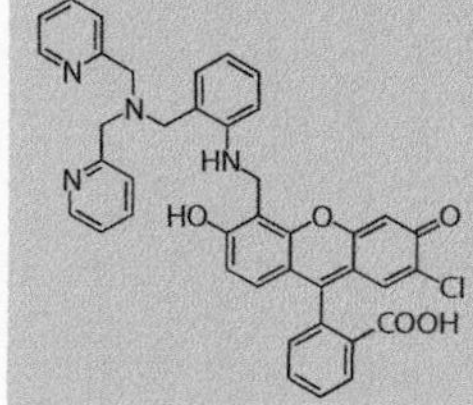	0.65 nM	5.2×10^6	3.4×10^{-3}	4.4	Burdette et al. (2003)

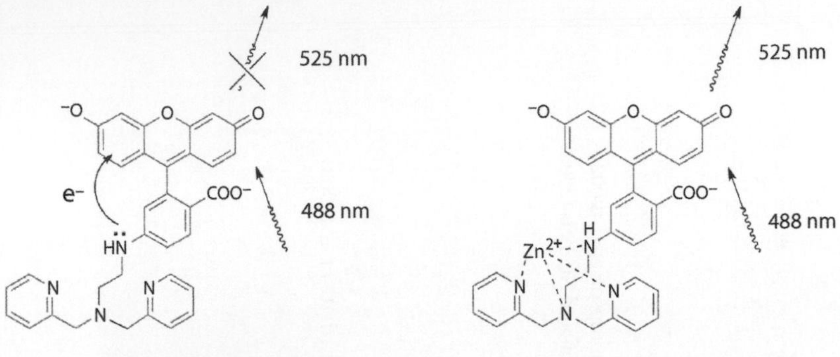

Figure 11.1 Design of fluorescent Zn^{2+} sensors based on the principle of photo-induced electron transfer, illustrated using ZnAF2 as an example.

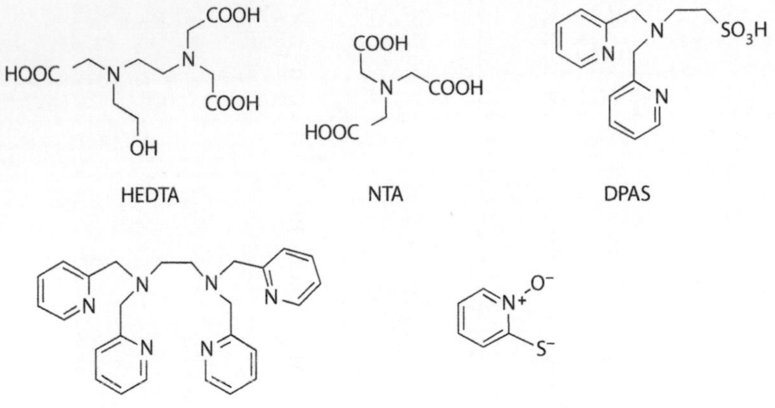

Figure 11.2 Examples of zinc chelators and ionophores for zinc titration *in vitro* or for calibrating signals from zinc sensors in cells. HEDTA, NTA, and DPAS are cell membrane impermeable, whereas TPEN and pyrithione are cell permeable.

of Zn^{2+}. Ca^{2+} and Mg^{2+}, for example, are two divalent cations abundantly present in biological fluids. Zn^{2+} indicators employing Zn^{2+} ligands made of pyridines and amines (DPA, for example) are generally selective against Ca^{2+} and Mg^{2+}, whereas those containing carboxylate ligands (IDA, for example) coordinate with Ca^{2+} or Mg^{2+} more strongly. FluoZin-3, for instance, binds to Zn^{2+} with a K_d of 15 nM, yet it is also somewhat sensitive to Ca^{2+} changes higher than tens of micromolar (albeit with much lower amplitude than its response to Zn^{2+}) (Gee et al. 2002). Interestingly, the Ca^{2+} sensitivity of FluoZin-3 does not seem to have a major effect on the ability of FluoZin-3 to monitor [Zn^{2+}] in the extracellular medium containing millimolar amounts of Ca^{2+} (Gee et al. 2002; Zhao et al. 2008).

In addition to thermodynamic stability, another important parameter for characterizing fluorescent Zn^{2+} sensors is their kinetics of Zn^{2+} binding. The Zn^{2+} association (k_{on}) and dissociation (k_{off}) rate constant of several fluorescent Zn^{2+} sensors have been determined using stopped-flow fluorescence measurements (Hirano et al. 2002; Komatsu et al. 2005; Nolan and Lippard 2009). Sensors based on the DPA chelating moiety bind Zn^{2+} rapidly, with k_{on} on the order of 10^6 M^{-1} s^{-1} (Table 11.2). However, once bound to Zn^{2+}, sensors with nanomolar binding affinity also appear to be fairly "sticky" and release Zn^{2+} only slowly, with k_{off} on the order of 10^{-3} s^{-1} to 10^{-2} s^{-1}—rates that are probably too sluggish for monitoring transient Zn^{2+} activity at second or subsecond timescales. DPA modifications that result in a reduction in the Zn^{2+} binding affinity may affect both k_{on} and k_{off} and can dramatically speed up the Zn^{2+} dissociation rate constant to above 1 s^{-1} (Table 11.2; ZnAF3, for example) (Komatsu et al. 2005; Nolan and Lippard 2009).

The majority of the synthetic Zn^{2+} sensors reported thus far, including those mentioned previously, change their emission intensity when [Zn^{2+}] fluctuates. In general, these intensity-based indicators display a larger dynamic range than the current ratiometric Zn^{2+} sensors that shift their excitation or emission wavelength when [Zn^{2+}] changes (Carol et al. 2007). However, ratiometric Zn^{2+} sensors are expected to be more reliable in quantifying cellular [Zn^{2+}] because variations in illumination intensity and/or the light collection efficiency of the imaging system, as well as dye loading, cell thickness, dye bleaching, and so forth, can be largely corrected when the fluorescence signals at two different wavelengths are determined ratiometrically. Few biological applications of these synthetic ratiometric Zn^{2+} sensors have been reported thus far, possibly owing to their tendency to aggregate in undefined compartments in cells. In addition to the fluorescence intensity or ratio, the fluorescence lifetime of Newport Green DCF has also been exploited as a readout of Zn^{2+} levels in biological fluids (Thompson et al. 2002).

11.2.3 Genetically Encoded Zn^{2+} Indicators

Genetically encoded fluorescent Zn^{2+} sensors are built using fluorescent proteins (FPs) (Vinkenborg et al. 2010). So far, the majority of these protein Zn^{2+} sensors have been constructed based on the principle of Förster/fluorescence resonance energy transfer (FRET). These fluorescent Zn^{2+} sensors contain a Zn^{2+}-binding motif that is fused with a pair of FRET partners, typically a cyan fluorescent protein (CFP) and a yellow fluorescent protein (YFP) or their enhanced versions. Zn^{2+} binding induces a conformational change that alters the intramolecular FRET efficiency, which can be read out ratiometrically (Figure 11.3, illustrated with eCALWY as an example). Different Zn^{2+}-binding motifs have been utilized to craft fluorescent Zn^{2+} sensors varying in Zn^{2+} affinity and dynamic range. In addition, because naturally occurring Zn^{2+}-binding motifs (or mutants of binding motifs of other metals [van Dongen et al. 2006]) typically manifest high binding specificity to Zn^{2+} against other metals present in biological fluids, these genetically encoded Zn^{2+} indicators also maintain high ion selectivity for sensing Zn^{2+}. Some representative examples are listed in Table 11.3.

Compared to small synthetic fluorescent dyes, genetically encoded fluorescent Zn^{2+} indicators offer several advantages for cellular imaging, including more specific and uniform cytoplasmic distribution, subcellular targetability when fused with peptide sequences encoding organelle-specific trafficking or

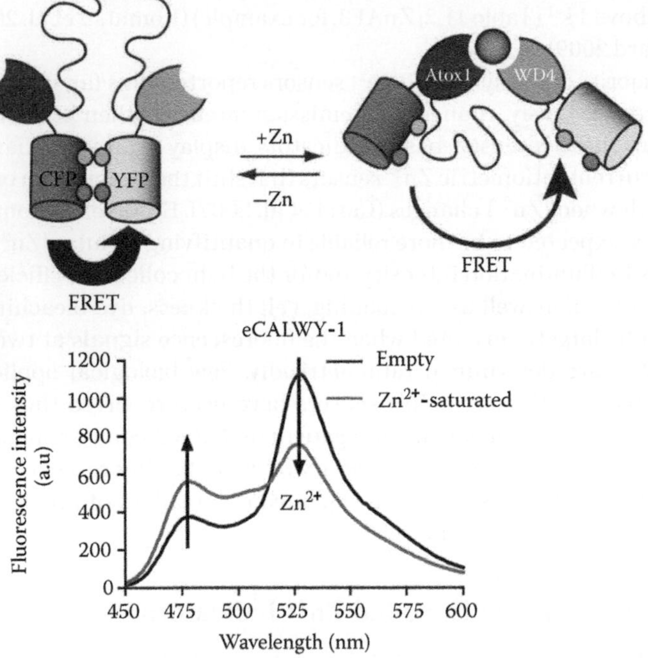

Figure 11.3 FRET-based protein Zn^{2+} sensors illustrated with eCALWY as an example. eCALWY sensors contain hydrophobic amino acids on the surface to enhance protein association. In the Zn^{2+}-free state, FRET efficiency is high. When $[Zn^{2+}]$ rises, the Zn^{2+}-binding motifs coordinate with Zn^{2+} to disrupt the interaction between the fluorophores and to reduce the FRET efficiency. (Adapted from Vinkenborg, J. L. et al., *Nat. Methods* 6:737–740, 2009. With permission.)

TABLE 11.3 GENETICALLY ENCODED FRET Zn^{2+} SENSORS

Sensor	Zn^{2+} Binding Motif	K_d (Zn^{2+})	Dynamic Range (*In Vitro*)	Reference
ZF1/2 ZapCY-1	ZF1/2 of Zap1	0.2 nM, 4 nM 2.5 pM	1.3-fold 4.2-fold	Qiao et al. (2006); Qin et al. (2011)
eCALWYs	Mutant of Atox1/WD 4	2 pM–2.9 nM	2–3-fold	Vinkenborg et al. (2009)
eZinCh	Engineered	8 μM	4-fold	Evers et al. (2007); Vinkenborg et al. (2009)

retention signals (Dittmer et al. 2009; Vinkenborg et al. 2009), and predictable expression in cells or tissues of interest through the use of appropriate promoters, etc. However, the dynamic ranges of protein-based Zn^{2+} sensors reported thus far are smaller than those of synthetic Zn^{2+} dyes (cf. Tables 11.2 and 11.3), so their sensitivity for detecting small-amplitude Zn^{2+} changes may be relatively limited. Moreover, because few data are currently available on the kinetics of Zn^{2+} binding or dissociation by these protein sensors, their utility for detecting rapid or pulsatile Zn^{2+} fluctuations remains to be demonstrated.

11.3 APPLICATIONS OF FLUORESCENT Zn^{2+} SENSORS

11.3.1 Measuring Intracellular Zn^{2+}

Total Zn^{2+} in biological samples is usually measured by atomic absorption spectroscopy or by inductively coupled plasma mass spectrometry (ICP-MS) (Taylor 2000; Shi and Chance 2008). Using these techniques, total cellular zinc concentrations have been determined to be in the range of hundreds of micromolar (Colvin et al. 2008, 2010). However, the vast majority of cellular Zn^{2+} is sequestered by Zn^{2+}-binding proteins, which results in a cytosolic free Zn^{2+} concentration that is orders of magnitude lower.

Fluorescent Zn^{2+} sensors have been applied to measure the concentration of free Zn^{2+} in the cytoplasm ($[Zn^{2+}]_i$) of eukaryotic cells (Table 11.4). Among small synthetic Zn^{2+} sensors, FluoZin-3 and ZnAF-2F display a fairly uniform cell distribution that is indicative of specific cytoplasmic labeling, so their signals should provide a more accurate assessment of free Zn^{2+} levels in the cytosol. Furthermore, FluoZin-3 and ZnAF-2F can be conveniently delivered into cells using neutral esters (acetoxymethyl ester or acetate) that are sensitive to intracellular esterases. To convert the measured fluorescence signal (F) to $[Zn^{2+}]_i$, the equation $[Zn^{2+}]_i = K_d(F - F_{min})/(F_{max} - F)$ is used, where K_d is the dissociation

TABLE 11.4 MEASURING CYTOPLASMIC FREE Zn^{2+} CONCENTRATION ($[Zn^{2+}]_i$) WITH FLUORESCENT Zn^{2+} SENSORS

$[Zn^{2+}]_i$ (pM)	Cell Type	Zn^{2+} Sensor	Reference
614	HT-29 (human colon cancer)	FluoZin-3	Krezel and Maret (2006)
1050	Rat primary cortical neuron	ZnAF-2F	Colvin et al. (2008)
970	PC12 (rat neuroendocrine tumor)	FluoZin-3	Li and Maret (2009)
400	INS-1 (rat insulinoma β cell) HEK293 (viral transformed human embryonic kidney cell)	eCALWY	Vinkenborg et al. (2009)
80	Hela (human cervical cancer)	ZapCY	Qin et al. (2011)

constant of the zinc sensor; F_{min} is the minimum cellular fluorescence signal of the sensor, measured in the presence of a cell-permeable Zn^{2+} chelating agent, TPEN (Figure 11.2), that depletes cellular free Zn^{2+}; and F_{max} is the maximum cellular fluorescence signal when the zinc sensor is saturated with Zn^{2+}, which is typically achieved by treating cells with a high level of Zn^{2+} and a Zn^{2+} ionophore such as pyrithione (Figure 11.2) (Zalewski et al. 1991).

Because FluoZin-3 (or any other Zn^{2+} sensor) is a Zn^{2+} chelator by itself and is usually loaded into cells at concentrations of micromolar or above, it also acts as a zinc buffer and can alter cellular zinc homeostasis. To correct this perturbation and to get a more accurate assessment of $[Zn^{2+}]_i$, Maret and coworkers introduced a method of collecting a series of FluoZin-3 images in a set of cells loaded with different amounts of FluoZin-3. Plotting the measured $[Zn^{2+}]_i$ against [FluoZin-3] and extrapolating the plot to a zero concentration of the zinc sensor yielded $[Zn^{2+}]_i$ values closer to its native level (Krezel and Maret 2006). This and other studies (Dineley et al. 2002) have highlighted the potential pitfall of loading excess zinc sensors into cells during quantification of $[Zn^{2+}]_i$, especially when the sensor becomes a major contributor to cellular Zn^{2+} buffering.

Fluorescence imaging using genetically encoded Zn^{2+} sensors, including eCALWY and ZapCY, also provided $[Zn^{2+}]_i$ values in the subnanomolar range (Table 11.4). These results are in agreement with earlier measurements using different techniques including biochemical assays (Peck and Ray 1971; Simons 1991) and ^{19}F-nuclear magnetic resonance (Benters et al. 1997). To determine the sensor K_d (Zn^{2+}) in the cytosolic environment, the protein-based Zn^{2+}

sensor can be calibrated *in situ* by limited permeabilization of the plasma membrane with α-toxin in buffers containing different free zinc concentrations (Vinkenborg et al. 2009).

In addition to measuring the steady-state $[Zn^{2+}]_i$ in resting cells, fluorescent Zn^{2+} sensors have also been applied to follow dynamic fluctuations of $[Zn^{2+}]_i$. In insulin-releasing MIN6 β-cells, membrane depolarization by different means resulted in a marked increase in the rate of intracellular Zn^{2+} accumulation, as measured by FluoZin-3 (Gyulkhandanyan et al. 2006). Pharmacological studies revealed that Zn^{2+} transport into β-cells was mediated at least in part through the L-type voltage-gated Ca^{2+} channel (L-VGCC). In another study using eCALWY-4 as the Zn^{2+} sensor, Rutter and co-workers found $[Zn^{2+}]_i$ nearly doubled in primary mouse β-cells that had been stimulated with high glucose (16.7 mM) for 24 h. Interestingly in this case, blocking L-VGCC had little effect on glucose-mediated $[Zn^{2+}]_i$ increase. Instead, it was proposed that glucose-associated increases in the level of zinc importers, including ZiP6, ZiP7, and ZiP8, likely played a role in $[Zn^{2+}]_i$ elevation (Bellomo et al. 2011). Another example of dynamic $[Zn^{2+}]_i$ fluctuation was again revealed by FluoZin-3 imaging and was related to the cell cycle: In PC-12 cells, $[Zn^{2+}]_i$ reached a peak value early in the G1 phase and late in the G1/S phase (Li and Maret 2009). These transient $[Zn^{2+}]_i$ fluctuations have been speculated to be involved in cell cycle control.

Using Newport Green DCF to monitor cellular Zn^{2+}, Yamasaki and co-workers reported the so-called zinc wave in mast cells following activation of the high-affinity immunoglobin E receptor: There was an increase in the Newport Green DCF signal starting from the perinuclear area, which gradually spread out to the rest of the cell within several minutes (Yamasaki et al. 2007). This observation was interpreted as free Zn^{2+} release into the cytoplasm from the intracellular stores. The authors further proposed that the released Zn^{2+} functioned as an intracellular second messenger to modulate the signaling events initiated by the immunoglobin E receptor. However, because Newport Green DCF has a $K_d(Zn^{2+})$ of 1.5 μM (Table 11.2), this observation would suggest that either $[Zn^{2+}]_i$ of mast cells was much higher than $[Zn^{2+}]_i$ of other mammalian cells, or the observed fluorescence enhancement of Newport Green DCF might result from fluctuations in Zn^{2+} activity from other cellular compartments. For instance, it has been reported that once inside cells, Newport Green DCF accumulated in cellular granules with high Zn^{2+} content, such as insulin granules (Lukowiak et al. 2001).

Besides monitoring $[Zn^{2+}]_i$, there have also been attempts to image free Zn^{2+} levels in different cellular organelles. Earlier efforts at measuring Zn^{2+} levels in mitochondria exploited the fact that RhodZin-3, a Zn^{2+} sensor containing a positive charge, tends to accumulate in mitochondria following its electrical gradient (Sensi et al. 2003a,b). A more general approach to targeting small synthetic dyes to different cellular compartments is to integrate chemistry with genetics. Lippard and co-workers applied the SNAP labeling strategy (Keppler et al. 2003) by expressing *O*[6]-alkylguaninetransferase in different organelles, including Golgi and mitochondria (Tomat et al. 2008). The subsequent addition of a membrane-permeable Zn^{2+} sensor, ZP1BG, to transfected cells led to the labeling of the expressed *O*[6]-alkylguaninetransferase. More recent efforts to quantify Zn^{2+} levels in cellular organelles have largely focused on expressing genetically encoded zinc sensors fused with appropriate targeting sequences (Vinkenborg et al. 2009; Qin et al. 2011; McCranor et al. 2012). These studies indicated that the resting levels of free Zn^{2+} in several organelles, including the endoplasmic reticulum, mitochondria, and Golgi, were orders of magnitude lower than $[Zn^{2+}]_i$.

11.3.2 Measuring Zn^{2+} in the Extracellular Fluid of Physiological Preparations

Total zinc in the blood plasma of mammals is in the range of 8–20 μM, as determined by atomic absorption spectroscopy (Magneson et al. 1987; Faure et al. 1990; Folin et al. 1994; Chen et al. 1995; Zhang and Allen 1995). Conventional enzymatic and biochemical assays have determined that almost all serum zinc is protein bound, resulting in subnanomolar free Zn^{2+} concentrations in the serum—more than four orders of magnitude lower than that of total zinc (Magneson et al. 1987; Zhang and Allen 1995). The majority of the plasma zinc is associated with albumin, and another sizable portion of Zn^{2+} is bound to α_2-macroglobulin (Foote and Delves 1984). Albumin and α_2-macroglobulin bind to Zn^{2+} with high affinity, displaying apparent stability constants of $3 \times 10^7\ M^{-1}$ and $1.3 \times 10^6\ M^{-1}$, respectively (Pratt and Pizzo 1984; Masuoka et al. 1993).

To measure free Zn^{2+} levels in the extracellular cerebrospinal fluid, Frederickson and co-workers applied a ratiometric fluorescent Zn^{2+} sensor composed of carbonic anhydrase and dansylamide (Frederickson et al. 2006). Among several species they examined, including rat, rabbit, and human, free $[Zn^{2+}]$ in the extracellular cerebrospinal fluid was found to be around 19 nM.

11.3.3 Imaging Zn^{2+} Secretion

A number of mammalian cells, including pancreatic islet β-cells, prostate epithelial cells, certain glutamatergic neurons in the central nervous system, Paneth cells, pituitary cells, and mast cells, contain a high level of Zn^{2+} in their secretory granules (Frederickson et al. 2005). During secretion, Zn^{2+} is co-released with the other contents of the secretory granule into the extracellular medium. The function of released Zn^{2+} and its role in mediating cell–cell signaling have been an intriguing topic, and there have been major efforts toward understanding the fate, mode of action, and targets of the secreted Zn^{2+} in two biological systems: pancreatic islet β-cells and certain glutamatergic neurons.

The insulin granules of pancreatic islet β-cells contain a high level of Zn^{2+}. Within the granule, six insulin molecules coordinate with two zinc ions to form a hexameric insulin$_6$–Zn_2 complex that is insoluble below pH 7 (Emdin et al. 1980; Dodson and Steiner 1998), thereby giving rise to an electron-dense crystal (dense core granule) (Brader and Dunn 1991; Dodson and Steiner 1998). Because the granular insulin content is greater than 70 mM (Matthews et al. 1982; Huang et al. 1995), the total Zn^{2+} in insulin granules may reach or even exceed 20 mM. The free Zn^{2+} level in insulin granules is not known, but it has been estimated to be between 1 and 100 μM (Vinkenborg et al. 2009). The high granular zinc content in islet β-cells has been exploited to purify β-cells from dissociated human islets by fluorescence-activated cell sorting (FACS) after labeling cells with Newport Green DCF (Lukowiak et al. 2001).

Islet β-cells release insulin in response to stimulation by a variety of secretagogues. Because of its essential role in maintaining blood glucose homeostasis, there have been major efforts in understanding the regulation of insulin secretion. When the insulin granule fuses with the plasma membrane, there is a rapid dissolution of the insulin–Zn crystal due to alkalization from extracellular solution. This leads to a local elevation of $[Zn^{2+}]$ at the site of exocytosis (Aspinwall et al. 1997). This phenomenon has been exploited to design imaging assays using Zn^{2+} as a surrogate to monitor insulin secretion in islet β-cells. Numerous fluorescent Zn^{2+} sensors, including Zinquin (Qian et al. 2000), FluoZin-3 (Gee et al. 2002), and RhodZin-3 (Michael et al. 2006), have been applied to monitor insulin/Zn^{2+} release in islet β-cells. Because these Zn^{2+} sensors are applied to the extracellular bath, and because common biological salines contain residual Zn^{2+} near micromolar concentration unless they are treated with the Chelex resin (Qian et al. 2003), the sensitivity of detecting local Zn^{2+} release near the plasma membrane is compromised by the background fluorescence from the bulk solution. To reject the bulk fluorescence signal, total internal reflection of fluorescence (TIRF) microscopy has been applied to study secretion at the interface between a cell and the underlying glass coverslip (Michael et al. 2006).

To enhance the sensitivity of detecting zinc secretion and to develop a robust imaging assay compatible with widefield epifluorescence detection, our laboratory developed an amphipathic fluorescent zinc indicator for monitoring induced exocytotic Zn^{2+} release (ZIMIR; Figure 11.4) (Li et al. 2011). ZIMIR contains a pair of dodecyl alkyl chains for membrane tethering, yet it is also highly charged near neutral pH so that it cannot diffuse across the cell membrane by itself. When ZIMIR is added to cells at low micromolar concentrations, it is rapidly taken up by cells through the hydrophobic interaction between the dodecyl chains and membrane lipids. The labeling is noninvasive

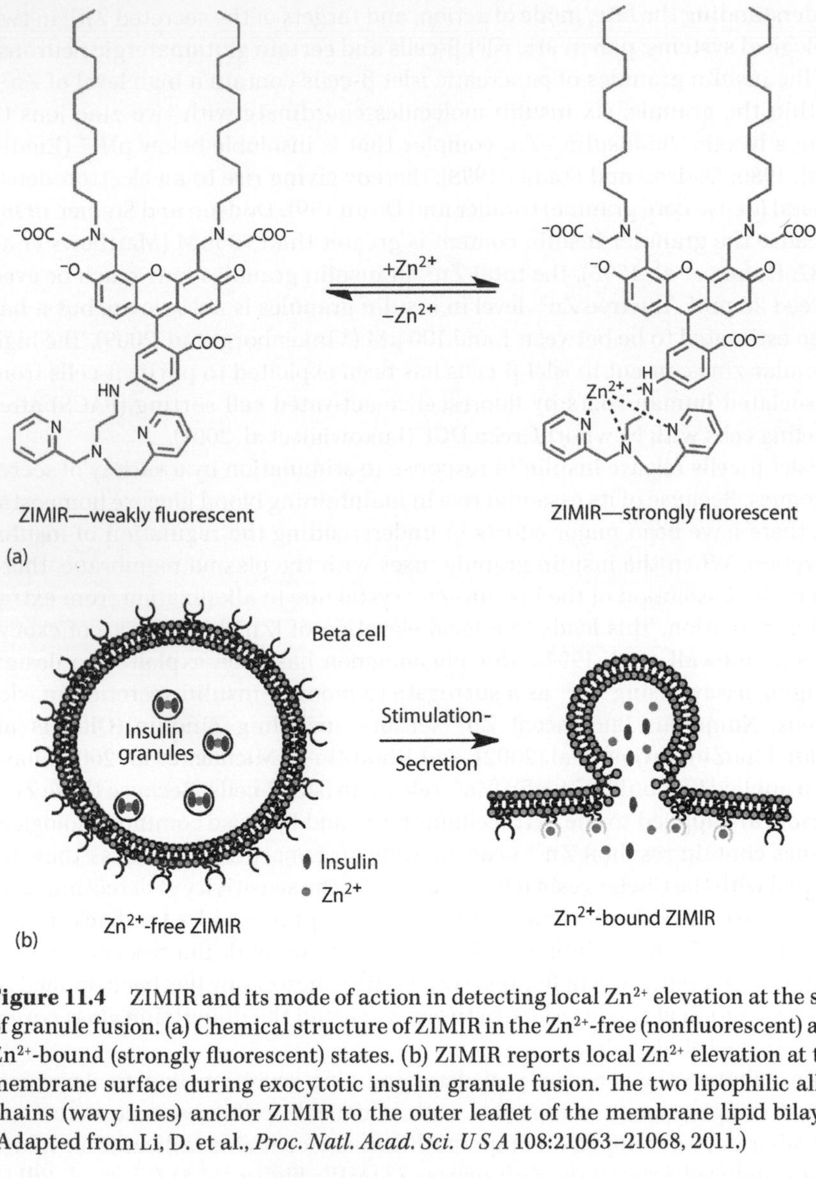

Figure 11.4 ZIMIR and its mode of action in detecting local Zn^{2+} elevation at the site of granule fusion. (a) Chemical structure of ZIMIR in the Zn^{2+}-free (nonfluorescent) and Zn^{2+}-bound (strongly fluorescent) states. (b) ZIMIR reports local Zn^{2+} elevation at the membrane surface during exocytotic insulin granule fusion. The two lipophilic alkyl chains (wavy lines) anchor ZIMIR to the outer leaflet of the membrane lipid bilayer. (Adapted from Li, D. et al., *Proc. Natl. Acad. Sci. U S A* 108:21063–21068, 2011.)

and does not appear to have any observable side effects on cell growth or function. Furthermore, because ZIMIR is anchored to the outer leaflet of the plasma membrane, it is most sensitive for detecting local Zn^{2+} increases at the sites of exocytosis (Figure 11.4b). ZIMIR binds Zn^{2+} with a K_d of 0.45 μM and displayed a 70-fold enhancement in fluorescence intensity on Zn^{2+} binding. A very useful feature of ZIMIR is that the sensor is able to diffuse through interstitial spaces to label both superficial and interior cells in tissues. This makes it possible to apply confocal laser scanning microscopy to image insulin/Zn^{2+} release dynamics in three dimensions throughout isolated islets with high spatial and temporal resolution. An imaging study in rat islets revealed that small clusters of adjacent islet β-cells release insulin in synchrony and that islet β-cells were fairly heterogeneous in their secretory responses to glucose stimulation (Li et al. 2011).

Numerous studies have shown that islet β-cells may manifest oscillatory insulin secretion activity (Henquin 2009). To reveal such a behavior by fluorescence imaging, we added a soluble, cell-membrane–impermeable Zn^{2+} chelator, dipicolylamine *N*-ethylsulfonate (DPAS, Figure 11.2), to the medium. The Zn^{2+} dissociation constant (k_{off}) of ZIMIR is estimated to be approximately 1 s^{-1}. In the presence of micromolar concentrations of DPAS, Zn^{2+} dissipation from the membrane into the bulk solution is sped up by about 10-fold (Li et al. 2011), likely resulting from the accelerated abstraction of Zn^{2+} by DPAS from the Zn^{2+}–ZIMIR complex. In cultured β-cells labeled with ZIMIR, simple epifluorescence imaging revealed oscillatory insulin release dynamics at subcellular resolution (Figure 11.5).

Like islet β-cells, a subset of glutamatergic nerve terminals in the central nervous system contain synaptic vesicles enriched with high levels of Zn^{2+} (Frederickson et al. 1983). The release of this vesicular zinc pool has been proposed to modulate synaptic transmission by different mechanisms, including the inhibitory regulation of postsynaptic glutamate receptors, principally *N*-methyl-D-aspartate receptor (NMDAR; Sensi et al. 2009). Attempts at imaging Zn^{2+} release at the synapse using ZnAF-2 or ZP4 were confounded by the report that these probes, despite containing negative charges and a hydrophilic dipicolylamine, were nevertheless cell membrane permeant when added to brain tissue (Kay and Toth 2006). This made it difficult to interpret the source of Zn^{2+} activity when changes in fluorescence intensity were spotted (Kay

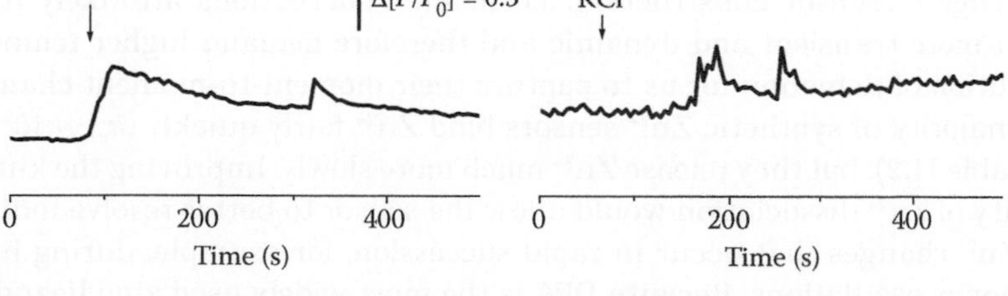

Figure 11.5 ZIMIR imaging in cultured MIN6 cells revealed oscillatory insulin/Zn^{2+} release when DPAS was included in the medium (right).

2006). In another study using cell membrane impermeant FluoZin-3, Qian and Noebels (2005) detected Zn^{2+} release at the hippocampal mossy fiber synapse that was highly correlated with glutamate release. Because the authors did not detect the FluoZin-3 signal increase in hippocampal neurons from the ZnT3 knockout mouse, they concluded that the observed Zn^{2+} release in wild-type animals was from the synaptic vesicles.

11.4 OUTLOOK

There has been substantial progress toward the engineering of fluorescent sensors for imaging Zn^{2+} dynamics in different biological systems over the past decade. Both small synthetic dyes and protein-based indicators have been developed to track Zn^{2+} levels within or outside of living cells. Although these innovations have enabled the examination of Zn^{2+} dynamics in living cells, imaging studies have also pointed out areas for future improvements that would be desirable when engineering the next generation Zn^{2+} sensors.

To study the signaling role of Zn^{2+} fluctuation or the cellular mechanisms regulating Zn^{2+} homeostasis, it would be necessary to follow rapid Zn^{2+} transients at subcellular and high temporal resolution while minimizing the buffering effect of zinc sensors. The resting level of $[Zn^{2+}]_i$ is in the subnanomolar range, and the amplitudes of Zn^{2+} fluctuations are likely to be in the same range. The challenge in detecting Zn^{2+} changes at this low level of abundance can be appreciated by contrasting it with cellular Ca^{2+} imaging, which typically monitors cellular $[Ca^{2+}]_i$ fluctuations near micromolar concentrations, a range containing 1000 times more analyte than free cellular Zn^{2+}. Thus, Zn^{2+} sensors showing robust contrast enhancement and drastically improved fluorescence brightness would be desirable. The latter requirement is particularly challenging and demands substantial investments in dye engineering. Once developed, such Zn^{2+} sensors can then be applied to cells at a much lower working concentration to minimize the perturbation of endogenous Zn^{2+} homeostasis.

Improving the kinetics of Zn^{2+} chelation and release represents another challenge in sensor construction. Local Zn^{2+} fluctuations are likely to be much more transient and dynamic and therefore demand higher temporal resolution of detection for us to capture their moment-to-moment changes. The majority of synthetic Zn^{2+} sensors bind Zn^{2+} fairly quickly ($k_{on} \sim 10^6$ M^{-1} s^{-1}, Table 11.2), but they release Zn^{2+} much more slowly. Improving the kinetic lability of Zn^{2+} dissociation would allow the sensor to better resolve individual Zn^{2+} changes that occur in rapid succession, for example, during high-frequency oscillations. Because DPA is the most widely used zinc ligand for designing zinc sensors, there have been systematic efforts to modulate its zinc-binding affinity and/or kinetics by modifying its structure. A number of

DPA derivatives varying in steric hindrance, linker length, electronegativity, composition, or the number of pyridine rings have been reported (Komatsu et al. 2005; Nolan and Lippard 2009; Wong et al. 2009). Several of these modifications result in an increase in k_{off} with a concomitant drop in K_d. For protein-based Zn^{2+} sensors, currently we know much less about the kinetics of their Zn^{2+} responses, so more studies are needed in this area to improve our understanding of the structural elements affecting kinetic behavior. The large number of naturally occurring Zn^{2+}-binding domains ought to provide a rich resource for protein engineering of sensors with fast Zn^{2+} responses. Proteins that bind Zn^{2+} with high thermodynamic stability and kinetic lability are potentially useful candidates, and examples of such proteins have been described (Jacob et al. 1998; Maret 2004; Maret and Li 2009). Furthermore, studies have also demonstrated that it is possible to mutate Zn^{2+}-binding proteins to drastically accelerate their Zn^{2+} dissociation rates while causing only a modest drop in their stability constants. For example, mutating a single amino acid in carbonic anhydrase from glutamate to alanine increases k_{off} by nearly 300-fold while reducing the zinc affinity only by 10-fold (Kiefer et al. 1995; Fierke and Thompson 2001).

Because the majority of synthetic Zn^{2+} sensors are known to distribute to more than one cellular compartment when applied to cells or tissues, protein-based Zn^{2+} sensors are superior with regard to the specificity of cellular localization, though currently their dynamic range is modest at best by comparison with small synthetic dyes. This limitation is not necessarily intrinsic to protein-based sensors. New strategies for sensor design and library construction, combined with robust and efficient screening, may eventually produce novel protein-based sensors with drastically improved responses to metal ions (including Zn^{2+}). For instance, recent work by Campbell and co-workers has demonstrated the feasibility of crafting genetically encoded Ca^{2+} sensors (GECIs) that rival or even surpass the performance of traditional synthetic Ca^{2+} indicators (Zhao et al. 2011).

Last but not least, because of the involvement of abnormal Zn^{2+} handling in a number of human diseases (Chasapis et al. 2012), monitoring Zn^{2+} in tissues in live animals represents a high-risk yet high-reward project that may have immediate implications in translational research and clinical diagnosis. Despite the challenges, we have recently seen some encouraging examples in this area. Moore, Lippard, Medarova, and co-workers have applied a fluorescent probe, ZPP1 (Zhang et al. 2008), to demonstrate the feasibility of using whole-body animal optical imaging to follow the progression of prostate cancer judged by the decreasing Zn^{2+} content in the prostates of tumor-bearing mice (Ghosh et al. 2010). Obviously, synthetic fluorescent Zn^{2+} sensors emitting at red or infrared wavelengths, showing robust Zn^{2+} responsivity, and ideally exhibiting appropriate cell or tissue targeting specificity would be future goals to enhance the application potential of this technique *in vivo*. To look even deeper into live animals, other imaging modalities such as magnetic resonance imaging (MRI) represent an alternative to optical methods (Lee et al. 2010; Lubag et al. 2011). Once again, realizing the application of these imaging techniques to acquire functional information on Zn^{2+} handling or cell states *in vivo* largely relies on the successful engineering of suitable Zn^{2+} probes or contrast agents that are biocompatible and that perform well in biological fluids under physiological conditions.

REFERENCES

Andreini, C., Banci, L., Bertini, I., and Rosato, A. (2006). Counting the zinc-proteins encoded in the human genome. *J Proteome Res 5*, 196–201.

Aslamkhan, A.G., Aslamkhan, A., and Ahearn, G.A. (2002). Preparation of metal ion buffers for biological experimentation: A methods approach with emphasis on iron and zinc. *J Exp Zool 292*, 507–522.

Aspinwall, C.A., Brooks, S.A., Kennedy, R.T., and Lakey, J.R. (1997). Effects of intravesicular H^+ and extracellular H^+ and Zn^{2+} on insulin secretion in pancreatic beta cells. *J Biol Chem 272*, 31308–31314.

Bellomo, E.A., Meur, G., and Rutter, G.A. (2011). Glucose regulates free cytosolic Zn(2) concentration, Slc39 (ZiP), and metallothionein gene expression in primary pancreatic islet beta-cells. *J Biol Chem 286*, 25778–25789.

Benters, J., Flogel, U., Schafer, T., Leibfritz, D., Hechtenberg, S., and Beyersmann, D. (1997). Study of the interactions of cadmium and zinc ions with cellular calcium homoeostasis using ^{19}F-NMR spectroscopy. *Biochem J 322(Pt 3)*, 793–799.

Brader, M.L., and Dunn, M.F. (1991). Insulin hexamers: New conformations and applications. *Trends Biochem Sci 16*, 341–345.

Burdette, S.C., Frederickson, C.J., Bu, W., and Lippard, S.J. (2003). ZP4, an improved neuronal Zn^{2+} sensor of the Zinpyr family. *J Am Chem Soc 125*, 1778–1787.

Carol, P., Sreejith, S., and Ajayaghosh, A. (2007). Ratiometric and near-infrared molecular probes for the detection and imaging of zinc ions. *Chem Asian J 2*, 338–348.

Chasapis, C.T., Loutsidou, A.C., Spiliopoulou, C.A., and Stefanidou, M.E. (2012). Zinc and human health: An update. *Arch Toxicol 86*, 521–534.

Chen, M.D., Lin, P.Y., Tsou, C.T., Wang, J.J., and Lin, W.H. (1995). Selected metals status in patients with noninsulin-dependent diabetes mellitus. *Biol Trace Elem Res 50*, 119–124.

Colvin, R.A., Bush, A.I., Volitakis, I., Fontaine, C.P., Thomas, D., Kikuchi, K., and Holmes, W.R. (2008). Insights into Zn^{2+} homeostasis in neurons from experimental and modeling studies. *Am J Physiol Cell Physiol 294*, C726–C742.

Colvin, R.A., Holmes, W.R., Fontaine, C.P., and Maret, W. (2010). Cytosolic zinc buffering and muffling: Their role in intracellular zinc homeostasis. *Metallomics 2*, 306–317.

Cotton, F.A., Wilkinson, G., Murillo, C.A., and Bochmann, M. (1999). *Advanced Inorganic Chemistry*, 6th ed. New York: Wiley-Interscience.

Cousins, R.J., Liuzzi, J.P., and Lichten, L.A. (2006). Mammalian zinc transport, trafficking, and signals. *J Biol Chem 281*, 24085–24089.

Dineley, K.E., Malaiyandi, L.M., and Reynolds, I.J. (2002). A reevaluation of neuronal zinc measurements: Artifacts associated with high intracellular dye concentration. *Mol Pharmacol 62*, 618–627.

Dittmer, P.J., Miranda, J.G., Gorski, J.A., and Palmer, A.E. (2009). Genetically encoded sensors to elucidate spatial distribution of cellular zinc. *J Biol Chem 284*, 16289–16297.

Dodson, G., and Steiner, D. (1998). The role of assembly in insulin's biosynthesis. *Curr Opin Struct Biol 8*, 189–194.

Emdin, S.O., Dodson, G.G., Cutfield, J.M., and Cutfield, S.M. (1980). Role of zinc in insulin biosynthesis: Some possible zinc-insulin interactions in the pancreatic B-cell. *Diabetologia 19*, 174–182.

Evers, T.H., Appelhof, M.A., de Graaf-Heuvelmans, P.T., Meijer, E.W., and Merkx, M. (2007). Ratiometric detection of Zn(II) using chelating fluorescent protein chimeras. *J Mol Biol 374*, 411–425.

Faure, H., Favier, A., Tripier, M., and Arnaud, J. (1990). Determination of the major zinc fractions in human serum by ultrafiltration. *Biol Trace Elem Res 24*, 25–37.

Fierke, C.A., and Thompson, R.B. (2001). Fluorescence-based biosensing of zinc using carbonic anhydrase. *Biometals 14*, 205–222.

Folin, M., Contiero, E., and Vaselli, G.M. (1994). Zinc content of normal human serum and its correlation with some hematic parameters. *Biometals 7*, 75–79.

Foote, J.W., and Delves, H.T. (1984). Albumin bound and alpha 2-macroglobulin bound zinc concentrations in the sera of healthy adults. *J Clin Pathol 37*, 1050–1054.

Frederickson, C.J., Giblin, L.J., Krezel, A., McAdoo, D.J., Mueller, R.N., Zeng, Y., Balaji, R.V., Masalha, R., Thompson, R.B., Fierke, C.A. et al. (2006). Concentrations of extracellular free zinc (pZn)e in the central nervous system during simple anesthetization, ischemia and reperfusion. *Exp Neurol 198*, 285–293.

Frederickson, C.J., Kasarskis, E.J., Ringo, D., and Frederickson, R.E. (1987). A quinoline fluorescence method for visualizing and assaying the histochemically reactive zinc (bouton zinc) in the brain. *J Neurosci Methods 20*, 91–103.

Frederickson, C.J., Klitenick, M.A., Manton, W.I., and Kirkpatrick, J.B. (1983). Cytoarchitectonic distribution of zinc in the hippocampus of man and the rat. *Brain Res 273*, 335–339.

Frederickson, C.J., Koh, J.Y., and Bush, A.I. (2005). The neurobiology of zinc in health and disease. *Nat Rev Neurosci 6*, 449–462.

Fukada, T., Yamasaki, S., Nishida, K., Murakami, M., and Hirano, T. (2011). Zinc homeostasis and signaling in health and diseases: Zinc signaling. *J Biol Inorg Chem 16*, 1123–1134.

Gee, K.R., Zhou, Z.L., Qian, W.J., and Kennedy, R. (2002). Detection and imaging of zinc secretion from pancreatic beta-cells using a new fluorescent zinc indicator. *J Am Chem Soc 124*, 776–778.

Ghosh, S.K., Kim, P., Zhang, X.A., Yun, S.H., Moore, A., Lippard, S.J., and Medarova, Z. (2010). A novel imaging approach for early detection of prostate cancer based on endogenous zinc sensing. *Cancer Res 70*, 6119–6127.

Gyulkhandanyan, A.V., Lee, S.C., Bikopoulos, G., Dai, F., and Wheeler, M.B. (2006). The Zn^{2+}-transporting pathways in pancreatic beta-cells: A role for the L-type voltage-gated Ca^{2+} channel. *J Biol Chem 281*, 9361–9372.

Henquin, J.C. (2009). Regulation of insulin secretion: A matter of phase control and amplitude modulation. *Diabetologia 52*, 739–751.

Hirano, T., Kikuchi, K., Urano, Y., Higuchi, T., and Nagano, T. (2000). Highly zinc-selective fluorescent sensor molecules suitable for biological applications. *J Am Chem Soc 122*, 12399–12400.

Hirano, T., Kikuchi, K., Urano, Y., and Nagano, T. (2002). Improvement and biological applications of fluorescent probes for zinc, ZnAFs. *J Am Chem Soc 124*, 6555–6562.

Huang, L., Shen, H., Atkinson, M.A., and Kennedy, R.T. (1995). Detection of exocytosis at individual pancreatic beta cells by amperometry at a chemically modified microelectrode. *Proc Natl Acad Sci U S A 92*, 9608–9612.

Jacob, C., Maret, W., and Vallee, B.L. (1998). Control of zinc transfer between thionein, metallothionein, and zinc proteins. *Proc Natl Acad Sci U S A 95*, 3489–3494.

Jiang, P.J., and Guo, Z.J. (2004). Fluorescent detection of zinc in biological systems: Recent development on the design of chemosensors and biosensors. *Coord Chem Rev 248*, 205–229.

Kay, A.R. (2006). Imaging synaptic zinc: Promises and perils. *Trends Neurosci 29*, 200–206.

Kay, A.R., and Toth, K. (2006). Influence of location of a fluorescent zinc probe in brain slices on its response to synaptic activation. *J Neurophysiol 95*, 1949–1956.

Keppler, A., Gendreizig, S., Gronemeyer, T., Pick, H., Vogel, H., and Johnsson, K. (2003). A general method for the covalent labeling of fusion proteins with small molecules in vivo. *Nat Biotechnol 21*, 86–89.

Kiefer, L.L., Paterno, S.A., and Fierke, C.A. (1995). Hydrogen-bond network in the metal-binding site of carbonic-anhydrase enhances zinc affinity and catalytic efficiency. *J Am Chem Soc 117*, 6831–6837.

Kikuchi, K., Komatsu, K., and Nagano, T. (2004). Zinc sensing for cellular application. *Curr Opin Chem Biol 8*, 182–191.

Klug, A. (2010). The discovery of zinc fingers and their applications in gene regulation and genome manipulation. *Annu Rev Biochem 79*, 213–231.

Koide, Y., Urano, Y., Hanaoka, K., Terai, T., and Nagano, T. (2011). Evolution of group 14 rhodamines as platforms for near-infrared fluorescence probes utilizing photo-induced electron transfer. *ACS Chem Biol 6*, 600–608.

Komatsu, K., Kikuchi, K., Kojima, H., Urano, Y., and Nagano, T. (2005). Selective zinc sensor molecules with various affinities for Zn^{2+}, revealing dynamics and regional distribution of synaptically released Zn^{2+} in hippocampal slices. *J Am Chem Soc 127*, 10197–10204.

Krezel, A., and Maret, W. (2006). Zinc-buffering capacity of a eukaryotic cell at physiological pZn. *J Biol Inorg Chem 11*, 1049–1062.

Laukens, D., Waeytens, A., De Bleser, P., Cuvelier, C., and De Vos, M. (2009). Human metallothionein expression under normal and pathological conditions: Mechanisms of gene regulation based on in silico promoter analysis. *Crit Rev Eukaryot Gene Expr 19*, 301–317.

Lee, T., Zhang, X.A., Dhar, S., Faas, H., Lippard, S.J., and Jasanoff, A. (2010). In vivo imaging with a cell-permeable porphyrin-based MRI contrast agent. *Chem Biol 17*, 665–673.

Li, D., Chen, S., Bellomo, E.A., Tarasov, A.I., Kaut, C., Rutter, G.A., and Li, W.H. (2011). Imaging dynamic insulin release using a fluorescent zinc indicator for monitoring induced exocytotic release (ZIMIR). *Proc Natl Acad Sci U S A 108*, 21063–21068.

Li, Y., and Maret, W. (2009). Transient fluctuations of intracellular zinc ions in cell proliferation. *Exp Cell Res 315*, 2463–2470.

Lichten, L.A., and Cousins, R.J. (2009). Mammalian zinc transporters: Nutritional and physiologic regulation. *Annu Rev Nutr 29*, 153–176.

Lubag, A.J., De Leon-Rodriguez, L.M., Burgess, S.C., and Sherry, A.D. (2011). Noninvasive MRI of beta-cell function using a Zn^{2+}-responsive contrast agent. *Proc Natl Acad Sci U S A 108*, 18400–18405.

Lukowiak, B., Vandewalle, B., Riachy, R., Kerr-Conte, J., Gmyr, V., Belaich, S., Lefebvre, J., and Pattou, F. (2001). Identification and purification of functional human beta-cells by a new specific zinc-fluorescent probe. *J Histochem Cytochem 49*, 519–528.

Magneson, G.R., Puvathingal, J.M., and Ray, W.J., Jr. (1987). The concentrations of free Mg^{2+} and free Zn^{2+} in equine blood plasma. *J Biol Chem 262*, 11140–11148.

Maret, W. (2004). Zinc and sulfur: A critical biological partnership. *Biochemistry 43*, 3301–3309.

Maret, W., and Li, Y. (2009). Coordination dynamics of zinc in proteins. *Chem Rev 109*, 4682–4707.

Martell, A.E. and Smith, R.M., 2003. NIST critically selected stability constants of metal complexes database. National Institute of Standards and Technology, Standard Reference Database 46.

Masuoka, J., Hegenauer, J., Van Dyke, B.R., and Saltman, P. (1993). Intrinsic stoichiometric equilibrium constants for the binding of zinc(II) and copper(II) to the high affinity site of serum albumin. *J Biol Chem 268*, 21533–21537.

Matthews, E.K., McKay, D.B., O'Connor, M.D., and Borowitz, J.L. (1982). Biochemical and biophysical characterization of insulin granules isolated from rat pancreatic islets by an iso-osmotic gradient. *Biochim Biophys Acta 715*, 80–89.

McCall, K.A., Huang, C., and Fierke, C.A. (2000). Function and mechanism of zinc metalloenzymes. *J Nutr 130*, 1437S–1446S.

McCranor, B.J., Bozym, R.A., Vitolo, M.I., Fierke, C.A., Bambrick, L., Polster, B.M., Fiskum, G., and Thompson, R.B. (2012). Quantitative imaging of mitochondrial and cytosolic free zinc levels in an in vitro model of ischemia/reperfusion. *J Bioenerg Biomembr 44*, 253–263.

Michael, D.J., Ritzel, R.A., Haataja, L., and Chow, R.H. (2006). Pancreatic beta-cells secrete insulin in fast- and slow-release forms. *Diabetes 55*, 600–607.

Newman, R.H., Fosbrink, M.D., and Zhang, J. (2011). Genetically encodable fluorescent biosensors for tracking signaling dynamics in living cells. *Chem Rev 111*, 3614–3666.

Nolan, E.M., and Lippard, S.J. (2009). Small-molecule fluorescent sensors for investigating zinc metalloneurochemistry. *Acc Chem Res 42*, 193–203.

Peck, E.J., Jr., and Ray, W.J., Jr. (1971). Metal complexes of phosphoglucomutase in vivo. Alterations induced by insulin. *J Biol Chem 246*, 1160–1167.

Pratt, C.W., and Pizzo, S.V. (1984). The effect of zinc and other divalent cations on the structure and function of human alpha 2-macroglobulin. *Biochim Biophys Acta 791*, 123–130.

Qian, J., and Noebels, J.L. (2005). Visualization of transmitter release with zinc fluorescence detection at the mouse hippocampal mossy fibre synapse. *J Physiol 566*, 747–758.

Qian, W.J., Aspinwall, C.A., Battiste, M.A., and Kennedy, R.T. (2000). Detection of secretion from single pancreatic beta-cells using extracellular fluorogenic reactions and confocal fluorescence microscopy. *Anal Chem 72*, 711–717.

Qian, W.J., Gee, K.R., and Kennedy, R.T. (2003). Imaging of Zn^{2+} release from pancreatic beta-cells at the level of single exocytotic events. *Anal Chem 75*, 3468–3475.

Qiao, W., Mooney, M., Bird, A.J., Winge, D.R., and Eide, D.J. (2006). Zinc binding to a regulatory zinc-sensing domain monitored in vivo by using FRET. *Proc Natl Acad Sci U S A 103*, 8674–8679.

Qin, Y., Dittmer, P.J., Park, J.G., Jansen, K.B., and Palmer, A.E. (2011). Measuring steady-state and dynamic endoplasmic reticulum and Golgi Zn^{2+} with genetically encoded sensors. *Proc Natl Acad Sci U S A 108*, 7351–7356.

Rungby, J. (2010). Zinc, zinc transporters and diabetes. *Diabetologia 53*, 1549–1551.

Sensi, S.L., Paoletti, P., Bush, A.I., and Sekler, I. (2009). Zinc in the physiology and pathology of the CNS. *Nat Rev Neurosci 10*, 780–791.

Sensi, S.L., Paoletti, P., Koh, J.Y., Aizenman, E., Bush, A.I., and Hershfinkel, M. (2011). The neurophysiology and pathology of brain zinc. *J Neurosci 31*, 16076–16085.

Sensi, S.L., Ton-That, D., Sullivan, P.G., Jonas, E.A., Gee, K.R., Kaczmarek, L.K., and Weiss, J.H. (2003a). Modulation of mitochondrial function by endogenous Zn^{2+} pools. *Proc Natl Acad Sci U S A 100*, 6157–6162.

Sensi, S.L., Ton-That, D., Weiss, J.H., Rothe, A., and Gee, K.R. (2003b). A new mitochondrial fluorescent zinc sensor. *Cell Calcium 34*, 281–284.

Shi, W., and Chance, M.R. (2008). Metallomics and metalloproteomics. *Cell Mol Life Sci 65*, 3040–3048.

Simons, T.J. (1991). Intracellular free zinc and zinc buffering in human red blood cells. *J Membr Biol 123*, 63–71.

Taylor, H.E. (2000). *Inductively Coupled Plasma-Mass Spectrometry: Practices and Techniques*. San Diego, CA: Academic Press.

Thompson, R.B., Peterson, D., Mahoney, W., Cramer, M., Maliwal, B.P., Suh, S.W., Frederickson, C., Fierke, C., and Herman, P. (2002). Fluorescent zinc indicators for neurobiology. *J Neurosci Methods 118*, 63–75.

Tomat, E., and Lippard, S.J. (2010). Imaging mobile zinc in biology. *Curr Opin Chem Biol 14*, 225–230.

Tomat, E., Nolan, E.M., Jaworski, J., and Lippard, S.J. (2008). Organelle-specific zinc detection using zinpyr-labeled fusion proteins in live cells. *J Am Chem Soc 130*, 15776–15777.

van Dongen, E.M., Dekkers, L.M., Spijker, K., Meijer, E.W., Klomp, L.W., and Merkx, M. (2006). Ratiometric fluorescent sensor proteins with subnanomolar affinity for Zn(II) based on copper chaperone domains. *J Am Chem Soc 128*, 10754–10762.

Vinkenborg, J.L., Koay, M.S., and Merkx, M. (2010). Fluorescent imaging of transition metal homeostasis using genetically encoded sensors. *Curr Opin Chem Biol 14*, 231–237.

Vinkenborg, J.L., Nicolson, T.J., Bellomo, E.A., Koay, M.S., Rutter, G.A., and Merkx, M. (2009). Genetically encoded FRET sensors to monitor intracellular Zn^{2+} homeostasis. *Nat Methods 6*, 737–740.

Wong, B.A., Friedle, S., and Lippard, S.J. (2009). Subtle modification of 2,2-dipicolylamine lowers the affinity and improves the turn-on of Zn(II)-selective fluorescent sensors. *Inorg Chem 48*, 7009–7011.

Yamasaki, S., Sakata-Sogawa, K., Hasegawa, A., Suzuki, T., Kabu, K., Sato, E., Kurosaki, T., Yamashita, S., Tokunaga, M., Nishida, K. et al. (2007). Zinc is a novel intracellular second messenger. *J Cell Biol 177*, 637–645.

Zalewski, P.D., Forbes, I.J., and Betts, W.H. (1993). Correlation of apoptosis with change in intracellular labile Zn(II) using zinquin [(2-methyl-8-*p*-toluenesulphonamido-6-quinolyloxy)acetic acid], a new specific fluorescent probe for Zn(II). *Biochem J 296(Pt 2)*, 403–408.

Zalewski, P.D., Forbes, I.J., and Giannakis, C. (1991). Physiological role for zinc in prevention of apoptosis (gene-directed death). *Biochem Int 24*, 1093–1101.

Zhang, P., and Allen, J.C. (1995). A novel dialysis procedure measuring free Zn^{2+} in bovine milk and plasma. *J Nutr 125*, 1904–1910.

Zhang, X.A., Hayes, D., Smith, S.J., Friedle, S., and Lippard, S.J. (2008). New strategy for quantifying biological zinc by a modified zinpyr fluorescence sensor. *J Am Chem Soc 130*, 15788–15789.

Zhao, J., Bertoglio, B.A., Gee, K.R., and Kay, A.R. (2008). The zinc indicator FluoZin-3 is not perturbed significantly by physiological levels of calcium or magnesium. *Cell Calcium 44*, 422–426.

Zhao, Y., Araki, S., Wu, J., Teramoto, T., Chang, Y.F., Nakano, M., Abdelfattah, A.S., Fujiwara, M., Ishihara, T., Nagai, T. et al. (2011). An expanded palette of genetically encoded Ca^{2+} indicators. *Science 333*, 1888–1891.

Chapter 12

Histone Modification Sensors in Living Cells

Hiroshi Kimura and Yuko Sato

CONTENTS

12.1 INTRODUCTION

In eukaryotic cell nuclei, DNA is wrapped around a histone octamer, consisting of two copies each of the core histones (H2A, H2B, H3, and H4), to form a single nucleosome, which is the fundamental unit of chromatin (Luger et al. 1997). Posttranslational modifications of these core histones play an important role in genome function, including the regulation of gene expression, DNA damage recovery, and accurate chromosome segregation (Figure 12.1) (Bhaumik et al. 2007; Bannister and Kouzarides 2011; Greer and Shi 2012). In particular, acetylation and methylation on specific lysine residues are important for epigenetic gene regulation. Although only a single acetyl group is added on the primary amine on lysine residues, methylation can occur at three different levels: monomethylation (me1), dimethylation (me2), and trimethylation (me3). These modifications are added and removed by modification and demodification

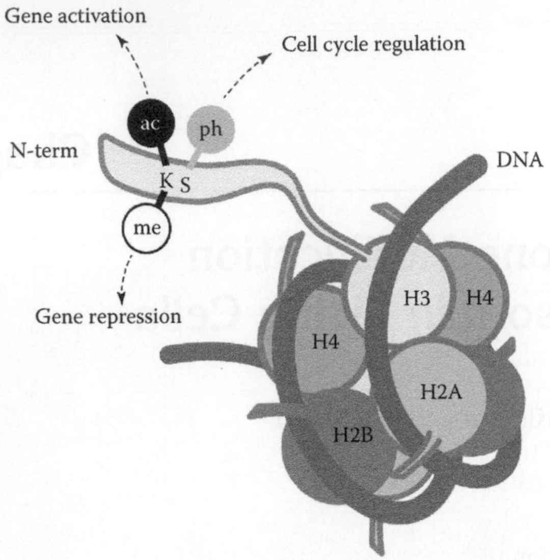

Figure 12.1 Nucleosome and histone modification. Schematic structure of a nucleosome, which consists of approximately 147 bp DNA and two copies of histones H2A, H2B, H3, and H4. Lysine residues on histone tails can be acetylated and methylated. In general, acetylation and methylation are associated with gene activation and repression, respectively, although methylations on some residues (such as H3K4) are associated with activation. Phosphorylation of serine and threonine residues is associated with cell cycle regulation and stress responses.

enzymes, respectively. For example, an acetyl group is added by histone acetyltransferase (HAT) and removed by histone deacetylase (HDAC) (Shahbazian and Grunstein 2007). Similarly, the methylation status is controlled by lysine (K) methyltransferase (KMT) and demethylase (KDM) (Dillon et al. 2005; Greer and Shi 2012). With the regulation of these enzymes, histone modifications change locally and globally during the cell cycle, development and differentiation, and in response to external stimuli. Some acetylations and phosphorylations turn over rapidly (Waterborg 2002; Clayton et al. 2006; Zheng et al. 2013), possibly playing a role in balancing transcription kinetics and signaling pathways. In contrast, some methylations are more stably maintained (Zee et al. 2010; Xu et al. 2012) and can be inheritable epigenetic markers, because core histones, particularly H3 and H4, bind stably with DNA over cell generations (Kimura and Cook 2001).

Acetylation of histone H3 occurs on many lysine residues, including K9 and K27, and is generally correlated with transcriptional activation together with methylation at K4, being localized to transcription start sites and/or enhancers of actively transcribed genes (Wang et al. 2008). In contrast, methylation on H3 at K9 and K27 is correlated with transcriptional repression (Barski et al. 2007). The effect of acetylation on histones could be due to the neutralization of positive charges, which may loosen the interaction with negatively charged DNA. Whereas some acetylation can subtly affect the biochemical properties of nucleosomes (Turner 1991; Hong et al. 1993), modifications on the histone N-terminal tails, which are extruded from the nucleosomes and are structurally disordered, can function as binding platforms for specific "reader" proteins (Yun et al. 2011). Indeed, many protein domains are known to bind to specifically modified forms of histones; for instance, bromodomains and chromodomains bind to acetylated and methylated histones, respectively (Filippakopoulos and Knapp 2012; Musselman et al. 2012). Proteins containing such domains mediate the transcriptional activation or repression of specific genes by binding to both modified histones and other regulatory proteins.

Because histone modifications mark epigenetic status, they are thought to be promising therapeutic targets (Biancotto et al. 2010; Mund and Lyko 2010; Zagni et al. 2013). In particular, tumorigenesis is often associated with mutations in and unregulated expression of genes involved in chromatin regulation, including histone modification and demodification enzymes. Indeed, two histone deacetylase inhibitors, Vorinostat (suberoylanilide hydroxamic acid [SAHA]) and Romidepsin (FK228), have been approved by the U.S. Food and Drug Administration as anticancer drugs for the treatment of cutaneous T-cell lymphoma, together with DNA methyltransferase inhibitors (New et al. 2012). Therefore, a high-throughput system to evaluate histone modification states would be beneficial for the development of future epigenome drugs.

These histone modifications are not necessarily prerequisite codes for determining the gene state for transcription, as some modifications are added as a consequence of transcription (Henikoff and Shilatifard 2011). Nevertheless, specific histone modifications can still serve as good epigenetic indicators of chromatin state (Kimura 2013). Thanks to the chromatin immunoprecipitation assay combined with genome-wide analysis based on microarray and deep sequencing, the epigenomic landscape of histone modifications has been uncovered in many different cell types (Barski et al. 2007; Wang 2008; The ENCODE Project Consortium 2012). However, the dynamics of histone modifications in single cells and living organisms has only just begun to emerge with the development of suitable monitoring methods (Kimura et al. 2010; Sasaki 2012). Unlike monitoring the general behavior of histone proteins, which can be marked by genetically fusing a fluorescent protein, tracking their modifications requires a separate probe. To monitor histone modifications in living cells, two types of probes have been developed (Table 12.1). One is for sensing the balance between modifying and demodifying enzymes, based on monitoring the modification state of a fluorescence/Förster resonance energy transfer (FRET) probe (Lin and Ting 2004; Lin et al. 2004; Sasaki et al. 2009; Ito 2011).

TABLE 12.1 SUMMARY OF DIFFERENT HISTONE MODIFICATION SENSORS

	Enzymatic Balance Sensor	Endogenous Modification Detector	
	FRET Probe	Labeled Fab	Mintbody
Genetically encoded	Yes	No	Yes
Requirements for probe development	Optimization of FRET efficiency	Specific IgG	Hybridoma (or ScFv library)
Spatial resolution	Low	Medium-high (depending on binding affinity)	Medium-high (depending on binding affinity)
Quantification method	FRET ratio change	FRAP, nucleus/ cytoplasm intensity ratio	FRAP, nucleus/ cytoplasm intensity ratio
Modifications detected	H3S28ph,[a] H3K9me3,[b] H3K27me3,[b] H4K5ac+K8ac,[c] H4K12ac[d]	H3S10ph,[e] H3K4me1,[f] H3K4me2,[f] H3K4me3,[f] H3K9ac,[f] H3K9me1,[f] H3K9me2,[f] H3K27ac,[f] H3K27me3[f]	H3K9ac[g]
Applications	Robust and sensitive detection of HDAC or HAT inhibitor activity in living cells	Spatiotemporal regulation of phosphorylation and acetylation in living cells and preimplantation embryos; Behavior of inactive X chromosomes during the cell cycle	Timing of acetylation in *Drosophila* embryos; Localizing acetylation on polytene chromosomes in *Drosophila* salivary glands

[a] Data from Lin, C. W., and Ting, A. Y., *Angew Chem. Int. Ed. Engl.* 43:2940–2943, 2004 (chromatin-free).
[b] Data from Lin, C. W. et al., *J Am Chem Soc.* 126:5982–5983, 2004 (chromatin-free).
[c] Data from Sasaki, K. et al., *Proc. Natl. Acad. Sci. U S A* 106:16257–16262, 2009 (chromatin-targeted).
[d] Data from Ito, T. et al., *Chem. Biol.* 18:495–507, 2011 (chromatin-targeted).
[e] Data from Hayashi-Takanaka, Y. et al., *J. Cell Biol.* 187:781–790, 2009.
[f] Data from Hayashi-Takanaka, Y. et al., *Nucleic Acids Res.* 39:6475–6488, 2011.
[g] Data from Sato, Y. et al., *Sci. Rep.* 3:2436, 2013.

Another is for detecting the modifications on endogenous histones using modification-specific antibody fragments (Hayashi-Takanaka et al. 2009, 2011; Sato et al. 2013).

12.2 FRET-BASED ENZYMATIC ACTIVITY SENSORS

By using FRET-based sensors, the levels of intracellular protein modification, or the balance between modifying and demodifying enzymes, can be monitored in living cells (VanEngelenburg and Palmer 2008; Aye-Han et al. 2009). These sensors use intramolecular FRET between a donor and an acceptor fluorescent protein (e.g., cyan fluorescent protein [CFP] and yellow fluorescent protein [YFP], respectively), which are typically connected with three components: a target amino acid sequence of a modifying enzyme, a flexible linker, and a modification-specific binder (Figure 12.2a). When the substrate amino acid is modified, it physically interacts with the specific binder, resulting in an intramolecular structural change that alters the distance and orientation of the two fluorescent proteins to induce or disturb FRET.

Several FRET sensors for histone modifications have been developed (Lin and Ting 2004; Lin et al. 2004; Sasaki et al. 2009; Ito 2011). In earlier studies, sensors for histone H3 phosphorylation and methylation were reported (Lin and Ting 2004; Lin et al. 2004). In the case of the H3 Ser28 phosphorylation sensor, the phospho-amino acid binding domain of 14-3-3 and the N-terminal 30 amino acids of H3 were inserted between CFP and YFP (Lin and Ting 2004). Similarly, histone H3 methylation sensors were generated by inserting a chromodomain (the methyl-histone binding domain of HP1 or Polycomb) and an H3 sequence containing the methylation site (K9 or K27) between two fluorescent proteins (Lin et al. 2004). The FRET signals from these sensors responded to the modification states both *in vitro* and in living cells, but these FRET sensors were not targeted to chromatin and their dynamic range was not very high.

Recently, chromatin-targeted histone H4 acetylation probes have also been developed (Figure 12.2b) (Sasaki et al. 2009). The original FRET sensor, named Histac, is designed for monitoring hyperacetylated H4 that harbors acetylation on both K5 and K8, providing a binding site for the bromodomain of BRDT (bromodomain testis-specific protein); the BRDT bromodomain and the full-length H4 (including the N-terminus containing K5 and K8) are inserted between CFP and Venus (a YFP variant with faster maturation kinetics). When K5 and K8 are acetylated, the BRDT bromodomain binds to the acetylated residues, resulting in an intramolecular conformational change. In this case, FRET signals were observed without acetylation and disappeared on acetylation (Figure 12.2b). This histone H4 hyperacetylation sensor has proven useful for monitoring the effects of HDAC and HAT inhibitors in living cells (Dancy et al. 2012). By using a

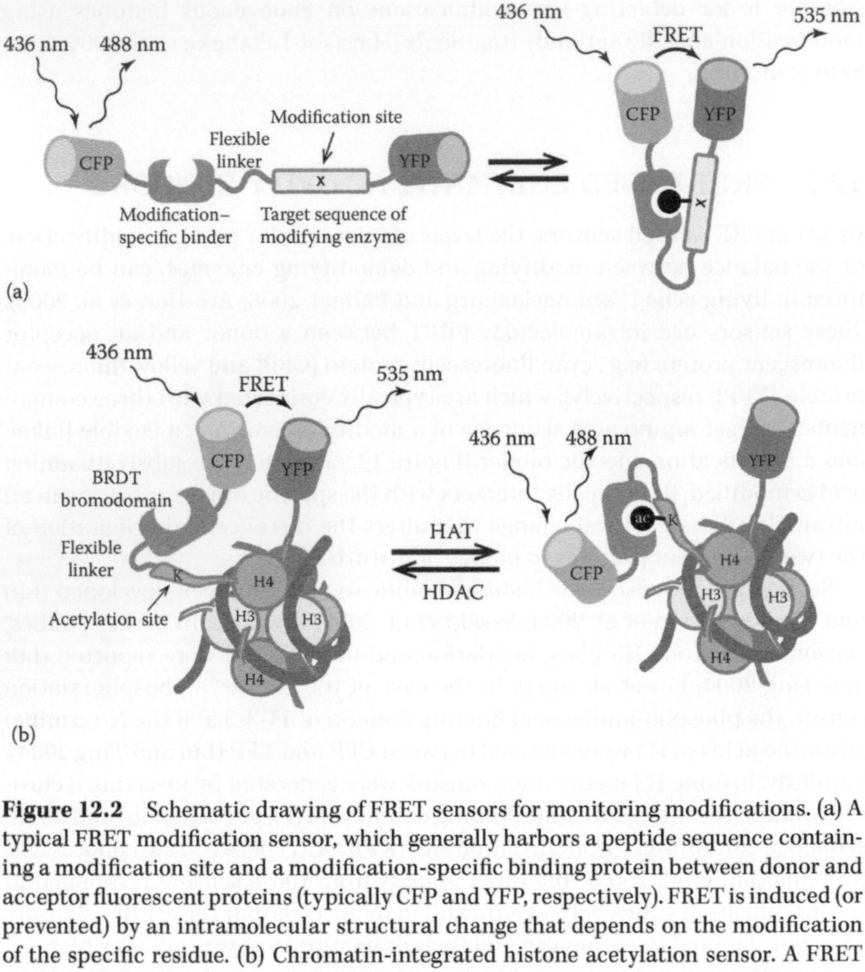

Figure 12.2 Schematic drawing of FRET sensors for monitoring modifications. (a) A typical FRET modification sensor, which generally harbors a peptide sequence containing a modification site and a modification-specific binding protein between donor and acceptor fluorescent proteins (typically CFP and YFP, respectively). FRET is induced (or prevented) by an intramolecular structural change that depends on the modification of the specific residue. (b) Chromatin-integrated histone acetylation sensor. A FRET sensor called Histac harbors a bromodomain (acetylation binder) and the full-length of histone H4 protein (containing acetylatable lysines) between CFP and YFP. In this case, FRET occurs in the absence of acetylation, and the interaction between the bromodomain and the acetylated lysine residue induces a configuration change that reduces FRET.

similar approach, another FRET sensor Histac-K12, which allows the monitoring of H4 K12 acetylation, has also been developed (Ito et al. 2011). In this probe, the bromodomain of BRD2 (bromodomain-containing 2), which preferentially binds to K12 acetylation, was used instead of the BRDT bromodomain. These FRET sensors contain the full-length histone H4 protein and are incorporated into chromatin as the endogenous H4, making it possible to monitor signals on chromatin. Importantly, these sensors are useful not only for monitoring histone modifications but also for screening small chemicals that bind to the bromodomain. While HAT and HDAC are potential targets for epigenome drugs, these enzymes generally show broad specificity and their inhibition may affect a number of downstream events. Specific inhibitors of bromodomain proteins, such as JQ1, I-BET, and PFI-1, can inhibit the expression of a subset of oncogenic genes more selectively and may form a new class of cancer drugs (Filippakopoulos et al. 2010; Delmore et al. 2011; Picaud et al. 2013).

12.3 SENSORS FOR ENDOGENOUS HISTONE MODIFICATIONS

12.3.1 Fab-Based Live Endogenous Modification Labeling

Endogenous histone modifications have been visualized in living cells using fluorescently labeled antigen binding fragments (Fabs) derived from modification-specific antibodies (Hayashi-Takanaka et al. 2009, 2011). In this Fab-based live endogenous modification labeling (FabLEM) technique, Fab fragments prepared from the whole immunoglobulin G (IgG) by protease digestion are labeled with a fluorescent dye and loaded into living cells (Figure 12.3a). Once loaded into cells, Fabs diffuse throughout the cytoplasm and enter into the nucleus via diffusion through the nuclear pore. If the target modification is present in the nucleus, Fab binds transiently to the modification. Although Fab repeatedly binds and unbinds to/from the target modification, its distribution reaches a steady state, being concentrated in regions where the modification is enriched over the diffuse background. As unbound Fabs can diffuse into the cytoplasm, the ratio of nuclear and cytoplasmic fluorescence intensity is correlated with the bound and unbound fractions (Figure 12.3b).

In principle, any histone modification can be monitored by FabLEM as long as the specific antibody is available. In addition to the specificity of the Fab, its binding affinity and the abundance of the target modifications are also critical for FabLEM. Empirically, antibodies with a dissociation constant (K_d) approximately 10^{-8} M *in vitro* at 25°C (as the IgG form) are adequate for FabLEM, but high-affinity antibodies with K_d of 10^{-11}–10^{-12} M will yield higher contrast images and will be generally more versatile (e.g., applicable to preimplantation embryo imaging; Hayashi-Takanaka et al. 2011). Fluorescence recovery after photobleaching (FRAP) experiments revealed that the residence time varies from <1 to >10 s, depending on the epitope binding affinity, in living cells. These binding times are shorter than *in vitro* dissociation rates at room temperature, probably because the high protein concentration (>100 mg/mL) in the nucleus can cause more frequent molecular collisions to force the dissociation of bound proteins.

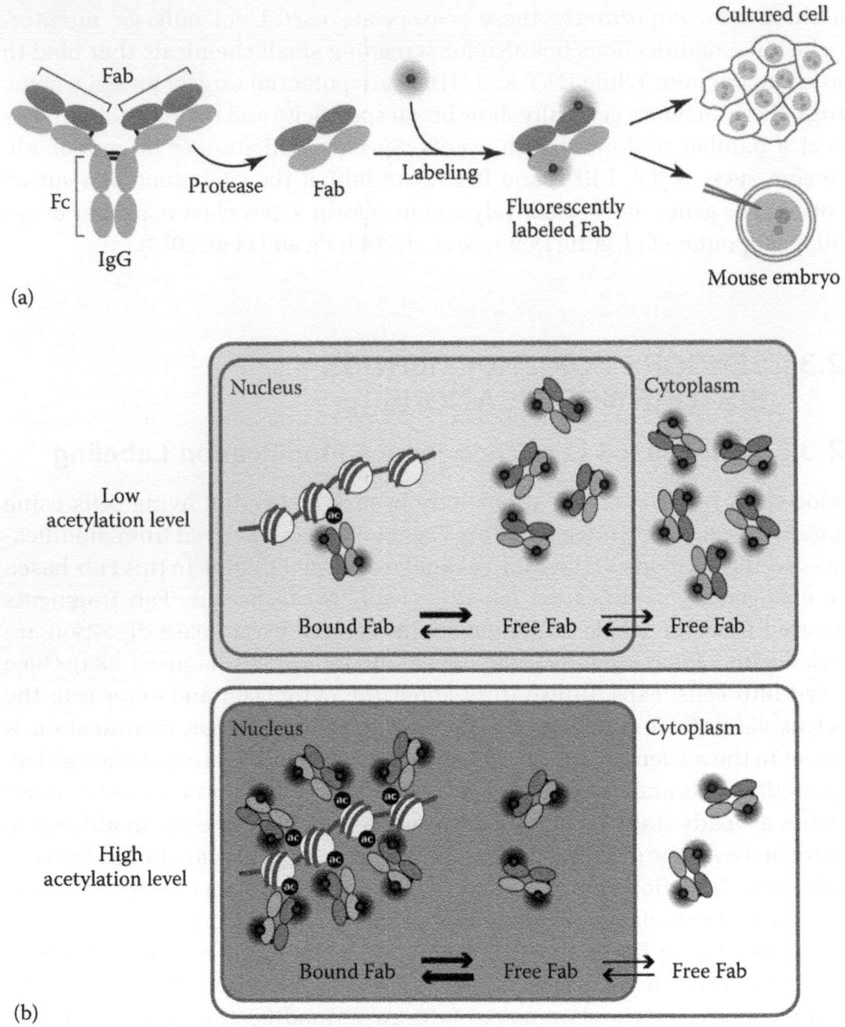

Figure 12.3 Fab-based modification sensors. (a) Schematic drawing of the Fab-based live endogenous modification labeling (FabLEM) technique. A fluorescently labeled Fab can be used as a modification sensor in living cells via direct loading into cultured cells and embryos. (b) Alterations in global modification levels can be monitored by FabLEM. The global modification level is correlated with the chromatin-bound fraction of a modification-specific Fab, and this can be measured by fluorescence recovery after photobleaching or by determining the nucleus/cytoplasm intensity ratio, as free Fab can diffuse into cytoplasm. When the modification level becomes high, the level of bound Fabs increases, resulting in a higher nucleus/cytoplasm intensity ratio.

(Continued)

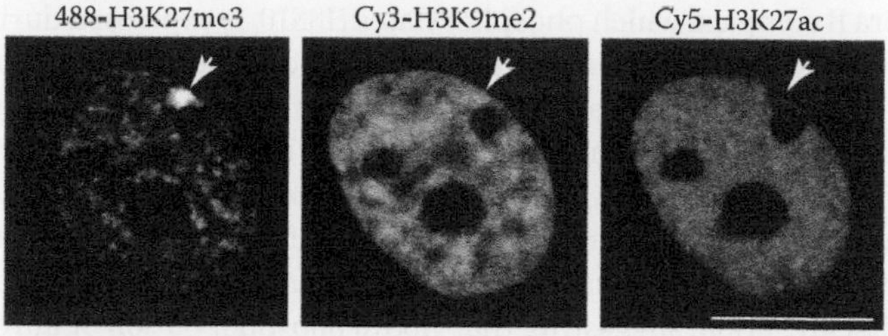

Figure 12.3 (Continued) Fab-based modification sensors. (c) Simultaneous detection of different histone modifications in single living cells. Telomerase-immortalized human epithelial cells were loaded with three Fabs (Alexa488-conjugated anti-H3K27me3, Cy3-conjugated anti-H3K9me2, and Cy5-conjugated anti-H3K27ac), and fluorescence images were collected using a confocal microscope. Distinct localizations are observed in living cells. Arrows indicate an inactive X chromosome, where H3K27me3 and H3K27ac are enriched and excluded, respectively. Bar = 10 μm.

Therefore, Fabs loaded into cells do not permanently block the access of endogenous binding proteins to the modification sites. Indeed, Fabs loaded into cells at approximately 1 μM (corresponding to approximately 3×10^6 molecules in a cell) did not disturb cell division (Hayashi-Takanaka et al. 2011). Moreover, mouse preimplantation embryos that were injected with Fabs and imaged under a fluorescence microscope developed normally to birth. Thus, the endogenous modifications can be tracked by FabLEM without affecting cell function.

As fluorescent dyes are directly conjugated to Fab fragments, the distribution of several different modifications could be visualized using multiple Fabs labeled with different dyes. For example, three different histone modifications can be visualized simultaneously using Alexa Fluor 488, Cy3, and Cy5 (Figure 12.3c). Once loaded into cells, Fab molecules are quite stable, and their localization can be followed for a few days until the signal becomes too diluted by cell division or is trapped in cytoplasmic vesicles.

12.3.2 Histone Modification Dynamics as Revealed by FabLEM

By using FabLEM, the spatiotemporal dynamics of histone H3 Ser10 (H3S10) phosphorylation were monitored in living human cells and mouse preimplantation embryos (Hayashi-Takanaka et al. 2009). H3S10 is known to be massively phosphorylated on mitotic chromosomes (Johansen and Johansen 2006). Consistent with this, H3S10 phosphorylation (H3S10ph)-specific Fabs were uniformly distributed in most interphase cells and became concentrated on condensed chromosomes in mitotic cells. Interestingly, the regulation of H3S10 phosphorylation before mitosis differs in normal and cancerous cells, whereas the aurora B complex, which phosphorylates H3S10, is expressed during S phase in both cell types. In normal cells, H3S10ph foci appear several hours before mitosis; in contrast, in cancer cells foci formed just before chromosome condensation. Treatments with kinase and phosphatase inhibitors suggest that the level of protein phosphatase activity is higher in cancerous cells than in normal diploid fibroblasts. Live-cell imaging using FabLEM further showed that transiently inhibiting aurora B in normal cells during interphase caused higher chromosome missegregation, suggesting that H3S10 phosphorylation or aurora B activity before mitosis is essential for accurate chromosome segregation.

FabLEM was also used to track the dynamics of inactive X chromosomes (Xi) during the cell cycle (Hayashi-Takanaka et al. 2011). Because Xi is enriched in histone H3 Lys27 trimethylation (H3K27me3), the concentration of H3K27me3-specific Fab serves as a good indicator of Xi in living cells (Chadwick and Willard 2003). Live imaging with a fluorescently labeled replication protein—proliferating cell nuclear antigen (PCNA; Leonhardt et al. 2000)—confirmed that human Xi replicates late in S phase. Although Xi was suggested to move dynamically from the nuclear periphery to the perinucleolus in association with DNA replication (Zhang et al. 2007), no such intracellular movement was visible by live imaging (Hayashi-Takanaka et al. 2011). This suggests that Xi replicates just like autosomes and its perinucleolar targeting is not essential for maintaining the inactive state.

Beyond the ability to detect the concentration of modifications in specific foci, FabLEM also allows for monitoring physiological changes in global modification levels at the single-cell level (Figure 12.3b) (Hayashi-Takanaka et al. 2011). If target modification levels increase, the chromatin-bound fraction of Fab also increases as a result of an enhanced binding association rate. This can be measured by FRAP, or the ratio of the nuclear to cytoplasmic fluorescence intensity, as the nuclear and cytoplasmic intensities are correlated with the amount of bound and freely diffusing unbound Fab, respectively. For example, when cells were treated with the HDAC inhibitor trichostatin A (TSA), which induces hyperacetylation, the nucleus/cytoplasm intensity ratio of histone H3 acetylation-specific Fab increased within 30 min. After the removal of TSA, Fab became diffuse throughout the cytoplasm, and the original distribution was restored. The effects of TSA on histone methylation were also monitored by various methylation-specific antibodies. Thus, changes in histone modification levels can be sensed with fluorescently labeled Fab probes.

12.3.3 Modification-Specific Intracellular Antibodies as *In Vivo* Sensors

FabLEM requires purified Fabs and their direct loading into cells, which prevents long-term and organism-level experiments, as well as high-throughput assays. To overcome these limitations due to the requirement for protein injection, a genetically encoded system using a single-chain variable fragment (scFv; Chames 2012) has been developed (Sato et al. 2013). An scFv was cloned from hybridoma cells producing the specific antibody against histone H3 Lys9 acetylation (H3K9ac) and then genetically fused with the enhanced green fluorescent protein (EGFP) to create a modification-specific intracellular antibody, or mintbody (Figure 12.4a).

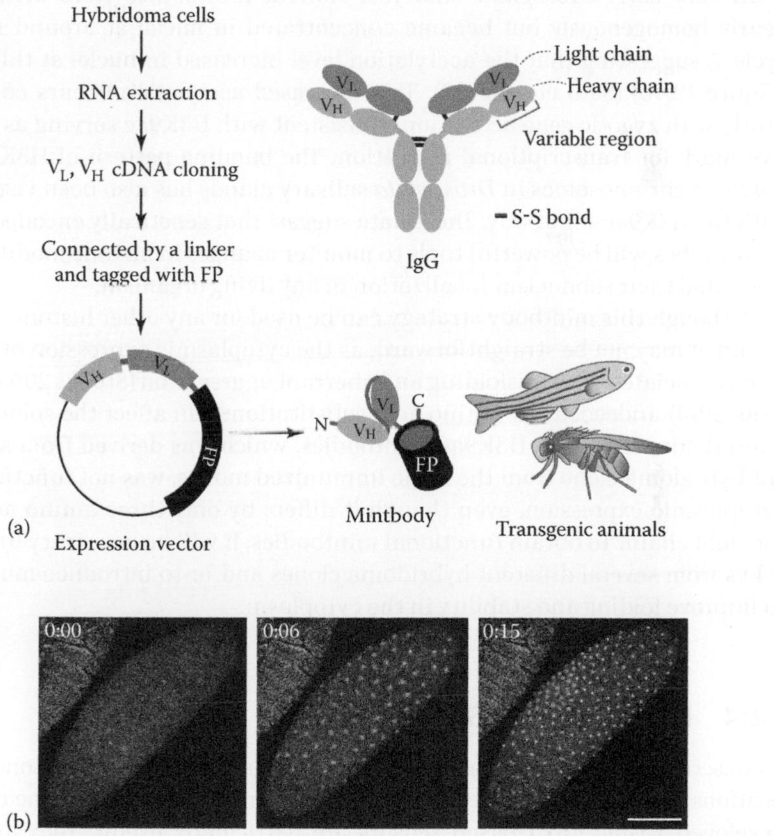

Figure 12.4 Modification-specific intracellular antibody (mintbody). (a) Genetically encoded system to track modifications in living cells and organisms. To construct an expression vector for a modification-specific intracellular antibody (mintbody), cDNA encoding the variable regions of the IgG heavy and light chains (V_H and V_L, respectively) are cloned and fused with a fluorescent protein (FP) gene. The resulting construct can be used to generate transgenic animals. (b) H3K9ac-mintbody imaging in live *Drosophila* embryos. Increases in H3K9ac levels can be monitored by nuclear enrichment of the H3K9ac-mintbody.

The H3K9ac-specific mintbody (H3K9ac-mintbody) has proven to be a useful acetylation sensor in living cells. The H3K9ac-mintbody binds to the target acetylation site, and changes in acetylation levels can be monitored by FRAP or through the nuclear/cytoplasmic intensity ratio, just like FabLEM. Moreover, transgenic animals that express the H3K9ac-mintbody have been generated. So far, *Drosophila* and zebrafish expressing the H3K9ac-mintbody have been shown to develop normally and remain fertile.

In very early *Drosophila* embryos, H3K9ac-mintbodies were distributed nearly homogenously but became concentrated in nuclei at around mitotic cycle 7, suggesting that the acetylation level increased in nuclei at this stage (Figure 12.4b) (Sato et al. 2013). This increased acetylation occurs concomitantly with zygotic gene activation, consistent with H3K9ac serving as a positive mark for transcriptional activation. The banding pattern of H3K9ac on polytene chromosomes in *Drosophila* salivary glands has also been visualized with the H3K9ac-mintbody. These data suggest that genetically encoded mintbody probes will be powerful tools to monitor changes in histone modification levels and their subnuclear localization in any living organism.

Although this mintbody strategy can be used for any other histone modification, it may not be straightforward, as the cytoplasmic expression of scFv is often associated with misfolding and aberrant aggregation (Stocks 2005; Kvam et al. 2010). Indeed, a few amino acid substitutions can affect the solubility of a mintbody; one of the H3K9ac-mintbodies, which was derived from a different hybridoma clone from the same immunized mouse, was not functional on cytoplasmic expression, even though it differs by only three amino acids on the light chain. To obtain functional mintbodies, it will be necessary to screen scFvs from several different hybridoma clones and/or to introduce mutations to improve folding and stability in the cytoplasm.

12.4 PERSPECTIVES

As described here, several sensors are now available to monitor histone modifications in living cells (Table 12.1). To monitor enzymatic balance, the recently developed Histac FRET-based sensors are particularly useful. They are integrated into chromatin and sensitively detect the effect of small molecule inhibitors against acetylases and deacetylases. To detect endogenous modifications directly, protein-based and genetically encoded systems (i.e., FabLEM and mintbody) are now also available. Direct injection of specific, fluorescently labeled Fab fragments is a versatile and convenient technique for monitoring changes in the localization and levels of specific modifications in cultured cells and in mouse preimplantation embryos. In addition, genetically encoded mintbodies have now also opened new opportunities to visualize these changes in living organisms.

These technologies could be used to monitor how epigenetic modifications are dynamically regulated both *in vitro* and *in vivo* under various conditions, such as during cell differentiation and dedifferentiation, and in response to stress. In addition, these sensors will also be valuable tools for screening and evaluating potential diagnostic inhibitors that affect epigenetic modifications. Further development of histone modification sensors will facilitate future studies on epigenetic regulation related to development, differentiation, and disease.

ACKNOWLEDGMENTS

We thank Timothy J. Stasevich and Yoko Hayashi-Takanaka for their useful comments on the manuscript. The authors' work was supported by grants-in-aid from the Ministry of Education, Culture, Sports, Science and Technology (MEXT) of Japan.

REFERENCES

Aye-Han, N.N., Ni, Q., and Zhang, J. 2009. Fluorescent biosensors for real-time tracking of post-translational modification dynamics. *Curr Opin Chem Biol*, **13**, 392–397.

Bannister, A.J., and Kouzarides, T. 2011. Regulation of chromatin by histone modifications. *Cell Res*, **21**, 381–395.

Barski, A., Cuddapah, S., Cui, K., Roh, T.Y., Schones, D.E., Wang, Z., Wei, G., Chepelev, I., and Zhao, K. 2007. High-resolution profiling of histone methylations in the human genome. *Cell*, **129**, 823–837.

Bhaumik, S.R., Smith, E., and Shilatifard, A. 2007. Covalent modifications of histones during development and disease pathogenesis. *Nat Struct Mol Biol*, **14**, 1008–1016.

Biancotto, C., Frigè, G., and Minucci, S. 2010. Histone modification therapy of cancer. *Adv Genet*, **70**, 341–386.

Chadwick, B.P., and Willard, H.F. 2003. Chromatin of the Barr body: Histone and non-histone proteins associated with or excluded from the inactive X chromosome. *Hum Mol Genet*, **12**, 2167–2178.

Chames, P. 2012. *Antibody Engineering: Methods and Protocols*, Second Edition. *Methods in Molecular Biology*, **907**. New York: Humana Press.

Clayton, A.L., Hazzalin, C.A., and Mahadevan, L.C. 2006. Enhanced histone acetylation and transcription: A dynamic perspective. *Mol Cell*, **23**, 289–296.

Dancy, B.M., Crump, N.T., Peterson, D.J., Mukherjee, C., Bowers, E.M., Ahn, Y.H., Yoshida, M., Zhang, J., Mahadevan, L.C., Meyers, D.J., Boeke, J.D., and Cole, P.A. 2012. Live-cell studies of p300/CBP histone acetyltransferase activity and inhibition. *ChemBioChem*, **13**, 2113–2121.

Delmore, J.E., Issa, G.C., Lemieux, M.E., Rahl, P.B., Shi, J., Jacobs, H.M., Kastritis, E., Gilpatrick, T., Paranal, R.M., Qi, J., Chesi, M., Schinzel, A.C., McKeown, M.R., Heffernan, T.P., Vakoc, C.R., Bergsagel, P.L., Ghobrial, I.M., Richardson, P.G., Young, R.A., Hahn, W.C.,

Anderson, K.C., Kung, A.L., Bradner, J.E., and Mitsiades, C.S. 2011. BET bromodomain inhibition as a therapeutic strategy to target c-Myc. *Cell*, **146**, 904–917.

Dillon, S.C., Zhang, X., Trievel, R.C., and Cheng, X. 2005. The SET-domain protein superfamily: Protein lysine methyltransferases. *Genome Biol*, **6**, 227.

Filippakopoulos, P., and Knapp, S. 2012. The bromodomain interaction module. *FEBS Lett*, **586**, 2692–2704.

Filippakopoulos, P., Qi, J., Picaud, S., Shen, Y., Smith, W.B., Fedorov, O., Morse, E.M., Keates, T., Hickman, T.T., Felletar, I., Philpott, M., Munro, S., McKeown, M.R., Wang, Y., Christie, A.L., West, N., Cameron, M.J., Schwartz, B., Heightman, T.D., La Thangue, N., French, C.A., Wiest, O., Kung, A.L., Knapp, S., and Bradner, J.E. 2010. Selective inhibition of BET bromodomains. *Nature*, **468**, 1067–1073.

Greer, E.L., and Shi, Y. 2012. Histone methylation: A dynamic mark in health, disease and inheritance. *Nat Rev Genet*, **13**, 343–357.

Hayashi-Takanaka, Y., Yamagata, K., Nozaki, N., and Kimura, H. 2009. Visualizing histone modifications in living cells: Spatiotemporal dynamics of H3 phosphorylation during interphase. *J Cell Biol*, **187**, 781–790.

Hayashi-Takanaka, Y., Yamagata, K., Wakayama, T., Stasevich, T.J., Kainuma, T., Tsurimoto, T., Tachibana, M., Shinkai, Y., Kurumizaka, H., Nozaki, N., and Kimura, H. 2011. Tracking epigenetic histone modifications in single cells using Fab-based live endogenous modification labeling. *Nucleic Acids Res*, **39**, 6475–6488.

Henikoff, S., and Shilatifard, A. 2011. Histone modification: Cause or cog? *Trends Genet*, **27**, 389–396.

Hong, L., Schroth, G.P., Matthews, H.R., Yau, P., and Bradbury, E.M. 1993. Studies of the DNA binding properties of histone H4 amino terminus. Thermal denaturation studies reveal that acetylation markedly reduces the binding constant of the H4 "tail" to DNA. *J Biol Chem*, **268**, 305–314.

Ito, T., Umehara, T., Sasaki, K., Nakamura, Y., Nishino, N., Terada, T., Shirouzu, M., Padmanabhan, B., Yokoyama, S., Ito, A., and Yoshida, M. 2011. Real-time imaging of histone H4K12-specific acetylation determines the modes of action of histone deacetylase and bromodomain inhibitors. *Chem Biol*, **18**, 495–507.

Johansen, K.M., and Johansen, J. 2006. Regulation of chromatin structure by histone H3S10 phosphorylation. *Chromosome Res*, **14**, 393–404.

Kimura, H. 2013. Histone modifications for human epigenome analysis. *J Hum Genet*, **58**, 439–445.

Kimura, H., and Cook, P.R. 2001. Kinetics of core histones in living human cells: Little exchange of H3 and H4 and some rapid exchange of H2B. *J Cell Biol*, **153**, 1341–1353.

Kimura, H., Hayashi-Takanaka, Y., and Yamagata, K. 2010. Visualization of DNA methylation and histone modifications in living cells. *Curr Opin Cell Biol*, **22**, 412–418.

Kvam, E., Sierks, M.R., Shoemaker, C.B., and Messer, A. 2010. Physico-chemical determinants of soluble intrabody expression in mammalian cell cytoplasm. *Protein Eng Des Sel*, **23**, 489–498.

Leonhardt, H., Rahn, H.P., Weinzierl, P., Sporbert, A., Cremer, T., Zink, D., and Cardoso, M.C. 2000. Dynamics of DNA replication factories in living cells. *J Cell Biol*, **149**, 271–280.

Lin, C.W., Jao, C.Y., and Ting, A.Y. 2004. Genetically encoded fluorescent reporters of histone methylation in living cells. *J Am Chem Soc*, **126**, 5982–5983.

Lin, C.W., and Ting, A.Y. 2004. A genetically encoded fluorescent reporter of histone phosphorylation in living cells. *Angew Chem Int Ed Engl*, **43**, 2940–2943.

Luger, K., Mäder, A.W., Richmond, R.K., Sargent, D.F., and Richmond, T.J. 1997. Crystal structure of the nucleosome core particle at 2.8 A resolution. *Nature*, **389**, 251–260.

Mund, C., and Lyko, F. 2010. Epigenetic cancer therapy: Proof of concept and remaining challenges. *Bioessays*, **32**, 949–957.

Musselman, C.A., Lalonde, M.E., Cote, J., and Kutateladze, T.G. 2012. Perceiving the epigenetic landscape through histone readers. *Nat Struct Mol Biol*, **19**, 1218–1227.

New, M., Olzscha, H., and La Thangue, N.B. 2012. HDAC inhibitor-based therapies: Can we interpret the code? *Mol Oncol*, **6**, 637–656.

Picaud, S., Da Costa, D., Thanasopoulou, A., Filippakopoulos, P., Fish, P.V., Philpott, M., Fedorov, O., Brennan, P., Bunnage, M.E., Owen, D.R., Bradner, J.E., Taniere, P., O'Sullivan, B., Müller, S., Schwaller, J., Stankovic, T., and Knapp, S. 2013. PFI-1: A highly selective protein interaction inhibitor, targeting BET Bromodomains. *Cancer Res*, **73**, 3336–3346.

Sasaki, K., Ito, A., and Yoshida, M. 2012. Development of live-cell imaging probes for monitoring histone modifications. *Bioorg Med Chem*, **20**, 1887–1892.

Sasaki, K., Ito, T., Nishino, N., Khochbin, S., and Yoshida, M. 2009. Real-time imaging of histone H4 hyperacetylation in living cells. *Proc Natl Acad Sci U S A*, **106**, 16257–16262.

Sato, Y., Mukai, M., Ueda, J., Muraki, M., Stasevich, T.J., Horikoshi, N., Kujirai, T., Kita, H., Kimura, T., Hira, S., Okada, Y., Hayashi-Takanaka, Y., Obuse, C., Kurumizaka, H., Kawahara, A., Yamagata, K., Nozaki, N., and Kimura, H. 2013. Genetically encoded system to track histone modification in vivo. *Sci Rep*, **3**, 2436.

Shahbazian, M.D., and Grunstein, M. 2007. Functions of site-specific histone acetylation and deacetylation. *Annu Rev Biochem*, **76**, 75–100.

Stocks, M. 2005. Intrabodies as drug discovery tools and therapeutics. *Curr Opin Chem Biol*, **9**, 359–365.

The ENCODE Project Consortium. 2012. An integrated encyclopedia of DNA elements in the human genome. *Nature*, **489**, 57–74.

Turner, B.M. 1991. Histone acetylation and control of gene expression. *J Cell Sci*, **99**, 13–20.

VanEngelenburg, S.B., and Palmer, A.E. 2008. Fluorescent biosensors of protein function. *Curr Opin Chem Biol*, **12**, 60–65.

Wang, Z., Zang, C., Rosenfeld, J.A., Schones, D.E., Barski, A., Cuddapah, S., Cui, K., Roh, T.Y., Peng, W., Zhang, M.Q., and Zhao, K. 2008. Combinatorial patterns of histone acetylations and methylations in the human genome. *Nat Genet*, **40**, 897–903.

Waterborg, J.H. 2002. Dynamics of histone acetylation in vivo. A function for acetylation turnover? *Biochem Cell Biol*, **80**, 363–378.

Xu, M., Wang, W., Chen, S., and Zhu, B. 2012. A model for mitotic inheritance of histone lysine methylation. *EMBO Rep*, **13**, 60–67.

Yun, M., Wu, J., Workman, J.L., and Li, B. 2011. Readers of histone modifications. *Cell Res*, **21**, 564–578.

Zagni, C., Chiacchio, U., and Rescifina, A. 2013. Histone methyltransferase inhibitors: Novel epigenetic agents for cancer treatment. *Curr Med Chem*, **20**, 167–185.

Zee, B.M., Levin, R.S., Xu, B., LeRoy, G., Wingreen, N.S., and Garcia, B.A. 2010. In vivo residue-specific histone methylation dynamics. *J Biol Chem*, **285**, 3341–3350.

Zhang, L.-F., Huynh, K.D. and Lee, J.T. 2007. Perinucleolar targeting of the inactive X during S phase: Evidence for a role in the maintenance of silencing. *Cell*, **129**, 693–706.

Zheng, Y., Thomas, P.M., and Kelleher, N.L. 2013. Measurement of acetylation turnover at distinct lysines in human histones identifies long-lived acetylation sites. *Nat Commun*, **4**, 2203.

Section III

Beyond Live-Cell Tracking

Sensors for Diverse Applications

Chapter 13

Chemical Probes for Fluorescence Imaging in Living Mice

Toshiyuki Kowada, Hiroki Maeda, and Kazuya Kikuchi

CONTENTS

13.1 INTRODUCTION

Recently, various fluorescent probes have been developed for noninvasive imaging in living mice. Among them, near-infrared (NIR) organic fluorescent dyes have been shown to exhibit fewer effects on viability than have fluorescent quantum dots, which can be cytotoxic (Yong et al. 2013). Indocyanine green is a clinically approved dye for diagnostic purposes. NIR light can penetrate fairly deep into tissue, because light absorption, light scattering, and autofluorescence in tissues are minimal in the NIR window (660–900 nm) (Weissleder 2001; Weissleder and Ntziachristos 2003). In addition to the appropriate emitting light wavelength, there are other characteristics for imaging probes, such as high aqueous solubility, high quantum yield, high photostability, and optimal clearance. Furthermore, probe delivery into target cells or tissues should be the most important factor for *in vivo* imaging. For this type of imaging, the target cells are required to express the characteristic molecules, enzymes, or receptors.

For *in vivo* tumor imaging, many types of NIR probes have been developed to date. Specifically, selective binding ligands for receptors, antibodies, or peptide sequences have been used to target characteristic markers that were overexpressed in tumor cells (Becker et al. 2001; Tung et al. 2002; Ke et al. 2003; Chen et al. 2004; Cheng et al. 2005; Garanger et al. 2005; Razkin et al. 2006; Wu et al. 2006; Barrett et al. 2007; Kovar et al. 2007, 2009; Ogawa et al. 2009a, b; Zhou et al. 2009; Kossodo et al. 2010; Uddin et al. 2010; Koide et al. 2012; Tafreshi et al. 2012) and in the angiogenic endothelium (Citrin et al. 2004; Backer et al. 2007). For instance, epidermal growth factor (Ke et al. 2003; Kovar et al. 2007), folic acid (Tung et al. 2002), RGD (Arg-Gly-Asp) peptides (Chen et al. 2004; Cheng et al. 2005; Garanger et al. 2005; Razkin et al. 2006; Wu et al. 2006), vascular endothelial growth factor (Backer et al. 2007), and 2-deoxyglucose (Kovar et al. 2009; Zhou et al. 2009) have been employed as targeting ligands. In other cases, phosphatidylserine, which is exposed on the outer cell membrane of apoptotic cells, was targeted for the visualization of tumor apoptosis using Cy5.5-labeled annexin V (Petrovsky et al. 2003; Ntziachristos et al. 2004).

In this chapter, we have omitted the reports on tumor imaging using these fluorophore–ligand conjugates (Luo et al. 2011), and we mainly describe the chemical fluorescent probes that target hydroxyapatite or enzymes functioning in living mice. In particular, we focused on activatable probes because of their advantages compared to conventional always-on imaging agents. For quantum dots or multimodal probe systems, readers are referred to other reviews (Louie 2010; Zrazhevskiy et al. 2010; Sapsford et al. 2013; Yong et al. 2013). In addition, a recent review summarizing the chemical probes for matrix metalloproteinases (MMPs) is also available (Knapinska and Fields 2012).

13.2 SELF-QUENCHED FLUORESCENT PROBES

Weissleder et al. have enthusiastically studied the development of self-quenched fluorescent probes for *in vivo* imaging (Figure 13.1). These probes are based on graft polymers, in which tethered fluorophores are essentially quenched owing to locally high concentrations of the fluorophores. In this system, additional quencher molecules are not required because the fluorophores themselves can act as quenchers. Thus, a simple molecular design seems to be quite feasible for developing *in vivo* imaging agents for a variety of applications.

A graft copolymer, consisting of a poly-L-lysine (PLL) backbone, methoxypolyethylene glycol (MPEG), and a Cy5.5 fluorophore, was used to image tumor-associated lysosomal protease activity in a xenograft mouse model (Figure 13.1a) (Weissleder et al. 1999). Before enzymatic cleavage of the polymer

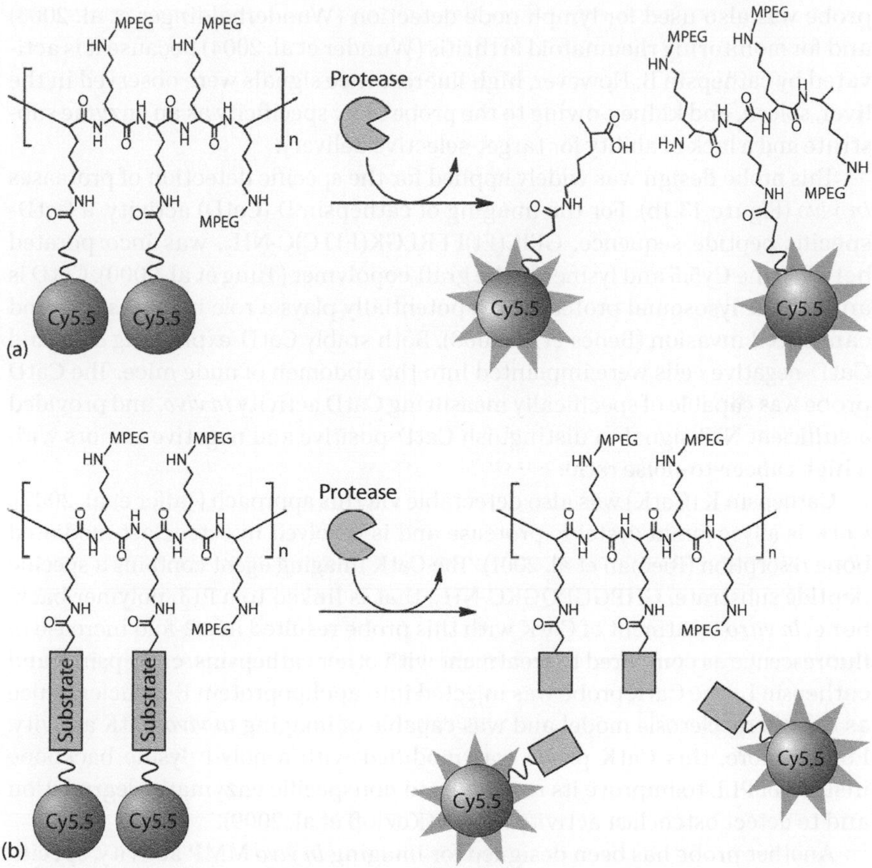

Figure 13.1 Schematic illustration of self-quenched fluorescent probes (a) without or (b) with additional protease substrate peptide linkers.

backbone, NIR fluorescence was self-quenched owing to the high local concentration of Cy5.5. *In vitro* treatment with trypsin cleaved the bonds next to the carboxylic end of the unmodified lysine and led to the release of the fluorophores. Furthermore, various cancer cell lines internalized the graft polymer by fluid-phase endocytosis. Enzyme inhibition experiments revealed that the NIR signal can be generated by active lysosomal cysteine/serine proteases such as cathepsin B, H, and L. At 24 h post-injection in a tumor-bearing mouse, 16% of the total injected graft copolymer was still present in the circulating blood. A high target-to-background ratio was achieved as a result of the protease-activatable property of the probe, despite its prolonged circulation time. This probe was also used for lymph node detection (Wunderbaldinger et al. 2003) and for monitoring rheumatoid arthritis (Wunder et al. 2004) because it is activated by cathepsin B. However, high fluorescence signals were observed in the liver, spleen, and kidney, owing to the probe's low specificity as an enzyme substrate and a lack of ability for target-selective delivery.

This probe design was widely applied for the specific detection of proteases *in vivo* (Figure 13.1b). For the imaging of cathepsin D (CatD) activity, a CatD-specific peptide sequence, GPIC(Et)FFRLGK(FITC)C-NH_2, was incorporated between the Cy5.5 and lysines of the graft copolymer (Tung et al. 2000). CatD is an aspartic lysosomal protease that potentially plays a role in metastasis and cancer cell invasion (Benes et al. 2008). Both stably CatD-expressing cells and CatD-negative cells were implanted into the abdomen of nude mice. The CatD probe was capable of specifically measuring CatD activity *in vivo*, and provided a sufficient NIR signal to distinguish CatD-positive and negative tumors with a high cancer-to-noise ratio.

Cathepsin K (CatK) was also detectable via this approach (Jaffer et al. 2007). CatK is a lysosomal cysteine protease and is involved in osteoclast-mediated bone resorption (Rieman et al. 2001). The CatK imaging agent contains a specific peptide substrate, GHPGGPQGKC-NH_2, that is linked to a PLL polymer backbone. *In vitro* treatment of CatK with this probe resulted in a 2-fold increase in fluorescence as compared to treatment with other cathepsins, cathepsin B and cathepsin L. The CatK probe was injected into apolipoprotein E–deficient mice as an atherosclerosis model and was capable of imaging *in vivo* CatK activity. Furthermore, this CatK probe was modified with a poly-D-lysine backbone instead of PLL to improve its resistance to nonspecific enzymatic degradation and to detect osteoclast activity *in vivo* (Kozloff et al. 2009).

Another probe has been designed for imaging *in vivo* MMP activity, specifically MMP-2 (Bremer et al. 2001a,b). This probe contained an MMP-2-cleavable peptide, GPLGVRGK(FITC)C-NH_2, which could also be cleaved to some extent by other MMPs, such as MMP-1, 7, 8, and 9. HT1080 human fibrosarcoma cells and BT20 tumor cells were used for *in vivo* experiments, because the former cells express high levels of MMP-2, while the latter cells express low levels of MMP-2. The MMP-sensitive probe was injected into HT1080- or BT20-tumor–bearing mice and resulted in intense fluorescence signals in the HT1080-tumor–bearing mouse. In addition, a control probe that contained a scrambled peptide sequence showed no significant fluorescence *in vivo*. It was further confirmed that the MMP-sensitive probe had the potential to assess the inhibitory effects of prinomastat, an MMP inhibitor.

The peptide substrate GWEHDGK(FITC)C-NH_2, which is selectively cleaved by caspase-1, was used to detect caspase-1 activity in living mice (Messerli et al. 2004). Caspase-1 belongs to a family of cysteinyl aspartate–specific proteases that process the inactive precursors of interleukin-1β and -18 and initiate a proinflammatory cell death program (Lavrik et al. 2005). The specificity of the probe for caspase-1 was verified by *in vitro* incubation with caspase-1 or caspase-3, as well as by an inhibitor assay. In an experiment using tumor-implanted mice, there was significantly increased fluorescence in caspase-1–expressing tumor tissue compared with tissues without caspase-1 expression.

A self-quenched probe with the specific peptide substrate GVSQNYPIVGK (FITC)C-NH_2 was also applicable for the detection of HIV-1 protease (PR) activity (Shah et al. 2004). HIV-1PR is an aspartyl protease that is required for viral infectivity (Navia et al. 1989). To image the HIV-1PR activity *in vivo*, human glioma cells were injected into mice abdomens, followed by intratumoral injection of the HIV-1PR amplicon vector. Greater than 4-fold higher NIR fluorescence was observed from HIV-1PR–expressing tumors compared with control tumors.

13.3 ACTIVITY-BASED PROBES

Activity-based probes (ABPs) covalently bind to the catalytic site of the target enzyme and have been widely applied in activity-based protein profiling (Cravatt et al. 2008; Heal et al. 2011; Deu et al. 2012; Li et al. 2012). Although ABPs are powerful tools to study proteases, owing to their binding specificities, a lack of signal amplification is the primary shortcoming for *in vivo* applications.

Recently, for the noninvasive monitoring of cathepsin activity *in vivo*, Bogyo and co-workers developed quenched near-infrared fluorescent ABPs (qNIRF-ABPs), which contained a dipeptide, a Cy5 fluorophore, a QSY7 quenching group, and an acyloxymethyl ketone (AOMK) warhead (Figure 13.2) (Blum et al. 2007, 2009). Cell studies have shown that the peptide AOMK scaffold is highly selective for cathepsin B and L (Blum et al. 2005). The qNIRF-ABP GB137 was intravenously injected into tumor-bearing mice via the tail vein. The probe fluorescence was detected in tumors at early time points and reached maximum levels by 6–8 h after probe injection. The rapid detectability was probably due to the quenching moiety, while the signal-to-background ratio of quenched GB137 was similar to that of the nonquenched probe. To confirm that the fluorescence signals in tumor tissues were derived from cathepsin activity, the cysteine protease inhibitor K11777 was used (Caffrey 2007). K11777 treatment reduced the probe signals in tumors. Thus, it is expected that GB137 can be applied to monitor the *in vivo* efficacy of protease inhibitors. Moreover, GB137 was compared with a commercially available probe, Prosense® 750, which is a polymer-based reagent and is also activated by cathepsins (Blum et al. 2009). As a result, GB137 exhibited more rapid and specific activation in tumor-bearing mice, while the fluorescence signals from ProSense 750 were detectable only at 8 h after probe injection.

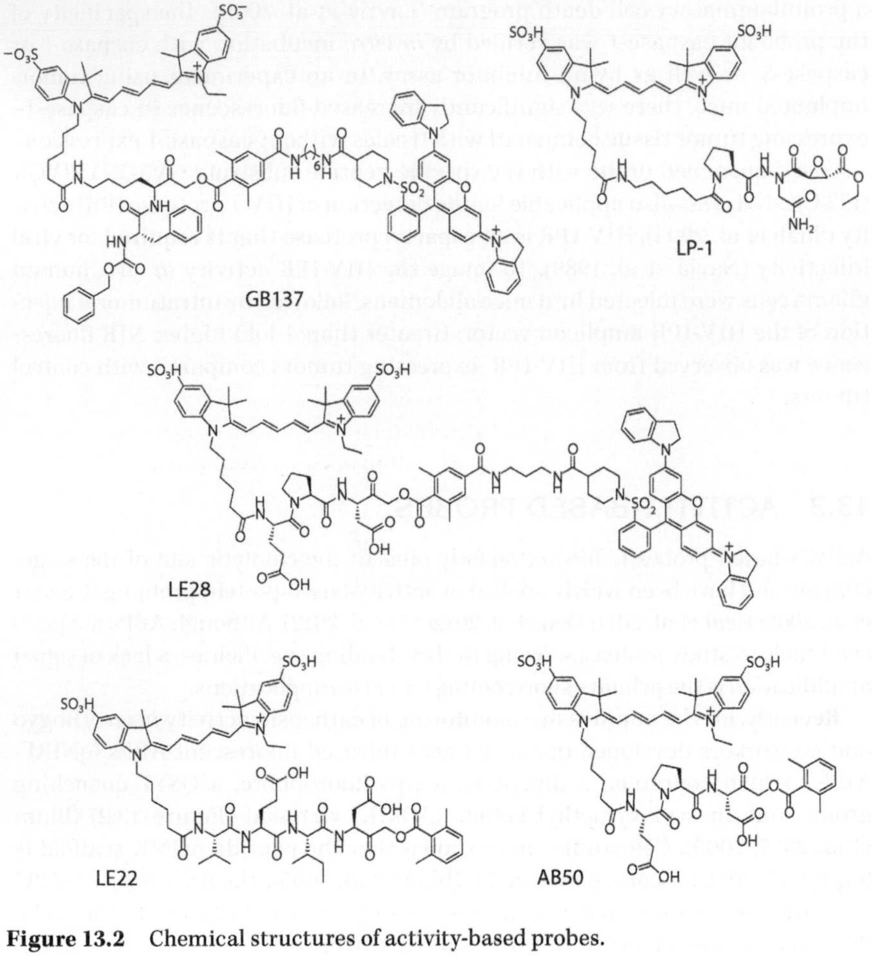

Figure 13.2 Chemical structures of activity-based probes.

Legumain is a lysosomal cysteine protease involved in antigen processing and matrix degradation. In addition, it is associated with atherosclerosis and tumor invasion (Liu et al. 2003; Mattock et al. 2010). LP-1, which contains a Pro-Asn-aza epoxide scaffold, was used for the noninvasive imaging of legumain (Figure 13.2) (Lee and Bogyo 2010). In an *in vitro* inhibition assay, LP-1 showed high selectivity toward legumain and had a higher rate constant than AOMK-based ABPs. However, LP-1 showed some cross-reactivity toward lysosomal cathepsins in macrophages. Tumor-xenografted mice were intravenously injected with LP-1, and the probe fluorescence was monitored noninvasively. After 90 min, the fluorescence signals in tumors reached a maximum signal-to-background ratio, whereas a control probe without the reactive epoxide group showed no accumulation in tumors. Furthermore, *ex vivo* imaging confirmed that LP-1 exhibited pronounced selectivity for legumain *in vivo*. An improved qNIRF-ABP, which was termed LE28 and contained a Glu-Pro-Asp-AOMK scaffold, has also been reported (Figure 13.2) (Edgington et al. 2013). LE28 is quenched before the formation of a covalent bond to legumain owing to the presence of a quencher molecule. Therefore, the unbound probes caused no background signals, leading to brighter tumor signals compared to LP-1. The brighter signals were also attributed to the enhanced cellular uptake of LE28. *In vitro* studies confirmed that LE28 exhibited specific labeling of legumain and less reactivity to caspases, because the bulky quenching moiety QSY21 favored uptake by the endocytic pathway in RAW cells.

Although annexin V–based fluorescent labeling has been widely and commonly used for the detection of apoptosis owing to its high affinity to phosphatidylserine, there are some limitations, such as slow clearance *in vivo* and the inability to detect early-stage apoptosis (Schutters and Reutelingsperger 2010). To circumvent these limitations, previous studies showed that the peptide AOMK scaffold was also applicable to the direct visualization of early-stage apoptosis *in vivo* by covalently labeling active caspases (Edgington et al. 2009). Caspases are a family of cysteine proteases that are involved in the early stages of apoptosis (Kurokawa and Kornbluth 2009). AB50-Cy5, which contains a Cy5 fluorophore and a Glu-Pro-Asp-AOMK sequence, showed labeling of caspase 3 and legumain *in vitro* without cathepsin B labeling (Figure 13.2). To monitor apoptosis in xenografted tumor tissues, the monoclonal antibody Apomab was injected into tumor-bearing mice. After various intervals, AB50-Cy5 was injected into these mice to enable the noninvasive imaging of apoptosis induced by Apomab. AB50-Cy5 showed specific labeling of caspases in tumor tissues and rapid clearance, which resulted in an ideal signal-to-background ratio. Bogyo and co-workers reported an improved probe, LE22, which exhibited higher potency toward caspases, with lower cross-reactivity (Figure 13.2) (Edgington et al. 2012). LE22 was sensitive to caspase-3 and -6 in RAW cells, whereas caspase-7 and legumain were scarcely labeled by LE22. After treating tumor-bearing mice with the anti-DR5 antibody, LE22 was administered intravenously. Consequently, the fluorescence signals from LE22-treated tumors were brighter than those from AB50-treated tumors.

13.4 ACTIVATABLE CELL-PENETRATING PEPTIDES

Cell-penetrating peptides (CPPs) have considerable potential for the intracellular delivery of various cargoes, such as imaging agents, siRNA, and proteins (Koren and Torchilin 2012). Tsien and co-workers have developed a system that regulates the membrane-permeability of CPPs by masking cationic polyarginine with polyanionic peptides (Figure 13.3) (Nguyen et al. 2010; Whitney et al. 2010, 2013; Chen et al. 2012; Olson et al. 2012; Savariar et al. 2013). In this system, particular enzyme activities or stimuli induce the release of polyanionic peptides and the recovery of the intrinsic membrane permeability of CPPs.

To image tumor tissues, MMP-cleavable ACPPs were developed (Jiang et al. 2004). Among various MMPs, the PLGLAG peptide sequence was used to target MMP-2 and MMP-9, which are overexpressed by many tumors. Cellular-uptake experiments using HT-1080 cells demonstrated that an ACPP composed of [11-kDa PEG]-X-e_9-XPLGLAG-r_9-X-k(Cy5), where X is a 6-aminohexanoyl group, exhibited 21- to 68-fold uptake after cleavage by MMP-2. In this ACPP, nonspecific proteolysis of the peptide linker *in vivo* was prevented by the incorporation of D-amino acids. In addition, PEG was effective at increasing solubility and reducing *in vivo* excretion rather than blocking the ability for cell penetration. The ACPP was injected into mice xenografted with HT-1080 cells, and *in vivo* imaging was then performed on tumor MMP-2 and MMP-9 activity. Significantly higher fluorescence signals were obtained using the ACPP compared to control probes composed of all D-amino acids or a scrambled recognition sequence. Furthermore, by changing the MMP recognition site, this probe design could be applied to the detection of different enzymes or disulfide bond reduction. The visualization of other cancer types using ACPPs was also confirmed (Olson et al. 2009; Nguyen et al. 2010; Whitney et al. 2010).

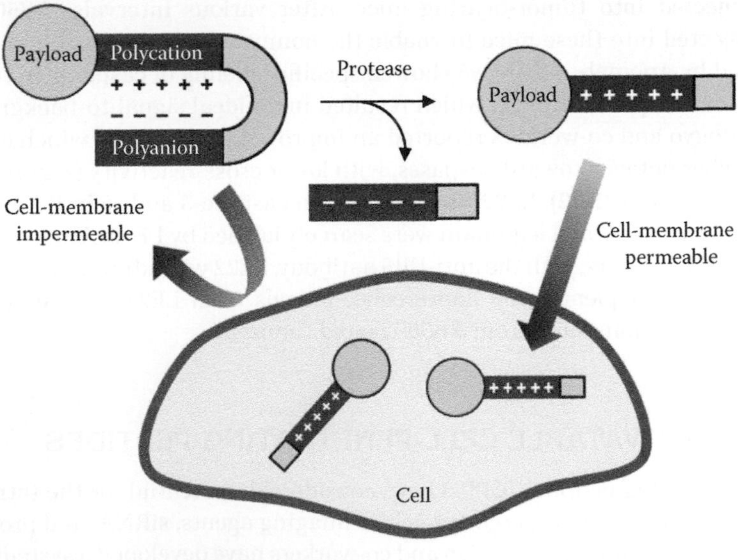

Figure 13.3 Schematic illustration of activatable cell-penetrating peptides.

ACPPs were also applied to image thrombin activity *in vivo* (Chen et al. 2012; Olson et al. 2012; Whitney et al. 2013). Thrombin is a serine protease that plays an important role in the early stages of blood coagulation (Huntington 2005). Ratiometric ACPPs, which contained a fluorescence donor (Cy5) and acceptor (Cy7) and a thrombin-cleavable sequence (PPRSFL) were developed, because ratiometric data acquisition has significant advantages over single wavelength intensity measurements (Figure 13.4) (Grynkiewicz et al. 1985). *In vitro* experiments using purified thrombin with $RACPP_{PPRSFL}$ resulted in a 34-fold ratio change. In addition, the ratio change was attributed to the selective cleavage of the linker by thrombin, because the k_{cat}/K_m value for thrombin was more than 14-fold higher than the values for plasmin and factor Xa. $RACPP_{PPRSFL}$ was then injected into the tail-clip model. An increase in the fluorescence ratio was observed locally around the wound within 10 min after intravenous injection of the probe. When the selective thrombin inhibitor hirudin was administered before probe injection, the increase in the ratio was inhibited by more than 90%. Thus, the fluorescence signal change was most likely due to *in vivo* thrombin activity. Furthermore, ratiometric Förster/fluorescence resonance energy transfer (FRET)-based ACPPs were used for the detection of tumor-associated MMP-2 and 9 (Savariar et al. 2013).

13.5 HYDROXYAPATITE-TARGETING FLUORESCENT PROBES

Hydroxyapatite (HA) is the main inorganic component of bone and is also associated with breast cancer and hypercalcemia. HA deposits are visualized *in vivo* by combining a fluorophore with functional groups that bind tightly to HA (Low and Kopeček 2012).

Bisphosphonates are analogs of pyrophosphate and are resistant to hydrolysis because of their P-C-P moiety. They exhibit a strong binding affinity to HA *in vivo*. To monitor micro-calcification in breast cancer, a near-infrared fluorescent probe containing pamidronate and IRDye 78 was developed and termed Pam78 (Figure 13.5) (Zaheer et al. 2001). After intravenous injection of Pam78 into nude mice, intense fluorescence was observed from the spine, ribs, paws, and knees. Sufficiently intense signals were observed even at a depth of 2.5 mm, although the intensity of the fluorescence signal depended on the distance from the skin surface and decreased exponentially. For application in large animals, a scaled-up synthesis of Pam800 was reported (Figure 13.5) (Bhushan et al. 2007). The HA-binding specificity of Pam800 was verified in binding assays by using other calcium salts, such as phosphate, oxalate, carbonate, and pyrophosphate, because Pam800 also contained pamidronate and the NIR fluorophore IRDye800 CW. When Pam800 was administered intravenously to 30-kg

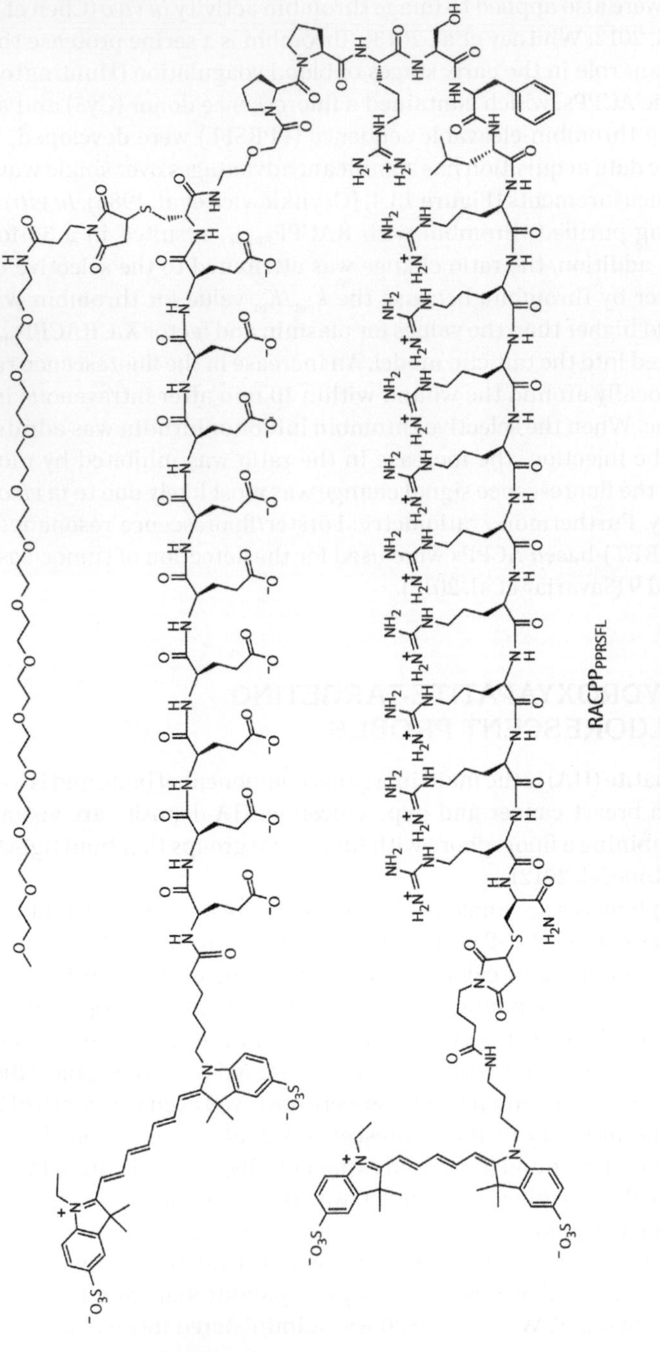

Figure 13.4 Chemical structures of activatable cell-penetrating probes.

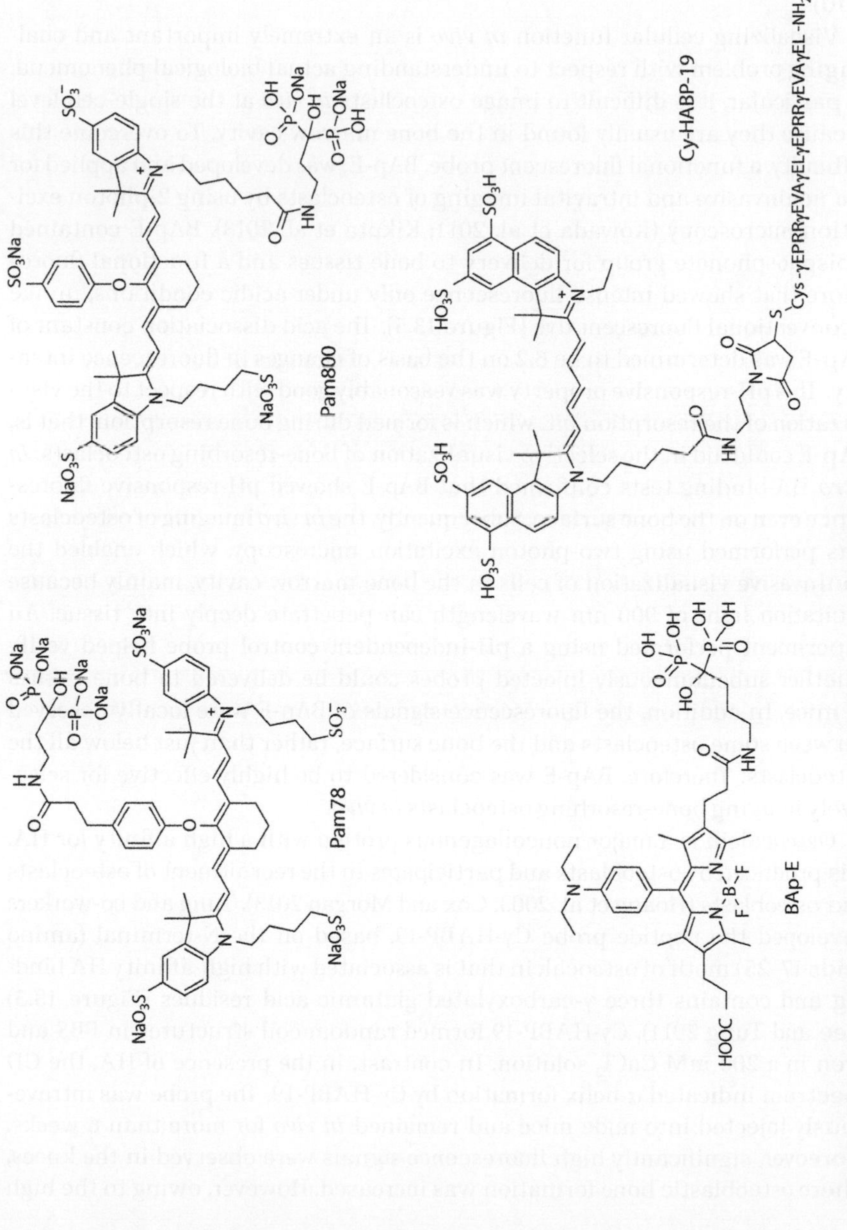

Figure 13.5 Chemical structures of hydroxyapatite-targeting probes.

Yorkshire pigs, fluorescence signals from Pam800 were observed in bones of the forearm. In addition, Pam800 was applicable to the image-guided surgery of subcutaneously embedded HA crystals. Moreover, a similar hydroxyapatite-targeting probe, OsteoSense®, is commercially available (Kozloff et al. 2007, 2010).

Visualizing cellular function *in vivo* is an extremely important and challenging problem with respect to understanding actual biological phenomena. In particular, it is difficult to image osteoclasts *in vivo* at the single-cell level because they are usually found in the bone marrow cavity. To overcome this difficulty, a functional fluorescent probe, BAp-E, was developed and applied for the noninvasive and intravital imaging of osteoclasts by using 2-photon excitation microscopy (Kowada et al. 2011; Kikuta et al. 2013). BAp-E contained a bisphosphonate group for delivery to bone tissues and a functional fluorophore that showed intense fluorescence only under acidic conditions, unlike a conventional fluorescent dye (Figure 13.5). The acid dissociation constant of BAp-E was determined to be 6.2 on the basis of changes in fluorescence intensity. This pH-responsive property was reasonably good with respect to the visualization of the resorption pit, which is formed during bone resorption; that is, BAp-E could aid in the selective visualization of bone-resorbing osteoclasts. *In vitro* HA-binding tests confirmed that BAp-E showed pH-responsive fluorescence even on the bone surface. Subsequently, the *in vivo* imaging of osteoclasts was performed using two-photon excitation microscopy, which enabled the noninvasive visualization of cells in the bone marrow cavity, mainly because excitation light of 900 nm wavelength can penetrate deeply into tissue. An experiment performed using a pH-independent control probe helped verify whether subcutaneously injected probes could be delivered to bone tissues in mice. In addition, the fluorescence signals of BAp-E were locally observed between some osteoclasts and the bone surface, rather than just below all the osteoclasts. Therefore, BAp-E was considered to be highly effective for selectively imaging bone-resorbing osteoclasts *in vivo*.

Osteocalcin is a major noncollagenous protein with a high affinity for HA. It is produced by osteoblasts and participates in the recruitment of osteoclasts and osteoblasts (Hoang et al. 2003; Cox and Morgan 2013). Tung and co-workers developed the peptide probe Cy-HABP-19, based on the N-terminal (amino acids 17–25) motif of osteocalcin that is associated with high-affinity HA binding and contains three γ-carboxylated glutamic acid residues (Figure 13.5) (Lee and Tung 2011). Cy-HABP-19 formed random coil structures in PBS and even in a 200 mM $CaCl_2$ solution. In contrast, in the presence of HA, the CD spectrum indicated α-helix formation by Cy-HABP-19. The probe was intravenously injected into nude mice and remained *in vivo* for more than 6 weeks. Moreover, significantly high fluorescence signals were observed in the knees, where osteoblastic bone formation was increased. However, owing to the high background fluorescence, the signal-to-noise ratio reached a maximum only after 2 weeks after probe injection.

To monitor bone deposition in mice, the IRDye 800CW was conjugated with a tetracycline derivative known to be a bone-chelating agent (Kovar et al. 2011). The probe was intraperitoneally injected into athymic nude mice, followed by noninvasive imaging 24 h post infection. High-intensity fluorescence signals were obtained from the tibia, femur, and vertebrae, with lower background signals in the kidney.

13.6 ACTIVATABLE PROBES FOR MONITORING REACTIVE OXYGEN SPECIES

Reactive oxygen species (ROS) function as intracellular signaling molecules, as well as components of oxidative stress (D'Autréaux and Toledano 2007). Thus, there has been much effort to understand where ROS are generated and how they act. With this aim, various fluorescent probes have been developed (Chen et al. 2011).

By using a simple one-step reaction from commercially available cyanine dyes, Hydro-Cy7 was developed for imaging ROS production in lipopolysaccharide (LPS)-stimulated mice (Figure 13.6) (Kundu et al. 2009). Superoxide, hydroxy radical, and *tert*-butyl peroxy radical could all oxidize Hydro-Cy7 into a fluorescent cyanine dye *in vitro*, and the fluorescence intensity increased up to 104-fold on oxidation. Hydro-Cy7 was highly stable in aqueous pH 7.4 buffer at 37°C, with a half-life of approximately 3 days. To image ROS production *in vivo*, mice were intraperitoneally injected with Hydro-Cy7 at 6 h after LPS injection. A 2-fold increase in fluorescence was measured, compared to the fluorescence observed after injection of saline.

FOSCY-1 consisted of a less-ROS-susceptible Cy5 dye conjugated with a highly ROS-susceptible Cy7 (Figure 13.6) (Oushiki et al. 2010). Its reactivity to ROS was confirmed *in vitro* by monitoring fluorescence intensity changes after the addition of ROS produced both chemically and enzymatically. In addition, owing to its membrane impermeability, FOSCY-1 could detect extracellular ROS, which are produced for intercellular signaling. For further investigation of ROS detection using FOSCY-1, *in vivo* imaging of ROS production was performed in a mouse model of peritonitis. The probe was intraperitoneally injected into mice at 4 h after zymogen injection. Significantly high-intensity fluorescence signals were observed compared to control mice and were partially suppressed in a concentration-dependent manner by NOX inhibitor treatment.

Recently, a new type of far-red to NIR fluorescent dye, Si-rhodamine, was developed by Nagano and co-workers. MMSiR exhibits a high degree of sensitivity and selectivity for the detection of HOCl compared to other ROS *in vitro* (Figure 13.6) (Koide et al. 2011). Through a reaction with HOCl, MMSiR was

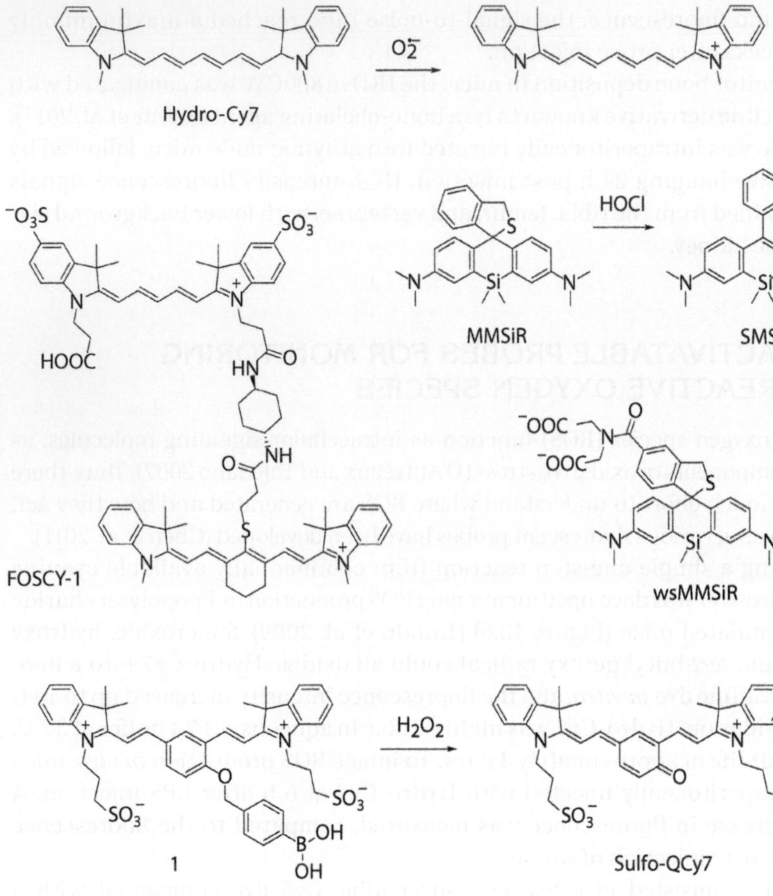

Figure 13.6 Chemical structures of ROS probes.

converted to SMSiR, which was highly fluorescent, photostable, and pH independent. For *in vivo* experiments, a highly hydrophilic derivative, wsMMSiR, was synthesized and intraperitoneally injected into phorbol myristate acetate-treated mice. Compared to unstimulated mice, a mouse model of peritonitis showed significant fluorescence enhancement.

Endogenously produced hydrogen peroxide was noninvasively detectable using a Cy7-like NIR probe (Karton-Lifshin et al. 2011). The probe was converted to sulfo-QCy7 by a reaction with H_2O_2, and was sensitive to H_2O_2 concentrations even below 1 μM *in vitro* (Figure 13.6, probe **1**). In LPS-stimulated mice, the probe showed an extremely high signal-to-noise ratio, approximately 10-fold higher compared to control mice without probe addition and LPS treatment.

Very recently, Lin and co-workers reported new types of NIR probes (Figure 13.6), probe 2 and NIR-H_2O_2, for imaging HClO or H_2O_2, respectively, *in vivo* (Yuan et al. 2012a,b). These probes appeared to be hybrids of Cy dyes and rhodamine or coumarin, respectively.

13.7 NIR FLUORESCENT PROBES FOR TUMOR HYPOXIA

The abnormal proliferation of tumor cells can induce the incomplete formation of vasculature, resulting in locally low concentrations of oxygen inside the tumor tissue. This hypoxic condition is a characteristic feature of various solid tumors (Wilson and Hay 2011).

Azobenzene derivatives can be reduced by thiols or reductases; therefore, Nagano and co-workers developed a FRET-based NIR probe for sensing hypoxia, which included Cy5 and BHQ-3 as a fluorophore and quencher, respectively (Figure 13.7) (Kiyose et al. 2010). Under hypoxic conditions *in vitro*, the azo moiety of QCy5 was readily reduced, and a significant enhancement of fluorescence was observed within 10 min. Furthermore, the fluorescence intensity increased immediately after portal vein ligation in QCy5-injected mice, resulting from the reduction of the azo moiety in the ischemic liver.

The 2-nitroimidazole moiety has been widely used as a hypoxia marker because it is reduced by nitroreductases under hypoxic conditions, with the resulting product able to bind to cellular nucleophilic macromolecules (Kizaka-Kondoh and Konse-Nagasawa 2009). Thus, GPU-167, which included Cy7 and two nitroimidazole moieties, was applied in the fluorescent imaging of tumor hypoxia *in vivo* (Figure 13.7) (Okuda et al. 2012). Studies confirmed that GPU-167 bound to intracellular components, although nonspecific binding could also occur under normoxic conditions. At 24 h after GPU-167 administration into tumor-bearing mice, high-intensity fluorescence signals were observed in the tumor. However, the most intense signals were observed in the liver, potentially disturbing selective tumor imaging.

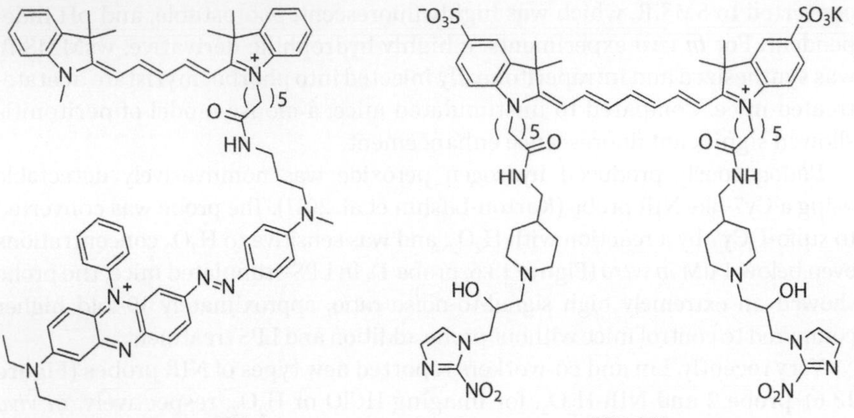

Figure 13.7 Chemical structures of hypoxia probes.

13.8 CONCLUSIONS

As described here, the fluorescence properties, targeting ability, aqueous solubility, and pharmacokinetics of chemical probes can be finely tuned by chemically modifying the probe structure. Therefore, numerous NIR fluorescent probes have been developed, and some of these are commercially available today. However, the present probes do not necessarily satisfy the requirements for diagnostic use or for understanding biological events. Thus, further improvements to each chemical probe can be expected to lead to target-selective imaging *in vivo*. Finally, we hope that this review will provide valuable insights into the development of *in vivo* fluorescent probes and for future developments.

REFERENCES

Backer, M. V., Levashova, Z., Patel, V., Jehning, B. T., Claffey, K., Blankenberg, F. G., Backer, J. M. *Nat. Med.* **2007**, *13*, 504–509.

Barrett, T., Koyama, Y., Hama, Y., Ravizzini, G., Shin, I. S., Jang, B.-S., Paik, C. H., Urano, Y., Choyke, P. L., Kobayashi, H. *Clin. Cancer Res.* **2007**, *13*, 6639–6648.

Becker, A., Hessenius, C., Licha, K., Ebert, B., Sukowski, U., Semmler, W., Wiedenmann, B., Grötzinger, C. *Nat. Biotechnol.* **2001**, *19*, 327–331.

Benes, P., Vetvicka, V., Fusek, M. *Crit. Rev. Oncol. Hematol.* **2008**, *68*, 12–28.

Bhushan, K. R., Tanaka, E., Frangioni, J. V. *Angew. Chem. Int. Ed.* **2007**, *46*, 7969–7971.

Blum, G., Mullins, S. R., Keren, K., Fonovič, M., Jedeszko, C., Rice, M. J., Sloane, B. F., Bogyo, M. *Nat. Chem. Biol.* **2005**, *1*, 203–209.

Blum, G., von Degenfeld, G., Merchant, M. J., Blau, H. M., Bogyo, M. *Nat. Chem. Biol.* **2007**, *3*, 668–677.

Blum, G., Weimer, R. M., Edgington, L. E., Adams, W., Bogyo, M. *PLoS One* **2009**, *4*, e6374.

Bremer, C., Bredow, S., Mahmood, U., Weissleder, R., Tung, C.-H. *Radiology* **2001a**, *221*, 523–529.

Bremer, C., Tung, C.-H., Weissleder, R. *Nat. Med.* **2001b**, *7*, 743–748.

Caffrey, C. R. *Curr. Opin. Chem. Biol.* **2007**, *11*, 433–439.

Chen, B., Friedman, B., Whitney, M. A., Van Winkle, J. A., Lei, I.-F., Olson, E. S., Cheng, Q., Pereira, B., Zhao, L., Tsien, R. Y., Lyden, P. D. *J. Neurosci.* **2012**, *32*, 7622–7631.

Chen, X., Conti, P. S., Moats, R. A. *Cancer Res.* **2004**, *64*, 8009–8014.

Chen, X., Tian, X., Shin, I., Yoon, J. *Chem. Soc. Rev.* **2011**, *40*, 4783–4804.

Cheng, Z., Wu, Y., Xiong, Z., Gambhir, S. S., Chen, X. *Bioconjug. Chem.* **2005**, *16*, 1433–1441.

Citrin, D., Lee, A. K., Scott, T., Sproull, M., Ménard, C., Tofilon, P. J., Camphausen, K. *Mol. Cancer Ther.* **2004**, *3*, 481–488.

Cox, R. F., Morgan, M. P. *Bone* **2013**, *53*, 437–450.

Cravatt, B. F., Wright, A. T., Kozarich, J. W. *Annu. Rev. Biochem.* **2008**, *77*, 383–414.

D'Autréaux, B., Toledano, M. B. *Nat. Rev. Mol. Cell Biol.* **2007**, *8*, 813–824.

Deu, E., Verdoes, M., Bogyo, M. *Nat. Struct. Mol. Biol.* **2012**, *19*, 9–16.

Edgington, L. E., Berger, A. B., Blum, G., Albrow, V. E., Paulick, M. G., Lineberry, N., Bogyo, M. *Nat. Med.* **2009**, *15*, 967–973.

Edgington, L. E., van Raam, B. J., Verdoes, M., Wierschem, C., Salvesen, G. S., Bogyo, M. *Chem. Biol.* **2012**, *19*, 340–352.

Edgington, L. E., Verdoes, M., Ortega, A., Withana, N. P., Lee, J., Syed, S., Bachmann, M. H., Blum, G., Bogyo, M. *J. Am. Chem. Soc.* **2013**, *135*, 174–182.

Garanger, E., Boturyn, D., Jin, Z., Dumy, P., Favrot, M.-C., Coll, J.-L. *Mol. Ther.* **2005**, *12*, 1168–1175.

Grynkiewicz, G., Poenie, M., Tsien, R. Y. *J. Biol. Chem.* **1985**, *260*, 3440–3450.

Heal, W. P., Dang, T. H. T., Tate, E. W. *Chem. Soc. Rev.* **2011**, *40*, 246–257.

Hoang, Q. Q., Sicheri, F., Howard, A. J., Yang, D. S. C. *Nature* **2003**, *425*, 977–980.

Huntington, J. A. *J. Thromb. Haemastas.* **2005**, *3*, 1861–1872.

Jaffer, F. A., Kim, D.-E., Quinti, L., Tung, C.-H., Aikawa, E., Pande, A. N., Kohler, R. H., Shi, G.-P., Libby, P., Weissleder, R. *Circulation* **2007**, *115*, 2292–2298.

Jiang, T., Olson, E. S., Nguyen, Q. T., Roy, M., Jennings, P. A., Tsien, R. Y. *Proc. Natl. Acad. Sci. U.S.A.* **2004**, *101*, 17867–17872.

Karton-Lifshin, N., Segal, E., Omer, L., Portnoy, M., Satchi-Fainaro, R., Shabat, D. *J. Am. Chem. Soc.* **2011**, *133*, 10960–10965.

Ke, S., Wen, X., Gurfinkel, M., Charnsangavej, C., Wallace, S., Sevick-Muraca, E. M., Li, C. *Cancer Res.* **2003**, *63*, 7870–7875.

Kikuta, J., Wada, Y., Kowada, T., Wang, Z., Sun-Wada, G.-H., Nishiyama, I., Mizukami, S., Maiya, N., Yasuda, H., Kumanogoh, A., Kikuchi, K., German, R. N., Ishii, M. *J. Clin. Invest.* **2013**, *123*, 866–873.

Kiyose, K., Hanaoka, K., Oushiki, D., Nakamura, T., Kajimura, M., Suematsu, M., Nishimatsu, H., Yamane, T., Terai, T., Hirata, Y., Nagano, T. *J. Am. Chem. Soc.* **2010**, *132*, 15846–15848.

Kizaka-Kondoh, S., Konse-Nagasawa, H. *Cancer Sci.* **2009**, *100*, 1366–1373.

Knapinska, A., Fields, G. B. *ChemBioChem* **2012**, *13*, 2002–2020.

Koide, Y., Urano, Y., Hanaoka, K., Terai, T., Nagano, T. *J. Am. Chem. Soc.* **2011**, *133*, 5680–5682.

Koide, Y., Urano, Y., Hanaoka, K., Piao, W., Kusakabe, M., Saito, N., Terai, T., Okabe, T., Nagano, T. *J. Am. Chem. Soc.* **2012**, *134*, 5029–5031.

Koren, E., Torchilin, V. P. *Trends Mol. Med.* **2012**, *18*, 385–393.

Kossodo, S., Pickarski, M., Lin, S.-A., Gleason, A., Gaspar, R., Buono, C., Ho, G., Blusztajn, A., Cuneo, G., Zhang, J., Jensen, J., Hargreaves, R., Coleman, P., Hartman, G., Rajopadhye, M., Duong, L. T., Sur, C., Yared, W., Peterson, J., Bednar, B. *Mol. Imaging Biol.* **2010**, *12*, 488–499.

Kovar, J. L., Volcheck, W. M., Chen, J., Simpson, M. A. *Anal. Biochem.* **2007**, *361*, 47–54.

Kovar, J. L., Volcheck, W., Sevick-Muraca, E., Simpson, M. A., Olive, D. M. *Anal. Biochem.* **2009**, *384*, 254–262.

Kovar, J. L., Xu, X., Draney, D., Cupp, A., Simpson, M. A., Olive, D. M. *Anal. Biochem.* **2011**, *416*, 167–173.

Kowada, T., Kikuta, J., Kubo, A., Ishii, M., Maeda, H., Mizukami, S., Kikuchi, K. *J. Am. Chem. Soc.* **2011**, *133*, 17772–17776.

Kozloff, K. M., Weissleder, R., Mahmood, U. *J. Bone. Miner. Res.* **2007**, *22*, 1208–1216.

Kozloff, K. M., Quinti, L., Patntirapong, S., Hauschka, P. V., Tung, C.-H., Weissleder, R., Mahmood, U. *Bone* **2009**, *44*, 190–198.

Kozloff, K. M., Volakis, L. I., Marini, J. C., Caird, M. S. *J. Bone. Miner. Res.* **2010**, *25*, 1748–1758.

Kundu, K., Knight, S. F., Willett, N., Lee, S., Taylor, W. R., Murthy, N. *Angew. Chem. Int. Ed.* **2009**, *48*, 299–303.

Kurokawa, M., Kornbluth, S. *Cell* **2009**, *138*, 838–854.

Lavrik, I. N., Golks, A., Krammer, P. H. *J. Clin. Invest.* **2005**, *115*, 2665–2672.

Lee, J., Bogyo, M. *ACS Chem. Biol.* **2010**, *5*, 233–243.

Lee, J. S., Tung, C.-H. *ChemBioChem* **2011**, *12*, 1669–1673.

Li, N., Overkleeft, H. S., Florea, B. I. *Curr. Opin. Chem. Biol.* **2012**, *16*, 227–233.

Liu, C., Sun, C., Huang, H., Janda, K., Edgington, T. *Cancer Res.* **2003**, *63*, 2957–2964.

Louie, A. *Chem. Rev.* **2010**, *110*, 3146–3195.

Low, S. A., Kopeček, J. *Adv. Drug Deliv. Rev.* **2012**, *64*, 1189–1204.

Luo, S., Zhang, E., Su, Y., Cheng, T., Shi, C. *Biomaterials* **2011**, *32*, 7127–7138.

Mattock, K. L., Gough, P. J., Humphries, J., Burnand, K., Patel, L., Suckling, K. E., Cuello, F., Watts, C., Gautel, M., Avkiran, M., Smith, A. *Atherosclerosis* **2010**, *208*, 83–89.

Messerli, S. M., Prabhakar, S., Tang, Y., Shah, K., Cortes, M. L., Murthy, V., Weissleder, R., Breakefield, X. O., Tung, C.-H. *Neoplasia* **2004**, *6*, 95–105.

Navia, M. A., Fitzgerald, P. M. D., McKeever, B. M., Leu, C.-T., Heimbach, J. C., Herber, W. K., Sigal, I. S., Darke, P. L., Springer, J. P. *Nature* **1989**, *337*, 615–620.

Nguyen, Q. T., Olson, E. S., Aguilera, T. A., Jiang, T., Scadeng, M., Ellies, L. G., Tsien, R. Y. *Proc. Natl. Acad. Sci. U.S.A.* **2010**, *107*, 4317–4322.

Ntziachristos, V., Schellenberger, E. A., Ripoll, J., Yessayan, D., Graves, E., Bogdanov, A., Jr., Josephson, L., Weissleder, R. *Proc. Natl. Acad. Sci. U.S.A.* **2004**, *101*, 12294–12299.

Ogawa, M., Kosaka, N., Choyke, P. L., Kobayashi, H. *Cancer Res.* **2009a**, *69*, 1268–1272.

Ogawa, M., Regino, C. A. S., Choyke, P. L., Kobayashi, H. *Mol. Cancer Ther.* **2009b**, *8*, 232–239.

Okuda, K., Okabe, Y., Kadonosono, T., Ueno, T., Youssif, B. G. M., Kizaka-Kondoh, S., Nagasawa, H. *Bioconjug. Chem.* **2012**, *23*, 324–329.

Olson, E. S., Aguilera, T. A., Jiang, T., Ellies, L. G., Nguyen, Q. T., Wong, E. H., Gross, L. A., Tsien, R. Y. *Integr. Biol.* **2009**, *1*, 382–393.

Olson, E. S., Whitney, M. A., Friedman, B., Aguilera, T. A., Crisp, J. L., Baik, F. M., Jiang, T., Baird, S. M., Tsimikas, S., Tsien, R. Y., Nguyen, Q. T. *Integr. Biol.* **2012**, *4*, 595–605.

Oushiki, D., Kojima, H., Terai, T., Arita, M., Hanaoka, K., Urano, Y., Nagano, T. *J. Am. Chem. Soc.* **2010**, *132*, 2795–2801.

Petrovsky, A., Schellenberger, E., Josephson, L., Weissleder, R., Bogdanov, A., Jr. *Cancer Res.* **2003**, *63*, 1936–1942.

Razkin, J., Josserand, V., Boturyn, D., Jin, Z.-H., Dumy, P., Favrot, M., Coll, J.-L., Texier, I. *ChemMedChem* **2006**, *1*, 1069–1072.

Rieman, D. J., Mcclung, H. A., Dodds, R. A., Hwang, S. M., Lark, M. W., Holmes, S., James, I. E., Drake, F. H., Gowen, M. *Bone* **2001**, *28*, 282–289.

Sapsford, K. E., Algar, W. R., Berti, L., Gemmill, K. B., Casey, B. J., Oh, E., Stewart, M. H., Medintz, I. L. *Chem. Rev.* **2013**, *113*, 1904–2074.

Savariar, E. N., Felsen, C. N., Nashi, N., Jiang, T., Ellies, L. G., Steinbach, P., Tsien, R. Y., Nguyen, Q. T. *Cancer Res.* **2013**, *73*, 855–864.

Schutters, K., Reutelingsperger, C. *Apoptosis* **2010**, *15*, 1072–1082.

Shah, K., Tung, C.-H., Chang, C.-H., Slootweg, E., O'Loughlin, T., Breakefield, X. O., Weissleder, R. *Cancer Res.* **2004**, *64*, 273–278.

Tafreshi, N. K., Huang, X., Moberg, V. E., Barkey, N. M., Sondak, V. K., Tian, H., Morse, D. L., Vagner, J. *Bioconjug. Chem.* **2012**, *23*, 2451–2459.

Tung, C.-H., Mahmood, U., Bredow, S., Weissleder, R. *Cancer Res.* **2000**, *60*, 4953–4958.

Tung, C.-H., Lin, Y., Moon, W. K., Weissleder, R. *ChemBioChem* **2002**, *3*, 784–786.

Uddin, Md. J., Crews, B. C., Blobaum, A. L., Kingsley, P. J., Gorden, D. L., McIntyre, J. O., Matrisian, L. M., Subbaramaiah, K., Dannenberg, A. J., Piston, D. W., Marnett, L. J. *Cancer Res.* **2010**, *70*, 3618–3627.

Weissleder, R., Tung, C.-H., Mahmood, U., Bogdanov, A., Jr. *Nat. Biotechnol.* **1999**, *17*, 375–378.

Weissleder, R. *Nat. Biotechnol.* **2001**, *19*, 316–317.

Weissleder, R., Ntziachristos, V. *Nat. Med.* **2003**, *9*, 123–128.

Whitney, M., Crisp, J. L., Olson, E. S., Aguilera, T. A., Gross, L. A., Ellies, L. G., Tsien, R. Y. *J. Biol. Chem.* **2010**, *285*, 22532–22541.

Whitney, M., Savariar, E. N., Friedman, B., Levin, R. A., Crisp, J. L., Glasgow, H. L., Lefkowitz, R., Adams, S. R., Steinbach, P., Nashi, N., Nguyen, Q. T., Tsien, R. Y. *Angew. Chem. Int. Ed.* **2013**, *52*, 325–330.

Wilson, W. R., Hay, M. P. *Nat. Rev. Cancer* **2011**, *11*, 393–410.

Wu, Y., Cai, W., Chen, X. *Mol. Imaging Biol.* **2006**, *8*, 226–236.

Wunder, A., Tung, C.-H., Müller-Ladner, U., Weissleder, R., Mahmood, U. *Arthritis Rheum.* **2004**, *50*, 2459–2465.

Wunderbaldinger, P., Turetschek, K., Bremer, C. *Eur. Radiol.* **2003**, *13*, 2206–2211.

Yong, K.-T., Law, W.-C., Hu, R., Ye, L., Liu, L., Swihart, M. T., Prasad, P. N. *Chem. Soc. Rev.* **2013**, *42*, 1236–1250.

Yuan, L., Lin, W., Yang, Y., Chen, H. *J. Am. Chem. Soc.* **2012a**, *134*, 1200–1211.

Yuan, L., Lin, W., Zhao, S., Gao, W., Chen, B., He, L., Zhu, S. *J. Am. Chem. Soc.* **2012b**, *134*, 13510–13523.

Zaheer, A., Lenkinski, R. E., Mahmood, A., Jones, A. G., Cantley, L. C., Frangioni, J. V. *Nat. Biotechnol.* **2001**, *19*, 1148–1154.

Zhou, H., Luby-Phelps, K., Mickey, B. E., Habib, A. A., Mason, R. P., Zhao, D. *PLoS One* **2009**, *4*, e8051.

Zrazhevskiy, P., Sena, M., Gao, X. *Chem. Soc. Rev.* **2010**, *39*, 4326–4354.

Chapter 14

Optical Probes for *In Vivo* Brain Imaging

Ksenia V. Kastanenka, Michal Arbel-Ornath, Eloise Hudry, Elena Galea, Hong Xie, and Brian J. Bacskai

CONTENTS

14.1 IMAGING THE COMPONENTS OF THE CENTRAL NERVOUS SYSTEM

Beginning with the pioneering work of Golgi and Ramon y Cajal, neuroscientists have focused on developing methodologies to reveal structure and function within the nervous system (Ramon y Cajal 1995). Visualization of the nervous system has allowed insight into its development, maturation, and disease. Until recently, however, structural studies were largely limited to *ex vivo* preparations. The advent of imaging techniques such as magnetic resonance imaging (MRI), positron emission tomography (PET), diffuse optical tomography (DOT), photoacoustic microscopy, bioluminescence tomography, and multiphoton microscopy has enabled *in vivo* imaging to flourish (Garcia-Alloza and Bacskai 2004; Luker and Luker 2008; Gibson and Dehghani 2009; Yao and Wang 2011). Optical techniques in particular have developed at a rapid pace, and a variety of imaging probes suitable for use in reduced preparations as well as in intact organisms have been generated. Here, we review fluorescent probes that have been used for *in vivo* imaging in the central nervous system (CNS) with high-resolution optical microscopy techniques. We have been implementing multiphoton microscopy combined with cranial window approaches to image structure and function in the brain. Multiphoton microscopy allows high-resolution imaging with a spectrum of fluorescent probes in living brain at depths of several hundred microns. Because the work in our laboratory is focused on Alzheimer's disease, we have applied many of these tools to transgenic mouse models of the disease, and begin with an introduction to imaging the pathological deposits themselves. Despite our focus on disease, the use of optical imaging probes for *in vivo* imaging has broad applicability to healthy brain and other neurological disorders.

14.1.1 Imaging Pathological Deposits in the Alzheimer Brain

Alzheimer's disease (AD) is the major cause of dementia in individuals over the age of 65. It is a progressive neurodegenerative disorder characterized by the presence of extracellular amyloid plaques composed of aggregated amyloid beta (Aβ) peptides, intracellular neurofibrillary tangles composed of hyperphosphorylated forms of tau protein, and neuronal cell loss in the brain (Holtzman et al. 2011). Plaque deposition and tangle formation precede neuronal loss and the onset of memory impairment. Therefore, considerable efforts have been invested in the design of tracers that would allow visualization of plaques and/or tangles to aid in the diagnosis of AD in patients or the visualization of AD pathology before the onset of symptoms. The development of Pittsburgh Compound-B (PiB) was a breakthrough for molecular imaging. It has allowed the labeling of amyloid plaques in humans and the visualization of amyloid burden with PET (Klunk et al. 2004).

Transgenic mouse models that overexpress human mutant amyloid precursor protein (APP) and/or presenilin were developed based on genetic evidence in familial Alzheimer's disease patients. Several genetic strains, including Tg2576 and APP^{swe}:PS1dE9 mouse models, were created to recapitulate plaque deposition (Hsiao et al. 1996; Borchelt et al. 1997; Jankowsky et al. 2001). Amyloid plaques can be visualized when labeled with PiB or methoxy-XO4, a Congo red derivative (Klunk et al. 2002) (Figures 14.1 and 14.2). Both of these compounds are fluorescent and can therefore be detected with multiphoton microscopy. The tracers cross the blood–brain barrier (BBB) and label plaques, which can be imaged acutely or longitudinally through cranial windows (Bacskai et al. 2003; Skoch et al. 2005; Spires-Jones et al. 2011). Also, plaque clearance can be assessed as a result of a drug treatment (Garcia-Alloza et al. 2006, 2007) and anti-Aβ immunotherapy (Bacskai et al. 2001; Lombardo et al. 2003). Amyloid plaques can also be imaged after topical application of a broad range of histological dyes that include thioflavin S (Christie et al. 2001), although this approach requires direct access to the brain and is limited by diffusion of the dyes.

A second hallmark of AD is neurofibrillary tangle formation. Tg4510 mice were engineered to model tangle formation (Santacruz et al. 2005). Topical application of thioflavin S can be used to image tangles in mice (Spires-Jones et al. 2008). Also, X-34 or FSB dyes, when administered systemically, can be used

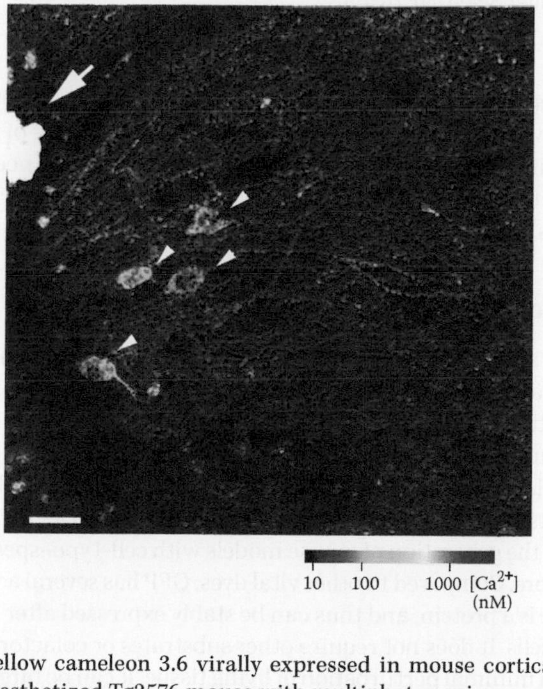

Figure 14.1 Yellow cameleon 3.6 virally expressed in mouse cortical neurons was imaged in an anesthetized Tg2576 mouse with multiphoton microscopy. Tg2576 mice exhibit plaque pathology due to expression of human amyloid precursor protein. An amyloid plaque is present in far left (arrow). Neuronal cell bodies expressing yellow cameleon 3.6 are demarcated by arrowheads. Scale bar 50 μm.

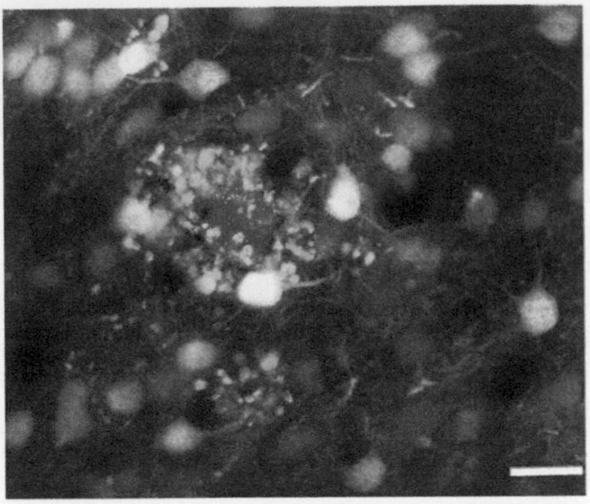

Figure 14.2 roGFP is expressed in the cortex of an APP:PS1 double transgenic mouse. Normal roGFP is in yellow, oxidized roGFP is in green, plaques are visualized with methoxy-X04 (blue). Scale bar 20 μm.

to label tangles repeatedly (Styren et al. 2000; Velasco et al. 2008; de Calignon et al. 2010). However, they still result in low-contrast labeling of the pathology, which can limit their use for *in vivo* imaging. There is a clear need for developing additional fluorescent dyes to image this pathology chronically with optical approaches and for PET imaging of tangles in humans.

14.1.2 Fluorescent Proteins

The discovery and isolation of green fluorescent protein (GFP) from jellyfish (Shimomura et al. 1962) and its exogenous expression in living organisms (Chalfie et al. 1994) revolutionized biological research. It provided a powerful approach for the direct detection of gene expression. Improved GFP photostability through genetic manipulations led to enhanced GFP (EGFP) (Heim et al. 1995), as well as a spectrum of color variants (Shaner et al. 2005). These advances led to its widespread use in living cells and the generation of mouse models with cell-type–specific expression of the fluorophore. Compared to other vital dyes, GFP has several advantages for *in vivo* imaging. It is a protein, and thus can be stably expressed after introduction of its cDNA into cells. It does not require other substrates or cofactors, and thus can be imaged with minimal perturbation of living tissue. It can be targeted to distinct cell types or specific cellular compartments through cell-specific or compartment-specific promoters. GFP fused to a protein of interest allows direct examination of subcellular localization of that protein in cells or organisms. GFP color variants, such as red (RFP), yellow (YFP), and cyan (CFP) fluorescent proteins, collectively referred to as XFPs (Feng et al. 2000), can be used to monitor the expression of several genes in a single organism. Since then, a second generation of monomeric fluorescent proteins, mFruits, have been developed that provide improved brightness and photostability (Shaner et al. 2004; Shu et al. 2006). This ultimately led to the generation of a "Brainbow mouse" (Livet et al. 2007), in which distinct neurons expressed different ratios of the four basic proteins, GFP, RFP, YFP, and CFP, resulting in brain structures exhibiting approximately 90 colors within their neurons. Furthermore, a lack of toxicity after repeated imaging of XFPs makes fluorescent proteins (FPs) suitable candidates for longitudinal *in vivo* studies.

Additional color variants of fluorescent proteins are based on reef coral fluorescent protein (RCFP). Similar to XFP, RCFPs come in multiple varieties, red (HcRed, AsRed, DsRed, mRFP) and cyan (AmCyan), for example (Hirrlinger et al. 2005). However, RCFPs tend to self-aggregate, and are thus limited in their application (Hirrlinger et al. 2005).

14.1.3 Cell-Type–Specific Promoters and Viruses

XFPs and RCFPs can be expressed ubiquitously or targeted with cell-type–specific promoters. The cytomegalovirus (CMV) promoter allows nonspecific gene expression (van den Pol and Ghosh 1998). Glial fibrillary acidic protein (GFAP) targets gene expression to astrocytes (Zhuo et al. 1997), while proteolipid protein promoter (PLP) allows targeting to oligodendrocytes (Fuss et al. 2000). Jung et al. (2000) labeled microglia with GFP by replacing the CX(3) CR1 gene in a knock-in model. The Thy1 promoter can be used to target FPs to neurons (Feng et al. 2000). Different promoters can be used to target specific populations of neurons. For instance, CamKII is used to target excitatory cortical neurons (Dittgen et al. 2004), while glutamic acid decarboxylase is used to target inhibitory interneurons (Tamamaki et al. 2003).

The introduction of genetic tracers, such as XFPs, into an organism can be achieved in one of two ways. The first is through the creation of transgenic mice, in which a tracer is inserted into the mouse genome. The second is by using viral delivery. There are several common viral vectors and each is suitable for specific applications (Papale et al. 2009). Adeno-associated viruses and lentiviruses are especially useful because they can be transduced into nondividing cells of the CNS. After introduction of an adeno-associated virus into the host, its DNA remains distinct from the host DNA (Papale et al. 2009), providing an advantage because it reduces the possibility of mutagenesis due to random insertion. The genetic material introduced via lentivirus, however, gets incorporated into the host DNA (Sakuma et al. 2012). The transgene delivered via a virus manifests protein expression within days to weeks and can continue to express over many months and even years.

Once a tracer is expressed in the CNS, it can be imaged in living animals through a cranial window over the brain (Garcia-Alloza and Bacskai 2004; Spires-Jones et al. 2007), thinned or intact skull (Helmchen and Denk 2005), or through an implanted chamber over a spinal cord (Farrar et al. 2012) using multiphoton microscopy.

14.2 IMAGING CELLULAR STRUCTURES

Mouse models such as Thyl-XFP transgenics can be used for high-resolution imaging of neurons. Specifically, these models allow imaging of structure in the transgenic lines where neurons are sparsely labeled (Feng et al. 2000). The XFP fills neurons in their entirety, including axons, dendrites, somas, nuclei, and dendritic spines.

Furthermore, FPs can be specifically targeted to different organelles (De Giorgi et al. 1999). Nuclear localization of GFP can be achieved by tagging GFP with a part of the glucocorticoid receptor, which is translocated into the nucleus on hormone binding (Picard and Yamamoto 1987) or tagging a protein of interest with a nuclear localization sequence (NLS) (Kalderon et al. 1984). To target GFP to mitochondria, GFP can be fused with a mitochondrial presequence (Hartl et al. 1989). Localization to endoplasmic reticulum can be achieved by fusing GFP to an N-terminal hydrophobic leader sequence, along with a "retention" signal, to sequester GFP within the endoplasmic reticulum once it arrives there (Munro and Pelham 1987). Golgi localization of GFP can be achieved by fusing a single transmembrane domain and sialyltransferase to GFP, thus retaining proteins in the Golgi (Schwientek et al. 1995). Plasma membrane can also be visualized when GFP is fused to a transmembrane protein on the cytosolic side of the membrane or fused to a protein that gets localized to the membrane, such as SNAP25 (Marsault et al. 1997). In addition, myr sequence can be used to localize a protein of interest to plasma membrane (Ruppert et al. 1995).

Astrocytic networks can be visualized using XFPs driven by a modified GFAP promoter in both transgenic animals (Zhuo et al. 1997) and after viral expression (Figure 14.3). Topical application of sulforhodamine 101 (SR101), a water-soluble red fluorescent dye, onto exposed brain before imaging also allows visualization of astrocytes (Nimmerjahn et al. 2004). It is hypothesized that the dye is distributed through gap junctions that connect individual astrocytes into astrocytic networks. However, because the application is topical, SR101 is best for acute experiments, whereas GFAP-driven XFPs are better suitable for longitudinal studies.

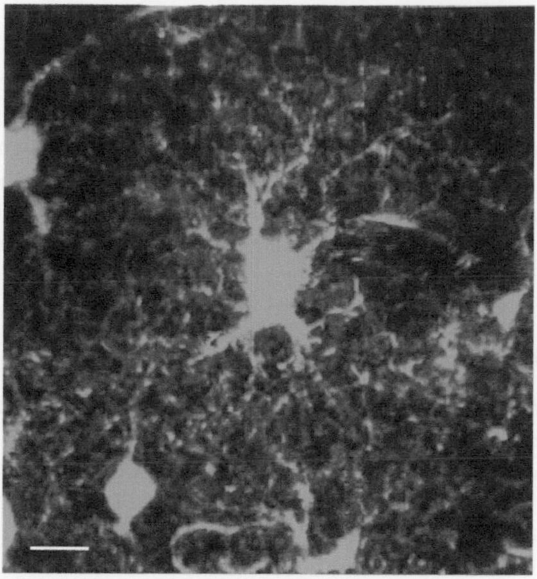

Figure 14.3 Image taken on a multiphoton microscope of cortical astrocytes expressing GFP under GFAP promoter. GFAP-GFP construct was expressed via adeno-associated virus in mouse cortex. Scale bar 7 μm.

14.2.1 Imaging Mitochondrial Mobility

Mitochondria provide the energy necessary for cellular function. They are also highly mobile structures within the cell. Mitochondria can be transported in the anterograde and retrograde direction along the microtubules lining axonal tracts by motor complexes composed of adaptor proteins and kinesins or dyneins, respectively. Mitochondria can also be mobilized along the actin filaments localized primarily to axonal terminals, dendrites, and growth cones via the motor complexes composed of myosin and adaptor proteins. Furthermore, mitochondria are physically stabilized via docking machinery (Sheng and Cai 2012).

Recently, genetically engineered mice were developed to allow imaging of mitochondrial mobility *in vivo*. These mice express mitochondrially targeted cyan fluorescent protein (CFP) and yellow fluorescent protein (YFP) specifically in neurons under the control of *Thy1* or *nse* (*Eno2*) regulatory elements (Misgeld et al. 2007). Another way to track mitochondria is by viral expression of mitoGFP (Figure 14.2). *In vivo* studies have confirmed that mitochondria can be transported bidirectionally along neurites or can be docked (Misgeld et al. 2007).

In summary, the advent of XFPs and their specific targeting has allowed the resolution and monitoring of cellular organelles as well as different cell types acutely and longitudinally in intact animals. This has led to greater insights into cellular structures and a deeper understanding of the organization of these structures. Monitoring changes in cellular architecture has become feasible and has allowed greater insight into the behavior of cellular structures and cells over time.

14.3 IMAGING NEURONAL FUNCTION

Sensation, locomotion, perception, and cognition involve the activity of elaborate neuronal networks. Understanding the function of these networks requires monitoring activity within large neuronal populations, not just single neurons. Until recently, *in vivo* studies were limited to multielectrode recordings and functional magnetic resonance imaging, which are limited by simultaneous recording from a small number of cells or by low resolution, respectively. Calcium imaging, voltage-sensitive dye (VSD) imaging, and other imaging methods described in Sections 14.3.1–14.3.4 overcome these limitations and allow the investigation of neuronal activity on a larger scale.

14.3.1 Calcium Imaging

Calcium is an omnipresent signaling ion in the body. It regulates a multitude of cellular processes involved in the cell cycle, ranging from cell division and proliferation to cellular death (Lu and Means 1993). Hence, intense efforts have been devoted to designing tools to monitor calcium dynamics to gain insight into processes regulated by this ion. Furthermore, electrical activity in the nervous system can be indirectly extrapolated from calcium imaging, as synaptic activity and the generation of action potentials are followed by an influx of calcium.

Currently, there are two classes of calcium indicators, small-molecule calcium dyes and genetically encoded calcium indicators, that allow the visualization of calcium dynamics *in vivo*. Small-molecule calcium dyes were designed in the laboratory of Roger Tsien (1980), who first developed the calcium-selective chelator BAPTA (1,2-bis[*o*-aminophenoxy]ethane-*N*,*N*,*N*′,*N*′-tetraacetic acid) and subsequently coupled it with fluorescent chromophores. Examples of small-molecule calcium dyes include fura-2 (Grynkiewicz et al. 1985), Oregon Green BAPTA, and fluo-4 dyes (Paredes et al. 2008). These small-molecule reporters have been successfully used for decades in cell culture, but their use required a specialized delivery approach to get them to load into cells in the intact brain (Stosick et al. 2003). After they are pressure injected into tissues, these dyes are taken up by cells and allow the visualization of changes in intracellular calcium using either spectral, lifetime, or brightness changes.

The second class of calcium indicators are genetically encoded calcium indicators (GECIs), which are available in two flavors, Förster/fluorescence resonance energy transfer (FRET)-based and single-fluorophore calcium indicators (see Chapter 3). Yellow Cameleon 3.6 (YC3.6) is a representative FRET-based calcium indicator (Nagai et al. 2004). It contains a donor, cyan fluorescent protein (CFP), and an acceptor, yellow fluorescent protein (YFP), connected by the calcium-binding protein calmodulin and the calmodulin-binding peptide M13 (Nagai et al. 2004) (Figure 14.1). In the absence of calcium, the predominant signal is from CFP. Calcium binding to the sensor results in more efficient FRET and stimulated emission by YFP. The calcium readout can be expressed as a ratio of YFP fluorescence to CFP fluorescence. Thus, YC3.6 is a ratiometric probe that allows for the measurement of absolute intracellular calcium concentrations and not just changes in calcium, as is the case for single-wavelength indicators (Section 14.3.1). Another advantage of using GECIs is the ability to simultaneously measure calcium concentration and visualize the structural morphology of the cells expressing the reporter, including individual neurites, dendritic spines, and astrocytic processes. This is difficult with small-molecule calcium indicators, because they are detected primarily in cell bodies. Lastly, YC3.6 allows *in vivo* chronic imaging over time.

An example of a genetically encoded single-fluorophore calcium indicator is GCaMP, which is composed of EGFP bound to calmodulin and the calmodulin-binding protein M13 (Nakai et al. 2001). Several iterations of this probe have led to dramatic improvements (Akerboom et al. 2012). Calcium binding to the GCaMP sensor results in a conformational change in the fluorophore environment that leads to an increase in fluorescence intensity (Nakai et al. 2001; Tian et al. 2009).

In vivo calcium imaging has been used in a variety of animal systems as well as in models of disease (Grienberger and Konnerth 2012). Both resting calcium and calcium transients underlying electrical activity can be imaged in neurons, while intracellular calcium waves can be imaged in glia. Our laboratory has recently discovered disruptions in resting calcium within a subset of neurites in transgenic animals with amyloid plaques, resulting in calcium overload in several animal models of Alzheimer's disease (Kuchibhotla et al. 2008). Furthermore, in addition to neuronal calcium disruptions, we observed elevations of resting calcium within the astrocytes. This resulted in more frequent calcium waves, which were not dependent on neuronal activity (Kuchibhotla et al. 2009). Other groups have monitored calcium transients in animal models of Alzheimer's disease and detected hyperactive and hypoactive neurons (Busche et al. 2008; Grienberger et al. 2012). These data provide further support for the calcium hypothesis of Alzheimer's disease that was originally formulated by Z. S. Khachaturian (Khachaturian 1984) and revised since.

Finally, there is substantial interest in imaging calcium in freely moving animals to correlate calcium dynamics and brain circuit activity with specific behavioral paradigms (Dombeck et al. 2009; Komiyama et al. 2010). This can be achieved by imaging calcium with two-photon microscopy of head-fixed animals or with head-mounted imaging devices (Andermann et al. 2010; Barretto and Schnitzer 2012).

14.3.2 Chloride Imaging

Similar to monitoring calcium influx to detect excitatory synaptic activity, it is now possible to monitor chloride influx to detect inhibitory synaptic activity. There are two classes of chloride indicators, small-molecule chloride dyes and genetically encoded chloride indicators. Quinoline-based chloride indicator dyes such as MQAE (*N*-(ethoxycarbonylmethyl)-6-methoxyquinolinium bromide), SPQ (6-methoxy-*N*-(3-sulfopropyl)quinolinium), and MEQ (6-methoxy-*N*-ethylquinolinium iodide) are small-molecule dyes (Verkman et al. 1989; Biwersi and Verkman 1991; Schwartz and Yu 1995) that can be applied topically to neural tissue in anesthetized animals. MQAE and MEQ are membrane permeable, whereas SPQ is not.

An example of a genetically encoded chloride-sensing probe is the FRET-based fusion protein Clomeleon, which is composed of a CFP donor and a YFP acceptor connected by a linker (Kuner and Augustine 2000). At low intracellular chloride concentrations, there is energy transfer between CFP and YFP, resulting in YFP emission. With increasing chloride concentrations, chloride ions bind YFP and quench its signal (Jayaraman et al. 2000; Markova et al. 2008). Thus, Clomeleon is a ratiometric probe that provides a read-out of chloride concentrations, which are proportional to the ratio of YFP to CFP. Clomeleon imaging in the brain can be performed with multiphoton imaging through a cranial window (Berglund et al. 2008). Particularly useful are the genetically engineered mice that express Clomeleon under a *thy1* promoter (Berglund et al. 2008). Several disadvantages of using Clomeleon and MQAE are their sensitivity to pH and susceptibility to bleaching (Kuner and Augustine 2000; Markova et al. 2008).

14.3.3 Voltage-Sensitive Dye Imaging

Similar to calcium imaging, voltage-sensitive dyes (VSDs; see Chapter 8) allow activity recordings of large populations of neurons with high spatial (on the order of micrometers) and temporal (with millisecond timescale) resolution (Grinvald and Hildesheim 2004; Chemla and Chavane 2010). Topical application of VSD results in their binding to cellular membranes and translates a change in membrane voltage into a fluorescent signal that can be detected with an imaging device. Fast neuronal activity is the major contributor to the fluorescent signal, while slow changes in glial activation play a minor part (Konnerth and Orkand 1986; Lev-Ram and Grinvald 1986). Because the change in fluorescence during VSD imaging is related to a change in membrane potential, VSD imaging provides a direct readout of neuronal circuit activity. This is in contrast to optical imaging based on slow signals such as intrinsic imaging that allows monitoring neural activity indirectly (Cohen et al. 1974). Furthermore, the millisecond time scale of VSD imaging offers higher temporal resolution compared to intrinsic imaging. However, the inability to resolve individual action potentials limits VSD imaging to monitoring macro-scale changes in neuronal activity. VSDs have been used in a number of preparations including dissociated neuronal cultures (Parsons et al. 1991), neuronal populations in brain slices (Grinvald et al. 1982; Sato et al. 1998), neuronal populations in spinal cord preparations (Momose-Sato et al. 2009), and recently, imaging in anesthetized as well as awake animals through cranial windows (Ferezou et al. 2006) with minimal side effects and phototoxicity. Imaging of cortical activity in freely behaving animals (Ferezou et al. 2006) is especially powerful, as it allows studying neural activity without the confounding effects of anesthesia and observing the activity of the neural networks that underlie behavior.

14.3.4 Measuring Immediate Early Gene Activation

Specific patterns of neuronal activity leading to lasting changes in connectivity, such as long-term potentiation (LTP), involve gene expression and the synthesis of new proteins (Davis and Squire 1984; Kandel 2001). The expression of immediate early genes (IEGs) is a first indication that this type of neuronal activity has occurred, before the activation of late response genes by the protein products of IEG expression (Nguyen et al. 1994). Thus, imaging IEG expression provides indirect evidence of where and when neuronal activity has occurred. A list of IEGs expressed in the brain is discussed elsewhere (Okuno 2011). Here, we focus on IEGs whose expression is localized to the mammalian cortex and can be imaged *in vivo*. One example is *Arc*, whose expression is dynamically regulated by activity. Until recently, its expression could be assessed only *ex vivo* using immunohistochemistry or *in situ* hybridization (Lyford et al. 1995; Guzowski et al. 1999). Recently, however, several Arc indicator mouse lines were engineered (Wang et al. 2006; Eguchi and Yamaguchi 2009; Grinevich et al. 2009), in which Arc expression is coupled to a destabilized reporter protein. One of these lines holds promise for imaging *in vivo*, as it expresses a sufficient number of copies of the fluorescent protein to allow protein visualization in live mice (Eguchi and Yamaguchi 2009; Rudinskiy et al. 2012).

IEG activation is used to monitor activity on the scale of minutes to hours compared to voltage sensors and ion indicators, which monitor activity within a millisecond-to-second range (Barth 2007).

Over the years, a multitude of imaging probes have been developed to study neuronal activity. These probes allow the visualization of activity from within individual cells, as is the case with calcium imaging, to the neuronal activity of entire populations of neurons made feasible with voltage-imaging dyes, thus allowing insights into the roles of individual neurons within a circuit or the study of neuronal bursting patterns underlying complex behaviors.

14.4 OTHER TYPES OF IMAGING

14.4.1 Imaging Oxidative Stress

Oxidation-sensitive variants of GFP allow monitoring of redox status within tissues in real time (Ostergaard et al. 2001; Dooley et al. 2004). These genetically encoded reporters can provide the oxidation status with high spatial and temporal resolution. The roGFP reporter exhibits a change in its excitation spectrum when oxidized, permitting ratiometric detection with dual excitation imaging (Dooley et al. 2004). Because this is a genetically encoded sensor, it can be targeted to specific cell types or organelles to be expressed for long periods of time, allowing chronic, minimally invasive imaging through cranial windows *in vivo* (Xie et al. 2013) (Figure 14.2).

Finally, the Amplex Red reagent has been used to detect extracellular oxidative stress. This colorless reagent reacts with hydrogen peroxide to generate the fluorescent product resorufin (Zhou et al. 1997). The probe provides a sensitive approach to measure hydrogen peroxide within the picomolar range (Zhou et al. 1997). Amplex Red can be applied topically to the brain to image free-radical production through cranial windows (McLellan et al. 2003; Garcia-Alloza et al. 2009).

14.4.2 Imaging Vascular Function

Most of the imaging techniques discussed in Sections 14.3.1 and 14.3.3, such as those involving voltage-sensitive dyes and small-molecule calcium indicators, use exogenous fluorophores. Hemodynamic changes in the brain can be detected with intrinsic signals that are related to changes in neuronal activity (Dunn et al. 2003, 2005). Thus, imaging based on hemodynamics (Kleinfeld et al. 1998), or blood oxygenation and deoxygenation levels (Bouchard et al. 2009), provides an indirect measure of neuronal activity. These hemodynamic events are thought to be the basis for the blood oxygen level dependent (BOLD) effect in functional magnetic resonance imaging (fMRI) (Kwong et al. 1992), which is also discussed further in

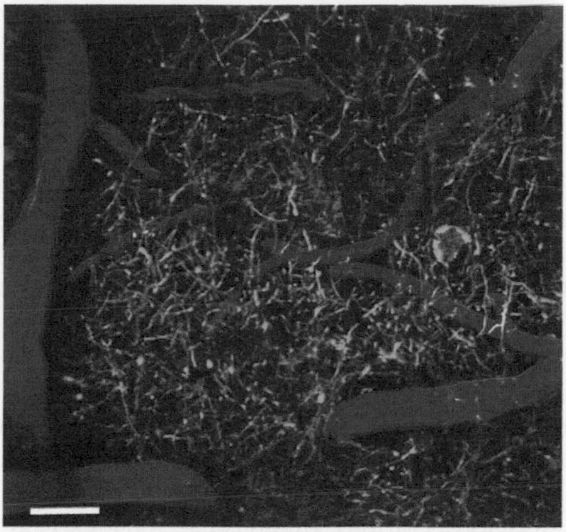

Figure 14.4 An image of AAV-mtGFP (in green) labeled mitochondria is shown in the living mouse brain. Vasculature was labeled with Texas red dextran (in red). Scale bar 20 μm.

Chapter 17. Recently, Petzold et al. (2008) determined that increases in synaptic activity were accompanied by increases in blood flow and the induction of calcium transients in astrocytic endfeet, suggesting that astrocytes might be the mediators coupling hemodynamic changes and neuronal activity in the brain.

In addition, the vasculature can be labeled for direct measurements of vessel diameter and blood flow. The intravenous administration of fluorescently labeled dextrans is routinely used for fluorescence angiography (Helmchen and Kleinfeld 2008) (Figure 14.4). These angiograms can be used as 3D maps to assist in returning to identified locations within the brain for repeated imaging (Bacskai et al. 2001, 2002).

High-resolution imaging also allows the measurement of instantaneous velocity at a single-vessel resolution, because the movement of unlabeled red blood cells (RBCs) through the vessels generates black streaks against the fluorescently labeled plasma (Kleinfeld et al. 1998). In addition, blood vessel diameter can be measured and used to monitor vascular reactivity over time (Driscoll et al. 2011).

14.4.3 Imaging Caspase Activity

Traditionally, caspases have been thought to play crucial roles in programmed cell death, or apoptosis, which is a mechanism for organisms to eliminate cellular debris and maintain homeostasis. Recently, caspases have also been implicated in non–cell death processes, such as neurite loss, spine loss, and synaptic pruning, that are necessary for normal development (Hyman 2011; Hyman and Yuan 2012). In addition to executing beneficial functions, caspases have been shown to play a role in the progression of a variety of disease processes, including Alzheimer's disease, where caspase activation has been linked to pathophysiology (D'Amelio et al. 2011). Therefore, there has been great interest in imaging caspase activity *in vivo*, where these processes can be studied.

To that end, fluorescent activity-based probes have been developed (Heal et al. 2011) that provide higher signal-to-noise ratios than previously designed fluorescent or radioactively labeled annexin V (Ntziachristos et al. 2004; Blankenberg et al. 2006) or small amphipathic molecules (Aloya et al. 2006). In addition, several small-molecule inhibitors, such as WC-II-89, M808 (Zhou et al. 2006), and fluorescently labeled peptide fluoromethyl ketones FLICA (fluorochrome-labeled inhibitors of caspases) (Bedner et al. 2000; Smolewski et al. 2001), can be used to monitor caspase activity *in vivo* (de Calignon et al. 2009; Xie et al. 2013).

14.5 CONCLUSION

The last several decades have brought substantial progress to the field of optical microscopy, whose uses have expanded far beyond the imaging of histological specimens and *in vitro* preparations. The discovery of GFP and the development of fluorescent probes revolutionized the way we look at cell biology, physiology, and pathophysiology. The ability to target specific organs, cells, organelles, and molecular targets has resulted in an ever increasing toolbox for biomedical research. We can now image specific structures within the intact nervous system as well as the function of individual components and their networks. In combination with sophisticated imaging technologies such as multiphoton microscopy, it is now possible to image the structure and function of almost every component of the CNS both acutely and over time. This has led to great insight into brain function in health and during disease, particularly in the progression of Alzheimer's disease in mouse models.

REFERENCES

Akerboom, J., Chen, T.W., Wardill, T.J., Tian, L., Marvin, J.S., Mutlu, S., Calderon, N.C., Esposti, F., Borghuis, B.G., Sun, X.R., Gordus, A., Orger, M.B., Portugues, R., Engert, F., Macklin, J.J., Filosa, A., Aggarwal, A., Kerr, R.A., Takagi, R., Kracun, S., Shigetomi, E., Khakh, B.S., Baier, H., Lagnado, L., Wang, S.S., Bargmann, C.I., Kimmel, B.E., Jayaraman, V., Svoboda, K., Kim, D.S., Schreiter, E.R., Looger, L.L. (2012). Optimization of a GCaMP calcium indicator for neural activity imaging. *J Neurosci* 32(40): 13819–13840.

Aloya, R., Shirvan, A., Grimberg, H., Reshef, A., Levin, G., Kidron, D., Cohen, A., Ziv, I. (2006). Molecular imaging of cell death in vivo by a novel small molecule probe. *Apoptosis* 11(12): 2089–2101.

Andermann, M.L., Kerlin, A.M., Reid, R.C. (2010). Chronic cellular imaging of mouse visual cortex during operant behavior and passive viewing. *Front Cell Neurosci* 4: 3.

Bacskai, B.J., Hickey, G.A., Skoch, J., Kajdasz, S.T., Wang, Y., Huang, G.F., Mathis, C.A., Klunk, W.E., Hyman, B.T. (2003). Four-dimensional multiphoton imaging of brain entry, amyloid binding, and clearance of an amyloid-beta ligand in transgenic mice. *Proc Natl Acad Sci USA* 100(21): 12462–12467.

Bacskai, B.J., Kajdasz, S.T., Christie, R.H., Carter, C., Games, D., Seubert, P., Schenk D., Hyman, B.T. (2001). Imaging of amyloid-beta deposits in brains of living mice permits direct observation of clearance of plaques with immunotherapy. *Nat Med* 7(3): 369–372.

Bacskai, B.J., Klunk, W.E., Mathis, C.A., Hyman, B.T. (2002). Imaging amyloid-beta deposits in vivo. *J Cereb Blood Flow Metab* 22(9): 1035–1041.

Barretto, R.P., Schnitzer, M.J. (2012). In vivo optical microendoscopy for imaging cells lying deep within live tissue. *Cold Spring Harb Protoc* 2012(10): 1029–1034.

Barth, A.L. (2007). Visualizing circuits and systems using transgenic reporters of neural activity. *Curr Opin Neurobiol* 17(5): 567–571.

Bedner, E., Smolewski, P., Amstad, P., Darzynkiewicz, Z. (2000). Activation of caspases measured in situ by binding of fluorochrome-labeled inhibitors of caspases (FLICA): Correlation with DNA fragmentation. *Exp Cell Res* 259(1): 308–313.

Berglund, K., Schleich, W., Wang, H., Feng, G., Hall, W.C., Kuner, T., Augustine, G.J. (2008). Imaging synaptic inhibition throughout the brain via genetically targeted Clomeleon. *Brain Cell Biol* 36(1–4): 101–118.

Biwersi, J., Verkman, A.S. (1991). Cell-permeable fluorescent indicator for cytosolic chloride. *Biochemistry* 30(32): 7879–7883.

Blankenberg, F.G., Vanderheyden, J.L., Strauss, H.W., Tait, J.F. (2006). Radiolabeling of HYNIC-annexin V with technetium-99m for in vivo imaging of apoptosis. *Nat Protoc* 1(1): 108–110.

Borchelt, D.R., Ratovitski, T., van Lare, J., Lee, M.K., Gonzales, V., Jenkins, N.A., Copeland, N.G., Price, D.L., Sisodia, S.S. (1997). Accelerated amyloid deposition in the brains of transgenic mice coexpressing mutant presenilin 1 and amyloid precursor proteins. *Neuron* 19(4): 939–945.

Bouchard, M.B., Chen, B.R., Burgess, S.A., Hillman, E.M. (2009). Ultra-fast multispectral optical imaging of cortical oxygenation, blood flow, and intracellular calcium dynamics. *Opt Express* 17(18): 15670–15678.

Busche, M.A., Eichhoff, G., Adelsberger, H., Abramowski, D., Wiederhold, K.H., Haass, C., Staufenbiel, M., Konnerth, A., Garaschuk, O. (2008). Clusters of hyperactive neurons near amyloid plaques in a mouse model of Alzheimer's disease. *Science* 321(5896): 1686–1689.

Chalfie, M., Tu, Y., Euskirchen, G., Ward, W.W., Prasher, D.C. (1994). Green fluorescent protein as a marker for gene expression. *Science* 263(5148): 802–805.

Chemla, S., Chavane, F. (2010). Voltage-sensitive dye imaging: Technique review and models. *J Physiol (Paris)* 104(1–2): 40–50.

Christie, R.H., Bacskai, B.J., Zipfel, W.R., Williams, R.M., Kajdasz, S.T., Webb, W.W., Hyman, B.T. (2001). Growth arrest of individual senile plaques in a model of Alzheimer's disease observed by in vivo multiphoton microscopy. *J Neurosci* 21(3): 858–864.

Cohen, L.B., Salzberg, B.M., Davila, H.V., Ross, W.N., Landowne, D., Waggoner, A.S., Wang, C.H. (1974). Changes in axon fluorescence during activity: Molecular probes of membrane potential. *J Membr Biol* 19(1): 1–36.

D'Amelio, M., Cavallucci, V., Middei, S., Marchetti, C., Pacioni, S., Ferri, A., Diamantini, A., De Zio, D., Carrara, P., Battistini, L., Moreno, S., Bacci, A., Ammassari-Teule, M., Marie, H., Cecconi, F. (2011). Caspase-3 triggers early synaptic dysfunction in a mouse model of Alzheimer's disease. *Nat Neurosci* 14(1): 69–76.

Davis, H.P., Squire, L.R. (1984). Protein synthesis and memory: A review. *Psychol Bull* 96(3): 518–559.

de Calignon, A., Fox, L.M., Pitstick, R., Carlson, G.A., Bacskai, B.J., Spires-Jones, T.L., Hyman, B.T. (2010). Caspase activation precedes and leads to tangles. *Nature* 464(7292): 1201–1204.

de Calignon, A., Spires-Jones, T.L., Pitstick, R., Carlson, G.A., Hyman, B.T. (2009). Tangle-bearing neurons survive despite disruption of membrane integrity in a mouse model of tauopathy. *J Neuropathol Exp Neurol* 68(7): 757–761.

De Giorgi, F., Ahmed, Z., Bastianutto, C., Brini, M., Jouaville, L.S., Marsault, R., Murgia, M., Pinton, P., Pozzan, T., Rizzuto, R. (1999). Targeting GFP to organelles. *Methods Cell Biol* 58: 75–85.

Dittgen, T., Nimmerjahn, A., Komai, S., Licznerski, P., Waters, J., Margrie, T.W., Helmchen, F., Denk, W., Brecht, M., Osten, P. (2004). Lentivirus-based genetic manipulations of cortical neurons and their optical and electrophysiological monitoring in vivo. *Proc Natl Acad Sci USA* 101(52): 18206–18211.

Dombeck, D.A., Graziano, M.S., Tank, D.W. (2009). Functional clustering of neurons in motor cortex determined by cellular resolution imaging in awake behaving mice. *J Neurosci* 29(44): 13751–13760.

Dooley, C.T., Dore, T.M., Hanson, G.T., Jackson, W.C., Remington, S.J., Tsien, R.Y. (2004). Imaging dynamic redox changes in mammalian cells with green fluorescent protein indicators. *J Biol Chem* 279(21): 22284–22293.

Driscoll, J.D., Shih, A.Y., Drew, P.J., Cauwenberghs, G., Kleinfeld, D. (2011). Two-photon imaging of blood flow in the rat cortex. In F. Helmchen and A. Konnerth (eds.), *Imaging in Neuroscience*. New York: Cold Spring Harbor Laboratory Press.

Dunn, A.K., Devor, A., Bolay, H., Andermann, M.L., Moskowitz, M.A., Dale, A.M., Boas, D.A. (2003). Simultaneous imaging of total cerebral hemoglobin concentration, oxygenation, and blood flow during functional activation. *Opt Lett* 28(1): 28–30.

Dunn, A.K., Devor, A., Dale, A.M., Boas, D.A. (2005). Spatial extent of oxygen metabolism and hemodynamic changes during functional activation of the rat somatosensory cortex. *NeuroImage* 27(2): 279–290.

Eguchi, M., Yamaguchi, S. (2009). In vivo and in vitro visualization of gene expression dynamics over extensive areas of the brain. *NeuroImage* 44(4): 1274–1283.

Farrar, M.J., Bernstein, I.M., Schlafer, D.H., Cleland, T.A., Fetcho, J.R., Schaffer, C.B. (2012). Chronic in vivo imaging in the mouse spinal cord using an implanted chamber. *Nat Methods* 9(3): 297–302.

Feng, G., Mellor, R.H., Bernstein, M., Keller-Peck, C., Nguyen, Q.T., Wallace, M., Nerbonne, J.M., Lichtman, J.W., Sanes, J.R. (2000). Imaging neuronal subsets in transgenic mice expressing multiple spectral variants of GFP. *Neuron* 28(1): 41–51.

Ferezou, I., Bolea, S., Petersen, C.C. (2006). Visualizing the cortical representation of whisker touch: Voltage-sensitive dye imaging in freely moving mice. *Neuron* 50(4): 617–629.

Fuss, B., Mallon, B., Phan, T., Ohlemeyer, C., Kirchhoff, F., Nishiyama, A., Macklin, W.B. (2000). Purification and analysis of in vivo-differentiated oligodendrocytes expressing the green fluorescent protein. *Dev Biol* 218(2): 259–274.

Garcia-Alloza, M., Bacskai, B.J. (2004). Techniques for brain imaging in vivo. *Neuromol Med* 6(1): 65–78.

Garcia-Alloza, M., Borrelli, L.A., Rozkalne, A., Hyman, B.T., Bacskai, B.J. (2007). Curcumin labels amyloid pathology in vivo, disrupts existing plaques, and partially restores distorted neurites in an Alzheimer mouse model. *J Neurochem* 102(4): 1095–1104.

Garcia-Alloza, M., Dodwell, S.A., Meyer-Luehmann, M., Hyman, B.T., Bacskai, B.J. (2006). Plaque-derived oxidative stress mediates distorted neurite trajectories in the Alzheimer mouse model. *J Neuropathol Exp Neurol* 65(11): 1082–1089.

Garcia-Alloza, M., Prada, C., Lattarulo, C., Fine, S., Borrelli, L.A., Betensky, R., Greenberg, S.M., Frosch, M.P., Bacskai, B.J. (2009). Matrix metalloproteinase inhibition reduces oxidative stress associated with cerebral amyloid angiopathy in vivo in transgenic mice. *J Neurochem* 109(6): 1636–1647.

Gibson, A., Dehghani, H. (2009). Diffuse optical imaging. *Philos Trans A Math Phys Eng Sci* 367(1900): 3055–3072.

Grienberger, C., Konnerth, A. (2012). Imaging calcium in neurons. *Neuron* 73(5): 862–885.

Grienberger, C., Rochefort, N.L., Adelsberger, H., Henning, H.A., Hill, D.N., Reichwald, J., Staufenbiel, M., Konnerth, A. (2012). Staged decline of neuronal function in vivo in an animal model of Alzheimer's disease. *Nat Commun* 3: 774.

Grinevich, V., Kolleker, A., Eliava, M., Takada, N., Takuma, H., Fukazawa, Y., Shigemoto, R., Kuhl, D., Waters, J., Seeburg, P.H., Osten, P. (2009). Fluorescent Arc/Arg3.1 indicator mice: A versatile tool to study brain activity changes in vitro and in vivo. *J Neurosci Methods* 184(1): 25–36.

Grinvald, A., Hildesheim, R. (2004). VSDI: A new era in functional imaging of cortical dynamics. *Nat Rev Neurosci* 5(11): 874–885.

Grinvald, A., Manker, A., Segal, M. (1982). Visualization of the spread of electrical activity in rat hippocampal slices by voltage-sensitive optical probes. *J Physiol* 333: 269–291.

Grynkiewicz, G., Poenie, M., Tsien, R.Y. (1985). A new generation of Ca^{2+} indicators with greatly improved fluorescence properties. *J Biol Chem* 260(6): 3440–3450.

Guzowski, J.F., McNaughton, B.L., Barnes, C.A., Worley, P.F. (1999). Environment-specific expression of the immediate-early gene Arc in hippocampal neuronal ensembles. *Nat Neurosci* 2(12): 1120–1124.

Hartl, F.U., Pfanner, N., Nicholson, D.W., Neupert, W. (1989). Mitochondrial protein import. *Biochim Biophys Acta* 988(1): 1–45.

Heal, W.P., Dang, T.H., Tate, E.W. (2011). Activity-based probes: Discovering new biology and new drug targets. *Chem Soc Rev* 40(1): 246–257.

Heim, R., Cubitt, A.B., Tsien, R.Y. (1995). Improved green fluorescence. *Nature* 373(6516): 663–664.

Helmchen, F., Denk, W. (2005). Deep tissue two-photon microscopy. *Nat Methods* 2(12): 932–940.

Helmchen, F., Kleinfeld, D. (2008). In vivo measurements of blood flow and glial cell function with two-photon laser-scanning microscopy, pp. 231–254. In *Methods in Enzymology*, Vol. 444. San Diego, CA: Academic Press.

Hirrlinger, P.G., Scheller, A., Braun, C., Quintela-Schneider, M., Fuss, B., Hirrlinger, J., Kirchhoff, F. (2005). Expression of reef coral fluorescent proteins in the central nervous system of transgenic mice. *Mol Cell Neurosci* 30(3): 291–303.

Holtzman, D.M., Goate, A., Kelly, J., Sperling, R. (2011). Mapping the road forward in Alzheimer's disease. *Sci Transl Med* 3(114): 114ps148.

Hsiao, K., Chapman, P., Nilsen, S., Eckman, C., Harigaya, Y., Younkin, S., Yang, F., Cole, G. (1996). Correlative memory deficits, Abeta elevation, and amyloid plaques in transgenic mice. *Science* 274(5284): 99–102.

Hyman, B.T. (2011). Caspase activation without apoptosis: Insight into Abeta initiation of neurodegeneration. *Nat Neurosci* 14(1): 5–6.

Hyman, B.T., Yuan, J. (2012). Apoptotic and non-apoptotic roles of caspases in neuronal physiology and pathophysiology. *Nat Rev Neurosci* 13(6): 395–406.

Jankowsky, J.L., Slunt, H.H., Ratovitski, T., Jenkins, N.A., Copeland, N.G., Borchelt, D.R. (2001). Co-expression of multiple transgenes in mouse CNS: A comparison of strategies. *Biomol Eng* 17(6): 157–165.

Jayaraman, S., Haggie, P., Wachter, R.M., Remington, S.J., Verkman, A.S. (2000). Mechanism and cellular applications of a green fluorescent protein-based halide sensor. *J Biol Chem* 275(9): 6047–6050.

Jung, S., Aliberti, J., Graemmel, P., Sunshine, M.J., Kreutzberg, G.W., Sher A., Littman, D.R. (2000). Analysis of fractalkine receptor CX(3)CR1 function by targeted deletion and green fluorescent protein reporter gene insertion. *Mol Cell Biol* 20(11): 4106–4114.

Kalderon, D., Roberts, B.L., Richardson, W.D., Smith, A.E. (1984). A short amino acid sequence able to specify nuclear location. *Cell* 39(3 Pt 2): 499–509.

Kandel, E.R. (2001). The molecular biology of memory storage: A dialogue between genes and synapses. *Science* 294(5544): 1030–1038.

Khachaturian, Z.S. (1984). Towards theories of brain aging. In D.W. Kilbourne and G.D. Burrows (eds.), *Handbook of Studies on Psychiatry and Old Age*. Amsterdam: Elsevier.

Kleinfeld, D., Mitra, P.P., Helmchen F., Denk, W. (1998). Fluctuations and stimulus-induced changes in blood flow observed in individual capillaries in layers 2 through 4 of rat neocortex. *Proc Natl Acad Sci USA* 95(26): 15741–15746.

Klunk, W.E., Bacskai, B.J., Mathis, C.A., Kajdasz, S.T., McLellan, M.E., Frosch, M.P., Debnath, M.L., Holt, D.P., Wang, Y., Hyman, B.T. (2002). Imaging Abeta plaques in living transgenic mice with multiphoton microscopy and methoxy-X04, a systemically administered Congo red derivative. *J Neuropathol Exp Neurol* 61(9): 797–805.

Klunk, W.E., Engler, H., Nordberg, A., Wang, Y., Blomqvist, G., Holt, D.P., Bergstrom, M., Savitcheva, I., Huang, G.F., Estrada, S., Ausen, B., Debnath, M.L., Barletta, J., Price, J.C., Sandell, J., Lopresti, B.J., Wall, A., Koivisto, P., Antoni, G., Mathis, C.A., Langstrom, B. (2004). Imaging brain amyloid in Alzheimer's disease with Pittsburgh Compound-B. *Ann Neurol* 55(3): 306–319.

Komiyama, T., Sato, T.R., O'Connor, D.H., Zhang, Y.X., Huber, D., Hooks, B.M., Gabitto, M., Svoboda, K. (2010). Learning-related fine-scale specificity imaged in motor cortex circuits of behaving mice. *Nature* 464(7292): 1182–1186.

Konnerth, A., Orkand, R.K. (1986). Voltage-sensitive dyes measure potential changes in axons and glia of the frog optic nerve. *Neurosci Lett* 66(1): 49–54.

Kuchibhotla, K.V., Goldman, S.T., Lattarulo, C.R., Wu, H.Y., Hyman, B.T., Bacskai, B.J. (2008). Abeta plaques lead to aberrant regulation of calcium homeostasis in vivo resulting in structural and functional disruption of neuronal networks. *Neuron* 59(2): 214–225.

Kuchibhotla, K.V., Lattarulo, C.R., Hyman, B.T., Bacskai, B.J. (2009). Synchronous hyperactivity and intercellular calcium waves in astrocytes in Alzheimer mice. *Science* 323(5918): 1211–1215.

Kuner, T., Augustine, G.J. (2000). A genetically encoded ratiometric indicator for chloride: Capturing chloride transients in cultured hippocampal neurons. *Neuron* 27(3): 447–459.

Kwong, K.K., Belliveau, J.W., Chesler, D.A., Goldberg, I.E., Weisskoff, R.M., Poncelet, B.P., Kennedy, D.N., Hoppel, B.E., Cohen, M.S., Turner, R. et al. (1992). Dynamic magnetic resonance imaging of human brain activity during primary sensory stimulation. *Proc Natl Acad Sci USA* 89(12): 5675–5679.

Lev-Ram, V., Grinvald, A. (1986). Ca^{2+}- and K^{+}-dependent communication between central nervous system myelinated axons and oligodendrocytes revealed by voltage-sensitive dyes. *Proc Natl Acad Sci USA* 83(17): 6651–6655.

Livet, J., Weissman, T.A., Kang, H., Draft, R.W., Lu, J., Bennis, R.A., Sanes, J.R., Lichtman, J.W. (2007). Transgenic strategies for combinatorial expression of fluorescent proteins in the nervous system. *Nature* 450(7166): 56–62.

Lombardo, J.A., Stern, E.A., McLellan, M.E., Kajdasz, S.T., Hickey, G.A., Bacskai, B.J., Hyman, B.T. (2003). Amyloid-beta antibody treatment leads to rapid normalization of plaque-induced neuritic alterations. *J Neurosci* 23(34): 10879–10883.

Lu, K.P., Means, A.R. (1993). Regulation of the cell cycle by calcium and calmodulin. *Endocr Rev* 14(1): 40–58.

Luker, G.D., Luker, K.E. (2008). Optical imaging: Current applications and future directions. *J Nucl Med* 49(1): 1–4.

Lyford, G.L., Yamagata, K., Kaufmann, W.E., Barnes, C.A., Sanders, L.K., Copeland, N.G., Gilbert, D.J., Jenkins, N.A., Lanahan, A.A., Worley, P.F. (1995). Arc, a growth factor and activity-regulated gene, encodes a novel cytoskeleton-associated protein that is enriched in neuronal dendrites. *Neuron* 14(2): 433–445.

Markova, O., Mukhtarov, M., Real, E., Jacob, Y., Bregestovski, P. (2008). Genetically encoded chloride indicator with improved sensitivity. *J Neurosci Methods* 170(1): 67–76.

Marsault, R., Murgia, M., Pozzan, T., Rizzuto, R. (1997). Domains of high Ca^{2+} beneath the plasma membrane of living A7r5 cells. *EMBO J* 16(7): 1575–1581.

McLellan, M.E., Kajdasz, S.T., Hyman, B.T., Bacskai, B.J. (2003). In vivo imaging of reactive oxygen species specifically associated with thioflavine S-positive amyloid plaques by multiphoton microscopy. *J Neurosci* 23(6): 2212–2217.

Misgeld, T., Kerschensteiner, M., Bareyre, F.M., Burgess, R.W., Lichtman, J.W. (2007). Imaging axonal transport of mitochondria in vivo. *Nat Methods* 4(7): 559–561.

Momose-Sato, Y., Mochida, H., Kinoshita, M. (2009). Origin of the earliest correlated neuronal activity in the chick embryo revealed by optical imaging with voltage-sensitive dyes. *Eur J Neurosci* 29(1): 1–13.

Munro, S., Pelham, H.R. (1987). A C-terminal signal prevents secretion of luminal ER proteins. *Cell* 48(5): 899–907.

Nagai, T., Yamada, S., Tominaga, T., Ichikawa, M., Miyawaki, A. (2004). Expanded dynamic range of fluorescent indicators for $Ca^{(2+)}$ by circularly permuted yellow fluorescent proteins. *Proc Natl Acad Sci USA* 101(29): 10554–10559.

Nakai, J., Ohkura, M., Imoto, K. (2001). A high signal-to-noise $Ca^{(2+)}$ probe composed of a single green fluorescent protein. *Nat Biotechnol* 19(2): 137–141.

Nguyen, P.V., Abel, T., Kandel, E.R. (1994). Requirement of a critical period of transcription for induction of a late phase of LTP. *Science* 265(5175): 1104–1107.

Nimmerjahn, A., Kirchhoff, F., Kerr, J.N., Helmchen, F. (2004). Sulforhodamine 101 as a specific marker of astroglia in the neocortex in vivo. *Nat Methods* 1(1): 31–37.

Ntziachristos, V., Schellenberger, E.A., Ripoll, J., Yessayan, D., Graves, E., Bogdanov, A., Jr., Josephson, L., Weissleder, R. (2004). Visualization of antitumor treatment by means of fluorescence molecular tomography with an annexin V-Cy5.5 conjugate. *Proc Natl Acad Sci USA* 101(33): 12294–12299.

Okuno, H. (2011). Regulation and function of immediate-early genes in the brain: Beyond neuronal activity markers. *Neurosci Res* 69(3): 175–186.

Ostergaard, H., Henriksen, A., Hansen, F.G., Winther, J.R. (2001). Shedding light on disulfide bond formation: Engineering a redox switch in green fluorescent protein. *EMBO J* 20(21): 5853–5862.

Papale, A., Cerovic, M., Brambilla, R. (2009). Viral vector approaches to modify gene expression in the brain. *J Neurosci Methods* 185(1): 1–14.

Paredes, R.M., Etzler, J.C., Watts, L.T., Zheng, W., Lechleiter, J.D. (2008). Chemical calcium indicators. *Methods* 46(3): 143–151.

Parsons, T.D., Salzberg, B.M., Obaid, A.L., Raccuia-Behling, F., Kleinfeld, D. (1991). Long-term optical recording of patterns of electrical activity in ensembles of cultured *Aplysia* neurons. *J Neurophysiol* 66(1): 316–333.

Petzold, G.C., Albeanu, D.F., Sato, T.F., Murthy, V.N. (2008). Coupling of neural activity to blood flow in olfactory glomeruli is mediated by astrocytic pathways. *Neuron* 58(6): 897–910.

Picard, D., Yamamoto, K.R. (1987). Two signals mediate hormone-dependent nuclear localization of the glucocorticoid receptor. *EMBO J* 6(11): 3333–3340.

Ramon y Cajal, S. (1995). *Histology of the Nervous System of Man and Vertebrates.* Oxford: Oxford University Press.

Rudinskiy, N., Hawkes, J.M., Betensky, R.A., Eguchi, M., Yamaguchi, S., Spires-Jones, T.L., Hyman, B.T. (2012). Orchestrated experience-driven Arc responses are disrupted in a mouse model of Alzheimer's disease. *Nat Neurosci* 15(10): 1422–1429.

Ruppert, C., Godel, J., Muller, R.T., Kroschewski, R., Reinhard, J., Bahler, M. (1995). Localization of the rat myosin I molecules myr 1 and myr 2 and in vivo targeting of their tail domains. *J Cell Sci* 108(Pt 12): 3775–3786.

Sakuma, T., Barry, M., Aikeda, Y. (2012). Lentiviral vectors: Basic to translational. *Biochem J* 443(3): 603–618.

Santacruz, K., Lewis, J., Spires, T., Paulson, J., Kotilinek, L., Ingelsson, M., Guimaraes, A., DeTure, M., Ramsden, M., McGowan, E., Forster, C., Yue, M., Orne, J., Janus, C., Mariash, A., Kuskowski, M., Hyman, B., Hutton, M., Ashe, K.H. (2005). Tau suppression in a neurodegenerative mouse model improves memory function. *Science* 309(5733): 476–481.

Sato, K., Momose-Sato, Y., Hirota, A., Sakai, T., Kamino, K. (1998). Optical mapping of neural responses in the embryonic rat brainstem with reference to the early functional organization of vagal nuclei. *J Neurosci* 18(4): 1345–1362.

Schwartz, R.D., Yu, X. (1995). Optical imaging of intracellular chloride in living brain slices. *J Neurosci Methods* 62(1–2): 185–192.

Schwientek, T., Lorenz, C., Ernst, J.F. (1995). Golgi localization in yeast is mediated by the membrane anchor region of rat liver sialyltransferase. *J Biol Chem* 270(10): 5483–5489.

Shaner, N.C., Campbell, R.E., Steinbach, P.A., Giepmans, B.N., Palmer, A.E., Tsien, R.Y. (2004). Improved monomeric red, orange and yellow fluorescent proteins derived from *Discosoma* sp. red fluorescent protein. *Nat Biotechnol* 22(12): 1567–1572.

Shaner, N.C., Steinbach, P.A., Tsien, R.Y. (2005). A guide to choosing fluorescent proteins. *Nat Methods* 2(12): 905–909.

Sheng, Z.H., Cai, Q. (2012). Mitochondrial transport in neurons: Impact on synaptic homeostasis and neurodegeneration. *Nat Rev Neurosci* 13(2): 77–93.

Shimomura, O., Johnson, F.H., Saiga, Y. (1962). Extraction, purification and properties of aequorin, a bioluminescent protein from the luminous hydromedusan, Aequorea. *J Cell Comp Physiol* 59: 223–239.

Shu, X., Shaner, N.C., Yarbrough, C.A., Tsien, R.Y., Remington, S.J. (2006). Novel chromophores and buried charges control color in mFruits. *Biochemistry* 45(32): 9639–9647.

Skoch, J., Hickey, G.A., Kajdasz, S.T., Hyman, B.T., Bacskai, B.J. (2005). In vivo imaging of amyloid-beta deposits in mouse brain with multiphoton microscopy. *Methods Mol Biol* 299: 349–363.

Smolewski, P., Bedner, E., Du, L., Hsieh, T.C., Wu, J.M., Phelps, D.J., Darzynkiewicz, Z. (2001). Detection of caspases activation by fluorochrome-labeled inhibitors: multiparameter analysis by laser scanning cytometry. *Cytometry* 44(1): 73–82.

Spires-Jones, T.L., de Calignon, A., Matsui, T., Zehr, C., Pitstick, R., Wu, H.Y., Osetek, J.D., Jones, P.B., Bacskai, B.J., Feany, M.B., Carlson, G.A., Ashe, K.H., Lewis, J., Hyman, B.T. (2008). In vivo imaging reveals dissociation between caspase activation and acute neuronal death in tangle-bearing neurons. *J Neurosci* 28(4): 862–867.

Spires-Jones, T.L., de Calignon, A., Meyer-Luehmann, M., Bacskai, B.J., Hyman, B.T. (2011). Monitoring protein aggregation and toxicity in Alzheimer's disease mouse models using in vivo imaging. *Methods* 53(3): 201–207.

Spires-Jones, T.L., Meyer-Luehmann, M., Osetek, J.D., Jones, P.B., Stern, E.A., Bacskai, B.J., Hyman, B.T. (2007). Impaired spine stability underlies plaque-related spine loss in an Alzheimer's disease mouse model. *Am J Pathol* 171(4): 1304–1311.

Stosiek, C., Garaschuk, O., Holthoff, K., Konnerth, A. (2003). In vivo two-photon calcium imaging of neuronal networks. *Proc Natl Acad Sci USA* 100(12): 7319–7324.

Styren, S.D., Hamilton, R.L., Styren, G.C., Klunk, W.E. (2000). X-34, a fluorescent derivative of Congo red: A novel histochemical stain for Alzheimer's disease pathology. *J Histochem Cytochem* 48(9): 1223–1232.

Tamamaki, N., Yanagawa, Y., Tomioka, R., Miyazaki, J., Obata, K., Kaneko, T. (2003). Green fluorescent protein expression and colocalization with calretinin, parvalbumin, and somatostatin in the GAD67–GFP knock-in mouse. *J Comp Neurol* 467(1): 60–79.

Tian, L., Hires, S.A., Mao, T., Huber, D., Chiappe, M.E., Chalasani, S.H., Petreanu, L., Akerboom, J., McKinney, S.A., Schreiter, E.R., Bargmann, C.I., Jayaraman, V., Svoboda, K., Looger, L.L. (2009). Imaging neural activity in worms, flies and mice with improved GCaMP calcium indicators. *Nat Methods* 6(12): 875–881.

Tsien, R.Y. (1980). New calcium indicators and buffers with high selectivity against magnesium and protons: Design, synthesis, and properties of prototype structures. *Biochemistry* 19(11): 2396–2404.

van den Pol, A.N., Ghosh, P.K. (1998). Selective neuronal expression of green fluorescent protein with cytomegalovirus promoter reveals entire neuronal arbor in transgenic mice. *J Neurosci* 18(24): 10640–10651.

Velasco, A., Fraser, G., Delobel, P., Ghetti, B., Lavenir, I., Goedert, M. (2008). Detection of filamentous tau inclusions by the fluorescent Congo red derivative FSB [(trans,trans)-1-fluoro-2,5-bis(3-hydroxycarbonyl-4-hydroxy)styrylbenzene]. *FEBS Lett* 582(6): 901–906.

Verkman, A.S., Sellers, M.C., Chao, A.C., Leung, T., Ketcham, R. (1989). Synthesis and characterization of improved chloride-sensitive fluorescent indicators for biological applications. *Anal Biochem* 178(2): 355–361.

Wang, K.H., Majewska, A., Schummers, J., Farley, B., Hu, C., Sur, M., Tonegawa, S. (2006). In vivo two-photon imaging reveals a role of arc in enhancing orientation specificity in visual cortex. *Cell* 126(2): 389–402.

Xie, H., Hou, S., Jiang, J., Sekutowicz, M., Kelly, J., Bacskai, B.J. (2013). Rapid cell death is preceded by amyloid plaque-mediated oxidative stress. *Proc Natl Acad Sci USA* 110(19): 7904–7909.

Yao, J., Wang, L.V. (2011). Photoacoustic tomography: Fundamentals, advances and prospects. *Contrast Media Mol Imaging* 6(5): 332–345.

Zhou, D., Chu, W., Rothfuss, J., Zeng, C., Xu, J., Jones, L., Welch, M.J., Mach, R.H. (2006). Synthesis, radiolabeling, and in vivo evaluation of an 18F-labeled isatin analog for imaging caspase-3 activation in apoptosis. *Bioorg Med Chem Lett* 16(19): 5041–5046.

Zhou, M., Diwu, Z., Panchuk-Voloshina, N., Haugland, R.P. (1997). A stable nonfluorescent derivative of resorufin for the fluorometric determination of trace hydrogen peroxide: Applications in detecting the activity of phagocyte NADPH oxidase and other oxidases. *Anal Biochem* 253(2): 162–168.

Zhuo, L., Sun, B., Zhang, C.L., Fine, A., Chiu, S.Y., Messing, A. (1997). Live astrocytes visualized by green fluorescent protein in transgenic mice. *Dev Biol* 187(1): 36–42.

Chapter 15

Smart Imaging Probes for the Study of Protease Function

Ehud Segal and Matthew Bogyo

CONTENTS

15.1 INTRODUCTION

Enzymes play essential functional roles in many biological and physiological processes in all organisms (Koblinski et al. 2000; Frank and Hargreaves 2003). Proteolytic enzymes, also known as proteases, are enzymes that catalyze the breakdown of proteins by hydrolyzing their peptide bonds. This type of post-translational regulation allows the cell to control the activation or inactivation of proteins. Because of the irreversibility of this process, as well as the importance of protease activity, an extensive regulatory network has evolved to ensure targeted spatial and temporal control of protease activity. Imbalance of this regulation is frequently associated with diverse diseases including cancer, inflammation, cardiovascular disease, and neurodegeneration, as well as viral and bacterial infections. Consequently, many proteases have been identified as promising drug targets, while others have been evaluated for their potential as biomarkers for various diseases (Clardy and Walsh 2004).

So far almost 600 proteases (about 2% of the human genome) have been identified and classified by their primary mechanism of action (Puente et al. 2005). However, deciphering the roles of specific proteases in a relevant biological environment and mapping the signaling pathways that control each proteolytic network remains a formidable challenge. One reason for this is the limited number of tools that allow the collection of real-time information regarding protease activity in a physiologically relevant environment. This valuable data can potentially lead to the identification and validation of proteases as relevant drug targets and can be utilized for the development of effective therapeutics against them. In this chapter, we describe various strategies of "smart" imaging reagents that allow sensitive, rapid, and high-resolution detection of proteases activity within cells and whole organisms.

15.2 THE BIOLOGICAL FUNCTION OF PROTEASES

The catalytic function of proteases is to hydrolyze amide bonds in proteins (Turk 2006). This process is irreversible and used by the cell to regulate the function and fate of many proteins (Chapman et al. 1997). Proteases mediate various processes including coagulation (Travis and Ferguson 1951), digestion (Neurath and Walsh 1976), the maturation of cytokines and pro-hormones (Lapidot and Petit 2002), apoptosis (Kutsyi et al. 1999), and the breakdown of intracellular proteins (Drag and Salvesen 2010). Consistent with the key roles of proteolytic events, aberrant proteolysis is often associated with diverse pathologies such as hemorrhagic disorders, immune disease, cancer, inflammation, arteriosclerosis, and neurodegeneration (Quesada et al. 2009). To prevent inappropriate proteolytic events, activity is tightly regulated by three main mechanisms: (1) the activation of inactive precursors (also known as zymogens), (2) control of access to substrates, and (3) the binding of inhibitors and cofactors (Kobe and Kemp 1999) (Figure 15.1). The activation of inactive protease precursors can be either autocatalytic or catalyzed by other proteases. In some cases, protease activation requires additional factors or the formation of large protein complexes such as the proteasome, involved in bulk protein turnover, and the apoptosome, which mediates the activation of pro-apoptotic caspases (Riedl and Salvesen 2007). Protease activity is often regulated by cofactors, proteins that reversibly bind to proteases and affect their final activity (Adam et al. 2004). Inhibition of protease activity can also be achieved by small molecules as well as protein based, endogenously expressed inhibitors that directly prevent access of substrates to the active site (i.e., canonical mechanism). Alternatively, inhibitors can bind to a region outside the active site, thus preventing substrate access without directly blocking the primary

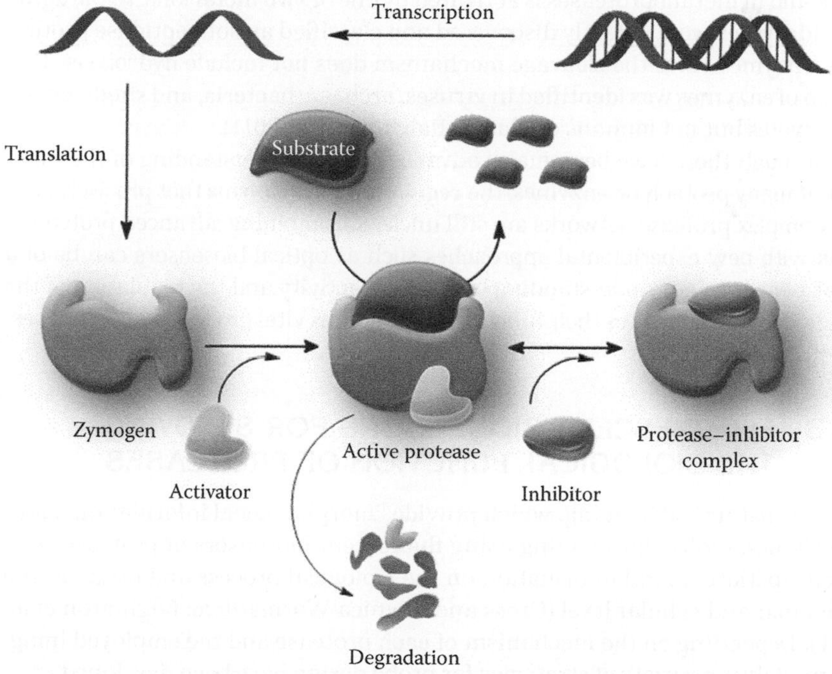

Figure 15.1 Regulation of protease activity. Protease activity is regulated at several different levels, including transcription and translation, the activation of zymogen forms, degradation, and the inactivation of the proteases by varying mechanisms. Once the protease is active, substrates and inhibitors compete for protease binding, and the outcome is defined by the local inhibitor concentration. (Adapted from Deu, E. et al., *Nat. Struct. Mol. Biol.* 19(1):9–16, 2012.)

catalytic residues (i.e., exosite-binding mechanism) (*Semin. Cell Dev. Biol.* 2009). A third group of protease inhibitors uses an intermediate mechanism based on a combination of the canonical and exosite-binding mechanisms. Other regulatory mechanisms include changes in protease gene expression, the control of mRNA stability, and substrate sequestration by inactive protease homologs and cellular internalization (Bode and Huber 2000).

Proteases are generally grouped into primary families—cysteine proteases, serine proteases, aspartyl proteases, metalloproteases, glutamate proteases, threonine and asparagine peptide lyases—based on their catalytic mechanisms (Lopez-Otin and Overall 2002). Serine, cysteine, and threonine proteases use the amino acid side chain of a primary catalytic residue for direct nucleophilic attack on a substrate, resulting in the formation of an acyl-enzyme intermediate. Aspartic and metallopeptidases, on the other hand, use water as the primary nucleophile, which in aspartic proteases is activated by two aspartates and in metalloproteases is activated by one or two metal ions. Asparagine peptide lyases were recently discovered and classified as nonpeptidase proteolytic enzymes since the cleavage mechanism does not include hydrolysis. This group of enzymes was identified in viruses, archaea, bacteria, and single-celled eukaryotes but not humans or plants (Rawlings et al. 2011).

Although there have been major advances in our understanding of the function of many proteolytic enzymes, the regulatory mechanisms that precisely control complex protease networks are still unclear. Combining advanced proteomic tools with new experimental approaches such as optical biosensors can be of a great benefit to our understanding of protease activity and the regulation of the biochemistry that relates their function to numerous vital processes and diseases.

15.3 FLUORESCENT BIOSENSORS FOR STUDYING THE BIOLOGICAL FUNCTION OF PROTEASES

Unlike anatomical imaging, which provides morphological information about live tissues, molecular imaging using fluorescent biosensors of proteases can reveal spatiotemporal information on the biological process and target at the molecular and cellular level (Gross and Piwnica-Worms 2006; Edgington et al. 2011). Depending on the mechanism of each protease and the employed imaging modality, a variety of strategies for probe design have been developed.

15.3.1 Fluorescent Protein Substrates

Fluorogenic substrates have been extensively used for kinetic measurements of enzyme activity *in vitro* (Boonacker and Van Noorden 2001). These reagents carry a fluorescent reporter group that, on enzymatic conversion to products, produces a fluorescent signal that can be monitored over time (Choe et al. 2006). The most widely used fluorogenic substrate probes consist of a peptide sequence attached at the C-terminus to a small organic fluorophore, such as an aminomethylcoumarin (AMC) (Los et al. 2000). In the presence of the active protease, the AMC is cleaved from the peptide, leading to a detectable shift in its fluorescent spectrum (Figure 15.2a). Other fluorochromes used in fluorogenic substrates include BODIPY (Jones et al. 1997), rhodamine (Claveau et al. 2000), cresyl violet (Van Noorden et al. 1997), Cy5.5 (Rochefort and Liaudet-Coopman 1999), fluorescein (Burchak et al. 2006), and resorufin (Coleman et al. 2007). Alternatively, it is also possible to make peptide substrates containing a fluorophore and quencher at opposite ends of the substrate. These substrates can then be cleaved to release fluorescent fragments. A different class of substrate-based probes for proteases uses two or more fluorophores that are self-quenched when in close proximity (Packard et al. 1997; Araujo et al. 1999). In addition, multiple fluorophores can be linked to graft polymers containing peptide substrate sequences. When these linkers are cleaved by the protease, free fluorescent monomers are released and the signal can be detected.

Fluorogenic substrates for proteases can be used not only for the determination of cell viability but also for metabolic studies. Juliano et al. (1999) synthesized fluorogenic substrates for serine, cysteine, and aspartyl proteases. The chromogenic and fluorogenic glycosylated and acetyl glycosylated peptides were found to have different kinetic characteristics when evaluated as substrates for bovine trypsin, papain, human tissue kallikrein, and rat tonin. Szabelski et al. (2005) developed internally quenched (by Förster/fluorescence resonance energy transfer [FRET]) substrates for cysteine proteases. The peptides were synthesized with diverse properties based on benzoxazol-5-yl-alanine derivatives and evaluated for their kinetic parameters of hydrolysis by papain and cathepsin B. The authors showed that the donor–acceptor pair used for FRET significantly affects the kinetic parameters of the enzymatic process. In another study, Tepel et al. (2000) developed a fluorogenic substrate for cathepsin K. The authors used benzyloxycarbonyl-glycylprolyl-arginine-4-methoxy-naphthylamide in combination with nitrosalicylaldehyde, which directly couples with the proteolytically released 4-methoxy-β-naphthylamine (4MβNA) to produce a fluorescent signal. Other examples include fluorogenic substrates for cathepsins B (Bleeker et al. 2000), D (Rochefort and Liaudet-Coopman 1999), H (Mahmood et al. 1999) and L (Assfalg-Machleidt et al. 1992), caspases (Zhang et al. 2003), matrix metalloproteinases (MMPs; Faust et al. 2009), and cytochrome P450 (Sukumaran et al. 2009).

In general, the use of fluorescent substrates for enzyme activity studies is fast and simple, and it is therefore mainly suitable for kinetic measurements of enzyme activity *in vitro* and for high-throughput screening. However, the specificity exhibited by many of the fluorogenic substrates is often difficult to control and may not be sufficient for use in complex cellular environments.

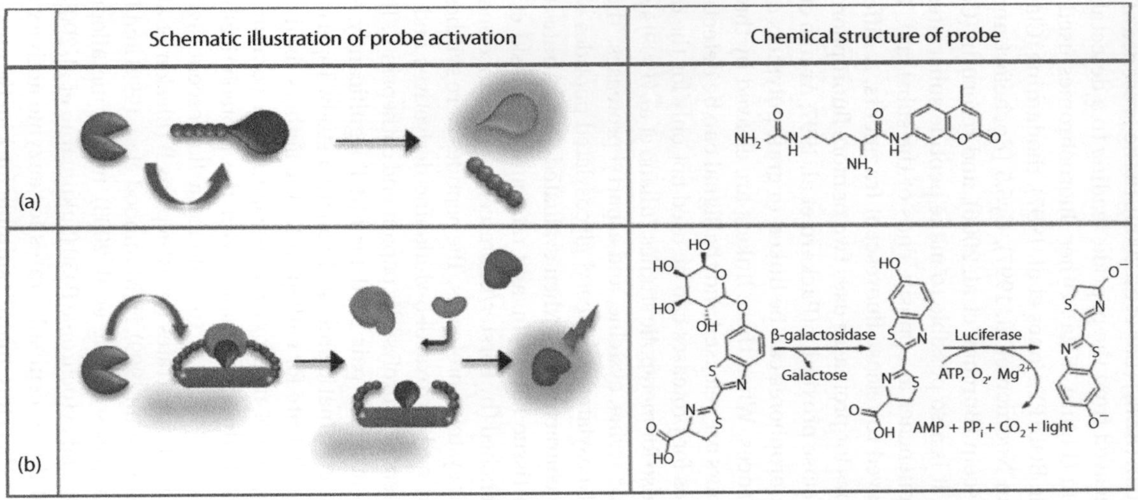

Figure 15.2 Fluorescent biosensors for studying the biological function of proteases. (a) Fluorogenic substrate. On enzymatic conversion, fluorescent group becomes active and can be monitored. (b) Macromolecular self-quenched probe and chemical structure of repeating graft copolymer segment. Arrows indicate quenching of fluorophores and enzymatic cleavage site (Mahmood et al. 1999).

(Continued)

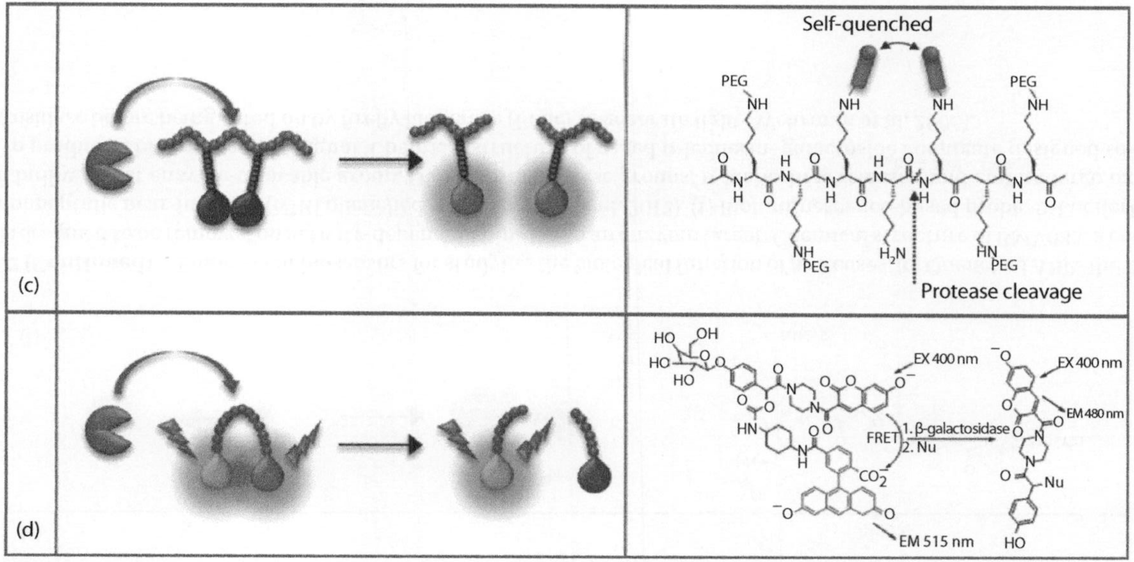

Figure 15.2 (Continued) Fluorescent biosensors for studying the biological function of proteases. (c) Fluorescence resonance energy transfer (FRET) probe based on donor and acceptor. Chemical structure shows a FRET-quenched β-galactosidase probe that, on cleavage, releases the donor (7-hydroxy-coumarin) and acceptor (fluorescein) fragments (Kwan et al. 2011). "Nu" represents a nucleophilic residue of an enzyme. (d) Activity-based probe (ABP). The highly selective probe binds covalently to an enzyme target using chemical interactions that are specific for the target enzyme. Chemical structure of the Cy5-labeled peptidic ABP GB123 for noninvasive optical imaging of cysteine protease activity (Blum et al. 2009).

(*Continued*)

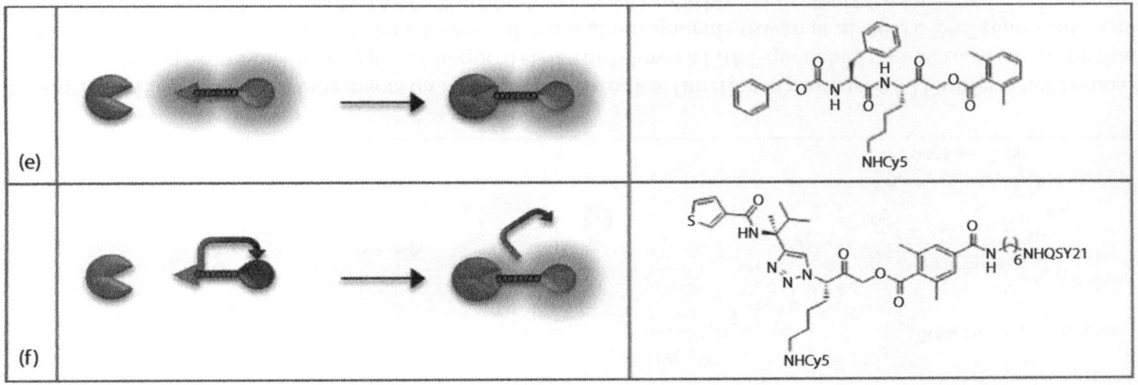

Figure 15.2 (Continued) Fluorescent biosensors for studying the biological function of proteases. (e) Quenched ABP. The fluorescent quencher is designed to be removed on activity-dependent binding to an enzyme target. Chemical structure of BMV083, a cathepsin S–directed, nonpeptidic near-infrared (NIR) quenched-ABP (Verdoes et al. 2012). (f) Bioluminescence-based probe. D-Luciferin is modified with a bulky, target enzyme-cleavable group. On removal of these groups, D-luciferin is released and can undergo oxidation by luciferase to produce a bioluminescent signal. Chemical structure of caged D-luciferin–galactoside conjugate designed to be cleaved by β-galactosidase before being acted on by firefly luciferase (FLuc) to generate light (Wehrman et al. 2006).

15.3.2 Bioluminescent Substrate-Based Probes

Most substrate-based probes utilize fluorescence as a reporter signal, but bioluminescence (BLI) is another modality that can be used. Compared with fluorescence, BLI has better tissue penetration and exceptionally low background. Nevertheless, there are significantly fewer reports of BLI substrate-based probes in the literature. One reason may be that most BLI studies use D-luciferin as the substrate, which is tolerant to only limited chemical modification (O'Brien et al. 2005; Shinde et al. 2006). In addition, in some cases D-luciferin can dramatically interfere with the target protease activity (Scherer et al. 2008). Laxman et al. (2002) have designed a luciferase-based reporter probe for the imaging of caspase-3 activity. This hybrid reporter comprises a luciferase protein flanked by two estrogen receptor (ER) regulatory domains that are linked through a caspase-3-peptide substrate (Figure 15.2b). The probe was used for real-time imaging of caspase activity in mice bearing reporter-expressing tumors. The induction of apoptosis resulted in a 3-fold increase in BLI compared with the un-induced control animals. In another study, Luker et al. designed a BLI ubiquitin–luciferase (Ub-Luc) fusion protein reporter to image the activity of the 26S proteasome (Luker et al. 2003). In cells transfected with this probe, a rapid attenuation of the BLI signal was observed as a result of substrate degradation. A comparison of the signal from construct-expressing cells with cells expressing wild-type luciferase provided a measure of signal loss caused by proteasome activity. The efficacy of the proteasome inhibitor bortezomib could be determined using this imaging approach, suggesting that it can be a powerful method for monitoring the efficacy of small-molecule drugs in a rapid and noninvasive manner. Wendt and colleagues used a different approach to develop a BLI probe for caspase-1 (Kindermann et al. 2010). This BLI probe was developed using a "reverse design" concept, in which chemically optimized protease inhibitors are converted into selective substrate probes (Watzke et al. 2008). By replacing the warhead of the inhibitor with a cleavable peptide bond and subsequently attaching amino-luciferin as the reporter group, a caspase-1 inhibitor was converted into a BLI probe for caspase-1 (CM-269) (Siegmund and Zeitz 2003). The probe exhibited selectivity for caspase-1 and an approximately 1000-fold increase in sensitivity compared to fluorogenic peptides substrates.

Although BLI probes are clearly valuable tools for *in vitro* and cell-based assays, they are generally not applicable for clinical studies because of the need to introduce the luciferase enzyme.

15.3.3 Near-Infrared Substrate-Based Probes

Photon penetration into living tissue is dependent on the absorption and scattering properties of tissue components (Ntziachristos et al. 2003). Therefore, the use of conventional fluorescent substrates emitting light in the visible range (e.g., coumarin, *p*-nitrophenol, and other analogues) is usually restricted to *in vitro* studies. The near-infrared (NIR) region of the spectrum offers significant advantages over the visible range, including low autofluorescence, higher penetration of photons through tissues, and improved signal-to-background ratios. As a consequence, the use of NIR reagents for *in vivo* imaging applications has expanded significantly in the last decade (Licha et al. 2008). The incorporation of NIR imaging reagents with "smart" molecules for enzyme activity allows noninvasive visualization of enzyme activity in whole organisms (Pham et al. 2004; Morimoto 2007). In addition, some NIR reagents have the ability to act as quenchers if located in close spatial proximity to one another. Weissleder and colleagues developed several NIR molecular probes for different applications (Tung et al. 1999, 2000; Jaffer et al. 2002). One interesting example is a NIR probe for imaging the serine protease thrombin. The NIR probe comprised a Cy5.5-labeled thrombin-cleavable peptide attached to a partially PEGylated poly-L-lysine scaffold. This results in a probe that is quenched due to the high density of fluorophore on the probe. On cleavage by the protease, free fluorophore is liberated. *In vivo* studies demonstrated that thrombin activity can be detected by the probe in a mouse model of intravascular thrombosis. Other examples of NIR substrate-based reporters include probes for caspase-1 (Messerli et al. 2004), caspase-3, and caspase-7 (Kim et al. 2006). A different design strategy for NIR substrate-based probes was recently introduced by Tsien and colleagues. The authors developed activatable cell-penetrating peptides (CPPs) for MMP-2 and MMP-9 (Jiang et al. 2004). The probes are composed of a fluorescently tagged peptide containing a cell-penetrating poly-cation stretch of amino acids followed by a poly-anion stretch separated by a cleavable linker. In the absence of proteases, the cationic region forms a strong charge interaction with the anionic region and the cellular uptake is largely blocked. Once the linker is cleaved by proteases, the poly-anionic inhibitory domain is lost and the fluorescently tagged poly-cation region is internalized into cells. The authors showed that probe uptake into HT-1080 cells is dependent on cleavage by MMP-2. In addition, following injection into HT-1080 tumor-bearing mice, the CPPs containing the MMP-cleavable linker showed 3-fold more accumulation in tumor cells than CPPs containing a scrambled linker.

15.3.4 Macromolecular Substrate-Based Probes

An alternative approach for developing optical biosensors for enzyme activity involves the use of macromolecular probes (Figure 15.2c). This class of biosensors mainly uses self-quenched extracellular matrix (ECM) substrates or other proteins (such as fluorescein-labeled collagen or BODIPY dye-labeled casein substrates) to visualize proteolysis *in vitro* or in living cells (Mahmood et al. 1999; Olson et al. 2012; Saravanakumar et al. 2012). In this approach, the substrates become fluorescent *in situ* via the action of proteases secreted into the ECM. To address the nonspecific cleavage of the macromolecules, more sophisticated probes were developed, which include a protease-specific linker between the polymeric backbone and the fluorophore (Gabriel et al. 2011). For example, a NIR probe specifically targeting cathepsin D was synthesized using the amino acid sequence Gly-Pro-Ile-Cys(Et)-Phe-Phe-Arg-Leu-Gly-Lys(FITC)-$CysNH_2$ as a linker between the polymeric carrier and the fluorophore (Tung et al. 2000). In its native state, this reporter probe is nonfluorescent at 700 nm due to resonance energy transfer among the quenched fluorochromes but becomes brightly fluorescent when the latter are released by cathepsin D. The autoquenched polymeric probe exhibited a long circulation time and high specificity toward cathepsin D in mice. In another study, Kim et al. (2006) reported a cell-permeable and biocompatible polymeric nanoparticle for apoptosis imaging using a Cy5.5-labeled caspase-sensitive Asp-Glu-Val-Asp motif attached to a branched poly-(ethylenimine) deoxycholic acid conjugate. Incubation of the polymeric nanoparticle with either caspase-3 or caspase-7 led to an approximately 7-fold increase in fluorescence intensity compared to background and could be abolished by coincubation with caspase inhibitors. In a step toward multimodal molecular imaging, Olson et al. (2010) reported the development of activatable cell-penetrating peptides linked to nanoparticles as dual probes for *in vivo* fluorescence and magnetic resonance imaging (MRI) of proteases. The authors were able to visualize MMP activities by MRI and by the fluorescence of dendrimeric nanoparticles coated with activatable cell-penetrating peptides (ACPPs) labeled with Cy5, gadolinium, or both.

Macromolecule protease probes can be further utilized by their passive targeting capacity due to the enhanced and retention permeability (EPR) effect in tumors and inflammatory lesions (Matsumura and Maeda 1986). However, the main limitation of macromolecule protease probes is the nonspecific cleavage of the linker between the polymeric carrier and the fluorophore. Depending on the application, this can be either of value or a significant disadvantage.

15.3.5 FRET Protein Reporters

FRET can be utilized to detect protease activity (Matayoshi et al. 1990) (Figure 15.2d). FRET is effective only when fluorophores are less than 100 Å apart (Selvin 2000). Therefore, the separation of two fluorophores as a result of protease activity can be easily analyzed. This makes FRET highly suitable for detecting the degradation of protease substrates. Two main classes of FRET-based probes have been reported: genetically encoded and small-molecule–based FRET probes. Cummings et al. (2002) designed several genetically encoded FRET-based reporters for *Bacillus anthracis* lethal factor (LF) protease. These reporters consist of an optimized FRET pair of fluorescent proteins, YPet and CyPet. CyPet and YPet were linked together by a flexible linker containing a consensus recognition site for LF protease flanked at both its N- and C-termini by several repeats of the flexible tripeptide Gly-Gly-Ser. Other samples include FRET systems for the measurement of hepatitis C virus NS3-4A protease activity (Sabariegos et al. 2009) and for caspases (Rehm et al. 2002). The second class of FRET-based probes consists of synthetic small-molecule probes. Gehrig S. et al. (2012) developed the ratiometric FRET reporter NEmo for the serine protease neutrophil elastase. The authors synthesized two versions of the FRET probes, NEmo-1, a peptide-based probe for measuring the soluble enzyme activity, and NEmo-2, a lipidated variant for measuring the membrane activity. The probes exhibited detection limits in the subnanomolar range when incubated with lung fluid and with neutrophils from a mouse model of acute neutrophilic lung inflammation. These FRET reporters, in combination with relevant small-molecule libraries, could provide a promising approach for high-throughput screening efforts using fluorescence-activated cell sorting (FACS).

15.4 ACTIVITY-BASED PROBES

Nearly all of the methods described in Section 15.3 use reporter substrates for the imaging of protease activity. Although many of these methods are useful for activity-based imaging, they all suffer from relatively poor selectivity for a specific enzyme. Furthermore, several of these methods require either the use of bulky cell-impermeable molecules or the introduction of reporter probes through the genetic manipulation of cells or organisms. An alternative method for visualizing enzyme activities uses small-molecule probes that covalently attach to an enzyme-target through a chemical reaction that is specific for the target enzyme(s) (Terai and Nagano 2008) (Figure 15.2d–f). These probes are called activity-based probes (ABPs) (Fonovic and Bogyo 2007). ABP-labeled proteomes can be analyzed using biochemical methods, thereby allowing specific identification of the individual active enzymes targeted by a probe. Furthermore, selectivity for individual or subsets of enzymes can be controlled by optimizing the probe backbone and the reactive group.

15.4.1 Structure and Reactivity

ABPs are small-molecule reporters designed to be highly selective for the catalytically active form of a protease or protease family (Cravatt et al. 2008). Typical ABPs consists of three main elements: (1) a reactive functional group (also termed the warhead) that covalently reacts with the active site of the enzyme; (2) a linker region that confers specificity, directs binding to the target, and prevents steric congestion; and (3) a tag used for the identification, purification, or direct visualization of the probe-labeled proteins (Deu et al. 2012) (Figure 15.3).

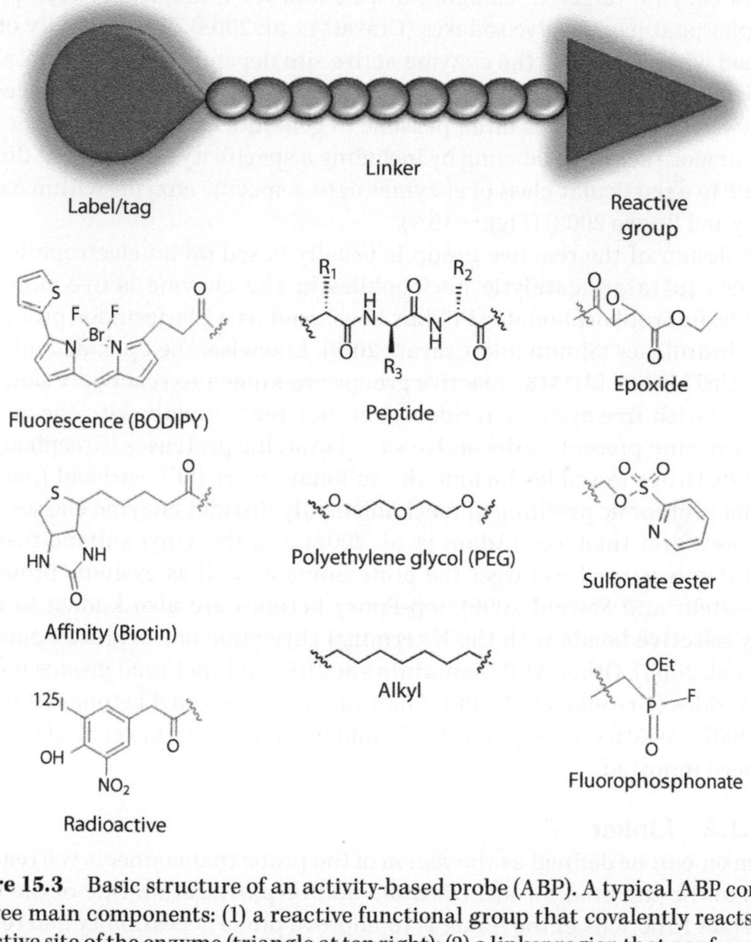

Figure 15.3 Basic structure of an activity-based probe (ABP). A typical ABP consists of three main components: (1) a reactive functional group that covalently reacts with the active site of the enzyme (triangle at top right); (2) a linker region that confers specificity, directs binding to the target, and prevents steric congestion (at middle); and (3) a tag used for the identification, purification, or direct visualization of the probe-labeled proteins (at left). Examples of chemical structures are shown in the columns beneath each of the 3 ABP elements. (Modified from Jeffery, D. A., and Bogyo, M., *Curr. Opin. Biotechnol.* 14(1):87–95, 2003.)

15.4.1.1 Reactive Group

The reactive group, or warhead, serves as the key functional element that directs the covalent, activity-dependent modification of a target enzyme or enzyme family (Nomura et al. 2010). It is usually an electrophile that specifically reacts with the nucleophile residue of the enzyme active site to form a stable covalent bond (Figure 15.3). The warhead typically defines the catalytic class of enzyme targeted, enabling distinction, for example, between proteases, phosphatases, or glycosidases (Cravatt et al. 2008). The reactivity of the warhead with respect to the enzyme active site depends on both affinity and chemical reactivity, and tuning this reactivity enables the control of selectivity between enzymes. It is often possible to generate an ABP with either very broad or more selective labeling by including a specificity element that directs the ABP to a particular class of enzymes or to a specific enzyme within a class (Jeffery and Bogyo 2003) (Figure 15.3).

The design of the reactive group is usually based on an electrophile that is known to target catalytic nucleophiles in the enzyme active sites. For example, fluorophosphonate (FP) has been used as a platform for probes for serine hydrolases (Simon and Cravatt 2010). Likewise, the epoxide and acyloxymethyl ketone (AOMK) reactive groups are known to react very slowly or not at all with free cysteine residues, but they react rapidly with the nucleophilic cysteine present in the active site of cysteine proteases (Greenbaum et al. 2000). Other examples include the sulfonate ester (SE) warhead that was used for proteomic profiling of mechanistically distinct enzyme classes such as kinases and thiolases (Adam et al. 2002) and the vinyl sulfone reactive group that was used to target the proteasome as well as cysteine proteases (Hagenstein and Sewald 2006). α,β-Epoxy ketones are also known to form highly selective bonds with the N-terminal threonine of the proteasome (Bo Kim et al. 2005). Other ABPs containing additional functional groups including epoxides (Grzonka et al. 2001; Shan 2004), diazomethyl ketones (Brouwer et al. 2007), AOMKs (Chong et al. 2011), and vinyl sulfones (Liu et al. 2008) have also been reported.

15.4.1.2 Linker

The linker can be defined as the region of the probe that connects the reactive group to the tag used for identification and/or purification. One of the most important functions of the linker is to improve probe accessibility and reduce steric hindrance between the tag and the reactive group (Fonovic and Bogyo 2008). For this purpose, the linker can be designed as an extended alkyl or polyethylene glycol spacer (Figure 15.3). Alternatively, the linker can be used to increase probe specificity through the addition of structural elements (i.e., peptides). For most protease probes, the linker is a short peptide sequence that mimics a true protein substrate. This peptide can be optimized to target distinct subfamilies of proteases by changing the amino acid sequence to match the substrate selectivity of the target protease(s) (Greenbaum et al. 2002a). For example, diphenyl phosphonates with a phenylalanine or lysine side chain are recognized by chymotrypsin- or trypsin-like proteases, respectively (Pan et al. 2006).

Most of the ABPs designed for molecular imaging typically carry a noncleavable linker that enables detection of the probe by different detection modalities. Alternatively, ABPs can be designed for the identification and isolation of target enzymes. In this case, the enrichment and purification of the targets are usually facilitated by biotinylated ABPs and immobilized streptavidin, after which mass spectrometry is used to identify the protein target (Speers and Cravatt 2005). A more recent improvement in linker design has been the development of a region that can be enzymatically or chemically cleaved to release probe-labeled proteins (Leriche et al. 2012). This simplifies the elution process and decreases background protein contamination during affinity purification. Cravatt and co-workers developed a cleavable linker that has a peptide recognition site for tobacco etch virus protease (TEV) and applied it to the identification of proteins that had been labeled with a sulfonate ester probe (Weerapana et al. 2007, 2010). Other types of cleavable linkers for various applications have been reported, including acid-cleavable linkers (van der Veken et al. 2005), hydrazones (Dirksen et al. 2010), silanes (Szychowski et al. 2010), photocleavable linkers (Orth et al. 2010), and a diazobenzene chemoselective cleavable linker for functional proteomic applications (Verhelst et al. 2007). An interesting recent example is a laser-light-cleavable mass tag for mass spectrometric applications. This probe was designed as a dendrimer with multiple mass tags that are attached to a reactive group for serine hydrolases (Yang et al. 2012). On irradiation of the labeled tissue by the laser beam, the mass tags are liberated and recorded by the mass analyzer. Consequently, the ion image of the mass tag reveals the distribution of active serine hydrolases in the tissue. This approach is attractive because it allows the detection of low-abundance proteins in tissues.

15.4.1.3 Tag

A key component that differentiates ABPs from common irreversible inhibitors is the presence of a tag or label (Sadaghiani et al. 2007). This element enables the visualization and/or purification of labeled proteins. Various tags have been incorporated into ABPs. These include isotope tags for biochemical and imaging applications, affinity tags for biochemical purification, stable isotope tags for mass spectrometry quantification, and fluorescent tags for cell biological and whole-body imaging applications (Figure 15.3).

15.4.1.3.1 Affinity Tags

Many ABPs contain tags for the enrichment of labeled proteins using affinity resins. The most commonly used affinity tag is biotin, which binds to avidin resins with high affinity (Freitag et al. 1999). The affinity of the tag for a resin is important when probes are used to label low-abundance proteins. The direct attachment of biotin to an ABP is often the most efficient method of tagging. However, biotin tags often suffer from a lack of cell permeability, which limits their use to labeling applications in cell and tissue extracts rather than for *in vivo* or cell-based labeling studies. To overcome this issue, a number of groups have developed orthogonal tagging methods in which the biotin or fluorescent tag is added after the probe has already modified the target enzyme (Speers and Cravatt 2005; Zhou et al. 2007; Nilvebrant et al. 2012). Thus, it is possible to use ABPs that have small, cell-permeable tags that are suitable for subsequent chemical ligation with a biotin affinity tag. Another alternative is the use of short peptide tags such as an HA or FLAG tag on an ABP. This approach has been used mainly for larger protein-based probes such as those designed to target ubiquitin-specific proteases (Yee et al. 2005; Tully and Cravatt 2010; Altun et al. 2011). Another example of an affinity tag for ABPs is short stretches of peptide nucleic acids (PNAs) (Winssinger et al. 2004; Winssinger and Harris 2005). This method was used to create probes in which the structure of the probe is "coded" by the PNA tag. The tag also serves as a way to purify the probe-labeled targets through direct hybridization to a DNA-containing chip. Finally, there have been a number of examples of the use of small-molecule probes that have been directly attached to a solid support (Hesek et al. 2006a, b). In this case, the tag is the resin bead. Although this is not optimal for covalent probes, as it would require a method to cleave the probe from the resin, this method has shown to be valuable for use in the isolation of kinase targets using reversible probe-bound resins.

15.4.1.4 Fluorophores

Various types of fluorophores with different photochemical and physical features have been used to tag ABPs. Commonly used fluorophores include fluorescein and rhodamine (Patricelli et al. 2001), BODIPY (Witte et al. 2011), cyanine dyes (Chan et al. 2004), dansyl (Berkers et al. 2005), and nitrobenz-2-oxa-1,3-diazole (NBD; Schmidinger et al. 2005). Naturally, each fluorophore has advantages and disadvantages. For instance, fluorescein has excellent quantum yield, high water solubility, and is relatively inexpensive, but on the other hand suffers from rapid photobleaching and poor cell permeability. BODIPY and cyanine dyes have excellent photostability, high fluorescence quantum yields, and excellent cell permeability. However, compared with fluorescein and rhodamine, they are expensive.

One of the main advantages of using a fluorescent tag is the ability to obtain a direct and rapid readout of probe labeling in sodium dodecyl sulfate-polyacrylamide gel electrophoresis (SDS-PAGE) gels. The probe-labeled proteomes can be analyzed using a fluorescence scanner and quantified while still in the glass plates used for electrophoresis. This enables the quantification of multiple gels and high-throughput screening. In addition, the hydrophobic character of some fluorophores can be utilized to improve the cellular uptake of the tagged probe. For example, DCG-04, a general papain family protease probe (Greenbaum et al. 2000), was previously tagged with BODIPY analogs. This allowed biochemical profiling and live-cell imaging (Greenbaum et al. 2002b) as well as the *in vivo* labeling of cysteine cathepsins in whole animals (Joyce et al. 2004).

Although fluorescent ABPs have been used for cell-based and *in vivo* imaging applications, the fluorescence of the unbound probes can affect both the sensitivity and specificity of optical imaging. To address this limitation, quenched activity-based probes (qABPs) have been developed (Blum et al. 2005). In this approach, a quencher group is placed near the fluorophore to prevent fluorescent emission. Once the probe covalently binds to the target enzyme, the quencher is lost and the unquenched probe emits the desired fluorescent signal. These probes have been used previously for several applications including dynamic imaging of cysteine protease activity in living cells and in animals. Blum et al. (2005) developed ABPs that contain a fluorescent quencher that is removed on activity-dependent binding to an enzyme target. In this study, a qABP, comprising a peptide AOMK containing a BODIPY fluorophore and quenching group, was used to image cathepsin activity in whole cells (Figure 15.2f). Verdoes et al. (2012) later developed a cathepsin S–directed, nonpeptidic NIR qABP. This probe was used for the classification of tumor-associated myeloid-derived cells based on *in vivo* protease activity. Noninvasive *in vivo* imaging showed high tumor-specific fluorescence in mice bearing syngeneic breast tumors.

15.4.2 Applications of ABPs

The ability to target various enzymes using ABPs has created a wide range of applications. Because ABPs can be used to assess the activity of a given target enzyme in the context of complex proteomic samples and even in whole organisms, they can be used effectively to evaluate such important parameters as drug efficacy, pharmacodynamics, and overall target selectivity of drug leads (Cravatt et al. 2008). Furthermore, *in vitro* and *in vivo* application of ABPs allows direct imaging of protease activity within the context of whole cells and living subjects. These applications are outlined in the following.

15.4.2.1 Target Identification

A key step when developing a new therapeutic agent is the selection of a target that is expected to modulate a pathway or disease of interest. ABPs can be used to evaluate the repertoire of associated target enzymes in a given disease model. The relevance of a specific enzyme or family of enzymes can be assessed by monitoring changes in activity levels during disease pathogenesis or by comparison to healthy subjects (Figure 15.4a). In addition, by using selective small-molecule inhibitors coupled with an ABP, it is possible to correlate the inhibition of specific targets with effects on disease progression to validate targets for drug-discovery efforts (Verhelst and Bogyo 2005). Some of the first attempts at global profiling of enzyme activity in disease-associated tissues and cells have focused on cancer. Several enzyme classes have been associated with tumor initiation and progression, including proteases (Findeisen and Neumaier 2012), lipases (Notarnicola et al. 2012), and kinases (Ip and Wong 2012). For example, the general serine hydrolase FP-biotin probe was used to profile the activity of serine hydrolases in NCI-60 melanoma and breast cancer cell lines (Jessani et al. 2002). Serine hydrolase samples obtained from cancer cell lines were labeled with rhodamine-tagged FP and used for in-gel fluorescence analysis and profiling. The different levels of active serine hydrolases generated unique signatures that allowed the classification of each one of the cell lines. In addition, the invasiveness profiles of the cells were associated with their proteomic signatures. In a different study, Hanahan and colleagues used ABPs and small-molecule inhibitors to investigate the functional role of the cathepsins in different stages of cancer (Joyce et al. 2004). Several cysteine cathepsins were up-regulated during multiple stages of tumorigenesis in a mouse model of pancreatic cancer. This study and others demonstrate that it is possible to combine small-molecule inhibitors with ABPs to correlate

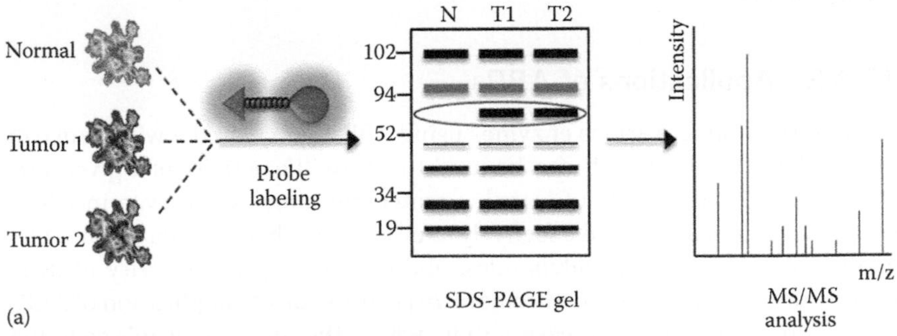

Figure 15.4 Applications of ABPs. (a) Target identification and validation by an ABP. The probe is used to label normal and tumor tissue lysates allowing visualization of fluorescently labeled samples by SDS-PAGE. The intensity of the labeled proteins is an indication of the levels of active enzymes in the tissues. Target proteins can be identified by affinity purification of probe-labeled proteins followed by mass spectrometry based identification.

(Continued)

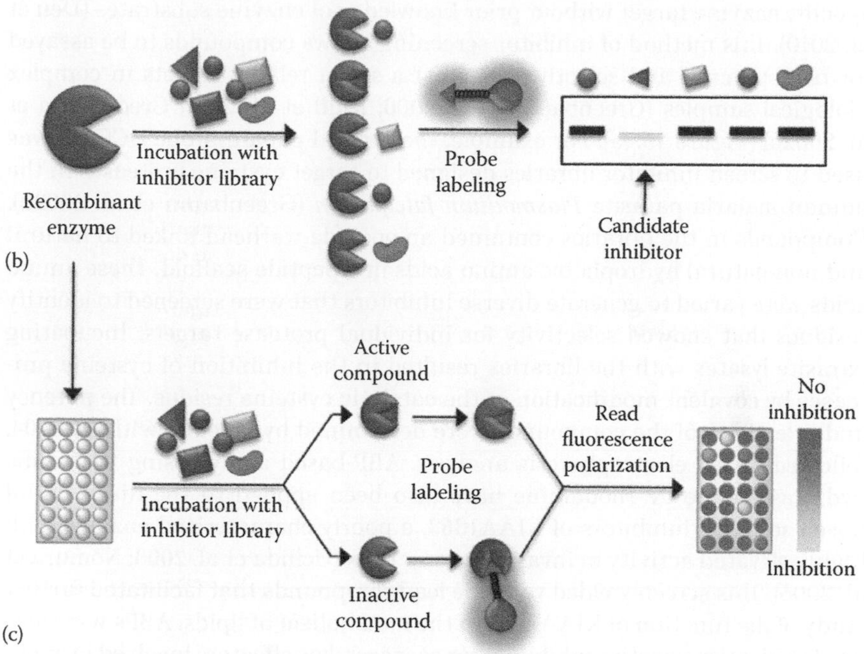

Figure 15.4 (Continued) Applications of ABPs. (b) Inhibitor screening and evaluation. An enzyme is incubated with an inhibitor library. The sample is then treated with a broad-spectrum ABP that labels multiple related enzyme targets. Samples are analyzed by SDS-PAGE, and labeled proteins are detected by scanning the gel with a flatbed laser scanner. When a candidate inhibitor binds in the active site of a target, it blocks labeling by the ABP and the signal is lost. (c) High-throughput screening using the fluorescence polarization (fluopol-ABPP) technique (Bachovchin et al. 2009). An enzyme is dispensed into a multi-well plate, and compounds are added to each well. An ABP is then dispensed to all wells, and the plate is incubated for a fixed time interval. The reaction of the probe with the enzyme target will increase the apparent mass of the probe, resulting in the induction of a strong fluorescence polarization signal.

the activities of proteases such as cathepsin with disease progression and that those proteases can serve as potential targets for anticancer therapy.

15.4.2.2 Inhibitor Screening

In traditional drug discovery, libraries of small molecules are screened *in vitro* against purified, often recombinant, protein targets to identify inhibitors (Inglese et al. 2007). However, these assays provide only limited information regarding the *in vivo* potency and selectivity of an inhibitor. ABPs can be used to perform high-throughput screens of small-molecule inhibitors for a specific enzyme target without prior knowledge of enzyme substrates (Deu et al. 2010). This method of inhibitor screening allows compounds to be assayed for both potency and selectivity against a set of related targets in complex biological samples (Greenbaum et al. 2000; Kidd et al. 2001; Greenbaum et al. 2002c) (Figure 15.4a). For example, the general papain probe DCG-04 was used to screen inhibitor libraries designed to target cysteine proteases in the human malaria parasite *Plasmodium falciparum* (Greenbaum et al. 2002b). Compounds in the libraries contained an epoxide warhead linked to natural and non-natural hydrophobic amino acids in a peptide scaffold. These amino acids were varied to generate diverse inhibitors that were screened to identify residues that showed selectivity for individual protease targets. Incubating parasite lysates with the libraries resulted in the inhibition of cysteine proteases by covalent modification of the catalytic cysteine residue. The potency and selectivity of the compounds were determined by labeling with DCG-04, followed by gel electrophoresis analysis. ABP-based assays using the serine hydrolase probe FP-rhodamine have also been applied to the discovery of novel, selective inhibitors of KIAA1363, a poorly characterized enzyme with highly elevated activity in invasive cancer cells (Uchida et al. 2003; Nomura et al. 2006). This screen yielded valuable lead compounds that facilitated further study of the function of KIAA1363 in the metabolism of lipids. ABPs were also used to identify selective inhibitors for caspases, key effectors involved in apoptosis (Berger et al. 2006). The same competition labeling approach can be used for purified target proteases (Figure 15.4b). For example, Berger et al. (2006) used a positional scanning combinatorial library (PSCL) approach to screen pools of peptide AOMKs containing both natural and non-natural amino acids for activity against several purified recombinant caspases. Based on the screening data, covalent inhibitors for caspases 3, 7, 8, and 9 were developed.

Most of the studies described above use one-dimensional SDS-PAGE gels as a readout, limiting throughput to compound libraries of relatively small size. Bachovchin et al. have addressed this limitation by developing a fluorescence polarization technology for competitive high-throughput ABP profiling (fluopol-ABPP) (Bachovchin et al. 2009; Knuckley et al. 2010). The principle of fluorescence polarization derives from the fact that the degree of polarization of a fluorophore is inversely related to its molecular rotation. Hence, fluorophores rotate quickly and emit low fluorescence polarization signals when free in solution; however, when bound to a large molecule (e.g., a protein), they rotate more slowly and emit high fluorescence polarization signals (Figure 15.4c). Fluopol-ABPP was used to discover selective inhibitors for two cancer-related enzyme targets, the hydrolytic enzyme retinoblastoma-binding protein-9 (RBBP9) and the thioltransferase glutathione *S*-transferase omega 1 (GSTO1) (Bachovchin et al. 2010).

15.4.2.3 *In Vivo* Imaging of Enzyme Activity

In vitro approaches are usually not capable of reproducing complex intracellular conditions, especially when attempting to decipher the complex regulatory network of proteolytic enzymes. The ability to monitor specific enzyme regulation in the context of the native cellular environment has tremendous clinical and basic science value. In addition to posttranslational modifications and interactions with cofactors or inhibitors, most enzymes are regulated by their localization and temporal expression within a cell or tissue (Shlomi et al. 2008). In addition, alterations in enzyme regulation and substrate proteolysis are contributing elements in the pathogenesis of many diseases. For instance, impaired proteolysis is a characteristic hallmark of neoplastic as well as inflammatory (Weaver et al. 1993), cardiovascular (Lutgens et al. 2007), neurodegenerative (Nijholt et al. 2011), bacterial (Johansson et al. 2008), viral (Nagai et al. 1976), and parasitic (Ingram et al. 2012) diseases. Insight into proteolytic events in living subjects accelerates the development of novel therapeutic agents and provides a more accurate means to monitor the effects of these agents (Figure 15.5).

Most imaging systems depend on physical parameters to generate image contrast. Consequently, specific molecular information is difficult to obtain. The development of new classes of quenched NIR ABPs has overcome these limitations and has created a number of possible applications for this technology (Blum et al. 2007; Puri and Bogyo 2009; Lee and Bogyo 2010; Lyo et al. 2012). These include real-time imaging applications in whole cells and the rapid readout of specific signals when used for in vivo imaging. NIR ABPs can also be used to noninvasively assess the efficacy of small-molecule drug leads. Several ABPs for different enzyme activities have been developed for *in vivo* imaging applications. Hanahan and colleagues used ABPs and class-specific small-molecule inhibitors to investigate the functional importance of the cathepsin cysteine proteases in pancreatic cancer (Joyce et al. 2004). The probes were injected into RIP1-Tag2 transgenic mice, and the dissected tissues were analyzed by SDS-PAGE and microscopy. Imaging cysteine cathepsin activity in tumor-bearing mice revealed pronounced up-regulation in precursor and neoplastic lesions of the endocrine pancreas and uterine cervix, with activity localized to the angiogenic vasculature. Pharmacological inhibition of cysteine cathepsin activity impaired angiogenic switching, tumor growth, and invasion in the pancreatic model. In a recent example with a cysteine cathepsin ABP, the effects of inhibiting target proteases using a general inhibitor were assessed by direct noninvasive imaging methods (Blum et al. 2007). Because the probes bind covalently, it was possible to measure the effects of the drug using noninvasive methods and to then collect tumor tissues and correlate the signals using SDS-PAGE analysis. Therefore, it was possible to interpret the noninvasive imaging data with confidence because of the link between the imaging data and the *ex vivo*

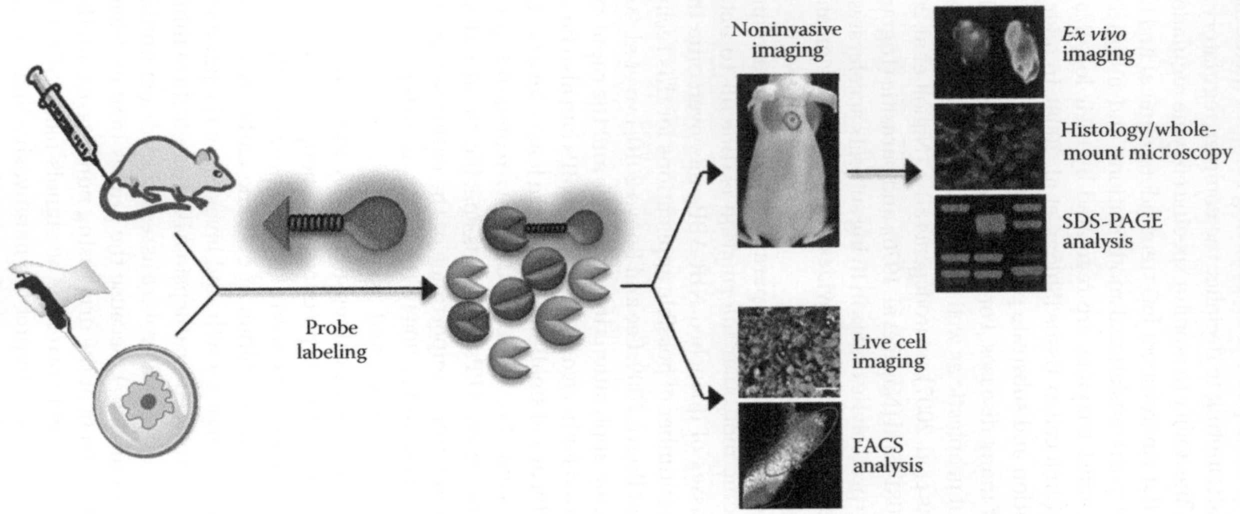

Figure 15.5 *In vitro* and *in vivo* imaging of enzyme activity using ABPs. ABPs can be used to profile enzyme activities in live cells as well as in whole animals. Cells in culture are incubated with a fluorescent ABP, the probe is removed, and the cell is washed to reduce signal from unbound probe. Live cells can be imaged using fluorescence microscopy to visualize the subcellular localization of active enzymes. Alternatively, cell samples can be analyzed by SDS-PAGE or FACS analysis (bottom). Alternatively, a fluorescent ABP is injected into a live subject and labeled enzymes can be monitored using noninvasive, whole-body imaging methods (top). Tissues can be collected, mounted, and analyzed by fluorescence microscopy. Active proteases within the given tissue can be visualized by the fluorescent signal or by histology analysis. Finally, tissues can be homogenized and analyzed biochemically using reducing SDS-PAGE and the labeled enzyme visualized by laser scanning of the fluorescent gel.

biochemical analysis. Lee at al. used a different class of ABPs for the lysosomal cysteine protease asparaginyl endopeptidase, also known as legumain (Lee and Bogyo 2010). The probes were labeled with a Cy5 fluorophore and also tagged with a series of cell-permeabilizing groups. In addition, the use of an aza-peptidyl scaffold and incorporation of P1 Asn resulted in rapid and highly selective binding of the probe to legumain, thus providing a high signal-to-noise ratio. This allowed the use of lower overall doses of probe and prevented extended circulation that can possibly cause cross-reactivity with other proteases. Most recently, an advanced and more effective version of an ABP for legumain was developed (Edgington et al. 2012). This new quenched NIR probe allowed tomographic imaging of legumain activity with high positional accuracy and selectivity. The use of the Cy5 fluorophore contributed to a relatively high resolution (1–2 mm) that enabled quantitative measurements of signals *in vivo*. In another example, Edgington et al. (2009) used NIR ABPs that label caspases to monitor the rates of apoptosis *in vivo*. In this study, the ABP was used to assess the efficacy of a clinical antibody that induces apoptosis in tumors via the ligation of death receptors on the tumor cell surface. The authors demonstrated that the NIR ABP could be used to monitor changes in the rate of apoptosis of tumors using noninvasive imaging methods. Furthermore, the accumulation of the probe in apoptotic tissues could be linked to the amount of labeled caspases that regulate the process of cell death. Verdoes et al. (2012) reported the synthesis and characterization of a cathepsin S–directed, nonpeptidic NIR qABP. The probe was used for noninvasive optical imaging in a syngeneic mouse model of breast cancer. In addition, FACS experiments identified specific subsets of myeloid-derived cells with an M2 macrophage phenotype as the *in vivo* cellular source of cysteine cathepsin activity responsible for probe fluorescence.

The route of administration of ABPs as well as their biodistribution and pharmacokinetics play a major role in influencing and determining both the quality and quantity of the signal. Most of the ABPs described above were administrated into live subjects by intravenous injection. However, quenched ABPs also have potential value for topical applications where the unbound probe cannot be washed away prior to imaging. Cutter et al. used a fluorescently quenched ABP as an intraoperative molecular imaging tool for the identification of tumor margins and infiltrating tumor cells in mice (Cutter et al. 2012). In this study, a quenched ABP was topically applied to the exposed brain tumor surface, and its activation was monitored by fluorescence imaging. The topically administered probe was activated within minutes and highlighted specific cathepsin activity in the tumor.

Taken together, ABPs designed for the dynamic monitoring of proteases in living subjects can serve as valuable tools for the development, evaluation, and dose-ranging of novel therapeutics *in vivo*. Moreover, they can be used to diagnose and potentially predict the outcome of a disease.

15.5 CONCLUSIONS AND FUTURE PROSPECTS

The field of activity-based biosensors has seen significant technological advances in the recent past that have led to the development of diverse applications for these agents. Both ABPs and substrate-based imaging probes have been applied to biologically and pathologically relevant fields. This includes the identification and evaluation of potential enzyme inhibitors in a complex proteome, drug lead identification, and the dynamic monitoring of enzymes in living subjects. The use of NIR quenched probes for noninvasive optical imaging has facilitated the process of validating pharmacological models and assessing the therapeutic efficacy of drug candidates. However, owing to physical limitations (i.e., limited tissue penetration, scattered photons, and low sensitivity), most current optical probes still require the development of adequate hardware to progress into the clinical setting. An alternative could be the use of other imaging modalities to generate new probes. Techniques such as positron emission tomography (PET) (Ray et al. 2003; Nahrendorf et al. 2009; Ren et al. 2011) and magnetic resonance imaging (MRI) have already begun to incorporate the use of "smart probes" to detect gene expression and molecular interactions (Olson et al. 2010). Furthermore, the integration of microbubbles into probes could result in ultrasound contrast agent as tools for the dynamic monitoring of enzyme activity *in vivo*. Such techniques and approaches will likely facilitate the identification and validation of new drug targets and the assessment of bioavailability and potency *in vivo* via the direct, noninvasive measurement of pharmacokinetic and pharmacodynamic parameters. Taken together, there is no doubt that the field will continue to grow in the coming years and additional biosensors will be added to the list of useful reagents for the study of protease biology.

ACKNOWLEDGMENTS

We thank the members of the Bogyo laboratory for helpful discussions and manuscript comments. This work was supported by funding from a grant from the National Institutes of Health R01-EB005011 (to M. B.).

REFERENCES

2009. Retraction notice to "Activation and silencing of matrix metalloproteinases" [*Semin. Cell Dev. Biol.* 19 (2008) 2–13]. *Seminars in Cell & Developmental Biology* 20 (3):375.

Adam, G. C., J. Burbaum, J. W. Kozarich, M. P. Patricelli, and B. F. Cravatt. 2004. Mapping enzyme active sites in complex proteomes. *Journal of the American Chemical Society* 126 (5):1363–8.

Adam, G. C., E. J. Sorensen, and B. F. Cravatt. 2002. Proteomic profiling of mechanistically distinct enzyme classes using a common chemotype. *Nature Biotechnology* 20 (8):805–9.

Altun, M., H. B. Kramer, L. I. Willems et al. 2011. Activity-based chemical proteomics accelerates inhibitor development for deubiquitylating enzymes. *Chemistry & Biology* 18 (11):1401–12.

Araujo, M. C., R. I. Melo, E. Del Nery et al. 1999. Internally quenched fluorogenic substrates for angiotensin I-converting enzyme. *Journal of Hypertension* 17 (5):665–72.

Assfalg-Machleidt, I., G. Rothe, S. Klingel et al. 1992. Membrane permeable fluorogenic rhodamine substrates for selective determination of cathepsin L. *Biological Chemistry Hoppe-Seyler* 373 (7):433–40.

Bachovchin, D. A., S. J. Brown, H. Rosen, and B. F. Cravatt. 2009. Identification of selective inhibitors of uncharacterized enzymes by high-throughput screening with fluorescent activity-based probes. *Nature Biotechnology* 27 (4):387–94.

Bachovchin, D. A., M. R. Wolfe, K. Masuda et al. 2010. Oxime esters as selective, covalent inhibitors of the serine hydrolase retinoblastoma-binding protein 9(RBBP9). *Bioorganic & Medicinal Chemistry Letters* 20 (7):2254–8.

Berger, A. B., K. B. Sexton, and M. Bogyo. 2006. Commonly used caspase inhibitors designed based on substrate specificity profiles lack selectivity. *Cell Research* 16 (12):961–3.

Berkers, C. R., M. Verdoes, E. Lichtman et al. 2005. Activity probe for in vivo profiling of the specificity of proteasome inhibitor bortezomib. *Nature Methods* 2 (5):357–62.

Bleeker, F. E., L. G. Hazen, A. Kohler, and C. J. Van Noorden. 2000. Direct comparison of the sensitivity of enzyme histochemical and immunohistochemical methods: Cathepsin B expression in human colorectal mucosa. *Acta Histochemica* 102 (3):247–57.

Blum, G., S. R. Mullins, K. Keren et al. 2005. Dynamic imaging of protease activity with fluorescently quenched activity-based probes. *Nature Chemical Biology* 1 (4):203–9.

Blum, G., G. von Degenfeld, M. J. Merchant, H. M. Blau, and M. Bogyo. 2007. Noninvasive optical imaging of cysteine protease activity using fluorescently quenched activity-based probes. *Nature Chemical Biology* 3 (10):668–77.

Blum, G., R. M. Weimer, L. E. Edgington, W. Adams, and M. Bogyo. 2009. Comparative assessment of substrates and activity based probes as tools for non-invasive optical imaging of cysteine protease activity. *PLoS One* 4 (7):e6374.

Bo Kim, K., F. N. Fonseca, and C. M. Crews. 2005. Development and characterization of proteasome inhibitors. *Methods in Enzymology* 399:585–609.

Bode, W., and R. Huber. 2000. Structural basis of the endoproteinase-protein inhibitor interaction. *Biochimica et Biophysica Acta* 1477 (1–2):241–52.

Boonacker, E., and C. J. Van Noorden. 2001. Enzyme cytochemical techniques for metabolic mapping in living cells, with special reference to proteolysis. *Journal of Histochemistry and Cytochemistry* 49 (12):1473–86.

Brouwer, A. J., A. Bunschoten, and R. M. Liskamp. 2007. Synthesis and evaluation of chloromethyl sulfoxides as a new class of selective irreversible cysteine protease inhibitors. *Bioorganic & Medicinal Chemistry* 15 (22):6985–93.

Burchak, O. N., L. Mugherli, F. Chatelain, and M. Y. Balakirev. 2006. Fluorescein-based amino acids for solid phase synthesis of fluorogenic protease substrates. *Bioorganic & Medicinal Chemistry* 14 (8):2559–68.

Chan, E. W., S. Chattopadhaya, R. C. Panicker, X. Huang, and S. Q. Yao. 2004. Developing photoactive affinity probes for proteomic profiling: Hydroxamate-based probes for metalloproteases. *Journal of the American Chemical Society* 126 (44):14435–46.

Chapman, H. A., R. J. Riese, and G. P. Shi. 1997. Emerging roles for cysteine proteases in human biology. *Annual Review of Physiology* 59:63–88.

Choe, Y., F. Leonetti, D. C. Greenbaum et al. 2006. Substrate profiling of cysteine proteases using a combinatorial peptide library identifies functionally unique specificities. *Journal of Biological Chemistry* 281 (18):12824–32.

Chong, C. M., S. Gao, B. Y. Chiang et al. 2011. An acyloxymethyl ketone-based probe to monitor the activity of glutathionylspermidine amidase in *Escherichia coli. ChemBioChem* 12 (15):2306–9.

Clardy, J., and C. Walsh. 2004. Lessons from natural molecules. *Nature* 432 (7019):829–37.

Claveau, D., D. Riendeau, and J. A. Mancini. 2000. Expression, maturation, and rhodamine-based fluorescence assay of human cathepsin K expressed in CHO cells. *Biochemical Pharmacology* 60 (6):759–69.

Coleman, D. J., M. J. Studler, and J. J. Naleway. 2007. A long-wavelength fluorescent substrate for continuous fluorometric determination of cellulase activity: Resorufin-beta-D-cellobioside. *Analytical Biochemistry* 371 (2):146–53.

Cravatt, B. F., A. T. Wright, and J. W. Kozarich. 2008. Activity-based protein profiling: From enzyme chemistry to proteomic chemistry. *Annual Review of Biochemistry* 77:383–414.

Cummings, R. T., S. P. Salowe, B. R. Cunningham et al. 2002. A peptide-based fluorescence resonance energy transfer assay for *Bacillus anthracis* lethal factor protease. *Proceedings of the National Academy of Sciences of the United States of America* 99 (10):6603–6.

Cutter, J. L., N. T. Cohen, J. Wang et al. 2012. Topical application of activity-based probes for visualization of brain tumor tissue. *PLoS One* 7 (3):e33060.

Deu, E., M. Verdoes, and M. Bogyo. 2012. New approaches for dissecting protease functions to improve probe development and drug discovery. *Nature Structural & Molecular Biology* 19 (1):9–16.

Deu, E., Z. Yang, F. Wang, M. Klemba, and M. Bogyo. 2010. Use of activity-based probes to develop high throughput screening assays that can be performed in complex cell extracts. *PLoS One* 5 (8):e11985.

Dirksen, A., S. Yegneswaran, and P. E. Dawson. 2010. Bisaryl hydrazones as exchangeable biocompatible linkers. *Angewandte Chemie* 49 (11):2023–7.

Drag, M., and G. S. Salvesen. 2010. Emerging principles in protease-based drug discovery. *Nature Reviews. Drug Discovery* 9 (9):690–701.

Edgington, L. E., A. B. Berger, G. Blum et al. 2009. Noninvasive optical imaging of apoptosis by caspase-targeted activity-based probes. *Nature Medicine* 15 (8):967–73.

Edgington, L. E., M. Verdoes, and M. Bogyo. 2011. Functional imaging of proteases: Recent advances in the design and application of substrate-based and activity-based probes. *Current Opinion in Chemical Biology* 15 (6):798–805.

Edgington, L. E., M. Verdoes, A. Ortega et al. 2012. Functional imaging of legumain in cancer using a new quenched activity-based probe. *Journal of the American Chemical Society* 135 (1):174–82.

Faust, A., B. Waschkau, J. Waldeck et al. 2009. Synthesis and evaluation of a novel hydroxamate based fluorescent photoprobe for imaging of matrix metalloproteinases. *Bioconjugate Chemistry* 20 (5):904–12.

Findeisen, P., and M. Neumaier. 2012. Functional protease profiling for diagnosis of malignant disease. *Proteomics. Clinical Applications* 6 (1–2):60–78.

Fonovic, M., and M. Bogyo. 2007. Activity based probes for proteases: Applications to biomarker discovery, molecular imaging and drug screening. *Current Pharmaceutical Design* 13 (3):253–61.

Fonovic, M., and M. Bogyo. 2008. Activity-based probes as a tool for functional proteomic analysis of proteases. *Expert Review of Proteomics* 5 (5):721–30.

Frank, R., and R. Hargreaves. 2003. Clinical biomarkers in drug discovery and development. *Nature Reviews. Drug Discovery* 2 (7):566–80.

Freitag, S., V. Chu, J. E. Penzotti et al. 1999. A structural snapshot of an intermediate on the streptavidin-biotin dissociation pathway. *Proceedings of the National Academy of Sciences of the United States of America* 96 (15):8384–9.

Gabriel, D., M. F. Zuluaga, and N. Lange. 2011. On the cutting edge: Protease-sensitive prodrugs for the delivery of photoactive compounds. *Photochemical & Photobiological Sciences: Official Journal of the European Photochemistry Association and the European Society for Photobiology* 10 (5):689–703.

Gehrig, S., M. A. Mall, and C. Schultz. 2012. Spatially resolved monitoring of neutrophil elastase activity with ratiometric fluorescent reporters. *Angewandte Chemie* 51 (25):6258–61.

Greenbaum, D. C., W. D. Arnold, F. Lu et al. 2002a. Small molecule affinity fingerprinting: A tool for enzyme family subclassification, target identification, and inhibitor design. *Chemistry & Biology* 9 (10):1085–94.

Greenbaum, D. C., A. Baruch, M. Grainger et al. 2002b. A role for the protease falcipain 1 in host cell invasion by the human malaria parasite. *Science* 298 (5600):2002–6.

Greenbaum, D., A. Baruch, L. Hayrapetian et al. 2002c. Chemical approaches for functionally probing the proteome. *Molecular & Cellular Proteomics: MCP* 1 (1):60–8.

Greenbaum, D., K. F. Medzihradszky, A. Burlingame, and M. Bogyo. 2000. Epoxide electrophiles as activity-dependent cysteine protease profiling and discovery tools. *Chemistry & Biology* 7 (8):569–81.

Gross, S., and D. Piwnica-Worms. 2006. Molecular imaging strategies for drug discovery and development. *Current Opinion in Chemical Biology* 10 (4):334–42.

Grzonka, Z., E. Jankowska, F. Kasprzykowski et al. 2001. Structural studies of cysteine proteases and their inhibitors. *Acta Biochimica Polonica* 48 (1):1–20.

Hagenstein, M. C., and N. Sewald. 2006. Chemical tools for activity-based proteomics. *Journal of Biotechnology* 124 (1):56–73.

Hesek, D., M. Toth, V. Krchnak, R. Fridman, and S. Mobashery. 2006a. Synthesis of an inhibitor-tethered resin for detection of active matrix metalloproteinases involved in disease. *The Journal of Organic Chemistry* 71 (16):5848–54.

Hesek, D., M. Toth, S. O. Meroueh et al. 2006b. Design and characterization of a metalloproteinase inhibitor-tethered resin for the detection of active MMPs in biological samples. *Chemistry & Biology* 13 (4):379–86.

Inglese, J., R. L. Johnson, A. Simeonov et al. 2007. High-throughput screening assays for the identification of chemical probes. *Nature Chemical Biology* 3 (8):466–79.

Ingram, J. R., S. B. Rafi, A. A. Eroy-Reveles et al. 2012. Investigation of the proteolytic functions of an expanded cercarial elastase gene family in *Schistosoma mansoni*. *PLoS Neglected Tropical Diseases* 6 (4):e1589.

Ip, C. K., and A. S. Wong. 2012. Exploiting p70 S6 kinase as a target for ovarian cancer. *Expert Opinion on Therapeutic Targets* 16 (6):619–30.

Jaffer, F. A., C. H. Tung, R. E. Gerszten, and R. Weissleder. 2002. In vivo imaging of thrombin activity in experimental thrombi with thrombin-sensitive near-infrared molecular probe. *Arteriosclerosis, Thrombosis, and Vascular Biology* 22 (11):1929–35.

Jeffery, D. A., and M. Bogyo. 2003. Chemical proteomics and its application to drug discovery. *Current Opinion in Biotechnology* 14 (1):87–95.

Jessani, N., Y. Liu, M. Humphrey, and B. F. Cravatt. 2002. Enzyme activity profiles of the secreted and membrane proteome that depict cancer cell invasiveness. *Proceedings of the National Academy of Sciences of the United States of America* 99 (16):10335–40.

Jiang, T., E. S. Olson, Q. T. Nguyen et al. 2004. Tumor imaging by means of proteolytic activation of cell-penetrating peptides. *Proceedings of the National Academy of Sciences of the United States of America* 101 (51):17867–72.

Johansson, B. P., O. Shannon, and L. Bjorck. 2008. IdeS: A bacterial proteolytic enzyme with therapeutic potential. *PLoS One* 3 (2):e1692.

Jones, L. J., R. H. Upson, R. P. Haugland, N. Panchuk-Voloshina, and M. Zhou. 1997. Quenched BODIPY dye-labeled casein substrates for the assay of protease activity by direct fluorescence measurement. *Analytical Biochemistry* 251 (2):144–52.

Joyce, J. A., A. Baruch, K. Chehade et al. 2004. Cathepsin cysteine proteases are effectors of invasive growth and angiogenesis during multistage tumorigenesis. *Cancer Cell* 5 (5):443–53.

Juliano, M. A., F. Filira, M. Gobbo et al. 1999. Chromogenic and fluorogenic glycosylated and acetylglycosylated peptides as substrates for serine, thiol and aspartyl proteases. *The Journal of Peptide Research* 53 (2):109–19.

Kidd, D., Y. Liu, and B. F. Cravatt. 2001. Profiling serine hydrolase activities in complex proteomes. *Biochemistry* 40 (13):4005–15.

Kim, K., M. Lee, H. Park et al. 2006. Cell-permeable and biocompatible polymeric nanoparticles for apoptosis imaging. *Journal of the American Chemical Society* 128 (11):3490–1.

Kindermann, M., H. Roschitzki-Voser, D. Caglic et al. 2010. Selective and sensitive monitoring of caspase-1 activity by a novel bioluminescent activity-based probe. *Chemistry & Biology* 17 (9):999–1007.

Knuckley, B., J. E. Jones, D. A. Bachovchin et al. 2010. A fluopol-ABPP HTS assay to identify PAD inhibitors. *Chemical Communications* 46 (38):7175–7.

Kobe, B., and B. E. Kemp. 1999. Active site-directed protein regulation. *Nature* 402 (6760):373–6.

Koblinski, J. E., M. Ahram, and B. F. Sloane. 2000. Unraveling the role of proteases in cancer. *Clinica Chimica Acta; International Journal of Clinical Chemistry* 291 (2):113–35.

Kutsyi, M. P., E. A. Kuznetsova, and A. I. Gaziev. 1999. Involvement of proteases in apoptosis. *Biochemistry. Biokhimiia* 64 (2):115–26.

Kwan, D. H., H. M. Chen, K. Ratananikom et al. 2011. Self-immobilizing fluorogenic imaging agents of enzyme activity. *Angewandte Chemie* 50 (1):300–3.

Lapidot, T., and I. Petit. 2002. Current understanding of stem cell mobilization: The roles of chemokines, proteolytic enzymes, adhesion molecules, cytokines, and stromal cells. *Experimental Hematology* 30 (9):973–81.

Laxman, B., D. E. Hall, M. S. Bhojani et al. 2002. Noninvasive real-time imaging of apoptosis. *Proceedings of the National Academy of Sciences of the United States of America* 99 (26):16551–5.

Lee, J., and M. Bogyo. 2010. Development of near-infrared fluorophore (NIRF)-labeled activity-based probes for in vivo imaging of legumain. *ACS Chemical Biology* 5 (2):233–43.

Leriche, G., L. Chisholm, and A. Wagner. 2012. Cleavable linkers in chemical biology. *Bioorganic & Medicinal Chemistry* 20 (2):571–82.

Licha, K., M. Schirner, and G. Henry. 2008. Optical agents. *Handbook of Experimental Pharmacology* 185 (Pt 1):203–22.

Liu, S., B. Zhou, H. Yang et al. 2008. Aryl vinyl sulfonates and sulfones as active site-directed and mechanism-based probes for protein tyrosine phosphatases. *Journal of the American Chemical Society* 130 (26):8251–60.

Lopez-Otin, C., and C. M. Overall. 2002. Protease degradomics: A new challenge for proteomics. *Nature Reviews. Molecular Cell Biology* 3 (7):509–19.

Los, M., H. Walczak, K. Schulze-Osthoff, and J. C. Reed. 2000. Fluorogenic substrates as detectors of caspase activity during natural killer cell-induced apoptosis. *Methods in Molecular Biology* 121:155–62.

Luker, G. D., C. M. Pica, J. Song, K. E. Luker, and D. Piwnica-Worms. 2003. Imaging 26S proteasome activity and inhibition in living mice. *Nature Medicine* 9 (7):969–73.

Lutgens, S. P., K. B. Cleutjens, M. J. Daemen, and S. Heeneman. 2007. Cathepsin cysteine proteases in cardiovascular disease. *FASEB Journal* 21 (12):3029–41.

Lyo, V., F. Cattaruzza, T. N. Kim et al. 2012. Active cathepsins B, L, and S in murine and human pancreatitis. *American Journal of Physiology. Gastrointestinal and Liver Physiology* 303 (8):G894–903.

Mahmood, U., C. H. Tung, A. Bogdanov, Jr., and R. Weissleder. 1999. Near-infrared optical imaging of protease activity for tumor detection. *Radiology* 213 (3):866–70.

Matayoshi, E. D., G. T. Wang, G. A. Krafft, and J. Erickson. 1990. Novel fluorogenic substrates for assaying retroviral proteases by resonance energy transfer. *Science* 247 (4945):954–8.

Matsumura, Y., and H. Maeda. 1986. A new concept for macromolecular therapeutics in cancer chemotherapy: Mechanism of tumoritropic accumulation of proteins and the antitumor agent smancs. *Cancer Research* 46 (12 Pt 1):6387–92.

Messerli, S. M., S. Prabhakar, Y. Tang et al. 2004. A novel method for imaging apoptosis using a caspase-1 near-infrared fluorescent probe. *Neoplasia* 6 (2):95–105.

Morimoto, S. 2007. In-vivo imaging of tumors with protease activated near-infrared fluorescent probes. *Tanpakushitsu kakusan koso. Protein, Nucleic Acid, Enzyme* 52 (13 Suppl):1774–5.

Nagai, Y., H. D. Klenk, and R. Rott. 1976. Proteolytic cleavage of the viral glycoproteins and its significance for the virulence of Newcastle disease virus. *Virology* 72 (2):494–508.

Nahrendorf, M., P. Waterman, G. Thurber et al. 2009. Hybrid in vivo FMT-CT imaging of protease activity in atherosclerosis with customized nanosensors. *Arteriosclerosis, Thrombosis, and Vascular Biology* 29 (10):1444–51.

Neurath, H., and K. A. Walsh. 1976. Role of proteolytic enzymes in biological regulation (a review). *Proceedings of the National Academy of Sciences of the United States of America* 73 (11):3825–32.

Nijholt, D. A., L. De Kimpe, H. L. Elfrink, J. J. Hoozemans, and W. Scheper. 2011. Removing protein aggregates: The role of proteolysis in neurodegeneration. *Current Medicinal Chemistry* 18 (16):2459–76.

Nilvebrant, J., T. Alm, and S. Hober. 2012. Orthogonal protein purification facilitated by a small bispecific affinity tag. *Journal of Visualized Experiments: JoVE* (59).

Nomura, D. K., M. M. Dix, and B. F. Cravatt. 2010. Activity-based protein profiling for biochemical pathway discovery in cancer. *Nature Reviews. Cancer* 10 (9):630–8.

Nomura, D. K., K. A. Durkin, K. P. Chiang et al. 2006. Serine hydrolase KIAA1363: Toxicological and structural features with emphasis on organophosphate interactions. *Chemical Research in Toxicology* 19 (9):1142–50.

Notarnicola, M., C. Messa, and M. G. Caruso. 2012. A significant role of lipogenic enzymes in colorectal cancer. *Anticancer Research* 32 (7):2585–90.

Ntziachristos, V., C. Bremer, and R. Weissleder. 2003. Fluorescence imaging with near-infrared light: New technological advances that enable in vivo molecular imaging. *European Radiology* 13 (1):195–208.

O'Brien, M. A., W. J. Daily, P. E. Hesselberth et al. 2005. Homogeneous, bioluminescent protease assays: Caspase-3 as a model. *Journal of Biomolecular Screening* 10 (2):137–48.

Olson, E. S., T. Jiang, T. A. Aguilera et al. 2010. Activatable cell penetrating peptides linked to nanoparticles as dual probes for in vivo fluorescence and MR imaging of proteases. *Proceedings of the National Academy of Sciences of the United States of America* 107 (9):4311–6.

Olson, E. S., M. A. Whitney, B. Friedman et al. 2012. In vivo fluorescence imaging of atherosclerotic plaques with activatable cell-penetrating peptides targeting thrombin activity. *Integrative Biology: Quantitative Biosciences from Nano to Macro* 4 (6):595–605.

Orth, R., T. Bottcher, and S. A. Sieber. 2010. The biological targets of acivicin inspired 3-chloro- and 3-bromodihydroisoxazole scaffolds. *Chemical Communications* 46 (44):8475–7.

Packard, B. Z., D. D. Toptygin, A. Komoriya, and L. Brand. 1997. Design of profluorescent protease substrates guided by exciton theory. *Methods in Enzymology* 278:15–23.

Pan, Z., D. A. Jeffery, K. Chehade et al. 2006. Development of activity-based probes for trypsin-family serine proteases. *Bioorganic & Medicinal Chemistry Letters* 16 (11):2882–5.

Patricelli, M. P., D. K. Giang, L. M. Stamp, and J. J. Burbaum. 2001. Direct visualization of serine hydrolase activities in complex proteomes using fluorescent active site-directed probes. *Proteomics* 1 (9):1067–71.

Pham, W., Y. Choi, R. Weissleder, and C. H. Tung. 2004. Developing a peptide-based near-infrared molecular probe for protease sensing. *Bioconjugate Chemistry* 15 (6):1403–7.

Puente, X. S., L. M. Sanchez, A. Gutierrez-Fernandez, G. Velasco, and C. Lopez-Otin. 2005. A genomic view of the complexity of mammalian proteolytic systems. *Biochemical Society Transactions* 33 (Pt 2):331–4.

Puri, A. W., and M. Bogyo. 2009. Using small molecules to dissect mechanisms of microbial pathogenesis. *ACS Chemical Biology* 4 (8):603–16.

Quesada, V., G. R. Ordonez, L. M. Sanchez, X. S. Puente, and C. Lopez-Otin. 2009. The Degradome database: Mammalian proteases and diseases of proteolysis. *Nucleic Acids Research* 37 (Database issue):D239–43.

Rawlings, N. D., A. J. Barrett, and A. Bateman. 2011. Asparagine peptide lyases: A seventh catalytic type of proteolytic enzymes. *The Journal of Biological Chemistry* 286 (44):38321–8.

Ray, P., A. M. Wu, and S. S. Gambhir. 2003. Optical bioluminescence and positron emission tomography imaging of a novel fusion reporter gene in tumor xenografts of living mice. *Cancer Research* 63 (6):1160–5.

Rehm, M., H. Dussmann, R. U. Janicke et al. 2002. Single-cell fluorescence resonance energy transfer analysis demonstrates that caspase activation during apoptosis is a rapid process. Role of caspase-3. *The Journal of Biological Chemistry* 277 (27):24506–14.

Ren, G., G. Blum, M. Verdoes et al. 2011. Non-invasive imaging of cysteine cathepsin activity in solid tumors using a 64Cu-labeled activity-based probe. *PLoS One* 6 (11):e28029.

Riedl, S. J., and G. S. Salvesen. 2007. The apoptosome: Signalling platform of cell death. *Nature Reviews. Molecular Cell Biology* 8 (5):405–13.

Rochefort, H., and E. Liaudet-Coopman. 1999. Cathepsin D in cancer metastasis: A protease and a ligand. *APMIS: Acta Pathologica, Microbiologica, et Immunologica Scandinavica* 107 (1):86–95.

Sabariegos, R., F. Picazo, B. Domingo et al. 2009. Fluorescence resonance energy transfer-based assay for characterization of hepatitis C virus NS3–4A protease activity in live cells. *Antimicrobial Agents and Chemotherapy* 53 (2):728–34.

Sadaghiani, A. M., S. H. Verhelst, and M. Bogyo. 2007. Tagging and detection strategies for activity-based proteomics. *Current Opinion in Chemical Biology* 11 (1):20–8.

Saravanakumar, G., D. G. Jo, and J. H. Park. 2012. Polysaccharide-based nanoparticles: A versatile platform for drug delivery and biomedical imaging. *Current Medicinal Chemistry* 19 (19):3212–29.

Scherer, R. L., J. O. McIntyre, and L. M. Matrisian. 2008. Imaging matrix metalloproteinases in cancer. *Cancer Metastasis Reviews* 27 (4):679–90.

Schmidinger, H., R. Birner-Gruenberger, G. Riesenhuber et al. 2005. Novel fluorescent phosphonic acid esters for discrimination of lipases and esterases. *ChemBioChem* 6 (10):1776–81.

Selvin, P. R. 2000. The renaissance of fluorescence resonance energy transfer. *Nature Structural Biology* 7 (9):730–4.

Shan, L. 2004. Cy5-labeled aza-peptidyl Pro-Asn epoxide. In *Molecular Imaging and Contrast Agent Database (MICAD)*. Bethesda, MD.

Shinde, R., J. Perkins, and C. H. Contag. 2006. Luciferin derivatives for enhanced in vitro and in vivo bioluminescence assays. *Biochemistry* 45 (37):11103–12.

Shlomi, T., M. N. Cabili, M. J. Herrgard, B. O. Palsson, and E. Ruppin. 2008. Network-based prediction of human tissue-specific metabolism. *Nature Biotechnology* 26 (9):1003–10.

Siegmund, B., and M. Zeitz. 2003. Pralnacasan (vertex pharmaceuticals). *IDrugs: The Investigational Drugs Journal* 6 (2):154–8.

Simon, G. M., and B. F. Cravatt. 2010. Activity-based proteomics of enzyme superfamilies: Serine hydrolases as a case study. *The Journal of Biological Chemistry* 285 (15):11051–5.

Speers, A. E., and B. F. Cravatt. 2005. A tandem orthogonal proteolysis strategy for high-content chemical proteomics. *Journal of the American Chemical Society* 127 (28):10018–9.

Sukumaran, S. M., B. Potsaid, M. Y. Lee, D. S. Clark, and J. S. Dordick. 2009. Development of a fluorescence-based, ultra high-throughput screening platform for nanoliter-scale cytochrome p450 microarrays. *Journal of Biomolecular Screening* 14 (6):668–78.

Szabelski, M., M. Rogiewicz, and W. Wiczk. 2005. Fluorogenic peptide substrates containing benzoxazol-5-yl-alanine derivatives for kinetic assay of cysteine proteases. *Analytical Biochemistry* 342 (1):20–7.

Szychowski, J., A. Mahdavi, J. J. Hodas et al. 2010. Cleavable biotin probes for labeling of biomolecules via azide-alkyne cycloaddition. *Journal of the American Chemical Society* 132 (51):18351–60.

Tepel, C., D. Bromme, V. Herzog, and K. Brix. 2000. Cathepsin K in thyroid epithelial cells: Sequence, localization and possible function in extracellular proteolysis of thyroglobulin. *Journal of Cell Science* 113 (Pt 24):4487–98.

Terai, T., and T. Nagano. 2008. Fluorescent probes for bioimaging applications. *Current Opinion in Chemical Biology* 12 (5):515–21.

Travis, B. L., and J. H. Ferguson. 1951. Proteolytic enzymes and platelets in relation to blood coagulation. *The Journal of Clinical Investigation* 30 (1):112–23.

Tully, S. E., and B. F. Cravatt. 2010. Activity-based probes that target functional subclasses of phospholipases in proteomes. *Journal of the American Chemical Society* 132 (10):3264–5.

Tung, C. H., S. Bredow, U. Mahmood, and R. Weissleder. 1999. Preparation of a cathepsin D sensitive near-infrared fluorescence probe for imaging. *Bioconjugate Chemistry* 10 (5):892–6.

Tung, C. H., U. Mahmood, S. Bredow, and R. Weissleder. 2000. In vivo imaging of proteolytic enzyme activity using a novel molecular reporter. *Cancer Research* 60 (17):4953–8.

Turk, B. 2006. Targeting proteases: Successes, failures and future prospects. *Nature Reviews. Drug Discovery* 5 (9):785–99.

Uchida, N., B. Dykstra, K. J. Lyons, F. Y. Leung, and C. J. Eaves. 2003. Different in vivo repopulating activities of purified hematopoietic stem cells before and after being stimulated to divide in vitro with the same kinetics. *Experimental Hematology* 31 (12):1338–47.

van der Veken, P., E. H. Dirksen, E. Ruijter et al. 2005. Development of a novel chemical probe for the selective enrichment of phosphorylated serine- and threonine-containing peptides. *ChemBioChem* 6 (12):2271–80.

Van Noorden, C. J., E. Boonacker, E. R. Bissell et al. 1997. Ala-Pro-cresyl violet, a synthetic fluorogenic substrate for the analysis of kinetic parameters of dipeptidyl peptidase IV (CD26) in individual living rat hepatocytes. *Analytical Biochemistry* 252 (1):71–7.

Verdoes, M., L. E. Edgington, F. A. Scheeren et al. 2012. A nonpeptidic cathepsin S activity based probe for noninvasive optical imaging of tumor-associated macrophages. *Chemistry & Biology* 19 (5):619–28.

Verhelst, S. H., and M. Bogyo. 2005. Chemical proteomics applied to target identification and drug discovery. *BioTechniques* 38 (2):175–7.

Verhelst, S. H., M. Fonovic, and M. Bogyo. 2007. A mild chemically cleavable linker system for functional proteomic applications. *Angewandte Chemie* 46 (8):1284–6.

Watzke, A., G. Kosec, M. Kindermann et al. 2008. Selective activity-based probes for cysteine cathepsins. *Angewandte Chemie* 47 (2):406–9.

Weaver, V. M., B. Lach, P. R. Walker, and M. Sikorska. 1993. Role of proteolysis in apoptosis: Involvement of serine proteases in internucleosomal DNA fragmentation in immature thymocytes. *Biochimie et Biologie Cellulaire* 71 (9–10):488–500.

Weerapana, E., A. E. Speers, and B. F. Cravatt. 2007. Tandem orthogonal proteolysis-activity-based protein profiling (TOP-ABPP): A general method for mapping sites of probe modification in proteomes. *Nature Protocols* 2 (6):1414–25.

Weerapana, E., C. Wang, G. M. Simon et al. 2010. Quantitative reactivity profiling predicts functional cysteines in proteomes. *Nature* 468 (7325):790–5.

Wehrman, T. S., G. von Degenfeld, P. O. Krutzik, G. P. Nolan, and H. M. Blau. 2006. Luminescent imaging of beta-galactosidase activity in living subjects using sequential reporter-enzyme luminescence. *Nature Methods* 3 (4):295–301.

Winssinger, N., R. Damoiseaux, D. C. Tully et al. 2004. PNA-encoded protease substrate microarrays. *Chemistry & Biology* 11 (10):1351–60.

Winssinger, N., and J. L. Harris. 2005. Microarray-based functional protein profiling using peptide nucleic acid-encoded libraries. *Expert Review of Proteomics* 2 (6):937–47.

Witte, M. D., M. T. Walvoort, K. Y. Li et al. 2011. Activity-based profiling of retaining beta-glucosidases: A comparative study. *ChemBioChem* 12 (8):1263–9.

Yang, J., P. Chaurand, J. L. Norris, N. A. Porter, and R. M. Caprioli. 2012. Activity-based probes linked with laser-cleavable mass tags for signal amplification in imaging mass spectrometry: Analysis of serine hydrolase enzymes in mammalian tissue. *Analytical Chemistry* 84 (8):3689–95.

Yee, M. C., S. C. Fas, M. M. Stohlmeyer, T. J. Wandless, and K. A. Cimprich. 2005. A cell-permeable, activity-based probe for protein and lipid kinases. *The Journal of Biological Chemistry* 280 (32):29053–9.

Zhang, H. Z., S. Kasibhatla, J. Guastella et al. 2003. N-Ac-DEVD-N′-(polyfluorobenzoyl)-R110: Novel cell-permeable fluorogenic caspase substrates for the detection of caspase activity and apoptosis. *Bioconjugate Chemistry* 14 (2):458–63.

Zhou, Z., P. Cironi, A. J. Lin et al. 2007. Genetically encoded short peptide tags for orthogonal protein labeling by Sfp and AcpS phosphopantetheinyl transferases. *ACS Chemical Biology* 2 (5):337–46.

Section IV

Emerging Techniques

Chapter 16

Super-Resolution Imaging

Robert K. Neely, Wim Vandenberg, and Peter Dedecker

CONTENTS

16.1 INTRODUCTION

The purpose of every microscope is to make visible the invisible; to magnify, measure, and better describe complex samples in minute detail. Put another way, every microscope constructs an image of a sample that allows its spatial distribution to be more clearly perceived. But any image is only as good as its fidelity, or the extent to which it provides an accurate picture. Even using the best techniques and instruments, any microscopist struggles against a range of issues that limit image fidelity on a daily basis. Particularly common complaints include low signal-to-noise ratios, incomplete or unspecific labeling of the sample, probe degradation, and sample motion (drift). Such problems can severely limit the imaging but none of those described are limiting in a fundamental sense; all are surmountable (though often not readily).

In this chapter, we are concerned with challenging the most fundamental limitation of far-field light microscopy—resolution—which limits the fidelity of the imaging process even in an ideal world where every aspect of the experiment works perfectly. The limited resolution in optical microscopy derives from the fact that we use light and are therefore bound by the laws of optics. In particular, the wave-like character of light means that it cannot be confined in an arbitrarily small volume, and it can be focused only to a point with dimensions of the same order of magnitude as its wavelength (typically a couple of hundred nanometers in diameter; Figure 16.1). In other words, on the detector of any microscope, even a point-like source will appear as a spot hundreds of nanometers in diameter.

This concept is known as point spreading because a single, infinitesimally small point source of light is spread out in space as the light passes through the imaging optics (Born 1999). The precise shape of this spreading is described using the point-spread function (PSF), which can be calculated or measured for a given optical system. The main determinants of the magnitude of the point spreading are the wavelength of the light and the spatial angle over which the light is

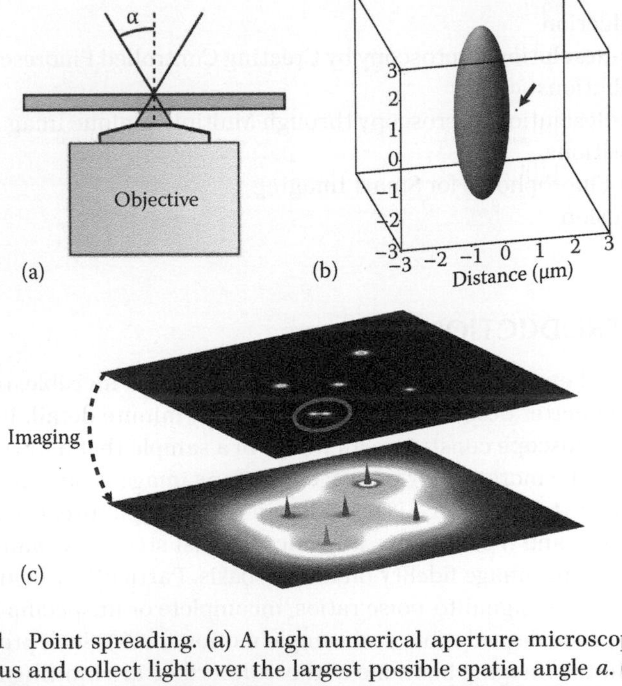

Figure 16.1 Point spreading. (a) A high numerical aperture microscope objective aims to focus and collect light over the largest possible spatial angle *a*. (b) Example point spread function for a high-numerical aperture confocal imaging system. The arrow points to a 10-times magnified depiction of a large (5 nm) fluorophore for scale. (c) Because of the spreading induced by the PSF, the emission patterns of closely spaced emitters will overlap, and information on closely spaced emitters is lost (circle). The cones show the location of the peak maxima. (Reprinted from *Mater. Today,* 11, Dedecker, P., Hofkens, J., and Hotta, J., Diffraction-unlimited optical microscopy, 12–21, Copyright 2008, with permission from Elsevier.)

focused or collected (represented by the numerical aperture of the optics, Figure 16.1a). As a rule of thumb, point spreading always occurs on the hundreds of nanometers length scale, regardless of the arrangement or quality of the optical system. The minimum area that is illuminated by a point source after light travels through the imaging apparatus is known as the "diffraction-limited region."

The consequences of point spreading are profound: in fluorescence microscopy, a single fluorescent dye, one or two nanometers in size, will be imaged on the detector as a blurry spot with a size of a few hundred nanometers (Figure 16.1b). Similarly, the excitation light, even if it originated from a point source, can be focused only to a diffraction-limited region. Fluorophores spaced within a few hundred nanometers cannot be excited independently from one another, and their emission will overlap to form a large and blurry intensity distribution. As a result, the ability to distinguish details smaller than the point spread function is severely compromised, and the information on length scales less than a few hundreds of nanometers is lost (Figure 16.1c). Importantly, this "diffraction limit" represents the best-case scenario: real-world issues such as low signal-to-noise ratio will reduce the resolution even further.

At this point, it is instructive to take a short detour into how fluorescence microscopy is typically performed. By and large, all fluorescence microscopes can be divided into two categories, based on whether the fluorescence is imaged using a single-point detector (a detector consisting of just a single "pixel"), or an imaging detector that consists of many pixels and can acquire a complete photograph at once. The advantage of the latter is that acquiring an image is both fast (typically limited by the read-out time of the camera) and the resulting movies can be contiguous because all points of the sample are imaged both constantly and simultaneously. This approach, known as widefield microscopy, requires that the entire sample be illuminated with near-uniform excitation light. In contrast to widefield imaging, confocal microscopy uses a point-by-point approach to sample illumination and, as a result, fluorescence detection. This necessitates the scanning of either the sample or the excitation beam to produce an image, though significant gains can be made in the imaging frame rate by scanning multiple points simultaneously, as in spinning disc and line scanning microscopies. One advantage of confocal microscopy is that the excitation light can be tightly focused to illuminate only a diffraction-limited spot. This contributes to a somewhat "sharper" fluorescence profile in the image. As a result, confocal imaging has a spatial resolution that is slightly higher than that of widefield imaging and also allows relatively high-resolution three-dimensional images to be acquired, though both systems remain limited to the hundreds-of-nanometers regime as a result of point spreading.

Point spreading is unavoidable, and although many attempts have been made to negate its effects, none of these holds the key to achieving super-resolution. As it turns out, the "secret" to breaking the diffraction limit is not to try to prevent or avoid point spreading, or resort to exotic optical strategies, but instead to exploit the properties of fluorophores. By doing so, we can break or distort the link between the excitation light applied to the sample and the fluorescence emission that we record. By carefully modifying this relationship, additional knowledge on the distribution of the fluorophores in the sample can be inferred. This can subsequently be used to improve the imaging resolution. In a sense, the fluorophore ceases to be a passive photon source and instead becomes an active partner in the experiment.

Fluorophores are defined by their interaction with light. The absorption of light by a fluorophore results in a redistribution of the electrons of the molecule. This state of the fluorophore, the excited state, is stable for a just few nanoseconds before the molecule relaxes back into its relatively stable ground state. In fluorophores, this relaxation is often achieved by emitting a photon of light; the process we know as fluorescence (Valeur 2002; Lakowicz 2006). For example, a fluorophore with a quantum yield of 90% returns a photon for, on average, 9 in 10 of the absorption events it undergoes. Nonradiative decay of the excited state is also possible, and is the dominant process in the case of nonfluorescent molecules. However, a wide range of other processes also occur and conspire to yield surprisingly dynamic fluorescence emission for all individual fluorophores or populations of fluorophores (Figure 16.2). These departures from the "normal" behavior of the fluorophore are relatively rare but can render the fluorophore "dark" for long periods. The key to all super-resolution microscopies

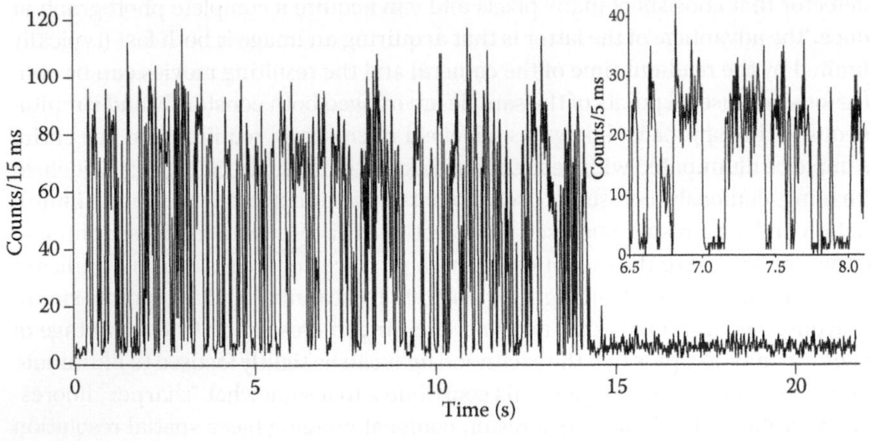

Figure 16.2 Fluorescence intensity time trace of an individual fluorophore (a green fluorescent protein) irradiated with continuous excitation light. The inset shows an expansion of the main trace. The emission from the molecule is continually fluctuating between bright and dark states, that is, blinking. At between 14 s and 15 s, the fluorophore bleaches irreversibly. (Reprinted from *Mater. Today*, 11, Dedecker, P., Hofkens, J., and Hotta, J., Diffraction-unlimited optical microscopy, 12–21, Copyright 2008, with permission from Elsevier.)

is being able to influence these processes and to make use of these dynamics. Broadly speaking, the super-resolution microscopies are accomplished through one of two approaches:

1. Creation of spatially controlled distributions of the emissive fluorophore population that differ from the distribution of the excitation light (typically by being sharper than the PSF allows).
2. Acquisition of a movie of the fluctuation of fluorophore emission over time, in which each movie frame provides a unique image of the sample. Super-resolution images can be derived through subsequent processing of these movies.

16.2 SUPER-RESOLUTION MICROSCOPY BY CREATING CONTROLLED FLUORESCENCE DISTRIBUTIONS

The capability of a microscope to convey structural details is really determined by the sharpness of its PSF; a sharper PSF is capable of distinguishing smaller details. We have already seen that diffraction limits the sharpness with which we can focus or pattern the excitation light. However, a fluorescence microscope is in the business of imaging fluorescence emission, not excitation light, so if we could somehow make sure that the fluorescence emission originates from a distribution that is sharper than the PSF, we can increase the resolution of the imaging system.

An excitation source can be focused onto the sample by the microscope objective. As a result of diffraction, the excitation light will be spread out according to the PSF, taking on a distribution similar to that shown in Figure 16.3. If the fluorescence intensity is entirely proportional to the power of the excitation light, the distribution of the emitted fluorescence is entirely similar to this distribution, and the imaging is subject to the traditional diffraction limit (i.e., the lowest profile in Figure 16.3).

However, there is an upper limit to the rate at which a single fluorophore can emit fluorescence. This is a direct result of the fact that the excited state of the fluorophore has a short but finite lifetime, on the order of a few nanoseconds. By shining enough light on the sample, the excited state can become "saturated," which occurs when the fluorophores are spending essentially all of their time in the excited state. At this point, increasing the excitation intensity will not result in an increase in fluorescence, but reaches a plateau beyond which no additional signal is possible. It is relatively straightforward to achieve saturation in practice by making use of high-power pulsed laser systems.

Figure 16.3 shows the effect that saturation has on the distribution of the fluorescence intensity of a uniform population of fluorophores at the focus of the microscope. By pushing closer and closer to saturation, we gain a distribution

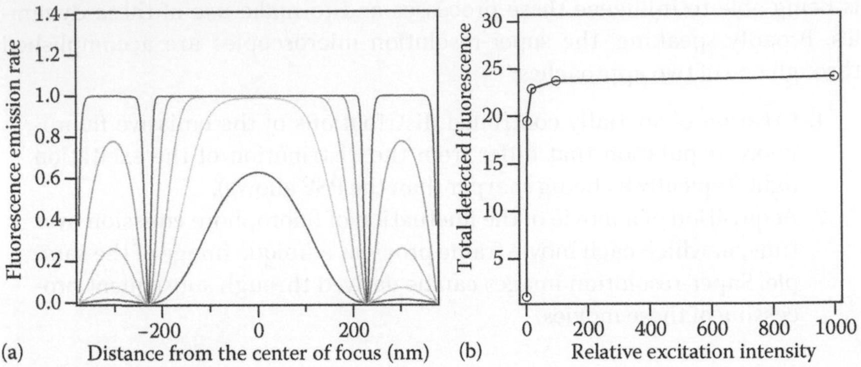

Figure 16.3 (a) Fluorescence emission rate for a uniform population of fluorophores as a function of the distance from the center of the focus, and at different total excitation powers. From lower to higher traces, the relative excitation powers are 1, 5, 20, 100, and 1000, respectively. (b) The total fluorescence intensity as it would be detected by a confocal detector.

with arbitrarily sharp edges. As a result, we can achieve arbitrarily high spatial information content at these locations, even though the excitation light remains diffraction-limited. In essence, the sharp edges arise because the fluorophores display a dynamic that is saturable (in the sense that there is a limit to how fast or how complete this process can be). At the same time, Figure 16.3 also shows hints of another aspect: achieving these distributions requires somewhat tougher imaging conditions (the high excitation power needed to achieve saturation).

By itself, saturation of the emission is sufficient to allow fluorescence imaging with a diffraction-unlimited spatial resolution. The most common implementation is saturated structured illumination microscopy (SSIM) (Gustafsson 2005). In SSIM, a widefield microscope is modified so that it illuminates the sample using a sinusoidal pattern, and the excitation intensity is increased sufficiently to induce arbitrarily sharp intensity distributions. The resulting fluorescence is then captured over the entire illuminated region using a sensitive camera. One of the main challenges in SSIM is that it produces an "inverse" image, in the sense that in standard imaging one typically tries to confine fluorescence to the smallest possible region when using patterned excitation, but SSIM instead confines the region in which no fluorescence is emitted (see Figure 16.3). In addition to requiring multiple acquisitions with shifted and rotated illumination patterns, this also implies that nontrivial calculations are required to reconstruct the "actual" fluorescence image from the recorded images.

Irradiation with light can also be used to suppress fluorescence emission from occurring. One way to achieve this is to make use of stimulated emission, which occurs when a molecule in the electronically excited state (following absorption of an excitation photon) interacts with a photon of light that "stimulates" the emission of a second photon, of the same wavelength, from the molecule. In so doing, the molecule is returned to its ground electronic state. As the name implies, stimulated emission does lead to the emission of additional photons, but these can be discarded easily using emission filters because their wavelength is identical to that of the light that triggered the stimulated emission. Any experiment that aims to use stimulated emission needs two light sources: one light pulse to excite the fluorophores, providing a pico- to nanosecond window during which the molecules are in the excited state, and a second light pulse to trigger stimulated emission (Lakowicz 2006; Dedecker et al. 2008). The two irradiation events are commonly referred to as the "excitation" and "depletion" pulses.

The most common way of using controlled fluorescence depletion (suppression) is implemented in confocal microscopy. However, to do so we have to open a new bag of tricks and start playing with the intensity distribution of the light at the focus point. Although light must always be focused to a distribution, the shape of this distribution can be modified by introducing controlled phase changes in the light path. By careful manipulation, different "modes" can be created. Of particular utility in this instance are the "donut modes," which are special because their distribution corresponds to an exact intensity zero at the center, surrounded by a bright ring of excitation light (Figure 16.4). If we were to focus the excitation light to a donut mode, a fluorophore at the precise center of the focus would experience no irradiation whatsoever. On the other hand, similar molecules a little further away would end up experiencing very large intensities, depending on the power of the laser beam used.

The concepts discussed in the preceding text, for example, stimulated emission of fluorophores and the donut modes of lasers, are those that underpin stimulated emission depletion (STED) microscopy (Hell and Wichmann 1994; Klar et al. 2000; Hell 2003), the first super-resolution technique proposed and successfully demonstrated. In STED, the microscope is reconfigured such that it uses two lasers: a "regular" excitation laser and a depletion laser that is tweaked so that it leads to a donut mode when focused on the sample. In an actual measurement, the sample is imaged one spot at a time, and at every position the fluorophores are first excited using a normal-mode excitation pulse, and then a subset of the fluorophores are rapidly "deactivated" using a donut-shaped depletion pulse. Because the power of the depletion pulse can be raised arbitrarily high by increasing the laser output, but is always zero at the exact center of the focus, the fluorescence emission can be confined to a region with sub-diffraction-limit dimensions. The size of this region depends simply on the power of the donut beam (the extent of the "saturation" of the stimulated emission transition), and this results in a theoretically unlimited and tunable spatial resolution (Figure 16.5).

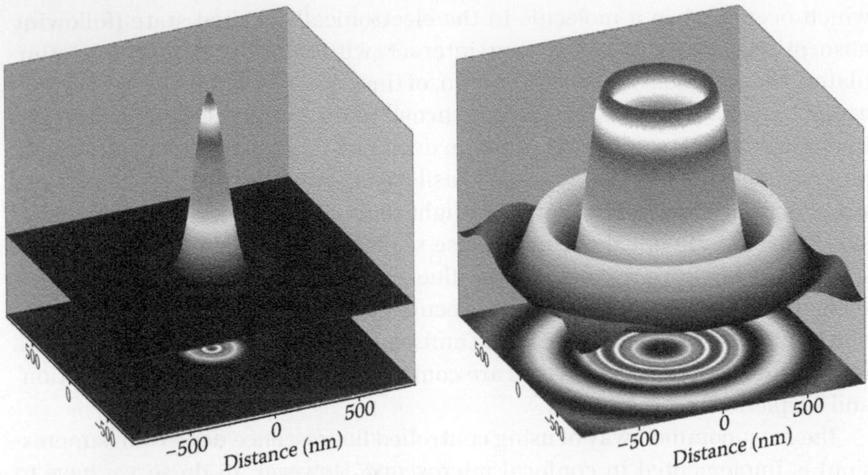

Figure 16.4 Different intensity distributions along the focal plane of a confocal microscope: A "standard" Airy disk shape and a "donut mode." (Reprinted from *Mater. Today*, 11, Dedecker, P., Hofkens, J., and Hotta, J., Diffraction-unlimited optical microscopy, 12–21, Copyright 2008, with permission from Elsevier.)

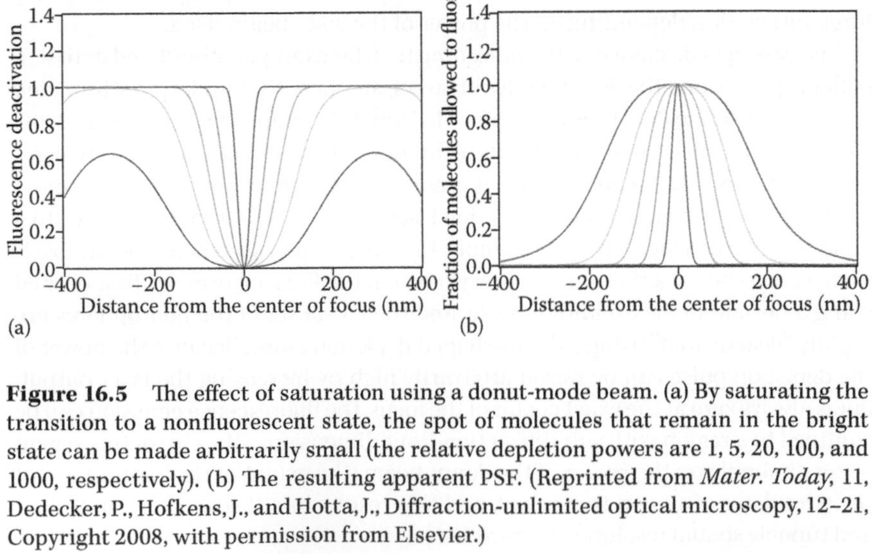

Figure 16.5 The effect of saturation using a donut-mode beam. (a) By saturating the transition to a nonfluorescent state, the spot of molecules that remain in the bright state can be made arbitrarily small (the relative depletion powers are 1, 5, 20, 100, and 1000, respectively). (b) The resulting apparent PSF. (Reprinted from *Mater. Today*, 11, Dedecker, P., Hofkens, J., and Hotta, J., Diffraction-unlimited optical microscopy, 12–21, Copyright 2008, with permission from Elsevier.)

Although STED microscopy is the most successful technique in this category at present, creating distributions that are much sharper than the PSF allows is not the only way to achieve a diffraction-unlimited resolution. For example, dynamic saturation optical microscopy (DSOM; Enderlein 2005) is a lesser known technique that requires fluorophores that can transition into a nonfluorescent state (such as the triplet state) when irradiated by light. Starting from an entirely fluorescent sample, exciting the fluorophores will cause the fluorescence to disappear as a result of the formation of this state. However, with normal-mode irradiation, fluorophores closer to the center of the focus will deactivate faster than fluorophores farther away, simply because these fluorophores experience higher excitation intensities. By monitoring the kinetics of the decrease in fluorescence and selectively extracting the fastest-decaying component (say, by fitting with a mixture of exponentials), the contribution of distant fluorophores can be eliminated and the spatial resolution increased. Although it has not achieved widespread use, DSOM is interesting because it relies on controlled spatial emission patterning, though not by creating sharper distributions but by relying on the temporal evolution of this pattern as a function of irradiation intensity. In Figure 16.6 examples of SSIM and STED imaging are shown.

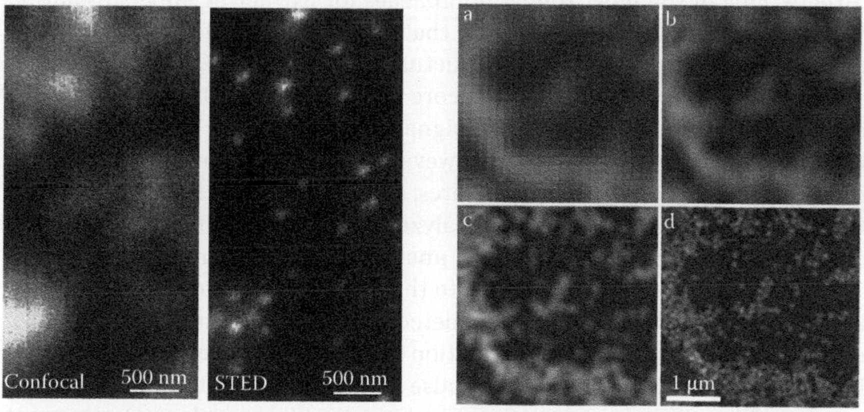

Figure 16.6 Examples of STED and nonlinear structured illumination. (Left) A comparison between a standard confocal image and a STED image of 20 nm fluorescent beads. (Right) A widefield image of 50 nm fluorescent beads. (a, b) Conventional and filtered image; (c) linear structured illumination; (d) nonlinear structured illumination. (Reprinted from *Mater. Today,* 11, Dedecker, P., Hofkens, J., and Hotta, J., Diffraction-unlimited optical microscopy, 12–21, Copyright 2008, with permission from Elsevier; Gustafsson, M. G. L., Nonlinear structured-illumination microscopy: Wide-field fluorescence imaging with theoretically unlimited resolution. *Proc. Natl. Acad. Sci. U. S. A.* 102:13081–13086. Copyright 2005 National Academy of Sciences U.S.A.)

16.3 SUPER-RESOLUTION MICROSCOPY THROUGH MULTIPLE, UNIQUE IMAGE ACQUISITIONS

Consider again the single-molecule time trace shown in Figure 16.2. If we were imaging many similarly behaved fluorophores using a sensitive microscope, we will observe each of the fluorophores blinking independently of all the others. This simultaneous blinking can be revealed using widefield microscopy, allowing images to be acquired on a very short timescale, typically tens of milliseconds. As a result of the blinking of individual fluorophores, none of these images will be identical because different fluorophores will be emitting at any given instant. In a sense, each image provides a complementary view of the sample, and the information that is encapsulated in this movie can ultimately be processed to yield super-resolution information. The main question, of course, is the optimal strategy to use in exploiting this information.

One of the complications is that the fluorescence dynamics can be well obscured. For example, depending on the acquisition speed, fluorophore concentration, and blinking kinetics, the resulting images can range from containing just a few emitters to containing a large number of overlapping fluorophores, making it impossible to distinguish individual blinking events directly. One of the most robust approaches for extracting super-resolution images from such data, in the sense that it can deal with this wide range of conditions, is stochastic optical fluctuation imaging (SOFI; Dertinger et al. 2009; Dedecker et al. 2012). The core assumption in SOFI is simply that, because of the emitter blinking, the signal in every detector pixel will contain fluctuations as a function of time. However, these fluctuations will be complex in the presence of multiple fluorophores, and impossible to interpret directly. Fortunately, these signals can be analyzed using an established mathematical technique known as correlation analysis, or its close relative cumulant analysis. These analyses are applied to the recorded fluorescence intensity as it fluctuates over time at each of the detector pixels. The result is a new image containing super-resolution information (Figure 16.7). Moreover, the imaging is entirely diffraction unlimited because a theoretically unlimited resolution can be achieved by extending the calculation to higher orders (Dertinger et al. 2009).

The main advantage of SOFI is that it is, experimentally, less demanding than any other super-resolution microscopy. It works over a relatively wide range of (active) fluorophore densities and produces results relatively rapidly. It is therefore capable of working with a wide range of samples, especially considering the ubiquity of fluorophore blinking. In addition, because the complexity is effectively contained in the algorithm used to analyze the data, of which public implementations are available, these experiments tend to be easy to

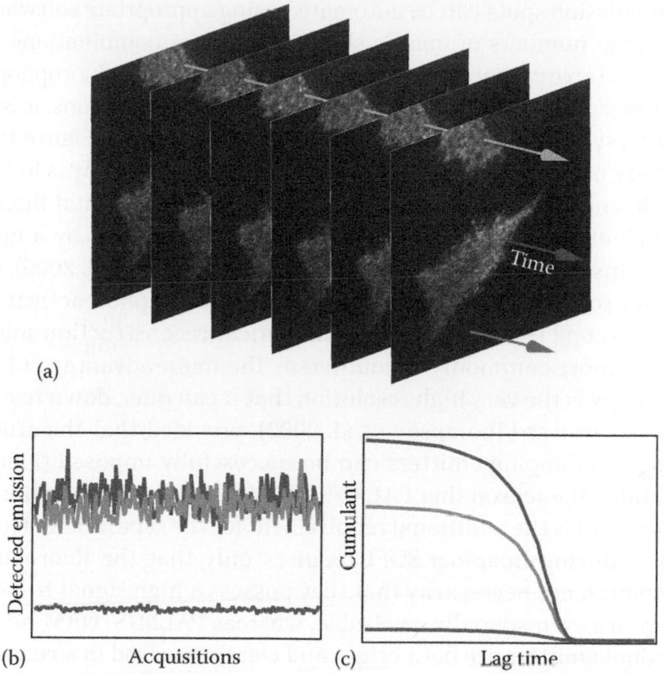

Figure 16.7 (a) In SOFI imaging, a large number of images are acquired on a sample showing fluorophore fluctuations. (b) A trajectory of the intensity values observed in every pixel (compare with the colors in a) is constructed and analyzed using a cumulant analysis. (c) The magnitude of the cumulant calculated for every pixel is then plotted to create the super-resolution image.

perform, and resolution enhancements of 2- to 3-fold can be routinely achieved (Dedecker et al. 2012).

Another super-resolution technique is based on the precise localization of individual fluorophores. This approach requires tight control over the number of active fluorophores in any frame of a given movie. The aim is typically that no more than a few tens of molecules are emitting in any acquired image and that all are detected with a high signal-to-noise ratio. In this case, we can have statistical "certainty" that the emission spots originate from nonoverlapping molecules simply because it would be unlikely that those few molecules that are active would just happen to overlap (we assume that the sample is more or less homogeneously labeled). If this holds, then the observed emission spot must originate from a single fluorophore located approximately at the center of the observed emission distribution. The identification and centroid analysis of each of the emission spots can be automated using appropriate software. When applied to large numbers of images showing different combinations of active fluorophores, a table containing the positions of all observed fluorophores in the sample can be constructed. From this list of fluorophore positions, it is possible to construct a synthetic super-resolution image of the sample (Figure 16.8).

The novelty of this approach to super-resolution microscopy is in the use of the dynamic emission of fluorescence to distinguish individual fluorophores from their identical peers. Developed nearly simultaneously by a number of research groups (Betzig et al. 2006; Hess et al. 2006; Rust et al. 2006), this concept has come to be known by a range of different names; photoactivation localization microscopy (PALM) or stochastic optical reconstruction microscopy (STORM) are most commonly encountered. The main advantage of localization microscopy is the very high resolution that it can offer, down to a few tens of nanometers or less (Thompson et al. 2002), provided that the crucial condition of nonoverlapping emitters can be successfully imposed (Figure 16.9). Fundamentally, the reason that PALM/STORM typically achieves higher resolutions than SOFI is the additional requirement for the separation of individual fluorophores during imaging: SOFI requires only that the fluorophores are blinking, though not necessarily that they possess a high signal-to-noise ratio or that they are even spatially resolvable, whereas PALM/STORM necessitates that individual emitters are both bright and clearly resolved in space. However, several groups have recently developed algorithms for localization microscopy

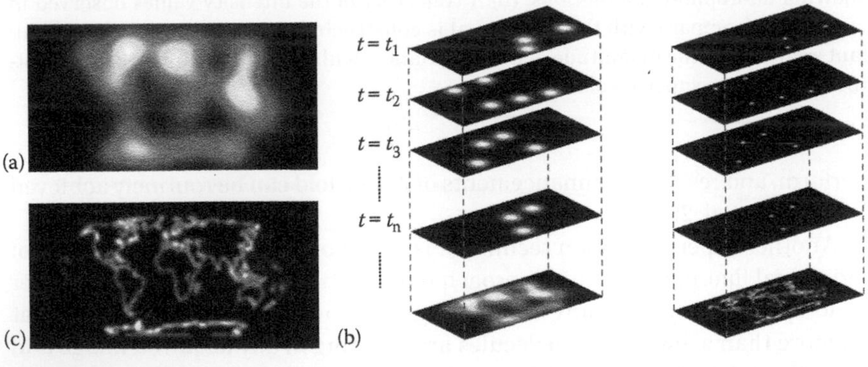

Figure 16.8 The principle behind localization imaging: By acquiring many images containing the emission of a low number of fluorophores, a synthetic super-resolution image can be constructed. (a) Simulated conventional image of the sample. (b) The localization imaging, showing the actual data (left) and computer-generated localization result (right). (c) The reconstructed picture. (Reprinted from *Mater. Today*, 11, Dedecker, P., Hofkens, J., and Hotta, J., Diffraction-unlimited optical microscopy, 12–21, Copyright 2008, with permission from Elsevier.)

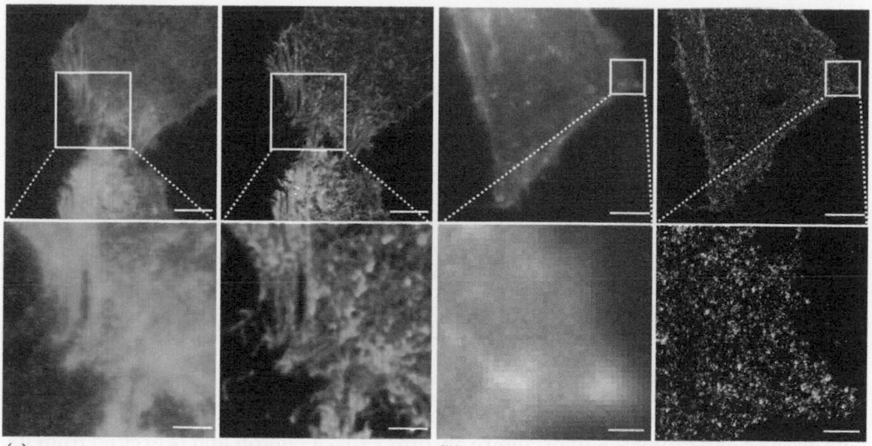

Figure 16.9 (a) Photochromic-SOFI (pcSOFI) imaging on a live HeLa cell labeled with the fluorescent protein Dronpa fused to actin. Scale bars correspond to 5 μm and 2 μm, respectively. (b) PALM imaging on a fixed HeLa cell labeled with the fluorescent protein Dendra-2 at the plasma membrane. Scale bars correspond to 5 μm and 500 nm, respectively.

that use additional information to somewhat reduce the requirement for non-overlapping emitters (Huang et al. 2011; Cox et al. 2012; Zhu et al. 2012). Although these strategies allow for considerably shorter acquisition times for localization microscopy, they still require data with a high signal-to-noise ratio.

The common element in all diffraction-unlimited imaging strategies is the reliance on specific fluorophore properties and dynamics, as well as the extent to which these properties can be induced in a given sample. Much of the challenge in super-resolution imaging is not in the optics or data analysis, but rather in the development and application of suitable fluorophores, probably the most crucial element in determining the performance of these techniques.

16.4 SMART FLUOROPHORES FOR SMART IMAGING

The key to successful super-resolution microscopy experiments is the fluorophore and the ability to achieve a reasonable degree of control over its labeling efficiency, distribution, and, crucially, its photophysical behavior. Figure 16.10 shows a pseudo-Jablonski diagram that incorporates the possible modalities that can be used to switch fluorophores between the emissive state and stable "dark" states (where "stable" depends on the specifics of the imaging but typically entails lifetimes that can reach from milliseconds to several hours in duration).

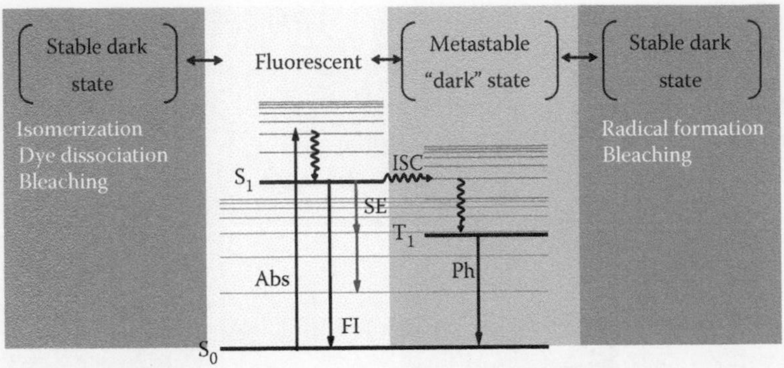

Figure 16.10 Jablonski diagram showing the transitions that occur in a "normal" fluorescence experiment. A fluorophore absorbs (Abs) a photon and is excited from its ground electronic state (S_0) to a vibronic level of a higher electronic state, typically in S_1. Vibrational relaxation (wavy vertical arrows) occurs rapidly (~10^{-12} s), relative to the lifetime of the excited state (~10^{-9} s). Relaxation back to the ground state is achieved by the emission of a photon of fluorescence (Fl). In STED, the depletion (donut) beam is used to stimulate the emission (SE) of a photon from the excited state, S_1. This depletes the excited state of the fluorophore (saturates the ground state), and hence prevents the emission of a photon of fluorescence from this area of the sample. In a very small percentage of excitation events intersystem crossing (ISC) can occur. ISC is the result of a change in the spin of one of the electrons in the S_1 state. The T_1 state, populated by ISC, is metastable and typically has a lifetime of microseconds to milliseconds. Transitions from both the S_1 state and T_1 state to stable (minutes to hours), non-emissive, or simply "invisible" (e.g., in the case of binding/dissociation) states are possible.

The specific requirements for the fluorophores depend on the imaging technique in question. STED microscopy requires only stimulated emission, which is inducible in all excited-state fluorophores. Despite this, however, only relatively few fluorophores are known to work well for STED imaging. The main determinant is likely the tolerance of the fluorophores to the high irradiation intensities that are required to achieve extensive fluorescence depletion. Other factors can play a role as well (Hotta et al. 2010), for instance the presence of an $S_1 \rightarrow S_n$ absorption band at the same wavelength as the emission spectrum of the dye. A range of dyes that are well suited to STED have been discovered or created, allowing STED to be performed at a variety of wavelengths. Similarly, it is possible to saturate the excited state of all fluorophores, and therefore SSIM theoretically can be applied to all fluorophores. However, as with STED, the main drawback of SSIM is the large irradiation intensity that must be applied to the sample to achieve significant saturation. This invariably leads to accelerated fluorophore photodestruction and thus has rather limited the broader application of SSIM.

At the next level of sophistication, we have fluorophores that display metastable dark states, which are characterized by light-induced formation and a lifetime that is determined by the fluorophore structure and environment. The typical example here is the triplet state that is present in organic and biological fluorophores, which typically has lifetimes on the order of micro- to milliseconds, though this can be extended by removing oxygen from the sample. While the molecule is in the triplet state, its absorption and emission are typically shifted from those used in the experiment, and these molecules are effectively "dark" in the imaging. In principle, the temporary "shelving" of fluorophores in the triplet state results in blinking that is compatible with SOFI or PALM imaging. However, the relatively short lifetime and low efficiency of intersystem crossing usually preclude these possibilities, unless the fluorophore is brought into a special environment. This shelving is used in ground-state depletion (GSD) (Bretschneider et al. 2007), a STED-variant that employs an intense donut mode laser in order to temporarily saturate the T_1 state. A less powerful excitation beam follows several milliseconds after the depletion beam, resulting in fluorescence only from the molecules at the center of the donut.

Other metastable dark states are well known. Fluorescent proteins, for example, are known to display fluorescence fluctuations that are caused by spontaneous (not light-induced) protonation/deprotonation of the fluorophore (Haupts et al. 1998). Like triplet states, however, these states are usually difficult to exploit in super-resolution experiments because of their short life times or low residency.

Perhaps one of the most significant advances for super-resolution imaging has been made as a result of the discovery and development of "smart fluorophores"—fluorophores with photophysical properties that can be actively and predictably controlled (Ha and Tinnefeld 2012). This group of fluorophores includes, for example, photochromic species, which can be rendered fluorescent with one wavelength of light and nonfluorescent with another, or photoactivatable species, which become fluorescent or change their emission wavelength on irradiation. Because the fluorescence dynamics of these fluorophores can be directly controlled by the wavelength and intensity of the incident light, thereby enabling detailed control over fluorophore emission, techniques such as SOFI and PALM can be applied under finely tuned conditions in a range of samples. These labels also allow the concepts behind STED to be applied using photochromism directly, by simply ensuring that the wavelength of the donut-mode depletion light induces off-switching of the fluorescence. This approach, known as reversibly saturable optically linear fluorescence transitions (RESOLFT) imaging (Hell 2003), achieves STED-like resolution enhancements using far lower incident light powers. The high degree of control obtained by using these "smart fluorophores" has furthermore allowed for highly parallelized variants of the RESOLFT technique allowing for fast acquisition (Chmyrov et al. 2013). Similarly, photochromism allows SSIM to be performed under milder imaging conditions and with impressive results (Rego et al. 2012).

Whereas some fluorophores are intrinsically smart, others can be made smart by employing appropriate environmental conditions (buffers, etc.). For example, the triplet state in most standard organic dyes can be readily reduced or oxidized to form a radical species in the presence of appropriate oxidizing or reducing agents (Heilemann et al. 2008, 2009; Cordes et al. 2009; Vogelsang et al. 2009). This radical species is non-emissive and can be stable for up to several hours, for example, in the case of the rhodamine dyes, which display high electron affinity (Figure 16.11), but the fluorescent species can also be rapidly recovered by irradiation with ultraviolet (UV) light (van de Linde et al. 2011a). The potential of this finding is tremendous as it allows super-resolution microscopy to be performed with many current and commercially available dyes.

In practice, it is possible to actively control the formation of the radical species, though the optimal conditions for achieving switching vary from one fluorophore to another. An overview of the optimal conditions for performing localization microscopy for 26 different fluorophores can be found in recent publications (Dempsey et al. 2011; van de Linde et al. 2011b). It should be noted that the optimal conditions for switching and imaging these fluorophores seem to vary across the literature. This is perhaps a reflection of the fact that the photophysical behavior of many (organic) fluorophores is sensitive to the atomic environment of the molecule. Considerable work will be needed before a more comprehensive and general picture can be established.

Some of the most fascinating fluorophores are the fluorescent proteins, genetically encoded labels that possess a wide range of photophysical behaviors.

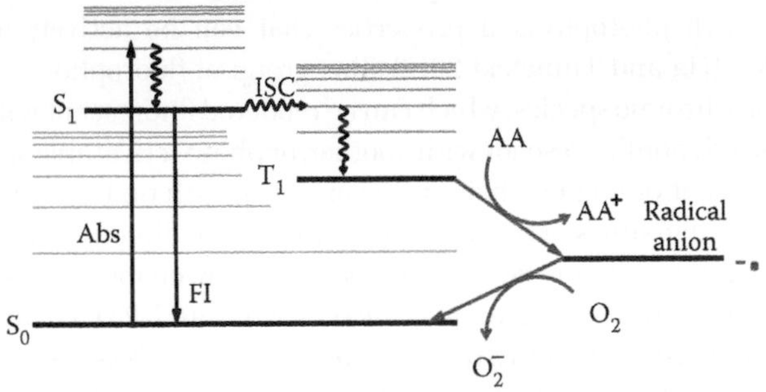

Figure 16.11 Schematic representation of the electronic transitions and transfer processes that lead to fluorescence blinking of a typical rhodamine dye. The cycle incorporates a critical reduction of the rhodamine triplet state to give a stable (hours) radical anion. The reducing agent is typically a thiol, present in low millimolar concentrations. The dye is returned to its fluorescent ground state by oxidation because of molecular oxygen dissolved in the solution. Striking the correct balance between these components allows some control over blinking, which is critical for the localization microscopies.

The photoswitching behavior of fluorescent proteins was first reported by Dickson et al. (1997), who showed that mutants of green fluorescent protein would switch to a non-emissive state after emitting about 10^6 photons. This dark state was stable on the timescale of (at least) several minutes, but remarkably, the fluorescent state could be recovered by irradiating the proteins with UV light (405 nm). Following this discovery, better-performing proteins that display higher contrast and higher fatigue resistance were discovered and were found to be excellent probes for SOFI (Dedecker et al. 2012; Moeyaert et al. 2014), RESOLFT (Hofmann et al. 2005; Dedecker et al. 2007; Grotjohann et al. 2011), PALM (Betzig et al. 2006; Hess et al. 2006), or SSIM (Rego et al. 2012) imaging. Even more remarkable is the fact that fluorescent proteins with photoactivation properties have been discovered as well (Lukyanov et al. 2005) and are tools of choice in PALM imaging. Surprisingly, the structure of the fluorophore is identical for many of these proteins, despite the large differences in their spectroscopic behavior, and consists of three amino acids that cyclize to form a single group embedded in the protein matrix. This remarkable finding shows that it is often the chromophore environment, not the covalent structure of the chromophore itself, that is responsible for many of the fluorescence dynamics known in smart fluorophores (Figure 16.12).

All of these properties essentially result in a fluorophore that can be rendered fluorescent or nonfluorescent as desired. While an intrinsic

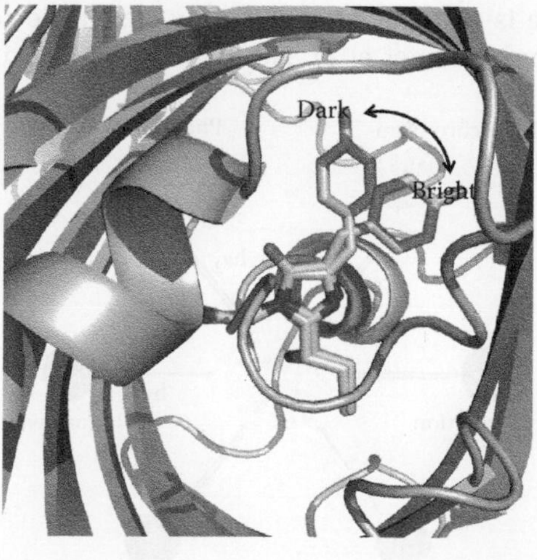

Figure 16.12 The chromophore in fluorescent proteins is formed from the cyclization of the three amino acids of the protein. In the fluorescent proteins the chromophore is embedded within a barrel-like structure. The anisotropic interactions of the fluorophore with the matrix are crucial in determining the structural flexibility and, hence, the photophysical properties of the chromophore.

propensity of the fluorophore towards fluorescence dynamics is the most convenient for broadly applicable super-resolution measurements, in some conditions other phenomena can also be used to induce pseudo-blinking behavior. We will discuss some examples of this in the text that follows (Figure 16.13).

A straightforward example is the use of transient fluorophore immobilization (binding). In widefield imaging, when fluorophores diffuse through the field of view, their emission "smears out" across multiple pixels and usually becomes near-invisible. However, fluorophores that bind become immobile and can be clearly perceived and subject to a SOFI or localization analysis. Any spatial distribution mapped out by the binding events can be revealed with sub-diffraction resolution. One of the first techniques to achieve this was points accumulation for imaging in nanoscale topography (PAINT) microscopy (Sharonov and Hochstrasser 2006), which demonstrated the possibility of using transient binding to a membrane to map out the spatial distribution of that membrane. Since then, a range of related techniques have been developed (Lew et al. 2011; Schoen et al. 2011).

A final option that we will consider here is the ability to perform super-resolution imaging using fluorogenic substrates, which are molecules that can be chemically converted from a nonfluorescent form to a fluorescent form. If this conversion can be mediated by a catalyst, the appearance of fluorescence will signal the location of the active sites on the catalyst and allow their mapping. One example is nanometer accuracy by stochastic catalytic reactions (NASCA) microscopy (Cremer et al. 2010), which has been used to map out

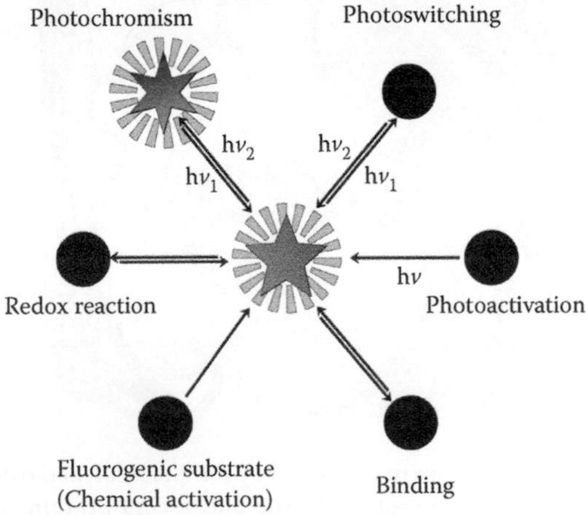

Figure 16.13 Schematic summary of the most commonly exploited forms of fluorophore "on–off" switching in super-resolution microscopy.

catalytic sites in zeolite crystals with very high detail. Other systems could be considered as well, including enzymes, as long as the fluorescent product remains associated with the immobilized catalyst for a sufficiently long duration, such that its emission can be clearly captured.

16.5 CONCLUSION

There is no doubt that diffraction-unlimited fluorescence microscopy is here to stay. Of all of the imaging approaches, fluorescence offers unparalleled selectivity and sensitivity, particularly in a complex environment such as a living cell. Breaking the diffraction limit represents a remarkable scientific and technical development that, because of its ground-breaking nature, will continue to yield exciting new results in the coming years. All of the super-resolution techniques that we have presented here have their own strengths and weaknesses, and hence will be more or less suitable for any particular experiment. Key to all super-resolution experiments is the fluorophore. Our discussion of the approaches that have been utilized to enable super-resolution imaging indicates one important feature, namely, that the fluorophore lies at the heart of all super-resolution experiments. Each super-resolution experiment requires a fluorophore with different attributes, be that stability or switchability. Further developments in the field will depend critically on both the development of new instrumental and analytical approaches as well as the development of fluorophores that enable these approaches. Super-resolution fluorescence microscopy has opened up an entirely new playing field for microscopists. The key is now to optimize and apply these approaches to problems of real significance in biology and the physical sciences.

REFERENCES

Betzig, E., Patterson, G.H., Sougrat, R., Lindwasser, O.W., Olenych, S., Bonifacino, J.S., Davidson, M.W., Lippincott-Schwartz, J., Hess, H.F., 2006. Imaging intracellular fluorescent proteins at nanometer resolution. *Science* 313, 1642–1645.

Born, M., 1999. *Principles of Optics: Electromagnetic Theory of Propagation, Interference and Diffraction of Light*, 7th expanded ed. Cambridge University Press, Cambridge, UK and New York.

Bretschneider, S., Eggeling, C., Hell, S.W., 2007. Breaking the diffraction barrier in fluorescence microscopy by optical shelving. *Phys. Rev. Lett.* 98, 218103.

Chmyrov, A., Keller, J., Grotjohann, T., Ratz, M., d' Este, E., Jakobs, S., Eggeling, C., Hell, S.W., 2013. Nanoscopy with more than 100,000 "doughnuts." *Nat. Methods* 10, 737–740.

Cordes, T., Vogelsang, J., Tinnefeld, P., 2009. On the mechanism of Trolox as antiblinking and antibleaching reagent. *J. Am. Chem. Soc.* 131, 5018–5019.

Cox, S., Rosten, E., Monypenny, J., Jovanovic-Talisman, T., Burnette, D.T., Lippincott-Schwartz, J., Jones, G.E., Heintzmann, R., 2012. Bayesian localization microscopy reveals nanoscale podosome dynamics. *Nat. Methods* 9, 195–200.

Cremer, G.D., Sels, B.F., Vos, D.E.D., Hofkens, J., Roeffaers, M.B.J., 2010. Fluorescence micro(spectro)scopy as a tool to study catalytic materials in action. *Chem. Soc. Rev.* 39, 4703–4717.

Dedecker, P., Hofkens, J., Hotta, J., 2008. Diffraction-unlimited optical microscopy. *Mater. Today* 11, 12–21.

Dedecker, P., Hotta, J., Flors, C., Sliwa, M., Uji-i, H., Roeffaers, M.B.J., Ando, R., Mizuno, H., Miyawaki, A., Hofkens, J., 2007. Subdiffraction imaging through the selective donut-mode depletion of thermally stable photoswitchable fluorophores: Numerical analysis and application to the fluorescent protein Dronpa. *J. Am. Chem. Soc.* 129, 16132–16141.

Dedecker, P., Mo, G.C.H., Dertinger, T., Zhang, J., 2012. Widely accessible method for superresolution fluorescence imaging of living systems. *Proc. Natl. Acad. Sci. U.S.A.* 109, 10909–10914.

Dempsey, G.T., Vaughan, J.C., Chen, K.H., Bates, M., Zhuang, X., 2011. Evaluation of fluorophores for optimal performance in localization-based super-resolution imaging. *Nat. Methods* 8, 1027–1036.

Dertinger, T., Colyer, R., Iyer, G., Weiss, S., Enderlein, J., 2009. Fast, background-free, 3D super-resolution optical fluctuation imaging (SOFI). *Proc. Natl. Acad. Sci. U.S.A.* 106, 22287–22292.

Dickson, R.M., Cubitt, A.B., Tsien, R.Y., Moerner, W.E., 1997. On/off blinking and switching behaviour of single molecules of green fluorescent protein. *Nature* 388, 355–358.

Enderlein, J., 2005. Breaking the diffraction limit with dynamic saturation optical microscopy. *Appl. Phys. Lett.* 87, 094105.

Grotjohann, T., Testa, I., Leutenegger, M., Bock, H., Urban, N.T., Lavoie-Cardinal, F., Willig, K.I., Eggeling, C., Jakobs, S., Hell, S.W., 2011. Diffraction-unlimited all-optical imaging and writing with a photochromic GFP. *Nature* 478, 204–208.

Gustafsson, M.G.L., 2005. Nonlinear structured-illumination microscopy: Wide-field fluorescence imaging with theoretically unlimited resolution. *Proc. Natl. Acad. Sci. U.S.A.* 102, 13081–13086.

Ha, T., Tinnefeld, P., 2012. Photophysics of fluorescent probes for single-molecule biophysics and super-resolution imaging. *Annu. Rev. Phys. Chem.* 63, 595–617.

Haupts, U., Maiti, S., Schwille, P., Webb, W.W., 1998. Dynamics of fluorescence fluctuations in green fluorescent protein observed by fluorescence correlation spectroscopy. *Proc. Natl. Acad. Sci. U.S.A.* 95, 13573–13578.

Heilemann, M., van de Linde, S., Mukherjee, A., Sauer, M., 2009. Super-resolution imaging with small organic fluorophores. *Angew. Chem. Int. Ed.* 48, 6903–6908.

Heilemann, M., van de Linde, S., Schüttpelz, M., Kasper, R., Seefeldt, B., Mukherjee, A., Tinnefeld, P., Sauer, M., 2008. Subdiffraction-resolution fluorescence imaging with conventional fluorescent probes. *Angew. Chem. Int. Ed.* 47, 6172–6176.

Hell, S.W., 2003. Toward fluorescence nanoscopy. *Nat. Biotechnol.* 21, 1347–1355.

Hell, S.W., Wichmann, J., 1994. Breaking the diffraction resolution limit by stimulated-emission: Stimulated-emission-depletion fluorescence microscopy. *Opt. Lett.* 19, 780–782.

Hess, S.T., Girirajan, T.P.K., Mason, M.D., 2006. Ultra-high resolution imaging by fluorescence photoactivation localization microscopy. *Biophys. J.* 91, 4258–4272.

Hofmann, M., Eggeling, C., Jakobs, S., Hell, S.W., 2005. Breaking the diffraction barrier in fluorescence microscopy at low light intensities by using reversibly photoswitchable proteins. *Proc. Natl. Acad. Sci. U.S.A.* 102, 17565–17569.

Hotta, J.-I., Fron, E., Dedecker, P., Janssen, K.P.F., Li, C., Müllen, K., Harke, B., Bückers, J., Hell, S.W., Hofkens, J., 2010. Spectroscopic rationale for efficient stimulated-emission depletion microscopy fluorophores. *J. Am. Chem. Soc.* 132, 5021–5023.

Huang, F., Schwartz, S.L., Byars, J.M., Lidke, K.A., 2011. Simultaneous multiple-emitter fitting for single molecule super-resolution imaging. *Biomed. Opt. Express* 2, 1377.

Klar, T.A., Jakobs, S., Dyba, M., Egner, A., Hell, S.W., 2000. Fluorescence microscopy with diffraction resolution barrier broken by stimulated emission. *Proc. Natl. Acad. Sci. U.S.A.* 97, 8206–8210.

Lakowicz, J.R., 2006. *Principles of Fluorescence Spectroscopy*. Springer, New York.

Lew, M.D., Lee, S.F., Ptacin, J.L., Lee, M.K., Twieg, R.J., Shapiro, L., Moerner, W.E., 2011. Three-dimensional superresolution colocalization of intracellular protein superstructures and the cell surface in live *Caulobacter crescentus*. *Proc. Natl. Acad. Sci. U.S.A.* 108, E1102–E1110.

Lukyanov, K.A., Chudakov, D.M., Lukyanov, S., Verkhusha, V.V., 2005. Innovation: Photoactivatable fluorescent proteins. *Nat. Rev. Mol. Cell Biol.* 6, 885–891.

Moeyaert, B., Nguyen Bich, N., De Zitter, E., Rocha, S., Clays, K., Mizuno, H., van Meervelt, L., Hofkens, J., Dedecker, P., 2014. Green-to-red photoconvertible Dronpa mutant for multimodal super-resolution fluorescence icroscopy. *ACS Nano* 8, 1664–1673.

Rego, E.H., Shao, L., Macklin, J.J., Winoto, L., Johansson, G.A., Kamps-Hughes, N., Davidson, M.W., Gustafsson, M.G.L., 2012. Nonlinear structured-illumination microscopy with a photoswitchable protein reveals cellular structures at 50-nm resolution. *Proc. Natl. Acad. Sci. U.S.A.* 109, E135–E143.

Rust, M.J., Bates, M., Zhuang, X., 2006. Sub-diffraction-limit imaging by stochastic optical reconstruction microscopy (STORM). *Nat. Methods* 3, 793–796.

Schoen, I., Ries, J., Klotzsch, E., Ewers, H., Vogel, V., 2011. Binding-activated localization microscopy of DNA structures. *Nano Lett.* 11, 4008–4011.

Sharonov, A., Hochstrasser, R.M., 2006. Wide-field subdiffraction imaging by accumulated binding of diffusing probes. *Proc. Natl. Acad. Sci. U.S.A.* 103, 18911–18916.

Thompson, R.E., Larson, D.R., Webb, W.W., 2002. Precise nanometer localization analysis for individual fluorescent probes. *Biophys. J.* 82, 2775–2783.

Valeur, B., 2002. *Molecular Fluorescence: Principles and Applications*. Wiley-VCH, Weinheim, New York.

van de Linde, S., Krstić, I., Prisner, T., Doose, S., Heilemann, M., Sauer, M., 2011a. Photoinduced formation of reversible dye radicals and their impact on super-resolution imaging. *Photochem. Photobiol. Sci.* 10, 499–506.

van de Linde, S., Löschberger, A., Klein, T., Heidbreder, M., Wolter, S., Heilemann, M., Sauer, M., 2011b. Direct stochastic optical reconstruction microscopy with standard fluorescent probes. *Nat. Protoc.* 6, 991–1009.

Vogelsang, J., Cordes, T., Forthmann, C., Steinhauer, C., Tinnefeld, P., 2009. Controlling the fluorescence of ordinary oxazine dyes for single-molecule switching and super-resolution microscopy. *Proc. Natl. Acad. Sci. U.S.A.* 106, 8107–8112.

Zhu, L., Zhang, W., Elnatan, D., Huang, B., 2012. Faster STORM using compressed sensing. *Nat. Methods* 9, 721–723.

Chapter 17

Imaging in Optogenetics

Xiaobo Wang, Li He, Carol A. Vandenberg, and Denise Montell

CONTENTS

17.1 OVERVIEW AND HISTORY OF OPTOGENETICS

The term optogenetics was coined in 2006 (Deisseroth et al. 2006) to describe a set of approaches that combine genetic and optical methods to control protein activities with light in living cells, tissues, and organisms (Deisseroth 2010, 2011). It was chosen as "method of the year 2010" by the journal *Nature Methods* (Editorial 2011), and was further highlighted in *Science* the same year (The New Staff 2010).

Francis Crick was the first to speculate that it might be possible to use light to control one type of cell in the brain without affecting other cells (Crick 1999). However, this idea languished owing to lack of a viable strategy to render specific cells responsive to light. Meanwhile, in another corner of the scientific world, microbiologists discovered proteins that function as light-activated ion pumps or channels, including bacteriorhodopsin, halorhodopsin, and channelrhodopsin (Oesterhelt and Stoeckenius 1971; Matsuno-Yagi and Mukohata 1977; Nagel et al. 2002). It took several years for these two fields to come together, in part because it did not seem particularly likely that microbial opsins would function properly in eukaryotic organisms. Instead, neurobiologists experimented with manipulating signaling cascades that respond to chemical stimulation. In 2002, Boris Zemelman and Gero Misenbock established a multiple-component optogenetic system (named "chARGe") in which coexpression of rhodopsin, a G protein–coupled light receptor, with its binding partners arrestin-2 and the α subunit of the heterotrimeric G protein, rendered non-photoreceptor *Drosophila* neurons responsive to visible light (Zemelman et al. 2002). Miesenbock and other groups later expressed TRPV1, a cation channel that is activated by capsaicin, or P2X2, an ATP-activated cation channel, heterologously in cultures of primary hippocampal neurons. They then used photo-uncaging of chemically caged agonists to activate the transfected neurons with light (Zemelman et al. 2003; Banghart et al. 2004; Lima and Miesenbock 2005).

Success in controlling neuronal activity with a bacterial opsin was first reported by Ed Boyden and Karl Deisseroth in their landmark study in 2005 using mammalian neurons (Boyden et al. 2005). This was the first single-component optogenetic system based on channelrhodopsin, a cation channel from algae that is normally activated by light. Channelrhodopsin works in mammalian cells because all vertebrate tissues contain the chromophore cofactor all-trans retinal, a vitamin A derivative that is the part of channelrhodopsin that absorbs light. Although all-trans retinal is absent from organisms such as fruit flies and worms, it can be easily added to their food. Thus channelrhodopsins and other microbial opsins such as halorhodopsins and bacteriorhodopsins have been used successfully to activate or inactivate neurons with light quickly, precisely, and safely in a variety of animals including worms, fruits flies, zebrafish, and mammals (Deisseroth et al. 2006; Adamantidis et al. 2007; Aravanis et al. 2007; Wyart et al. 2009; Arrenberg et al. 2010; Stirman et al. 2010, 2011; Witten et al. 2010; Stierl et al. 2011; Bundschuh et al. 2012; Honjo et al. 2012). Among these microbial opsins, channelrhodopsin family members excite neurons by allowing cations into the cell, and thus depolarizing it, in response to light. In contrast, halorhodopsins and bacteriorhodopsins serve as neuron silencers by pumping anions into the cell or pumping protons out of the cell, respectively, resulting in hyperpolarization (Chow et al. 2010).

Optogenetics has improved our fundamental understanding of neurobiology and has even opened the door to advances in clinical treatments for Parkinson's disease, blindness, and neuropsychiatric disorders (Nagel et al. 2003; Cardin et al. 2009; Gradinaru et al. 2009; Sohal et al. 2009; Tsai et al. 2009; Witten et al. 2010; Peron and Svoboda 2011). One particularly striking example is the use of halorhodopsin and channelrhodopsin to explore which cells and circuits are important in producing the therapeutic benefits of deep brain stimulation in Parkinson's disease. Although this therapy is clearly effective, it was unclear whether the benefits were due to inhibition of neuronal activity in specific regions or activation of other circuits. In an elegant study, Gradinaru et al. (2009) used optogenetics to systematically test different possibilities and discovered that high-frequency stimulation of neurons that innervate the subthalamic nucleus mimics the therapeutic benefit of general deep brain stimulation in this region. This study brought to fruition the promise of using optogenetics to dissect the circuitry that has gone awry in a disease such as Parkinson's and gain insight into the mechanism by which a poorly understood but effective treatment actually works.

This single-component optogenetic control of neuron activity has two important characteristics. First, it can achieve temporal control on a millisecond timescale, in keeping with the speed of endogenous neuronal signaling. Second, when expressed in specific subsets of neurons this technique enables precise temporal and spatial control of specific subtypes of cells even when they are sparsely distributed within the central nervous system of intact animals (Nagel et al. 2003; Peron and Svoboda 2011). This combination achieves temporal and spatial precision that is a significant advance over traditional methods (Peron and Svoboda 2011).

Optimal tools for measuring the consequences of activating or inactivating a particular neuron should match the temporal resolution of the approach used to manipulate cellular activity. Performing optogenetics experiments in a living animal requires delivery of both the light and the sensor into specific cells. Often, readouts are processed via electrical recordings using optrodes, which are a combined light source plus electrode. Light is commonly delivered with a fiber-coupled diode, such as optical fibers or light-emitting diodes (LEDs). Genetically encoded sensors are typically introduced by a commercially available viral infection system (Adamantidis et al. 2007; Aravanis et al. 2007; Gradinaru et al. 2007). Biosensors, such as fluorescent fusion proteins that sense changes in calcium or voltage, can also be used in place of electrical recordings as a readout of neuronal activity (Sakai et al. 2001; Lundby et al. 2008).

Although the first optogenetics tools were developed for neurobiology, and the underlying concept was to use natural light-sensitive channels to artificially excite or inhibit neurons, more recently the concept has been extended to control of other biochemical activities within cells (see Figure 17.1). Several groups have established strategies to use light to control different signals including small GTPases, adenylyl cyclases, kinases, and transcriptional and epigenetic regulators in cultured cells (Levskaya et al. 2005, 2009; Wu et al. 2009; Yazawa et al. 2009; Ryu et al. 2010; Stierl et al. 2011; Konermann et al. 2013).

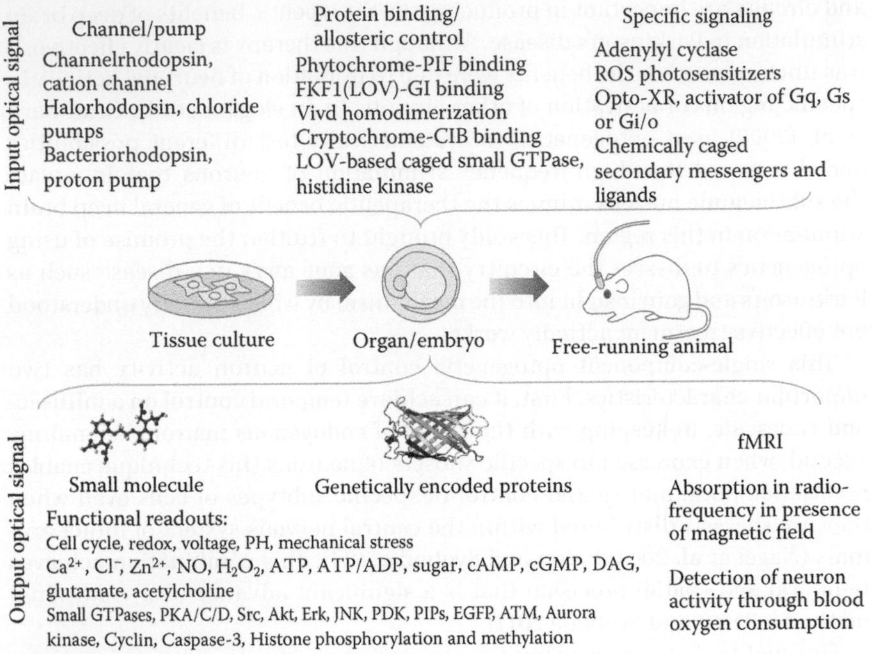

Figure 17.1 Summary of tools for optogenetic control and functional readout through imaging. Here we list the major systems reported for optogenetic control of different aspects of physiological signals. We also list most of the functional readouts through imaging by either small molecule indicators or genetically encoded protein reporters. The majority of them have not yet been combined together with optogenetic manipulation, which is a promising resource for future study.

Optogenetic activation of second messenger signaling pathways via G protein–coupled receptors can complement modulation of cell membrane potential with microbial rhodopsins. Airan et al. (2009) developed chimeric optoXR proteins composed of a vertebrate rhodopsin light sensing domain fused with intracellular domains from other G protein–coupled receptors such that these chimeras recruit and target different G protein signaling pathways in response to light. Remarkably these experiments were carried out in freely moving animals.

An alternative strategy to optically activate receptors or channels, pioneered by the Isacoff, Kramer, and Trauner laboratories, is to covalently modify the proteins with a tethered ligand (agonist, antagonist or blocker), such that light of different wavelengths alters the conformation of the tether and thus reversibly affects the ability of the ligand to bind the receptor or channel

(Kramer et al. 2013; Reiner and Isacoff 2014). Thus light-regulatable receptors and ion channels can be expressed and chemically modified to generate optogenetic proteins, or photo-switchable ligands can be introduced to modify endogenous channels (Banghart et al. 2004; Volgraf et al. 2006; Levitz et al. 2013).

In a series of remarkable applications of this technique, expression of light-activated glutamate receptor (LiGluR) in retinal ganglion cells was able to restore some visual responses in rodent models of inherited blindness (Caporale et al. 2011). More recently, modification of endogenous K^+ channels with photosensitive ligands was shown to confer visual responses in blind animals by chemical modification without the need to express heterologous proteins (Polosukhina et al. 2012; Tochitsky et al. 2014). The emerging application of optogenetic tools in biochemistry and cell biology will allow researchers ever more precise temporal and spatial control of diverse biochemical activities and cellular functions.

Several improvements are ongoing to expand the optogenetic toolkit even further. Mutagenesis-based modification of microbial opsins provided the working foundation of optogenetics early in its history and is used to isolate variants with enhanced properties (Berndt et al. 2009; Lin et al. 2009; Gunaydin et al. 2010; Hegemann and Moglich 2011). For example, mutation of one particular amino acid in channelrhodopsin 2 yields a protein with delayed inactivation kinetics so that a single pulse of light produces a long-lasting ion flux, greatly reducing the amount of light needed to activate a cell for a given amount of time. Other mutations and swapping domains between opsins have enabled high-frequency excitation or altered the wavelength of light required to activate the channel. Red-shifted channelrhodopsin derivatives allow enhanced photoactivation of cells deeper in tissue as a result of lower light scattering and reduced absorption of longer wavelengths by endogenous chromophores (Lin et al. 2013). Using traditional channelrhodopsins, photoactivation of deep brain structures is usually achieved by insertion of a thin optical fiber to illuminate the target tissue. With the new red-shifted ReChR channelrhodopsin, Tsien, Kleinfeld, and colleagues were able to circumvent the need for an invasive optical fiber, and instead could stimulate brainstem neurons through the skin and bone of the skull using noninvasive placement of an optical fiber in the external auditory canal (Lin et al. 2013). Rapid and reliable stimulation of whisker movement was elicited noninvasively through deep brain stimulation. Improvements under development, in addition to modification of the opsins, include increasing expression levels, subcellular targeting of the probes and sensors to specific organelles or the plasma membrane, and expanding the toolkit to gain light control of additional kinases and transcription factors (Deisseroth 2011). Of course, use of more and more optogenetic tools will be extended from cultured cells to tissues and to freely behaving animals (Airan et al. 2009; Sohal et al. 2009; Tsai et al. 2009; Lee et al. 2010).

Here we have briefly summarized the history, principles, and applications of optogenetics in life science and medical research. In the following section, we focus on the application of imaging in the detection and readout of optogenetic manipulation in cultured cells as well as in freely behaving animals.

17.2 IMAGING IN OPTOGENETICS

Imaging is critical in optogenetics experiments for two main reasons: confirming light-activation and monitoring its consequences. It is technically convenient to be able to use the same equipment (optical fibers or microscope) for both purposes. In neurons the consequences are changes in electrical activity whereas in nonexcitable cells, biochemical and morphological changes are key.

17.2.1 Imaging Neural Activity

To precisely monitor extremely rapid changes in neural activity controlled by channelrhodopsin and its derivatives, electrophysiological recordings are the time-honored gold standard. However, there are severe limitations to this approach: It is technically challenging and requires extensive training and expertise; placing electrodes into the brain or inside neurons is invasive; and intracellular recordings can usually sample only a small number of cells whereas field potentials average the signals from many cells, severely limiting the spatial resolution of the signals that are detected. Therefore additional approaches have been and continue to be developed including imaging of fluorescent biosensors and monitoring brain activity by functional magnetic resonance imaging (fMRI).

17.2.1.1 Neural Signals Recorded by Fluorescent Biosensors

One method of imaging patterns of neuronal activity is to use fluorescent imaging of calcium indicators including BAPTA (1,2-bis(*o*-aminophenoxy) ethane-*N*,*N*,*N*′,*N*′-tetraacetic acid)-based small molecules, such as Fura-2, indo-1, and fluo-4, or genetically encoded proteins like GCaMP and Cameleons (Perez Koldenkova and Nagai 2013; Terai et al. 2013) to record calcium signals before and after optogenetic stimulation. The advantage of calcium imaging is that the sensors are robust, sensitive, and easy to apply in tissue culture. For example, the Diesseroth group demonstrated that light exposure can strongly induce calcium signals in HEK cells expressing optoXRs (Airan et al. 2009). However, changes in intracellular calcium concentration are slow relative to changes in membrane potential and are distributed throughout the cell. And the loading of the dye *in vivo* is much more challenging and lacks of selectivity for particular neurons of interest.

A genetically encoded family of calcium reporters known as GCaMPs are fusion proteins of green fluorescent protein (GFP) with the calcium-binding domain of calmodulin. These sensors now span a range of calcium affinities and kinetics, and are capable of detecting single action potentials as well as localized calcium changes in individual synaptic spines in dendrites in the brain of living animals (Chen et al. 2013). Highlighting the ability of new GCaMPs to monitor neuronal activity *in vivo*, a study in zebrafish larvae Muto et al. (2013) demonstrated that visual perception of a swimming paramecium evokes direction-sensitive neuronal activity in the optic tectum of larval fish and correlated with prey capture activity in free-swimming fish.

In a tour de force of imaging, Ahrens et al. (2013) have used lightsheet microscopy together with the genetically encoded calcium sensor GCaMP5G to record electrical activity from the whole brain of larval zebrafish. They were able to capture activity at 0.8 Hz from more than 80% of the neurons in the brain at single-cell resolution. Using this method they demonstrated the feasibility of identifying correlated activity patterns of neuronal circuits within the whole brain. Engert and collaborators have also demonstrated the power of this approach by imaging the activity of individual neurons in behaving zebrafish exposed to a changing virtual environment (Ahrens et al. 2012).

As an alternative to calcium imaging, a great deal of effort has been invested in developing genetically encoded fluorescent sensors of voltage changes (reviewed in Peterka et al. 2011; Patti and Isacoff 2013). Voltage-sensitive fluorescent proteins were originally designed by inserting a fluorescent protein reporter within a sodium or potassium channel. The voltage-sensing domain in these fusion proteins generates a conformational change in response to a voltage difference, and this change modulates fluorescence emission (Sakai et al. 2001). In a significant design improvement, the voltage-sensing domain from a voltage-sensitive phosphatase enzyme replaced the one from an ion channel (Akemann et al. 2009). With this improvement, the voltage-sensitive biosensor reported cortical electrical responses to single sensory stimuli *in vivo* in pyramidal cells of the mouse somatosensory cortex in brain slices and even in living mice (Akemann et al. 2010).

Methods for optically recording membrane voltage are advancing rapidly. For example, one approach takes advantage of a voltage-sensitive archaebacterial rhodopsin protein Arch, the natural function of which is to convert light into proton pumping activity, allowing the organism to capture the energy of the sun. When Arch is heterologously expressed, it can silence neurons in response to light (Chow et al. 2010) in the now-traditional manner of optogenetics. However, Kralj et al. (2012) cleverly imagined that they could "run this protein in reverse," that is, that a voltage change might lead to a change in the spectral properties of the Arch protein. This successful approach led to detection of individual action potentials in mammalian neurons and changes in fluorescence that responded linearly to changes in voltage. One drawback of this particular protein is that the wild type version generates a hyperpolarizing current. Therefore the authors identified a point mutation that eliminated that current but this mutation also slowed the response significantly. These voltage indicators are very sensitive but quantum yield and signal intensity are still low (Cohen and Hochbaum 2014). Further improvements are likely to come along soon in this rapidly evolving field.

17.2.1.2 Imaging Animal Behaviors

The goal of optogenetics experiments is not only to observe subcellular and cellular responses but also to determine the behavioral effects of activating or inactivating specific neurons and thus to map the circuitry responsible for specific behaviors. *Drosophila* and zebrafish larvae, as well as *Caenorhabditis elegans* adults, are amenable to such studies because they are transparent. Therefore it is relatively straightforward to express channelrhodopsin-2 (ChR2) or other light activated ion channels in specific subsets of neurons or muscle cells using specific enhancer and promoter sequences and then to record animal movements in the absence or presence of whole-body illumination of the appropriate wavelength. When ChR2 was expressed in muscles in *C. elegans*, light-dependent muscle contractions lasted as long as the illumination and when ChR2 was expressed in mechanosensory neurons, light effectively generated behavioral responses normally seen in response to mechanical stimulation (Nagel et al. 2005).

Optogenetic methods have been used to investigate conditional learning paradigms. In *Drosophila* larvae, activating dopaminergic neurons were demonstrated to be sufficient to substitute for an aversive stimulus in *Drosophila* larvae olfactory learning whereas activating octopamine/tyramine neurons were sufficient to substitute for a reward. This study demonstrated directly and conclusively that distinct neurotransmitter pathways exert opposite neuromodulatory effects on olfactory learning (Schroll et al. 2006). In a further step up in sophistication, Stirman et al. (2011) were able to illuminate specific neurons or muscle cells in *C. elegans* rather than the entire animal while simultaneously recording the animals' movements. This required simultaneous multicolor illumination, specifically a three-color imaging system, to separate and probe different nodes in a neuromuscular networks with spatial, temporal and spectral precision (Stirman et al. 2011).

In a tour de force, Deisseroth's group was able to stimulate neurons expressing different optoXRs and then track the movements of mice to study the causal effects of activating different biochemical signaling pathways in freely moving animals (Airan et al. 2009). They injected lentiviral vectors carrying the optoXR genes into the nucleus accumbens of adult mice and used the fluorescent protein tag mCherry to follow their expression. Optical stimulation deep in the brain was achieved via an implanted fiber optic cable. Then they used time-lapse video to follow the preferred position of mice before and after light illumination. Through this video recording, they found that the effects of optoXRs are more related to reward behavior than direct control of locomotor activity or anxiety-related behaviors (Airan et al. 2009).

17.2.1.3 Tracking of Brain Activity by fMRI

Several recent studies using fMRI combined with optogenetic tools have revealed that a specific neuronal circuit can regulate local blood flow in living animals (Lee et al. 2010; Desai et al. 2011; Kahn et al. 2011). All studies used a blood oxygenation level-dependent fMRI (BOLD fMRI), which is a commonly used technology for noninvasive whole brain imaging (Ogawa et al. 1992). fMRI is a technique using an MRI scanner to measure brain activity by detecting changes in blood flow. BOLD fMRI, developed by Seiji Ogawa, is an fMRI based on a basic measure of the BOLD contrast, which is the change in magnetization between oxygen-rich and oxygen-poor blood (Ogawa et al. 1992). An fMRI graphically represents brain activity by color-defined neuronal activity across a whole brain or a particular brain region, with millimeter spatial resolution and kinetic resolution of seconds (Kim and Ogawa 2002).

BOLD fMRI can reflect complicated but poorly understood changes of brain activity in several aspects such as cerebral blood volume cerebral blood flow, and cerebral metabolic rate of oxygen consumption (Kim and Ogawa 2002). However, because of technical limitation, it has been difficult to confirm whether BOLD signals in a specific region are caused by activation of local excitatory neurons (Logothetis et al. 2001). Deisseroth's group integrated optogenetics with high-field fMRI signal readouts to confirm that the light stimulation of local CaMKIIα in excitatory neocortex or thalamus neurons could significantly result in positive BOLD signals at the stimulus site (Lee et al. 2010). This combination permitted interrogation of the causal effects of specific cell types by the axonal projection target. Finally, optogenetic fMRI (opto-fMRI) within the intact living mouse brain revealed that the downstream targets and the BOLD stimulus have distant signals (Lee et al. 2010), indicating that this potent tool may be suitable for functional circuit analysis as well as global phenotyping of dysfunctional circuitry. Later work from Boyden's group used the light-gated ion channel ChR2 with fMRI to investigate brain activity within different distributed regions in the brains of awake and anesthetized mice (Desai et al. 2011). And they demonstrated that opto-fMRI can be used to characterize a distributed network downstream of a defined cell class in an awake brain. Moreover, Moore's group applied opto-fMRI in mice to test the linearity of BOLD signals driven by locally induced excitatory activity (Kahn et al. 2011). More recently, Rossier's group used photostimulation of two specific subsets of neurons that express the calcium binding protein parvalbumin, together with fMRI, and monitored the pattern of vasoconstriction (Urban et al. 2012). They revealed a complex pattern consistent with increased local blood flow accompanied by vasoconstriction in deeper arterioles.

17.2.2 Imaging in Study of Optogenetics Control of Biochemical Processes

In addition to imaging neural activity induced by optogenetics, a number of laboratories also used imaging to study the feasibility and readout results of different optogenetic tools, which have been used to control small GTPases and other biochemical activities. Different from rhodopsins, most light-sensitive protein domains used in control of biochemical activities are not from integral membrane proteins, but from soluble intracellular proteins. Here, we describe, in detail, the different mechanisms of optogenetic biochemical tools and their corresponding imaging methods.

17.2.2.1 Designs of Optogenetic Biochemical Tools

There are several major protein domains that have been successfully engineered as optogenetic tools to be responsive to different wavelengths of light, which are the light-oxygen-voltage (LOV) domain, phytochrome photoreceptor, and the cryptochrome 2–CIB1 system.

17.2.2.1.1 Optogenetic Tools Based on the LOV Domain

The LOV domain is found in bacterial, fungal, and plant proteins belonging to a subgroup of the Per-ARNT-Sim (PAS) family, which is named after three proteins: period circadian rhythm (Per), single-minded (Sim), and aryl hydrocarbon receptor nuclear transporter (ARNT; Taylor and Zhulin 1999; Moglich et al. 2009b). The LOV domain has a conserved central core containing five strands of antiparallel β sheet and several α helices, within which resides a flavin, typically a flavin mononucleotide (FMN; Moglich et al. 2009b). On illumination with 400–500 nm light, excitation of the flavin molecule results in the formation of a covalent linkage between a thiol from a conserved cysteine residue in the LOV domain and the C4 (a) atom in the flavin. This reaction is reversible with a half-life of seconds to minutes in the dark, depending on the specific protein (Swartz et al. 2001). Attached to the LOV domain in plant protein known as phototropin is a helix called the Jα, which is docked onto the β-sheet of the LOV domain (Harper et al. 2003). On illumination, conformational changes occur throughout the LOV domain, leading to dissociation and unwinding of the Jα helix.

Several optogenetic tools have been designed, which capitalize on this light-induced conformational change. Sosnick's group fused the LOV domain via its Jα helix linker to the N-terminus of Trp repressor protein (TrpR), so that the

DNA binding activity of TrpR can be controlled by light (Strickland et al. 2008). This construct has been named LOV- and tryptophan-activated *protein* (*LOV–TAP*). Their recent work modified the LOV domain conformation by amino acid point mutation so that the regulatory effect of light on DNA binding activity of LOV–TAP was improved more than 60-fold (Strickland et al. 2010). Using this tool, target gene transcription can be controlled by light.

In another ground-breaking study, Hahn's group used the LOV domain of phototropin to design a photoactivatable analog of the small GTPase Rac1 (PA-Rac) (Wu et al. 2009). In PA–Rac, the LOV2 domain and Jα helix were fused to the N-terminus of Rac1. In the dark, the LOV2 domain binds to Rac1 and thereby occludes effector binding; on illumination with blue light, the Jα helix unwinds and the LOV domain dissociates from Rac1, thus allowing Rac1 to bind its downstream effector proteins such as the serine/threonine kinase PAK. Illumination increases the binding affinity of the LOV–Rac1 fusion protein for PAK 10-fold. This was sufficient to direct cell migration *in vitro* and *in vivo*: it can drive the movement of fibroblasts in 2D cultured conditions (Wu et al. 2009), guide the movement of neutrophil cells in the embryos of developing zebrafish (Yoo et al. 2010), and repolarize the protrusion and migration of collectively migratory border cells in the *Drosophila* ovary (Wang et al. 2010).

Moffat's group used the LOV domain to engineer the light-sensitive version of the histidine kinase FixL (Moglich et al. 2009a), one component of the FixL/FixJ sensor/response regulatory system. In this two-component system, FixL is an oxygen-binding kinase that can control the phosphorylation and activation of the transcriptional activator FixJ, and thus the induction of nitrogen fixation (*nif* and *fix*) genes in response to changing oxygen concentrations (Monson et al. 1995). Replacement of the PAS A/B domains in FixL with the LOV domain can change the signal response specificity from oxygen to blue light, without affecting the catalytic efficiency of FixL kinase (Moglich and Moffat 2007; Moglich et al. 2009a). This engineered FixL/FixJ represses target gene expression 70-fold on light illumination (Moglich et al. 2009a). More recently, a similar idea has been adapted by Jaeger's group to control the activity of lipase A by blue light (Krauss et al. 2010).

In addition, the Dolmetsch group established a blue-light–sensitive protein–protein interaction system based on the interaction of two proteins—FKF1, which possesses a LOV domain, and GIGANTEA (GI) (Yazawa et al. 2009). They used this strategy to achieve two different biochemical controls. First, they fused the small GTPase Rac1 with FKF1 and a membrane anchoring CAAX motif with GI, such that absorption of a photon relocates Rac1 to the plasma membrane, bringing it into proximity with its exchange factors, thus resulting in Rac activation, actin polymerization, and cell movement in light-treated fibroblasts (Yazawa et al. 2009). Second, they fused GI to the DNA binding domain of Gal4 and a fragment of FKF to the transactivation domain of VP16.

This pair can induce the expression level of a reporter gene approximately 5-fold upon illumination (Yazawa et al. 2009). More recently, Yang's group used Vivid (VVD), the smallest protein containing the LOV domain, which dimerizes in response to light, and fused it with a modified Gal4 lacking the dimerization domain—Gal4(65). Illumination with blue light enhances binding to the UASG sequence more than 200-fold in cultured cells (Wang et al. 2012). More strikingly, they transferred this light-switchable system vectors into diabetic mice, and then used light to induce insulin expression, which significantly decreased blood glucose, suggesting a promising strategy for treating diabetes (Wang et al. 2012).

17.2.2.1.2 Optogenetic Tools Based on the Phytochrome Receptors

Phytochrome receptors are a group of proteins that covalently bind linear tetrapyrrole chromophores, which absorb red/far-red light in bacteria, fungi, and plants. Phytochromes from different species bind various linear tetrapyrroles including biliverdin, phycocyanobilin, phytochromobilin, or others (Rockwell and Lagarias 2010; Rockwell et al. 2006). Phytochromes in plants and bacteria are composed of three individual domains named PAS, GAF (named after the three proteins it is found in: cGMP-specific phosphodiesterases, adenylyl cyclases, and FhlA), and PHY (Ho et al. 2000; Wagner et al. 2005; Essen et al. 2008; Yang et al. 2008; Moglich et al. 2009b). Phytochromes experience a reversible photocycle transition between the Pr (red-light absorption) and Pfr (far red-light absorption) spectroscopic states, mediated at the molecular level via the isomerization of the bilin chromophore around the C15=C16 double bond. Based on this isomerization, phytochromes can interact with or dissociate from different phytochrome-interacting factors (Ni et al. 1998) in a light-dependent manner, which forms the basis for the design of phytochrome-based optogenetic tools.

Quail's group first established a light-switchable gene promoter system using this light-sensitive interaction between plant phytochromes PhyA and PhyB with the phytochrome-interacting factor (PIF) 3 (Shimizu-Sato et al. 2002). They fused a DNA binding domain with PhyB and a transactivation domain with PIF3. This idea was conceptually similar to the later work of the Dolmetsch group (Yazawa et al. 2009). This switchable promoter system can produce a more than 1000-fold induction of gene expression by red light, with the added advantage that it can be quickly turned off by far-red light (Shimizu-Sato et al. 2002), enhancing both temporal and spatial resolution.

After this success of light-switchable systems, similar approaches have been used to control protein interactions or localization, and thus biochemical function, with light. Rosen's group set up a PHY/PIF–mediated interaction between the small GTPase Cdc42 with its downstream effector protein WASP, so that they can use red light to induce actin polymerization (Leung et al.

2008). Similarly, Muir's group established a moderate red-light activatable protein-splicing system (Tyszkiewicz and Muir 2008). Subsequently Voigt's group linked two nucleotide exchange factors, Tiam1 and intersectin, to the plasma membrane to control the activation of Rac1 and Cdc42 by red light, which can promote light-dependent membrane protrusion and fibroblast cell movement (Levskaya et al. 2009).

In addition to the application of plant phytochromes, several groups also engineered bacteriophytochromes or xanthopsin photoreceptor in the design of optogenetic tools. For example, Voigt's group fused the phytochrome sensor of the histidine kinase Cph1 to the histidine kinase region of EnvZ generating a red-light–inducible gene expression system (Levskaya et al. 2005). This design is conceptually similar to the aforementioned YF1 engineered to be sensitive to blue light (Moglich et al. 2009a). Woolley's group integrated a photoactive yellow protein (PYP), a member of the xanthopsin family, with the leucine zipper protein GCN4 to control gene expression with yellow light (Morgan et al. 2010). Although phytochromes have been widely applied in the optogenetic control of gene expression and small GTPase activity in cultured cells or yeast, currently there is not any successful report about its application in fruit fly, zebrafish, and transgenic mouse. This might stem from the characteristic of phytochromes: compared with the widely distributed flavin molecule among different species, the tetrapyrroles phycocyanobilin or phytochromobilin are not available in most animals. This could be a limiting factor for this application in living animals.

17.2.2.1.3 Optogenetic Tools Based on Cryptochromes

The *Arabidopsis* blue light–inducible binding interaction between cryptochrome 2 and its interacting partner CIB1 is gaining utility as an optical tool for promoting protein interactions owing to its robustness and response time. Cryptochromes are flavoproteins that are found in plants and animals. The light-responsive cofactors of cryptochrome 2 are universally abundant and the protein interactions can be induced rapidly and reversibly with light. This system was first used by Kennedy et al. (2010) to control gene expression and Cre-mediated recombination. Later, Koonermann et al. (2013) further employed this partnership to develop light-inducible transcriptional activation or inhibition. Upon illumination, CIB1 fused to a transcriptional regulatory domain is recruited to a customizable DNA-binding domain fused to cryptochrome 2. The approach is modular, allowing a variety of DNA binding elements (e.g., TALE domains or CRISPR-Cas domains) to recruit various effectors to either modulate transcription directly or epigenetically alter gene activity with histone modifying enzymes. Because CRISPR-Cas is easily modified to recognize different DNA targets, the approach offers rapid customization to target various genetic loci.

17.2.2.2 Imaging Systems Used in Optogenetic Biochemical Studies

Both nonfluorescent and fluorescent imaging systems have been used to visualize the downstream cellular changes that occur in response to optogenetic stimulation of biochemical probes. The choice of imaging modality depends on the availability of fluorescent proteins to track cell responses.

17.2.2.2.1 Nonfluorescent Widefield Imaging

For visualization and analysis of membrane ruffles and other types of cellular protrusions that can be controlled by the LOV domain fused to Cdc42 or Rac, nonfluorescent widefield microscopy imaging is sufficient. This tracking system is favorable for cultured mammalian cells. Hahn's group used this approach to visualize cell membrane changes involved in ectopic ruffle formation, membrane protrusion or retraction, protrusion polarity direction, and for analyzing cell migration speed (Wu et al. 2009). They found that light-induced Rac activity could control, within a few minutes, the formation of ectopic membrane ruffles and protrusions in different types of mammalian cells, and the effect was extremely local occurring directly in the 7 μm region of illumination. Conversely, light-induced Rac inhibition led to local membrane retraction. With longer illumination times, light-activated Rac could efficiently drive prominent cell movement of fibroblasts approximately 50 μm in 90 min along a fibronectin-coated surface. Although generally useful for cells cultured in 2D dishes, nonfluorescent imaging methods are normally unsuitable for cells cultured in 3D matrigel or hydrogel matrices, or for cells deep in endogenous microenvironments in living tissues owing to poor cell contrast in thick tissues (Guo et al. 2012).

17.2.2.2.2 Fluorescent Imaging

Fluorescent imaging is a powerful tool to monitor the effects of optogenetic tools both in cultured cells and in samples *in vivo*. These imaging studies will be divided into two categories: testing the direct effects of optogenetic tools and monitoring the downstream changes induced by light.

17.2.2.2.2.1 Imaging to Assess and Optimize Optogenetic Tools

When activating a protein activity with light, it is necessary to determine the amount of laser energy to deliver to the sample and the duration and frequency of stimulation for maximal efficacy and minimal tissue damage. This is easily achieved by the measurement of the induced protrusion area. To assess the precision of the spatial effect of PA-Rac, Klaus Hahn's group designed fusion proteins composed of the PA-Rac and PAGFP or mVenus. By analysis of fluorescence recovery after photobleaching (FRAP) or photoactivatable GFP, they demonstrated that the diffusion rate of PA-Rac1 is much slower than cytosolic proteins, possibly due to its membrane binding. Using the estimated half-life of photoactivation is approximately 43 s at room temperature, they simulated that the 10-μm spot adjacent to the illuminated region achieves at most 7.5% of the activation within the illuminated region, indicating that PA-Rac1 is a robust caged control in living cells. The limitation of this approach however had been the difficulty adapting it to other proteins.

To analyze the efficiency of the PHY–PIF photoswitchable membrane recruitment system, Christopher Voigt's group linked YFP to PIF (Levskaya et al. 2009). Following illumination with 650 nm light, dynamic imaging of YFP fluorescence revealed the translocation of PIF–YFP to the plasma membrane via its light-induced physical interaction with membrane-anchored PHY. Through this dynamic imaging analysis, Voigt's group demonstrated that the induced PHY–PIF membrane recruitment system achieves rapid kinetics near the theoretical limit for diffusion. They were able to identify optimized PHY–PIF variants that support rapid binding that is reversible upon illumination with 750 nm light (Levskaya et al. 2009). Similarly, Dolmetsch's group fused two different fluorescent proteins YFP and mCherry to FKF1 and membrane-anchored GI respectively, and then used the fluorescent imaging to track the redistribution of FKF1 or FKF1-linked Rac1 from the cytosol to the plasma membrane (Yazawa et al. 2009).

Yang's group used the expression of mCherry as a reporter to determine the light energy required to maximally activate Gal4 (65)–VVD in cultured cells and in whole animals (Wang et al. 2012).

17.2.2.2.2.2 Imaging Downstream Consequences of Optogenetic Manipulations

Once the optogenetic method of choice has been optimized, the next step is to use it to control a cellular effect. In their studies of PA–Rac (Wu et al. 2009), the Hahn laboratory monitored phosphorylation of PAK, a Rac effector, using a specific antibody and indirect immunofluorescence of fixed cells after local illumination. This was possible because light-activated PA–Rac diffused slowly out of the illumination sites, with a diffusion coefficient of 0.55 $\mu m^2 s^{-1}$. To assess cellular behavioral responses to local Rac activation, Wu et al. quantified protrusion and retraction area changes. Fluorescent imaging of fibroblasts expressing mVenus-tagged PA–Rac allowed precise measurement of the displacement of the cellular margin before and after local illumination. The Hahn group also followed the rapid local inhibition of RhoA activity using a RhoA Förster/fluorescence resonance energy transfer (FRET)-based biosensor caused by local PA–Rac illumination (Wu et al. 2009). However, a limitation to combining photoactivation and FRET is that the local sites of illumination can cause photobleaching of the FRET FP pairs (Wu et al. 2009).

Levskaya et al. used total internal reflection of fluorescence (TIRF) microscopy and mCherry-tagged GTPase-binding domains (GBD) from either PAK or WASP as biosensors to follow downstream effects of Rac or Cdc42 photoactivation (Levskaya et al. 2009). By comparing membrane protrusion dynamics to biosensor membrane translocation and enrichment, they found that GTPase activation occurs within seconds, whereas membrane protrusion takes a few minutes, indicating that a series of subsequent downstream steps likely occur to cause membrane protrusion (Levskaya et al. 2009).

Following the consequences of PA–Rac in living tissues and organisms presents additional challenges that have been met in two systems: *Drosophila* border cells and zebrafish neutrophils. Our group generated transgenic flies expressing PA–Rac under control of the Gal4/UAS system so expression can be limited to specific cell types of interest. We then successfully used the Hahn laboratory's mCherry-tagged PA–Rac in *Drosophila* border cells, a model for collective epithelial cell migration. Photoactivation of Rac in one cell led to local membrane ruffling and protrusion and altered the direction and speed of migration of the entire cluster of cells. The mCherry tag was useful for following migration speed, membrane area changes, and border cell protrusion and retraction (Wang et al. 2010). Based on migration speed, PA–Rac was able to rescue the migration defect caused by loss of guidance receptor activity, indicating that Rac functions downstream of the receptors, which was confirmed by following Rac activity with an *in vivo* FRET biosensor (Wang et al. 2010). Further, the protrusion changes monitored by mCherry–PA–Rac were useful in demonstrating that when Rac activity in one border cell is elevated, the other cells retract protrusions and follow the cell with the highest Rac activity, demonstrating collective direction-sensing (Wang et al. 2010). Concurrently, Huttenlocher's group used the same PA–Rac construct to manipulate the migrations of individual neutrophils in zebrafish fins (Yoo et al. 2010). They injected the transposase mRNA and PA–Rac plasmid together into the cytoplasm of one-cell stage embryos in order to generate late-stage zebrafish embryos expressing PA–Rac. They demonstrated that blue light can easily drive neutrophil migration, and PA–Rac can be used to rescue the protrusion defects, but not the migration defects or rounded tail morphology induced by the inhibition of PI (3)K signaling pathways (Yoo et al. 2010), which indicates the specific role for Rac in control of protrusion in neutrophil migration.

17.3 FUTURE DIRECTIONS

The future of optogenetics is bright. Rapid advances optogeneticol should result in improvements to both the tools themselves and the ability to measure downstream effects *in vitro* and *in vivo*. Use of optogenetic approaches is expanding and may become standard assays, rather than one of proof-of-principle demonstrations. Besides increasing the robustness of the current photoactivation tools, a substantive improvement would be to multiplex different optical input signals and combine them with more downstream readouts, such as FRET reporters. The expanding palette of fluorescent proteins should reduce current problems with spectral overlap and photobleaching. The future will likely bring improvements to methods for activating photoactivatable proteins deep in opaque tissues. The "holy grail" for the field would be to find a truly generalizable strategy to render any protein sensitive to light in any cell type, tissue, or whole organism.

REFERENCES

Adamantidis, A.R., Zhang, F., Aravanis, A.M., Deisseroth, K., and de Lecea, L. (2007). Neural substrates of awakening probed with optogenetic control of hypocretin neurons. *Nature 450*, 420–424.

Ahrens, M.B., Li, J.M., Orger, M.B., Robson, D.N., Schier, A.F., Engert, F., and Portugues, R. (2012). Brain-wide neuronal dynamics during motor adaptation in zebrafish. *Nature 485*, 471–477.

Ahrens, M.B., Orger, M.B., Robson, D.N., Li, J.M., and Keller, P.J. (2013). Whole-brain functional imaging at cellular resolution using light-sheet microscopy. *Nat Methods 10*, 413–420.

Airan, R.D., Thompson, K.R., Fenno, L.E., Bernstein, H., and Deisseroth, K. (2009). Temporally precise in vivo control of intracellular signalling. *Nature 458*, 1025–1029.

Akemann, W., Lundby, A., Mutoh, H., and Knopfel, T. (2009). Effect of voltage sensitive fluorescent proteins on neuronal excitability. *Biophys J 96*, 3959–3976.

Akemann, W., Mutoh, H., Perron, A., Rossier, J., and Knopfel, T. (2010). Imaging brain electric signals with genetically targeted voltage-sensitive fluorescent proteins. *Nat Methods 7*, 643–649.

Aravanis, A.M., Wang, L.P., Zhang, F., Meltzer, L.A., Mogri, M.Z., Schneider, M.B., and Deisseroth, K. (2007). An optical neural interface: In vivo control of rodent motor cortex with integrated fiberoptic and optogenetic technology. *J Neural Eng 4*, S143–S156.

Arrenberg, A.B., Stainier, D.Y., Baier, H., and Huisken, J. (2010). Optogenetic control of cardiac function. *Science 330*, 971–974.

Banghart, M., Borges, K., Isacoff, E., Trauner, D., and Kramer, R.H. (2004). Light-activated ion channels for remote control of neuronal firing. *Nat Neurosci 7*, 1381–1386.

Berndt, A., Yizhar, O., Gunaydin, L.A., Hegemann, P., and Deisseroth, K. (2009). Bi-stable neural state switches. *Nat Neurosci 12*, 229–234.

Boyden, E.S., Zhang, F., Bamberg, E., Nagel, G., and Deisseroth, K. (2005). Millisecond-timescale, genetically targeted optical control of neural activity. *Nat Neurosci 8*, 1263–1268.

Bundschuh, S.T., Zhu, P., Scharer, Y.P., and Friedrich, R.W. (2012). Dopaminergic modulation of mitral cells and odor responses in the zebrafish olfactory bulb. *J Neurosci 32*, 6830–6840.

Caporale, N., Kolstad, K.D., Lee, T., Tochitsky, I., Dalkara, D., Trauner, D., Kramer, R., Dan, Y., Isacoff, E.Y., and Flannery, J.G. (2011). LiGluR restores visual responses in rodent models of inherited blindness. *Mol Ther 19*, 1212–1219.

Cardin, J.A., Carlen, M., Meletis, K., Knoblich, U., Zhang, F., Deisseroth, K., Tsai, L.H., and Moore, C.I. (2009). Driving fast-spiking cells induces gamma rhythm and controls sensory responses. *Nature 459*, 663–667.

Chen, T.W., Wardill, T.J., Sun, Y., Pulver, S.R., Renninger, S.L., Baohan, A., Schreiter, E.R., Kerr, R.A., Orger, M.B., Jayaraman, V. et al. (2013). Ultrasensitive fluorescent proteins for imaging neuronal activity. *Nature 499*, 295–300.

Chow, B.Y., Han, X., Dobry, A.S., Qian, X., Chuong, A.S., Li, M., Henninger, M.A., Belfort, G.M., Lin, Y., Monahan, P.E. et al. (2010). High-performance genetically targetable optical neural silencing by light-driven proton pumps. *Nature 463*, 98–102.

Cohen, A.E., and Hochbaum, D.R. (2014). Measuring membrane voltage with microbial rhodopsins. *Methods Mol Biol 1071*, 97–108.

Crick, F. (1999). The impact of molecular biology on neuroscience. *Philos Trans R Soc Lond B Biol Sci 354*, 2021–2025.

Deisseroth, K. (2010). Controlling the brain with light. *Sci Am 303*, 48–55.

Deisseroth, K. (2011). Optogenetics. *Nat Methods 8*, 26–29.

Deisseroth, K., Feng, G., Majewska, A.K., Miesenbock, G., Ting, A., and Schnitzer, M.J. (2006). Next-generation optical technologies for illuminating genetically targeted brain circuits. *J Neurosci 26*, 10380–10386.

Desai, M., Kahn, I., Knoblich, U., Bernstein, J., Atallah, H., Yang, A., Kopell, N., Buckner, R.L., Graybiel, A.M., Moore, C.I. et al. (2011). Mapping brain networks in awake mice using combined optical neural control and fMRI. *J Neurophysiol 105*, 1393–1405.

Editorial (2010). Stepping away from the trees for a look at the forest. *Science 330*, 1612–1613.

Essen, L.O., Mailliet, J., and Hughes, J. (2008). The structure of a complete phytochrome sensory module in the Pr ground state. *Proc Natl Acad Sci USA 105*, 14709–14714.

Gradinaru, V., Mogri, M., Thompson, K.R., Henderson, J.M., and Deisseroth, K. (2009). Optical deconstruction of parkinsonian neural circuitry. *Science 324*, 354–359.

Gradinaru, V., Thompson, K.R., Zhang, F., Mogri, M., Kay, K., Schneider, M.B., and Deisseroth, K. (2007). Targeting and readout strategies for fast optical neural control in vitro and in vivo. *J Neurosci 27*, 14231–14238.

Gunaydin, L.A., Yizhar, O., Berndt, A., Sohal, V.S., Deisseroth, K., and Hegemann, P. (2010). Ultrafast optogenetic control. *Nat Neurosci 13*, 387–392.

Guo, Q., Wang, X., Knoblich, U., Tibbitt, M.K., Anseth, K.S., Montell, D.J., and Elisseeff, J.H. (2012). Light activated cell migration in synthetic extracellular matrices. *Biomaterials 33*, 8040–8046.

Harper, S.M., Neil, L.C., and Gardner, K.H. (2003). Structural basis of a phototropin light switch. *Science 301*, 1541–1544.

Hegemann, P., and Moglich, A. (2011). Channelrhodopsin engineering and exploration of new optogenetic tools. *Nat Methods 8*, 39–42.

Ho, Y.S., Burden, L.M., and Hurley, J.H. (2000). Structure of the GAF domain, a ubiquitous signaling motif and a new class of cyclic GMP receptor. *EMBO J 19*, 5288–5299.

Honjo, K., Hwang, R.Y., and Tracey, W.D., Jr. (2012). Optogenetic manipulation of neural circuits and behavior in *Drosophila* larvae. *Nat Protoc 7*, 1470–1478.

Kahn, I., Desai, M., Knoblich, U., Bernstein, J., Henninger, M., Graybiel, A.M., Boyden, E.S., Buckner, R.L., and Moore, C.I. (2011). Characterization of the functional MRI response temporal linearity via optical control of neocortical pyramidal neurons. *J Neurosci 31*, 15086–15091.

Kennedy, M.J., Hughes, R.M., Peteya, L.A., Schwartz, J.W., Ehlers, M.D., and Tucker, C.L. (2010). Rapid blue-light-mediated induction of protein interactions in living cells. *Nat Methods 7*, 973–975.

Kim, S.G., and Ogawa, S. (2002). Insights into new techniques for high resolution functional MRI. *Curr Opin Neurobiol 12*, 607–615.

Konermann, S., Brigham, M.D., Trevino, A.E., Hsu, P.D., Heidenreich, M., Cong, L., Platt, R.J., Scott, D.A., Church, G.M., and Zhang, F. (2013). Optical control of mammalian endogenous transcription and epigenetic states. *Nature 500*, 472–476.

Kralj, J.M., Douglass, A.D., Hochbaum, D.R., Maclaurin, D., and Cohen, A.E. (2012). Optical recording of action potentials in mammalian neurons using a microbial rhodopsin. *Nat Methods 9*, 90–95.

Kramer, R.H., Mourot, A., and Adesnik, H. (2013). Optogenetic pharmacology for control of native neuronal signaling proteins. *Nat Neurosci 16*, 816–823.

Krauss, U., Lee, J., Benkovic, S.J., and Jaeger, K.E. (2010). LOVely enzymes—Towards engineering light-controllable biocatalysts. *Microb Biotechnol 3*, 15–23.

Lee, J.H., Durand, R., Gradinaru, V., Zhang, F., Goshen, I., Kim, D.S., Fenno, L.E., Ramakrishnan, C., and Deisseroth, K. (2010). Global and local fMRI signals driven by neurons defined optogenetically by type and wiring. *Nature 465*, 788–792.

Leung, D.W., Otomo, C., Chory, J., and Rosen, M.K. (2008). Genetically encoded photoswitching of actin assembly through the Cdc42-WASP-Arp2/3 complex pathway. *Proc Natl Acad Sci USA 105*, 12797–12802.

Levitz, J., Pantoja, C., Gaub, B., Janovjak, H., Reiner, A., Hoagland, A., Schoppik, D., Kane, B., Stawski, P., Schier, A.F. et al. (2013). Optical control of metabotropic glutamate receptors. *Nat Neurosci 16*, 507–516.

Levskaya, A., Chevalier, A.A., Tabor, J.J., Simpson, Z.B., Lavery, L.A., Levy, M., Davidson, E.A., Scouras, A., Ellington, A.D., Marcotte, E.M. et al. (2005). Synthetic biology: Engineering *Escherichia coli* to see light. *Nature 438*, 441–442.

Levskaya, A., Weiner, O.D., Lim, W.A., and Voigt, C.A. (2009). Spatiotemporal control of cell signalling using a light-switchable protein interaction. *Nature 461*, 997–1001.

Lima, S.Q., and Miesenbock, G. (2005). Remote control of behavior through genetically targeted photostimulation of neurons. *Cell 121*, 141–152.

Lin, J.Y., Knutsen, P.M., Muller, A., Kleinfeld, D., and Tsien, R.Y. (2013). ReaChR: A red-shifted variant of channelrhodopsin enables deep transcranial optogenetic excitation. *Nat Neurosci 16*, 1499–1508.

Lin, J.Y., Lin, M.Z., Steinbach, P., and Tsien, R.Y. (2009). Characterization of engineered channelrhodopsin variants with improved properties and kinetics. *Biophys J 96*, 1803–1814.

Logothetis, N.K., Pauls, J., Augath, M., Trinath, T., and Oeltermann, A. (2001). Neurophysiological investigation of the basis of the fMRI signal. *Nature 412*, 150–157.

Lundby, A., Mutoh, H., Dimitrov, D., Akemann, W., and Knopfel, T. (2008). Engineering of a genetically encodable fluorescent voltage sensor exploiting fast Ci-VSP voltage-sensing movements. *PLoS One 3*, e2514.

Matsuno-Yagi, A., and Mukohata, Y. (1977). Two possible roles of bacteriorhodopsin; a comparative study of strains of *Halobacterium halobium* differing in pigmentation. *Biochem Biophys Res Commun 78*, 237–243.

Moglich, A., Ayers, R.A., and Moffat, K. (2009a). Design and signaling mechanism of light-regulated histidine kinases. *J Mol Biol 385*, 1433–1444.

Moglich, A., Ayers, R.A., and Moffat, K. (2009b). Structure and signaling mechanism of Per-ARNT-Sim domains. *Structure 17*, 1282–1294.

Moglich, A., and Moffat, K. (2007). Structural basis for light-dependent signaling in the dimeric LOV domain of the photosensor YtvA. *J Mol Biol 373*, 112–126.

Monson, E.K., Ditta, G.S., and Helinski, D.R. (1995). The oxygen sensor protein, FixL, of *Rhizobium meliloti*. Role of histidine residues in heme binding, phosphorylation, and signal transduction. *J Biol Chem 270*, 5243–5250.

Morgan, S.A., Al-Abdul-Wahid, S., and Woolley, G.A. (2010). Structure-based design of a photocontrolled DNA binding protein. *J Mol Biol 399*, 94–112.

Muto, A., Ohkura, M., Abe, G., Nakai, J., and Kawakami, K. (2013). Real-time visualization of neuronal activity during perception. *Curr Biol 23*, 307–311.

Nagel, G., Brauner, M., Liewald, J.F., Adeishvili, N., Bamberg, E., and Gottschalk, A. (2005). Light activation of channelrhodopsin-2 in excitable cells of *Caenorhabditis elegans* triggers rapid behavioral responses. *Curr Biol 15*, 2279–2284.

Nagel, G., Ollig, D., Fuhrmann, M., Kateriya, S., Musti, A.M., Bamberg, E., and Hegemann, P. (2002). Channelrhodopsin-1: A light-gated proton channel in green algae. *Science 296*, 2395–2398.

Nagel, G., Szellas, T., Huhn, W., Kateriya, S., Adeishvili, N., Berthold, P., Ollig, D., Hegemann, P., and Bamberg, E. (2003). Channelrhodopsin-2: A directly light-gated cation-selective membrane channel. *Proc Natl Acad Sci USA 100*, 13940–13945.

The News Staff (2011). Method of the year 2010. *Nat Methods 8*, 1.

Ni, M., Tepperman, J.M., and Quail, P.H. (1998). PIF3: A phytochrome-interacting factor necessary for normal photoinduced signal transduction, is a novel basic helix-loop-helix protein. *Cell 95*, 657–667.

Oesterhelt, D., and Stoeckenius, W. (1971). Rhodopsin-like protein from the purple membrane of *Halobacterium halobium*. *Nat New Biol 233*, 149–152.

Ogawa, S., Tank, D.W., Menon, R., Ellermann, J.M., Kim, S.G., Merkle, H., and Ugurbil, K. (1992). Intrinsic signal changes accompanying sensory stimulation: Functional brain mapping with magnetic resonance imaging. *Proc Natl Acad Sci USA 89*, 5951–5955.

Patti, J., and Isacoff, E.Y. (2013). Measuring membrane voltage with fluorescent proteins. *Cold Spring Harb Protoc 2013*, 606–613.

Perez Koldenkova, V., and Nagai, T. (2013). Genetically encoded Ca($^{2+}$) indicators: Properties and evaluation. *Biochim Biophys Acta 1833*, 1787–1797.

Peron, S., and Svoboda, K. (2011). From cudgel to scalpel: Toward precise neural control with optogenetics. *Nat Methods 8*, 30–34.

Peterka, D.S., Takahashi, H., and Yuste, R. (2011). Imaging voltage in neurons. *Neuron 69*, 9–21.

Polosukhina, A., Litt, J., Tochitsky, I., Nemargut, J., Sychev, Y., De Kouchkovsky, I., Huang, T., Borges, K., Trauner, D., Van Gelder, R.N. et al. (2012). Photochemical restoration of visual responses in blind mice. *Neuron 75*, 271–282.

Reiner, A., and Isacoff, E.Y. (2014). Photoswitching of cell surface receptors using tethered ligands. *Methods Mol Biol 1148*, 45–68.

Rockwell, N.C., and Lagarias, J.C. (2010). A brief history of phytochromes. *ChemPhysChem 11*, 1172–1180.

Rockwell, N.C., Su, Y.S., and Lagarias, J.C. (2006). Phytochrome structure and signaling mechanisms. *Annu Rev Plant Biol 57*, 837–858.

Ryu, M.H., Moskvin, O.V., Siltberg-Liberles, J., and Gomelsky, M. (2010). Natural and engineered photoactivated nucleotidyl cyclases for optogenetic applications. *J Biol Chem 285*, 41501–41508.

Sakai, R., Repunte-Canonigo, V., Raj, C.D., and Knopfel, T. (2001). Design and characterization of a DNA-encoded, voltage-sensitive fluorescent protein. *Eur J Neurosci 13*, 2314–2318.

Schroll, C., Riemensperger, T., Bucher, D., Ehmer, J., Voller, T., Erbguth, K., Gerber, B., Hendel, T., Nagel, G., Buchner, E. et al. (2006). Light-induced activation of distinct modulatory neurons triggers appetitive or aversive learning in *Drosophila* larvae. *Curr Biol 16*, 1741–1747.

Shimizu-Sato, S., Huq, E., Tepperman, J.M., and Quail, P.H. (2002). A light-switchable gene promoter system. *Nat Biotechnol 20*, 1041–1044.

Sohal, V.S., Zhang, F., Yizhar, O., and Deisseroth, K. (2009). Parvalbumin neurons and gamma rhythms enhance cortical circuit performance. *Nature 459*, 698–702.

Stierl, M., Stumpf, P., Udwari, D., Gueta, R., Hagedorn, R., Losi, A., Gartner, W., Petereit, L., Efetova, M., Schwarzel, M. et al. (2011). Light modulation of cellular cAMP by a small bacterial photoactivated adenylyl cyclase, bPAC, of the soil bacterium Beggiatoa. *J Biol Chem 286*, 1181–1188.

Stirman, J.N., Brauner, M., Gottschalk, A., and Lu, H. (2010). High-throughput study of synaptic transmission at the neuromuscular junction enabled by optogenetics and microfluidics. *J Neurosci Methods 191*, 90–93.

Stirman, J.N., Crane, M.M., Husson, S.J., Wabnig, S., Schultheis, C., Gottschalk, A., and Lu, H. (2011). Real-time multimodal optical control of neurons and muscles in freely behaving *Caenorhabditis elegans*. *Nat Methods 8*, 153–158.

Strickland, D., Moffat, K., and Sosnick, T.R. (2008). Light-activated DNA binding in a designed allosteric protein. *Proc Natl Acad Sci USA 105*, 10709–10714.

Strickland, D., Yao, X., Gawlak, G., Rosen, M.K., Gardner, K.H., and Sosnick, T.R. (2010). Rationally improving LOV domain-based photoswitches. *Nat Methods 7*, 623–626.

Swartz, T.E., Corchnoy, S.B., Christie, J.M., Lewis, J.W., Szundi, I., Briggs, W.R., and Bogomolni, R.A. (2001). The photocycle of a flavin-binding domain of the blue light photoreceptor phototropin. *J Biol Chem 276*, 36493–36500.

Taylor, B.L., and Zhulin, I.B. (1999). PAS domains: Internal sensors of oxygen, redox potential, and light. *Microbiol Mol Biol Rev 63*, 479–506.

Terai, T., Tomiyasu, R., Ota, T., Ueno, T., Komatsu, T., Hanaoka, K., Urano, Y., and Nagano, T. (2013). Tokyo Green derivatives as specific and practical fluorescent probes for UDP-glucuronosyltransferase (UGT) 1A1. *Chem Commun (Camb) 49*, 3101–3103.

Tochitsky, I., Polosukhina, A., Degtyar, V.E., Gallerani, N., Smith, C.M., Friedman, A., Van Gelder, R.N., Trauner, D., Kaufer, D., and Kramer, R.H. (2014). Restoring visual function to blind mice with a photoswitch that exploits electrophysiological remodeling of retinal ganglion cells. *Neuron 81*, 800–813.

Tsai, H.C., Zhang, F., Adamantidis, A., Stuber, G.D., Bonci, A., de Lecea, L., and Deisseroth, K. (2009). Phasic firing in dopaminergic neurons is sufficient for behavioral conditioning. *Science 324*, 1080–1084.

Tyszkiewicz, A.B., and Muir, T.W. (2008). Activation of protein splicing with light in yeast. *Nat Methods 5*, 303–305.

Urban, A., Rancillac, A., Martinez, L., and Rossier, J. (2012). Deciphering the neuronal circuitry controlling local blood flow in the cerebral cortex with optogenetics in PV::Cre transgenic mice. *Front Pharmacol 3*, 105.

Volgraf, M., Gorostiza, P., Numano, R., Kramer, R.H., Isacoff, E.Y., and Trauner, D. (2006). Allosteric control of an ionotropic glutamate receptor with an optical switch. *Nat Chem Biol 2*, 47–52.

Wagner, J.R., Brunzelle, J.S., Forest, K.T., and Vierstra, R.D. (2005). A light-sensing knot revealed by the structure of the chromophore-binding domain of phytochrome. *Nature 438*, 325–331.

Wang, X., Chen, X., and Yang, Y. (2012). Spatiotemporal control of gene expression by a light-switchable transgene system. *Nat Methods 9*, 266–269.

Wang, X., He, L., Wu, Y.I., Hahn, K.M., and Montell, D.J. (2010). Light-mediated activation reveals a key role for Rac in collective guidance of cell movement in vivo. *Nat Cell Biol 12*, 591–597.

Witten, I.B., Lin, S.C., Brodsky, M., Prakash, R., Diester, I., Anikeeva, P., Gradinaru, V., Ramakrishnan, C., and Deisseroth, K. (2010). Cholinergic interneurons control local circuit activity and cocaine conditioning. *Science 330*, 1677–1681.

Wu, Y.I., Frey, D., Lungu, O.I., Jaehrig, A., Schlichting, I., Kuhlman, B., and Hahn, K.M. (2009). A genetically encoded photoactivatable Rac controls the motility of living cells. *Nature 461*, 104–108.

Wyart, C., Del Bene, F., Warp, E., Scott, E.K., Trauner, D., Baier, H., and Isacoff, E.Y. (2009). Optogenetic dissection of a behavioural module in the vertebrate spinal cord. *Nature 461*, 407–410.

Yang, X., Kuk, J., and Moffat, K. (2008). Crystal structure of *Pseudomonas aeruginosa* bacteriophytochrome: Photoconversion and signal transduction. *Proc Natl Acad Sci USA 105*, 14715–14720.

Yazawa, M., Sadaghiani, A.M., Hsueh, B., and Dolmetsch, R.E. (2009). Induction of protein-protein interactions in live cells using light. *Nat Biotechnol 27*, 941–945.

Yoo, S.K., Deng, Q., Cavnar, P.J., Wu, Y.I., Hahn, K.M., and Huttenlocher, A. (2010). Differential regulation of protrusion and polarity by PI3K during neutrophil motility in live zebrafish. *Dev Cell 18*, 226–236.

Zemelman, B.V., Lee, G.A., Ng, M., and Miesenbock, G. (2002). Selective photostimulation of genetically chARGed neurons. *Neuron 33*, 15–22.

Zemelman, B.V., Nesnas, N., Lee, G.A., and Miesenbock, G. (2003). Photochemical gating of heterologous ion channels: Remote control over genetically designated populations of neurons. *Proc Natl Acad Sci USA 100*, 1352–1357.

Index

Page numbers followed by f and t indicate figures and tables, respectively.

B

F

N

O

Q

R

X

Y

Z

Printed and bound by CPI Group (UK) Ltd, Croydon, CR0 4YY
21/10/2024
01777044-0020